Neural Cell Behavior and Fuzzy Logic

Uziel Sandler • Lev Tsitolovsky

Neural Cell Behavior and Fuzzy Logic

Springer

Editors
Uziel Sandler
Jerusalem College Technology
91 160 Jerusalem
Israel

Lev Tsitolovsky
Bar-Ilan University
Faculty of Life Sciences
52 900 Ramat-Gan
Israel

ISBN: 978-1-4419-3492-5 e-ISBN: 978-0-387-09543-1

Printed on acid-free paper

9 8 7 6 5 4 3 2 1

springer.com

To Tamar Sandler, my wife and friend in life and science

(U.Sandler)

Contents

Part II Mathematics of feeling

Introduction: Brain as a unique object

Brain is the most complex, puzzling and attractive object in the universe. Everybody knows that brain is responsible for control of our learning, memory, motivations, consciousness, etc. Moreover, a brain by a certain mysterious way produces the Subject (or Self, Person, Ego, Observer - whatever you like), which tries to look into brain and understand the Subject itself. The means for the objective observation of a subjective world are absent, but we can observe the consequences of objective existence. Neurobiology is the only science that considers the subtle facet between substance and mind.

Besides the fact that brain generates goals and does not need external programming, it possesses many particular properties that make it an appealing object for research. Brain works using fuzzy laws on the one hand and produces fuzzy logic on the other hand [1269]. The brain has the highest metabolic rate of all bodily organs and depends predominantly on oxidative metabolism as a source of energy [755]. The brain spends 20% of total body oxygen consumption, although its mass is less than 2% of the total body and a brain does not produces mechanical efforts, as does a muscle, does not synthesize enzymes for food digestion, as does a liver and does not pump overlarge volumes of blood, as does a heart. Energy-consuming processes include maintenance of ion equilibriums, generation of the basal electrical activity, neurotransmitter uptake, etc. There are two main consumers of cellular energy, protein synthesis (55%) and Na^+, K^+-ATPase (45%), but during hypoxia Na^+, K^+-ATPase becomes dominant (80%) [158]. The Na^+, K^+-ATPase, or the sodium pump, is a transmembrane protein accountable for maintaining electrochemical gradients across the membrane in all cells. However, even in the resting awake state, around 80% of energy used by the brain supports events associated with cycling of glutamate and gamma-aminobutyric acid (GABA) neurotransmitters [1139, 541], the main transmitters of excitation and inhibition, connected also with cell damage and protection. Therefore, equilibrium between neuronal damage and protection plays an important role in the normal brain function.

In the brain, at any one moment, a lot of neurons, although not every one, are active. Brainy activity is perceptible even when it appears to be doing absolutely nothing. The basic diversity of chemical reactions in a body also belongs to brain. Properties of neurons are much more complex than for any other cell. This is partially concerned with the excitability of neurons. Brain has infinite memory in a finite volume (in the sense that nobody sees as memory come to an end). Physical damage of brain structure causes only subtle impairment of memory, which is usually less for remote, important and learned-by-heart memory [1252, 1255]. By some evaluations, a massive parallel processing capacity permits our visual system to successfully decode complex images in 100 ms, and our brain to store information that many times exceeds the text contained in the US Library of Congress [794]. Neuronal memory is not reversible, but brain scarcely contains any empty storage medium. The capability to recognize exceeds the capability to recollect. Sometimes, one cannot recall required information from one's past experience, but images in the long-term memory are never destroyed: for example, one doesn't remember only half of the face of a friend, and one doesn't remember a vacation in black and white but in color. Brain does not contain moving parts or valves, works exclusively and consistently, and memory stays intact after a short-term shut down of all dynamic processes. Signals in the brain are spread slowly (ten million times slower than electrical signals), but time response for many tasks is exceptionally short. Brain contains billions of neurons, ten or more trillions of synaptic connections, and processes information during milliseconds. It has a highly parallel organization, but does not suffer from the dictates of a few processors.

Brain is a unique object. However, is brain a subject? Are you your brain? Is one the activity of his brain? When brain produces thinking, neurons and glial cells generate electrical fields, the distribution of ionic concentrations alters, various substances are synthesized and degraded, numerous enzymes are activated or inhibited, gases spread through nerve tissue, etc. How, when and where is physiology converted into sense and what kind of physiological activity is decisive? This is the crucial problem. Can we some day solve it? For sure, problems that are inaccessible to our brain do exist. Just as a crocodile cannot be timed and a dog can not speak, the human mind also has borders. Our brain can't read two texts simultaneously, can't remember its prenatal life, can't imagine how the electron runs through two holes at once and maybe never can comprehend how the outer environment is converted into an inside, subjective word. In fact, there are environmental features that our brain can't perceive, our memory is not ideal and imagination is not infinite. However, our capability to realize is exceptionally powerful. For instance, we can't imagine four-dimensional space but can comprehend that in four-dimensional space, the loss of one's keys from your safe is not worrying: one always can penetrate through the fourth dimension. If we never understand consciousness, this will be the first problem that humanity cannot manage.

The foremost mission of neurobiology is the cognition of subjective phenomena and this includes the disproving of supernatural or mystic explanations for this extremely hard problem. Thermodynamics teaches us that disorder in closed-loop systems increases, but the brain gives a kind of an object, which introduces order in the world. One may imagine Maxwell's daemon, (that little homunculus!) controlling microscopic damper in the vessel. When a high-speed molecule in the environment draws near to the vessel, the homunculus opens slightly the damper and accumulates energy, thus violating thermodynamic laws. If we mark on the cerebral cortex the points receiving information from the body (face, knees, belly, etc.) we will observe an ugly dwarf at the surface of the cortex with its huge jaws, widely spread fingers and colossal genitals (the more sensitivity, the more cortical representation). This homunculus has only technical, ancillary meaning. However, sometimes one imagined one's Self as the Homunculus living in the brain. Any problem is easy to explain by means of a homunculus, but this is an inadmissible explanation in science. If one says that the theory may work only with the contribution of the homunculus (that is as being run by a Homunculus inside), this means that the theory is not scientific theory.

At present, we poorly comprehend a brain. Sometimes comprehension of an object is identified with the skill to recreate and/or to improve this object. This is not our goal. Animals recreate brain naturally and improve it in evolution without any comprehension. To be more accurate, we want to understand some brain functions. Among them there are some so intricate that even vague explanations are absent.

Neurobiologists run into several hard problems:

1. After learning, the reaction to a specific signal depends upon its significance, which was acquired during learning. Reactions become specific to input, and one signal may increase its effect, whereas another signal turns out to be ineffective. Why are some signals preferred? It is not the rule that a stronger impact evokes a stronger reaction. Gentle footsteps in one's bare apartment may exert a stronger impression than the harsh noise of one's TV. A specific reaction to a given signal may be both innate and acquired during experience. Thus, a task is related to the nature of memory: reaction depends on past experience. Moreover, is the physical appearance of memory elements predetermined by their real appearance in the environment? If the response is "yes" and similar images or events retain similar material traces of memory in different brains, a one-to-one correspondence exists between possible events and memory elements, even if you will never meet these events in your life: for instance, the image corresponding to the face of my grandmother preexists in any brain in a passive form ("grandmother" cells, synapses or pathways). It is scarcely likely that whole world's diversity may be squeezed into the brain. However, if the response is "no" and the memory elements are specific for each brain, who decides that an activated memory trace corresponds to a specific image or event? Therefore, the puzzle is how neurons recognize the representations they keep. Should we suppose the existence of a certain mysterious homunculus, living at the skull, which observes the environment through our senses and executes his wishes by means of our muscles? Some people think that we are our homunculi.

2. The same signal may evoke one or another reaction, depending upon circumstances. Reactions are specific for an output and brain somehow decides what action is desirable. But there is the question about how the reaction is chosen if each neuron has only one output? Let an organism choose the correct response to the signal, say, to run in or out. Its neurons can generate different reactions in response to dissimilar stimuli, but are not necessarily able to send the right signal in the right direction. For a single neuronal output, where an axon may be branched, a neuron does not have the means to send a signal to the specific branch. Even if axonal output is not symmetric, real choice is impossible, as the degree of non-symmetry is predetermined. A neuron may only send or not send a spike in the axon. Who chooses the decision? Even if we will suppose that our reactions are chosen by a homunculus, it does not have the means in order to accomplish its will: when it has chosen decision, it cannot send it to a correct target (if we try to proceed from neuron's doctrine).

3. Brain responds to an image by the formation of electrochemical representation within neural tissue. It is easy to imagine that each image corresponds to the specific constellation of activated and inhibited neurons. How does brain know that a current neuronal constellation correctly matches a given picture? Perhaps, changes of these pictures in time produce successive frames, such as in the cinema. If so, who looks at the film? One's homunculus? Or, maybe, odd elements of brain's picture interact and are integrated in a new entity? Each image is characterized by size, form, time, color, smell, time, etc. How are spatially and temporally different processes united in the awareness? This problem is known as the "binding problem".

4. An aware performance is always the sole act and currently active brain elements are somehow amalgamated in decision-making. An actual action is dominant. If scattered brain neurons generate various subthreshold actions, they ought not to prevent the actual action. If there is no homunculus in the skull, why do the decisions in various brain areas have a tendency to turn out similarly? Moreover, if a central processor is absent in the brain, the actions in brain areas must be synchronized within a narrow time window and this time window is shorter than necessary for a neural excitation transmission through the brain.

5. If action is predetermined by the environment and by the state of the brain, goal-directed behavior can be treated as an ordinary reflex, i.e. as a reaction to stimuli (inner or outer), based on memory and heredity. This predetermination would preclude voluntary action as the object of physiological description. The issue of predetermination of action has been the subject of philosophical and psychological analysis, as free will, but has not yet been addressed in depth by neurobiologists. The problem remaining unsolved is how behavior can be both unpredictable and goal-directed, if we do not consider there to be a homunculus in the head.

6. Living beings generate aware actions. How is neuronal activity transformed into qualia (saltiness, redness, pleasure, painfulness, etc.)? We are capable of understanding physiological processes and neuronal computations,

but how does Self begin, as a tangle of desires and expectations? Is this, once again, the trace of the notorious homunculus? Does a homunculus have its own control center or must we, at this point, stop our efforts? This is the same hard problem. No one can suggest an experiment for examination of a working hypothesis, because nobody may propose any hypothesis and even vouch for the scientific validity of the problem itself. By the way, Niels Bohr, Erwin Schroedinger, Wolfgang Pauli and Werner Heisenberg, the great quantum physicists of the twenty century, all admitted that consciousness is as primary as a substance [1152]. They believed that the explanatory gap between the subjectivity and the physicochemical activity of neurons is analogous to wave-particle duality. The smallest units of matter are not substance objects in the ordinary sense; they are not hard and not soft, do not have dimension, are impossible to see, to touch, to smell and they do not reside at a given time in a given place. They display themselves only during the process of observation and have an aspect of subjectivity: one cannot determine the coordinates of the electron until carrying to completion the measurements. There are forms and ideas which can be expressed only in mathematical language. Equally, it may be that both the material universe and the human mind are equally 'objective'. The great quantum physicists were unanimous in considering that consciousness arises before humanity.

These hard problems are rather far from being solved. We do not promise to settle them, but let us try. Our task is to appreciate some, still unexplained, brain functions, or at any rate, to outline pathways to their understanding. Some of our explanations will be preliminary. We are considering as unacceptable only two extremes: when any explanation is absent and when there are several equivalent explanations.

Which "logic" does the brain use for description of an environment and for decision-making? Numerous literature's sources and our original experiments considered in this book, allow us to believe that capability of perceiving could take place already on a neuron level. So, if we want to develop an adequate mathematical description of neuron information processing, we are, in a certain sense, compelled to search an apparatus, which could operate with "perceptions" as with mathematical objects. In this book we show that the subjective attitude of the brain to an expected event can be the reason for the advent of the brain's "logic of perception". Another reason to develop a phenomenological theory of neuron's behavior, which is based on notation of perception, is the nature of the experimental data for the real neurons. In fact, in these experiments only a few parameters of a system are observable and controllable, while a number of the other ones are "hidden" or remain out of control. For a complex phenomenon it leads to considerable variability in experimental results from trial to trial and makes it difficult to estimate accuracy of the obtained values. For such a phenomenon fair description of a system behavior is based on our "percept" of observed tendencies rather than on precise numerical values of the experimental data.

Mathematical basis for "computing of perceptions" was proposed by L.Zadeh almost half century ago and one was named as "fuzzy logic". In the following decades fuzzy logic has been intensively developed and applied to numerous applications in several thousand articles and books. It has been shown that fuzzy logic is amazingly effective in processing of information with high level of uncertainty. A real neuron is a complex dynamic system, so for description of it behavior we need an extension of fuzzy logic on evolutionary processes: "evolution of perceptions". Such a theory called "fuzzy dynamics" has been developed and studied during the last decade. Fuzzy dynamics can be successfully used in the situation, where both the system states and the dynamics laws have considerable uncertainty, which is typical for description of the systems on a "perceptive level". Actually, these laws are naturally formulated in terms of "neuron activity", "neuron damage", "level of compensation" and "expectation of punishment", which can be considered as phenomenological variables describing our "perception" of a neuron state. Of course, such variables are very qualitative and have considerable level of uncertainty, but they fair reflect our real knowledge about the considered system.

It is important, that fuzzy dynamics can directly operate with above mentioned variables and dynamics laws without additional arbitrary assumptions and hardly verified hypotheses. In this book we explain how fuzzy dynamics equations can be obtained. An undoubted advantage of this approach is an ability to use well-developed apparatus of the theoretical physics for analysis and solution of the dynamics equations. It will be shown that the fuzzy dynamics approach to a neuron behavior leads to good compatibility with experimental observations and allows understanding some basic features of the neuron being. In particular, it predicts strong sudden, non-stochastic alterations in the neuron's activity. Such alterations are typical for a real neuron's behavior and it has an effect on macro-behavior of an animal. It is well known that even when it know a good solution to a given problem, an animal from tries time-to-time to find a new solution and if the new solution is a worse then the old one the animal returns to its previous behavior. Such "researcher's instinct" is very beneficial, since it enables the animal to effectively optimize its behavior in continuously changing environmental conditions. It should be emphasized, that in fuzzy dynamics approach such behavior is neither the consequence of random inner influences of the neural system nor only the result of the sudden changes in environment, but rather a fundamental feature of neurons. It seems very likely, that the inner logic of the neuron's behavior is close to fuzzy logic.

The idea of using some of the features of physiological processes in cybernetics is very old, but it is still attractive. For example, a concept of self-consciousness and emotion for robotic systems has been discussed recently in set of international conferences and congresses. Because fuzzy logic is easily computerized, it is feasible to design new kinds of artificial neurons: "motivational artificial neurons", which seem to be very promising as elements of the "brain of the feeling robot". Behavior of such robot will be initiated by a few artificial drives included energy recovering, avoidance of injury and aspiration to survive. So, in a "feeling robot" performance of the main task, trial-and-error learning, aspiration to survive and "instinct of researcher" could be naturally combined.

Reading this book, the readers will encounter a number of molecular, electrical and morphological devices that brain uses in its performance. The reader is not committed to understanding the fine dynamics of their functioning. We have used these data for a purpose. Simplified explanations for some brain functions sometime contradict easy explanations of other functions as well as the occurrence of more complex material substratum. However, it is productive to compare in which functions (sometimes rather different) the same material device participates. This may reveal relationships between these outwardly different functions. Therefore, for us it was more essential to delineate what function performs a given molecular device, than to linger on the fine details of its performance. Obviously, this approach cannot lead us to recreation of the working brain scheme. However, we hope to advance in this direction.

We don't assume that the readers are familiar with the fuzzy logic paradigm, so short introduction to the fuzzy sets theory and the fuzzy logic will be presented. Note that the Sections are designated by star (∗) contain more specialized material and can be omitted by a reader, which is interesting, mainly, in general information.

ACKNOWLEDGMENTS. The contributions of the following individuals in the experimental work: Drs. Pivovarov A., Kipor G., Tsatyrjan O., Babkina N. and Shvedov A. and Prof. Friedman Y. in the fuzzy dynamics theory are greatly appreciated. The author would like to take this opportunity to thank Drs. Kovalenko M, Rossokhin A. and Saakjan Ju. for computer modelling. The authors gratefully acknowledge Prof. Kraevsky A. for his fruitful advises in the field of biochemistry and to Profs. Butnario D., Meisar R., Pap E. and Diamond P. for their fruitful discussion in the field of fuzzy logic and fuzzy set theory. One of us (LT) especially thank his wife Dr. Ludmila Tsitolovskaya for constructive discussions, valuable insights and patience. The authors acknowledge Judy Hanaani and Evelin Lourie for aid in editing the manuscript.

Part I

The being of neural cells

1

The operation of memory (a single neuron can learn)

1.1 Brain straightforward sight

Till the present, an enormous number of facts concerning brain have been collected, while advancement of our understanding of the mechanisms of its activity remains insignificant. The set of known experimental data and the quantity of the scientific reports reaches a disturbing threshold, with old facts being "rediscovered" using new methods. For example, during conditional reflex elaboration, the reaction to a signal acquiring important significance usually increases, while response to the insignificant signals decreases. This change in the responses is highly specific. During learning, only those responses, whose significance was adjusted at the time of learning, are modified, whereas responses to foreign signals do not change. This axiom is true both for animal behavior [947] and for neuronal reactions [530, 19, 503], but it continues to surprise modern scientists [371, 1282, 1087].

Motivation is a leading factor in goal-directed behavior; it alters readiness to action and is the most accessible phenomenon for study of a subjective feeling. Brain evaluates the magnitudes of inner and external variables and compares present conditions with past experience and also evaluates the possible consequences of its own reactions to the alteration of these variables. Put simply, sensation affects inner environment and generates movements, which alter outer environment and thus change sensations. We usually examine brain behavior like that of any other object, but the brain itself also studies the environment and uses memory in order to control behavior.

Neuronal structure of the brain is equally complex and well-ordered. Human brain contains a number of neural centers. Certain centers are stratified, like the neocortex, hippocampus and cerebellum, while others are dispersed nuclei without strict structure. Large centers are divided into subdivisions; for example, the neocortex is separated to more than one hundred fields.

The cortex has extraordinarily specific connections between individual cells and cell classes. Brain pathways and cortical regions that are established during early development are partially genetically determined, but depend also on their synaptic inputs and behavioral instruction [1196]. In several categories of animals, such as some insects, mollusks and fishes, there are genetically specific neurons that are distinct from any other neurons, but they are identical in different creatures of the same species. Environment also plays a role in brain morphology and intelligence, particularly in humans, but the predominant influence appears to genetic (70%-80%). Monozygotic twins reared apart are more alike with respect to cognitive and morphologic properties than heterozygotic twins. Heritability does not imply inevitability because the environment can determine the relative impact of genetic variation. Between 20%-35% of the observed population differences in intellectual capability are due to differences in family environments. Also, heritability of intelligence increases with age: as twins grow older, phenotype reflects genotype more closely. A strictly environmental theory would predict the opposite [1233]. This means that morphological structure of the neural system may be predetermined. Nevertheless, this predetermined structure is not the only possible and necessary one for correct performance. If, during early individual development, brain structure is to a marked degree mechanically injured, behavior is usually recovered so completely that it is difficult to distinguish from the behavior of a normal uninjured brain. This is not the case for destructive factors affecting the whole brain, such as stress and energy deficiency. Extirpations of neural tissue are well compensated, while stress in early life leaves a negative trace and an enriched environment leaves a positive trace [1040].

Brain structure has changed dramatically during evolution, but properties of a neuron have remained conservative in the animal world. Some properties of behavior have also been conserved through evolution and these properties are, possibly, determined by the peculiarity of neurons and not by brain structure: the capability to accept signals; excitability; transportation of electrical message into chemical output and plasticity in accordance with a change of circumstances. A complication of brain structure in evolution is correlated with the property of behavior, which has been perfected during evolution. This perfection is, before all, the capability of simultaneous maintenance, in an active state, a large quantity of information ordered in time. This capability is essential for development of consciousness.

Although neurons are not complicated by evolution, they are complex themselves. In the past, textbooks of neuroscience sometimes illustrated a neuron as a primitive balloon-like construction while a brain was shown as an electronic scheme consisting of such simple neurons. Electrical signals, it was thought, are the only means of brain operation. The intracellular environment of a neuron (K^+ rich and Na^+ poor) has a negative membrane potential around 70 mV - and is separated from the extracellular environment by means of a microscopic neuronal lipid membrane [984]. An action potential (AP, spike, neural impulse) is a short electrical wave travelling along the cellular membrane of a neuron. Neuronal membranes contain pores, special proteins in synaptic areas (chemoreceptors) and ion channels in non-synaptic membrane. Potential-dependent (K^+, Na^+, Ca^{2+} and some others) and ligand-dependent channels control membrane permeability for corresponding ions. For example, Na^+-channels contain a selectivity filter, which is made of negatively charged amino acid residues. They attract the positive Na^+ ions, while the larger K^+ ions cannot fit through this area. Potential-dependence means that the probability of the channel opening for the corresponding ion increases, when, as the rule, the membrane potential decreases (during depolarization). Ligand-dependent channels change the ion permeability as a result of chemical influence. In the standard scenario, depolarization at the beginning opens some sodium channels. Na^+ ions penetrate the neuron and augment depolarization. As a result, the probability of the next sodium channels opening increases. There is a critical level of depolarization, the threshold, when membrane permeability for Na^+ grows to maximum and the membrane potential reaches an equilibrium potential for sodium (around +40 mV): thus an AP is generated (Fig. 1.1).

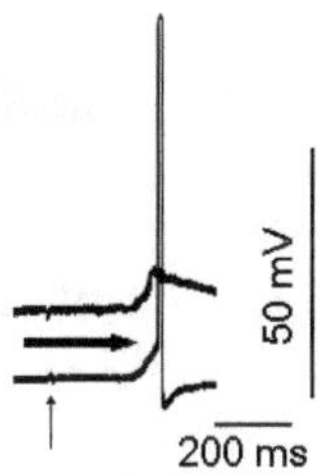

Fig. 1.1. Simultaneous recording of intracellular activity of two mollusk neurons. At the top, a neuron generates excitatory postsynaptic potential (EPSP), but fails to generate an action potential (AP). Vertical arrow presentation of tactile stimulus. At the bottom, a neuron generates the AP. Bold arrow, level of AP generation. Calibrations are pointed out at the figure.

Membrane potential returns to normal level, -70 mV, when depolarization initiates opening of the potassium channels. This requires higher depolarization than the opening of sodium channels. Ions of K^+ leave the neuron, the membrane acquires more negative charge and membrane potential recovers. During generation of each AP, the neuron loses some K^+ ions and loads Na^+ ions. The proportion between inner K^+ and outer Na^+ ions is sufficient for generation of hundreds or thousands of spikes, but each neuron has a special tool for ion equilibrium recovery: protein Na^+, K^+-ATPase, which transmits Na^+ out and K^+ in, spending energy as a molecule of ATP. Na^+, K^+-ATPase activity takes approximately half of the energy consumed by the brain [802].

Neural cells are a relatively new achievement of evolution. Basic metabolic pathways and mechanisms of any cell are preserved in a neural cell, but excitability and potential-dependent Na^+ channels are qualitatively new attributes of a neuron. Quantitative distinctions of a neuron are high levels of energetic and protein metabolism and low possibilities for displacement, although motility of neuronal processes is observed. Once arisen, neural cells have conservative properties in primitive multicellular animals and human.

Although properties of neural cells feebly change during evolution, they do change during individual development. Biophysical characteristics of cell membranes are, evidently, essential for brain function, since a decrease in the rate of learning from infancy to old age is accompanied by modifications in characteristic neuronal membranes. Input resistance, the duration of action potentials, the membrane time constant of many brain neurons, the amplitude of the peak of whole cell K^+ current, the excitability of identified neurons, the frequency of impulse generation, the rate of repolarization of the action potential, the synaptic efficacy and the electrical coupling between neurons all diminish with age, while the peak of whole cell Ca^{2+} current, the intracellular Ca^{2+} concentrations and the threshold for action potential generation all increase [212, 1379]. At the same time, parallels between neuronal function and development of neural tissue in the philo- and ontogenesis may be superficial.

Brain contains a lot of neural centers with incompletely known functions. In their turn, neural centers consist of neurons and glial cells, connected in sometimes strictly predetermined, but sometimes unpredetermined networks. Neurons are a much more favorite object of investigation than glial cells, since neural cells produce activity similar to a digital device and this has allowed the hope to understand brain. Neurons affect each others using specific substances, called neurotransmitters. Chemical signals are impossible to transmit at large distances by means of diffusion; this is a too slow and undirected process. Signals between cells for large distances are transmitted by, action potentials. A presynaptic neuron sends an AP, as its message into its axons to all neurons to which it is connected and triggers a chemical message, a neurotransmitter, at the synaptic cleft in the next neuron (see Fig. 1.2). Postsynaptic neurons usually receive input signals through synapses to their dendrites or soma. A transmitter interacts with a chemoreceptor that is specific for this transmitter, and transitory changes in membrane conductance and excitatory or inhibitory postsynaptic potential arises (dependently on the type of the chemoreceptor). The postsynaptic potential changes the membrane potential of a neuron by the flow of positive or negative ions through the membrane. A postsynaptic potential is excitatory if it promotes generation of an action potential, while it is inhibitory, if it prevents spike generation. When many excitatory synapses unite their action on a cellular membrane, the impact to a neuron may exceed threshold, the neuron generates spikes spreading into its axon and the next postsynaptic neuron will increase its firing rate. An axon is not electrical wire. Action potential spreads along an axon in a regenerative way, as does the Bickford fuse, while local potential-dependent channels are activated.

Neurotransmitters, which are large proteins, counteract with chemoreceptors in the cellular membrane and this counteraction leads to change in permeability of membrane to some ions (ionotropic receptors), or affect intra-

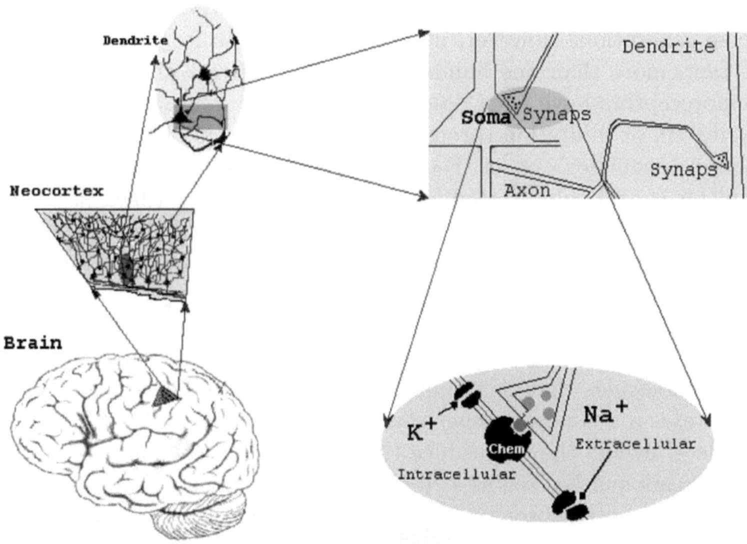

Fig. 1.2. Schematic representation of brain structure in different scales: whole brain; part of neocortex; morphology of cortex neurons; structure of soma, dendrite, axon and synaptic connections; potential-dependent channels (K^+, Na^+) and chemosensitive channel (Chem).

cellular biochemical reactions (metabotropic receptors, through one of the G proteins). Usually, one neurotransmitter can interact with several chemoreceptors, as, for example, GABA interacts either with $GABA_A$ or $GABA_B$ receptors and glutamate counteracts with α-amino-3-hydroxy-5-methyl-4- isoxazolepropionic acid (AMPA), N-methyl-D-aspartic acid (NMDA), kainite or metabotropic receptors, $dopamine_1$ and $dopamine_2$ receptors accept the effects of dopamine (there are some more kind of these receptors) and there are μ-, δ-, and κ- opioid receptors. In experimental conditions, one may affect selectively to specific kind of chemoreceptors by means of selective agonists.

Some neurons are connected with their neighbors interneurons while others are connected with foreign neural centers principal neurons. At any one moment, a lot of neurons in our brain are active, although not each one. A single interneuron influences thousands of postsynaptic principal cells. Interneurons having short axons usually inhibit postsynaptic cells, while principal cells with long axons excite them, though this is not a strict rule. Such a broad nonspecific dispersion of interneuronal influences indicates, rather, their impact on neuronal conditions of existence than a transmission of actual information.

Such an uncomplicated representation of neuronal activity could be explained by means of the existence of two neurotransmitters (for excitation and inhibition) and two channels in cellular membrane (Na^+ and K^+ channels) for spike generation. However, in reality the neuron is much more complex. At present, more than one hundred neurotransmitters and several hundreds of chemoreceptors have been described. The same neurotransmitter may exert excitation or inhibition through different receptors. The receptor dictates the neurotransmitter's effect. Facilitation and inhibition by the same neurotransmitter is probably exerted via different metabolic pathways. Sometimes, two different substances release at the same synapse. If the only function of synaptic transmission is a change in membrane potential, such vast diversity of chemical signals appears to be redundant.

Meanwhile, chemical interaction with a chemoreceptor may be continued within a neuron by means of second messengers. Second messengers are small diffusible molecules that relay a signal within a cell. The signaling molecule that activates a second messenger system does not enter the cell, but utilizes a cascade of events that transduce the signal into a cellular volume. Second messengers are synthesized as a result of external signals that are received by a transmembrane receptor protein and other membrane-associated proteins, such as G protein (GTP-binding proteins, that sense substances outside the cell and activate metabolic pathways inside the cell), etc. Their production and degradation can be localized, allowing the cell to shorten the time of signal transmission. The second messengers, cyclic adenosine monophosphate (cyclic AMP) and another cyclic nucleotide, cyclic GMP, Ca^{2+} and inositol trisphosphate that mobilizes Ca^{2+} from intracellular storages and regulates cellular reactions) unite metabolic pathways of a cell into an entire system. Second messenger the inositol trisphosphate regulates Ca^{2+} influx from intracellular stores and controls cell damage. Intracellular signaling is a very complex and diverse process, which enables cells to respond to a variety of stimuli in their interior milieu and environment.

The G protein coupled receptors are an important means for these reactions, which are the largest and most versatile protein family in the mammalian genome. G proteins activate many intracellular signaling pathways and modulate ion channel activity. There are about 1000 G protein-coupled receptors in neurons alone [1333]. Besides the G proteins, neurons may exert their influences apart from synapses through retrograde messengers, which are gases dissolved in the intra- or extracellular liquid. Little molecules of retrograde messengers penetrate in and out of neurons through lipid membrane and do not need chemoreceptors for transmission of influences between neurons. These reactions are served by energy-rich molecules, adenosine triphosphate (ATP), by means ATPases, ATP-dependent converters of energy, K^+ and Ca^{2+} channels, and by G proteins that change production of the second and retrograde messengers and modulate cellular metabolism and gap junctions, connecting the cytoplasm of adjacent cells. Gap junctions allow small ions, molecules and electrical currents to pass freely between cells. Therefore, coupling cells via gap junctions or via synaptic connections surely has different meaning. The retrograde messengers (the cannabinoids, arachidonic acid, nitric oxide (NO) and the carbon monoxide) united cells into a temporary organ. The complexity of neuronal behavior is growing as new functions are frequently being discovered [480].

Properties of ion channels are also much more complicated than we have described above. This is partially related to the excitability of neurons. Ion channels comprise a large family of transmembrane proteins. Most ion channels are multi-subunit proteins that undergo post-translation modification. They regulate the movement of ions across the cellular membrane and can be divided into groups according to their ion specificity. Only neurons and some muscle cells are excitable and produce Na^+ channels in the membranes, which are responsible for the rising phase of the AP. In order to generate spikes over and over, Na^+ channel gating must be fast and reliable. In fact Na^+ channel are transiently inactivated within a few milliseconds of opening. Na^+ channels are heteromeric protein complexes which open pores for Na^+ ions when the membrane potential decreases. There are at least nine mammalian genes encoding for Na^+ channel [219, 204, 32, 38]. Voltage sensitive Ca^{2+} channels (five types in the central nervous system) are also exclusively expressed in excitable cells such as neurons, skeletal, cardiac, smooth muscle and endocrine cells [40]. The voltage-gated K^+ channels account for a large group, and they are expressed in any cell in the body and stand out with 40 different genes [480]. Electrogenesis depends also upon Cl^- channels, a hyperpolarization-activated inward current, the persistent Na^+ current, the low threshold Ca^{2+} current, a transient outward current, etc. Mammalian central neurons typically express more than a dozen of potential-dependent ion channels [93].

Functional activity of neurons correlates not only with electro-chemical processes, but also with the micro displacement of filaments, dendrite's spines and macromolecules within the cellular membrane. Molecules and organelles move also along axonal branches with axoplasma flow, and, besides, axonal transport can be a two-way street [687]. Information is processed in the brain by both neurons and glial cells: astrocytes, microglia, etc. Astrocytes have been implicated in the dynamic regulation of neuron production, synaptic network formation, neuron electrical activity and specific neurological diseases [27].

When we are comparing the complexity of brain function with the complexity of the brain structure, we may do so with absolute satisfaction and conclude that one corresponds to the other: both are out of the range of our understanding. However, why is a neuron too complex? If its basic function is to transmit a signal to the target neuron while the only sense of this signal is exceed its input excitation over threshold, what is the purpose of so many intracellular electrical, chemical and mechanical events?

We, frequently, say "neurons send messages". This is true, but not all the truth. According to this premise, M. Stemmler and C. Koch [1178] have formulated the principle of a neuron's operation: a neuron seeks to minimize the metabolic energy expenditure due to spiking while still transmitting as much information as possible. Here immediately arises one of the plentiful questions that we meet each time we muse upon brain. A neuron does not seek to transfer information and it does not know about the existence of other neurons and knows almost nothing about the brain. Therefore, although brain can interpret action potentials, as information, a neuron somehow uses an AP for its own needs. However, even if we have rather lofty opinion concerning the neuron's inward world, the action of any living cell may carry out a few main senses: ingestion, aggression or flight. The ingestion function is doubtful for neurons having long axon, because proposed supplies are too extended from the cell soma. Great excitation of principal cells may damage adjacent neurons, whereas the inhibitory action of interneurons protects postsynaptic cells [797, 1390, 832, 1389]. Identifying mental and information processes has had a positive influence in neuroscience: they take us far from the idea of a homunculus. However, this may bring us to another extreme position: the association of motivation with thermostats and the association of consciousness with a computer [980].

So, are neuronal signals governed solely by the structure of their anatomical connections and the distribution of their excitabilities, or a neuron itself is a self-contained system, as is represented in Fig. 1.3? Many theories consider the brain as a complex network of neurons, which are approximated by simple elements that make a summation of excitations and generate an output reaction in accordance with steady activation function. Such an idealization is far from the properties of a real neuron; however, this is a very steady superstition. 35 yeas ago, one from us discussed this problem with the great Russian mathematician Andrey Kolmogorov. He explained that when he conceives of complex networks consisting of simple neurons, he feels giddiness and inspiration. However, when he thinks that, in addition, every neuron may be complex, he feels giddy too, because he feels sick. Nevertheless, the individual neurons constituting neural networks are much more complex elements than generally accepted and are involved in cognitive functions [1126, 181, 43, 1261, 1255, 1141, 58].

Plentiful attempts have been undertaken to determine what properties are responsible for the intellectual capabilities of humans. Certainly, sometimes, the advantages of our brain over that of animals are exaggerated. In some circumstances, chimpanzee memory may be superior to human memory. Young chimpanzees have an extraordinary working memory capability for numerical recollection better even than that of human adults [578]. Surely, memorizing of instantaneous visual images is not evidence of exceptional mental power. However, taken all together, diverse lines of evidence suggest that human intellect far exceeds the animal intellect. Therefore, it has seemed that comparing different brains we can appreciate the mystery of mind. For instance, brains of some of the great scientists, writers and politicians have been thoroughly investigated in order to find out distinction what distinguishes the great minds from brains of ordinary people. Different brains are always slightly different and eminent brains had distinctive features, although these features were not

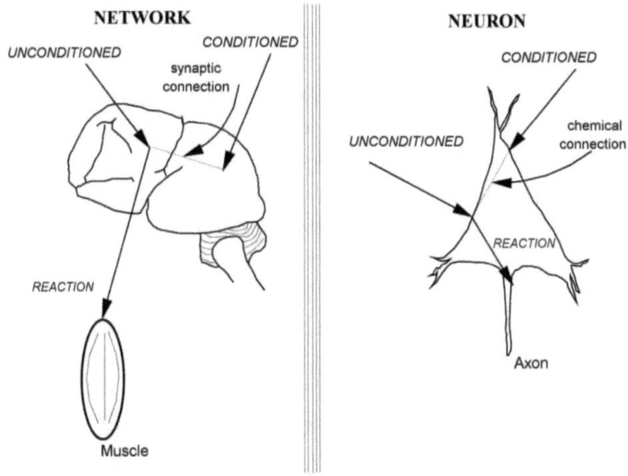

Fig. 1.3. Two hypothetic mechanisms of memory: establishment of new connections in the space of brain (left) and formation of chemical specificity. Before learning, an unconditioned stimulus produced an output reaction (muscle contraction or action potential), while a conditioned stimulus does not evoke reaction. After learning, a conditioned stimulus began to produce an output reaction with the participation of a newly emerging bond (dotted lines). This bond is spatial in a network hypothesis and chemical in a neuronal hypothesis.

the same for various eminent brains. This failure is not surprising, since we still do not know what characteristics of human brain are responsible for the superiority of our mind in the animal kingdom. Brain structure is perfected in evolution and its mass increases. Nevertheless, this is not a robust law. Large sea mammalians have larger brains than those of humans, but human brain has a larger fraction of body mass. Yet this fraction also is not an excellent indicator of brain excellence. A mouse brain is larger than human brain with respect to body mass. Humans, perhaps, take first place in the proportion of the square of brain mass to body mass, but nobody knows what this means.

An absence of quantitative differences between human and animal brains is not too breathtaking. But, by the way, there are not too many qualitative features distinguishing the human organism from animals and not every distinction determines the intellectual power of humans. For instance, only do human head hairs grow endlessly. An animal could not survive in the wild, if it would need a periodical hair-cut. Nevertheless, some important distinctions concerning the neocortex do exist. The primate cortex contains more neuronal layers, with an escalation of the form and function of cortical astrocytes and with an enlarged proportion of inhibitory $GABA_A$ergic interneurons among all other neurons [902]. The increased complexity of cortical astrocytes contrasts with the relatively limited changes to individual cortical neurons during phylogeny [683]. Frontal brain regions have rapidly expanded in recent primate evolution, consistent with their role in reasoning and intellectual function

[1233]. Various cortical areas, recently appearing in evolution, differ by their morphological organization, chemical properties, function, etc. [1196]. The monkey visual cortex contains more than 30 separate areas and the organization of the human and monkey visual cortex is remarkably similar. However, human cortex contains regions that are specialized for the processing of visual letters and words. The mystery is how cells "know" which image they hold, particularly because object representation must be constantly changing after acquiring new memories [1287].

However, only one property of the human brain is prominent. Our brain is the most complicated. Could it be that the enigma of brain is a function of complexity? We were personally acquainted with the enthusiast who, in Moscow during the period 1950-1980 tried to examine this hypothesis practically. He constructed an electronic scheme in a 20 m^2 room. He did not use any system and tried to create something large and complex. His model occupied more than two-thirds of the room's volume. One could see this giant scheme its early parts, constructed of large vacuum lamps, miniature lamps (included later), resistances, thermocouples, diodes, triodes, capacitances, pentodes, transistors, microminiature modules, and analog and digital elements that were included recently, etc. At a glance, this scheme consisted of around 10^6 elements, and when monster was in an operative state, light pipes blinked, inductances buzzed, resisters radiated heat and relays were switched on without any visible regularity in space and time. Of course, one could not observe any intellectual activity and although this does not prove that the initial idea was poor, we know that the neural system of a nematode worm contains only 302 neurons. For example, all the behavioral steps, neuronal events and gene products leading to copulation and sperm transfer are known [149]. The knowledge now is available, but what does it mean and what does the nematode gain by its particular sexual behavior? The nematode demonstrates associative learning and exhibits social behaviors. Does brain constitute a neuronal construction, or neuronal society? Correspondingly, what is changed after learning, the construction or the chemical reactions inside of neurons (and may be glia)? It is doubtful that even if we could create a precise electronic model of the brain it would work properly.

A vast diversity of brain properties will be logically ordered, when general principles can be highlighted. Right now, it is impossible to squeeze each particular fact into united theory. On the other hand, it is high time to understand where we are at present. We will try to generate some general principles that will have an explanatory force and will be in accord with some particular facts and will not contradict others. As an example, the calcium-binding proteins calretinin, calbindin and parvalbumin have important roles: they prevent Ca^{2+} enhancement within cells and protect cells from damage. However, in the visual and auditory systems of the bottlenose dolphin, calretinin and calbindin are the prevalent calcium-binding proteins, whereas parvalbumin is present in very few neurons. At the same time, in both auditory and visual systems of the macaque monkey, the parvalbumin-immunoreactive neurons [479] are present in comparable or higher densities than the calretinin and calbindin-immunoreactive neurons. It is rather possible, that different calcium-binding proteins play specific, still unknown roles in dolphins and primates. However, in our consideration one may confine ourselves to the general effects of intracellular Ca^{2+}.

1.2 Prediction of future events after learning

When an organism is faced with a new environment, it is compelled to adjust its behavior and adopt itself to altering characteristics. It memorizes an outer event, its impressions from the event, its own behavior and impression from the events that were a consequence of this behavior. When afterward similar circumstances will come, the memory will be recalled and will make easier overcoming of difficulties. This nontrivial process is called learning. Primitive forms of learning are feature of all living beings, even in bacteria and plants. A bacterium exhibits sensory adaptation in response to the continuous presence of a nonsaturating stimulus, such as the alcohol [1131] and the roots of soybean or potato form giant cells when larvae of a nematode invades the root and reach the developing cylinder [132]. The larvae feed exclusively from the giant cells, developing in the plants under control of the worm. However, these are non-associative forms of learning.

Learning leads to formation of new memory, which is usually divided into two forms: procedural (implicit, motor or unconscious) memory and declarative (explicit or aware) memory. Motor learning may be completely unconscious, but aware learning always includes an unconscious component. A declarative memory is concerned with the ability to remember or recognize objects, such as sounds, smells, visual scenes, etc. Images of declarative memory can be retained in a latent state unattainable to awareness. A procedural memory connects with perceptual and motor procedures, leading to a goal. A goal may be conscious, while a peculiarity of the procedure is usually unconscious. Both forms of memory are usual for animals and humans, but procedural memory is easier to study in animal experiments, while declarative memory is easier to study in humans, because of their strong tendency to acquire information as conscious knowledge [86]. Therefore, procedural memory has been well investigated in animals but is poorly studied in humans. Investigation of declarative memory in animals is usually restricted to a study of long-term potentiation [20, 845] although this form of neuronal plasticity evidently is not an example of memory at all (we later will return to this problem), if we do not consider any trace of a past event (like a scar, for example), as a memory.

1.2.1 Basic types of learning at the neuronal level

Primitive animals, such as invertebrates and cold-blooded vertebrates have been likened to automatons machine-like creatures whose behavior appears motorized and stereotypical. Some forms of behavior in high animals are also automatic, and may be simple, such as palpitation of the heart and breathing or relatively complex, such as swimming.

Animal behavior looks goal-directed and even "conscious", because it is not based on an exhaustive programming of a sequence of actions controlled by an external supervisor providing all probable strategies of behavior depending upon possible states of the environment. An animal behaves as if a supervisor is within it having its own needs and goals. On the other hand, humans also have the capacity for a gradual trial-and-error learning that operates outside awareness. In special circumstances, after training, people do not understand why they make the actions they do. "It seems to be automatic. My mind just seemed to tell me, "just pick it up, it's the right one" [91]. Automatic behavior in humans may be moderately complex, such as walking with the coordinated participation of many muscles, or extremely complex such as virtuoso piano playing. By the way, some forms of behavior cannot be acquired only with the help of conscious learning but without the participation of a procedural system. This is known to anybody who ever tried to learn to swim or cycle using textbook and without motor training.

The simplest forms of learning are habituation and sensitization and they are not reduced to fatigue and excitement, which are weakly selective phenomena. Habituation is a selective process. The neural system of any living being reacts powerfully to unexpected signals and attenuates its responsiveness to more frequent input, while its responsiveness to rarely delivered stimuli showed a marked average increase. The amplification of the response to rare stimuli required the presence of the other, more frequent stimulation source [1267, 371]. Response to a habitual signal recovers after an interruption in stimulation, but if this stimulus is repeated anew the decrease will be faster. Sensitization is, to some extent, an opposite process and responses to repeated painful influences are augmented [1242, 117]. Sensitization arises in response to frequent stimulation by a weak stimulus.

Associative learning is based on the pairing in time of indifferent and important events (unconditioned stimulus, US). Both the signal from an organism and from the environment (conditioned stimulus CS^+) may be used in the capacity of the indifferent event. In experiments, similar to the CS^+, other indifferent event, a discriminated stimulus (CS^-) that is never associated with the US is used as a control of specificity of learning-induced changes. Sensory information affects future movements and the movement of a sensory organ (for example, whiskers of rats) scans the environment, thus detecting the object [952, 1385]. New information enters the brain through receptors grouping into several informational channels. The environment may exert a subtle effect to the body and indirectly supply information or may directly affect the state of the body through casual inputs (US). Causal events exert important actions to an organism and may be positive (food, water, narcotics, etc.) or negative (pain, inedible food, or to threat of damage and death). Correspondingly, learning induces either attractive change towards or deflective changes from some characteristic: animals avoid negative influences and aspire to attractive ones. There are two basic forms of associative learning, classical and instrumental (operant) conditioning. Classical conditioning is characterized by the contingent reinforcement of a specific signal (CS^+), and instrumental conditioning by the contingent reinforcement of a specific type of behavior. During classical conditioning, the original response to the CS^+ is modified and may acquire the features of the unconditioned response. Instrumental conditioning, as a rule, changes the frequency of appearance of same motor action, which may include either autonomic responses such as the heart rate and almost any prominent physiological parameter [841] or it may be completely conscious. It is necessary to remark that although for description of the current behavior we use the terms conditioned stimulus, discriminative stimulus, unconditional stimulus, habitual stimulus, output action, etc. they are not concrete signals entering an animal during life. An animal meets continuous signals and generates reactions, which continuously turn into a next action. Entering these events into an experiment is a way to make an investigation easier and this may be considered as a permissible approximation of a behavior. Classical and instrumental conditioning are models of natural behavior in an artificially simplified environment. This means that continuous signals turn into discrete ones and only one action is considered.

The next step in studying natural behavior is the consideration of the complex chains of conditioned reflexes that occur when an animal consecutively executes several, generally speaking, different instrumental reactions, in order to receive reinforcement at the end of the reaction chain. During instrumental learning, an animal searches for the true signal by means of the trial-and-error method, that is, it generates by chance of a few determined actions. One from them later turns out to be correct, while the others are erroneous.

When during complex learning animals use the trial-and-error method, generation of erroneous movements does not arise "by chance" and they does not decrease after learning in accordance with the exponential low. Rather, they appear suddenly at definite phases of learning and afterwards suddenly disappear. In the complex (although discrete) environment an animal tries to accomplish some simple variants of complex chains of reflexes, in order to reach the reinforcement. Such erroneous movements can more exactly be considered as "probing" movements of accidental choice from a set of preferable actions [1251]. Sometimes, mistakes during trial-and-error behavior happen because of a shortage of reliable information from the environment or due to failures in memory for information necessary to perform the task successfully. However, this is not always the case; animals sometimes make "probing" movements despite apparently being able to remember the appropriate information [1340].

In parallel with learning, neurons in various brain regions change their activity correspondingly to the learning procedure. When habituation or sensitization are observed at the levels of behavior, responses of neurons to corresponding stimuli also decrease or increase in various areas of neural system. Therefore, traces of indifferent signals are weak and their distribution is restricted. On the other hand, neuronal activity, concerned with causal, important inputs is strong and broadly distributed in many brain areas. Nevertheless, when during learning indifferent signals becomes CS^+, their traces spread in the brain. Reorganization of neuronal activity was found in relation to any component of behavioral conditioned response: a preparatory increase in activity in response to the CS^+, between responses to the CS^+ and US, in response to the US, with an aftereffect and even change in the level of spontaneous activity. Usually, neurons re-order preexisting activities or, not often, generate activity anew. During the acquisition of classical conditioning, neural traces remaining after the CS^+ interact with the neural traces of the US. Neural analysis of this interaction is relatively straightforward because the researcher controls the onset of both the CS^+ and US. By contrast, during the acquisition of instrumental conditioning, a part of the neural system that generated the conditioned behavior has to interact with the neural traces that remain after the US. This makes neural analysis more difficult in that the researcher cannot directly know when the internal system initiates the generation of the conditioned behavior. CS^+ appearance during instrumental behavior is not obligatory, but CS^+ may be also an attribute of instrumental conditioning, for instance as the signal of the right time for instrumental reaction. This gives the investigator a modicum of control of cued operant conditioning. The CS^+ initiates the beginning of the behavioral task, and gives a reference point for time indication. On the other hand, when a CS^+ is introduced, instrumental reaction is the specific response to the CS^+. This remains the modified response to the CS^+ during classical conditioning.

We do not know whether the neural generation of the classical and instrumental reflexes has the same mechanism. In spite of a vast behavioral literature in comparative psychology concerning these forms of learning, we know very little of how they differ at the neural level. At the beginning of learning, the animal still does not have any information as to which form of learning will be presented. An animal immediately perceives that, for example, danger appears in the environment. Only during the training process can an animal determine which stimulus is a CS^+ (that is, predicts appearance of the US), whether something depends on its own actions and which action is profitable. We tried to analyze how an animal determines regularities in the experimental procedure. Decision-making is a complex multi-step process. We sharply simplify our task if we take into account only the magnitude of action in the given moment and neglect the temporal structure of behavior.

Brain exploits both fast electrical activity (10^{-3}-10^{-1} sec) and slow chemical signaling (10^{-2}-10^{3} sec). Therefore, mechanisms of higher neural functions are necessary an investigation of levels of chemical and electrical events during various forms of behavior. The most widely used object for investigations of neuronal mechanisms of behavior is the neural system of invertebrates (especially of mollusks) that posses a number of large identified neurons and a relatively simple neural system. The setup in Fig. 1.4. shows a design for experimental analysis of intracellular electrical activity during the execution of such neural functions as habituation, classical and instrumental conditioning. The roles of a specific metabolic system in neural function might be revealed by means of a chemical blockage or an augmentation of this systems. In the mollusk, classical conditioned reflexes [442, 712, 1170, 1258] and instrumental reflexes [261, 769, 881, 1270] have been described. During our own experiments, we received either classical conditioning, or the mollusk showed a modification of the probability of a specific neural pattern that occurs when it is contingently reinforced during instrumental actions [1260].

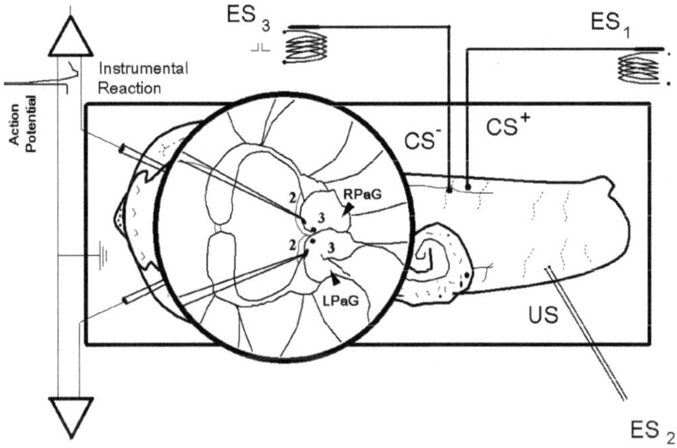

Fig. 1.4. Schematic representation of elaboration of a local instrumental reflex in Helix after infusion of pharmacological drugs. The central nervous system in semi-intact Helix preparation is augmented (circle). The preparation is used for intracellular recording from central neurons (with the central nervous system intact and connected to the periphery). The mollusk receives whole-body tactile stimuli, CS^+ and CS^-, from an electrical stimulator ES_1 and ES_3 by means of inductance and whole-body painful US from an electrical stimulator ES_2. During operant conditioning the mollusk receives an US in those cases in which beforehand chosen trained neuron does not generate an instrumental action potential. Appearance of US did not depend on activity of simultaneously recorded control neuron. During classical conditioning, unconditioned stimulus during acquisition is presented each time after the CS^+, independently of the response of either neuron. LPaG and RPaG, left and right parietal ganglion. Identified neurons LPa2, LPa3, RPa2 and RPa3 are shown by the points. Intracellular activity was augmented and recorded in digital mode. Drugs were administrated by means of extra - or intracellular microiontophoresis, or by means of perfusion.

In order to compare classical and instrumental conditioning, it is important to have data concerning the modification of responses to the CS^+ during classical and cued instrumental conditioning under similar circumstances. We compared generation of the neuronal analogs of classical and instrumental conditioned of defensive reflexes in two related neurons in the defensive system of the snail Helix [1260]. Fig. 1.5 exhibits the representative example of neuronal activity during classical conditioning and reacquisition after extinction.

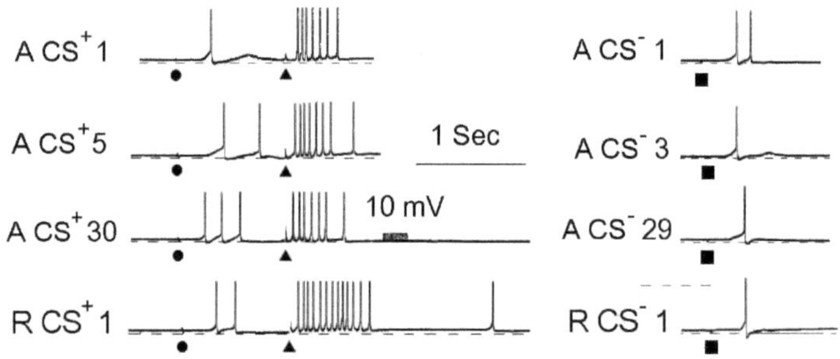

Fig. 1.5. Schematic representation of elaboration of a local instrumental reflex in Helix after infusion of pharmalogical drugs. The central nervous system in semi-intact Helix preparation is augmented (circle). The preparation is used for intracellular recording from central neurons (with the central nervous system intact and connected to the periphery). The mollusk receives whole-body tactile stimuli, CS^+ and CS^-, from an electrical stimulator ES_1 and ES_3 by means of inductance and whole-body painful US from an electrical stimulator ES_2. During operant conditioning the mollusk receives an US in those cases in which beforehand chosen trained neuron does not generate an instrumental action potential. Appearance of US did not depend on activity of simultaneously recorded control neuron. During classical conditioning, unconditioned stimulus during acquisition is presented each time after the CS^+, independently of the response of either neuron. LPaG and RPaG, left and right parietal ganglion. Identified neurons LPa2, LPa3, RPa2 and RPa3 are shown by the points. Intracellular activity was augmented and recorded in digital mode.

The dynamics of neuronal responses to the CS^+ and CS^- in our experiments demonstrate that neuronal analogs of classical conditioning satisfy the basic properties known from behavioral experiments. Averaged data demonstrates that the conditioned response increased during acquisition of classical conditioning, decreased during extinction sessions and recovered rapidly during reacquisition (Fig. 1.6). This was not the case for responses to the CS^-, which decreased to a steady level during acquisition, slightly decreased during extinction and almost did not change during reacquisition.

A short break in stimulation after the acquisition series led to augmentation of the responses to the CS^+. Similarly, after extinction, although the USs were omitted, response to the CS^+ partially recovered by itself after the break. This is an important general property of the classical conditioned reflex [947].

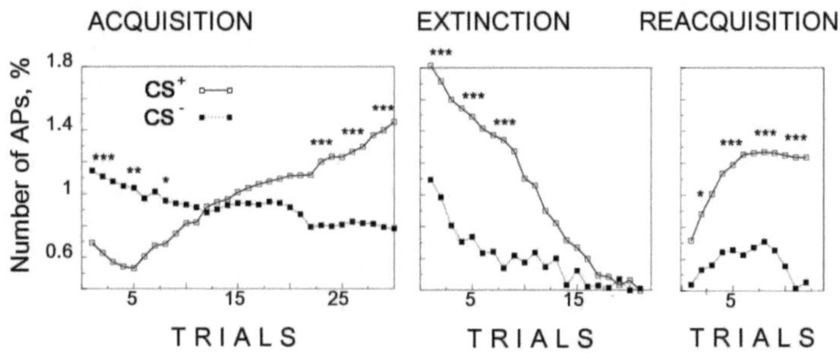

Fig. 1.6. Neuronal activity during acquisition, extinction and reacquisition of classical conditioning in Helix neurons LPa2, LPa3, PPa2 and PPa3. Ordinate the number of APs (relative units) in the responses to the conditioned (closed symbols) and discriminated (open symbols) stimuli v.s. trial number. Medians and significance in the difference between responses to the CS^+ and CS^- are shown at the top (Mann-Whitney U test, $*P < 0.05$; $**P < 0.01$; $***P < 0.001$). The training procedure consisted of the acquisition (25-35 combinations of the CS^+ and US), an extinction series (15-20 presentations of the isolated CS^+ after 5-10 min break), and a second acquisition stage consisting of repeated development of the conditioned reflex following a 20-minute break. The development of the associative connection was judged by the change in the electrical activity of defense command neurons.

Regularities for the instrumental conditioning were more complex. To ensure that the instrumental reaction occurred within the recorded neuron, the basic reinforcement schedule was delivered to only a single target neuron. Difference in the responses to the tactile stimuli CS^+ and CS^- served as the neuronal indication for the quality of the neuronal model of instrumental conditioning. US was delivered to the snail only if the preliminary selected experimental (trained) neuron failed to fire an AP within 1.5-3 seconds after the CS^+ (Fig. 1.7.). The appearance of a US did not depend on spike generation in the control neuron or on firing of any neuron to the CS^-. During training, the animal learned to determine which neuronal discharge was essential for avoidance of punishment. In this way, it was clear which neuron was responsible for the instrumental reaction.

Dynamics of changes in the response to the CS^+ during classical conditioning and during instrumental conditioning were rather different. At the beginning of classical conditioning, the animal, for a short period of time (around 7 pairings), collected data concerning the experimental procedure and, afterwards, response to the CS^+ gradually increased to the steady state. At the beginning of instrumental conditioning, response of the trained neuron to the CS^+ decreased, the animal received punishment and, after this, response of the trained neuron to the CS^+ recovered and took on the role of the instrumental reaction (Fig. 1.7). Responses of the control neuron to the CS^+ and

CS^- decreased after training (Fig. 1.8). Dynamics of responses to the CS^+ and CS^- in the control neuron were similar at the end of learning (Fig. 1.8). At the same time, responses to the CS^+ and CS^- in the trained neuron were significantly different (Fig. 1.9). Value of the trained neuron responses to the CS^+ before and after training was approximately the same, but response to the CS^- decreased. Difference between these responses demonstrates specificity of instrumental conditioning in respect to input. After learning, response of the trained neuron to the CS^+ exceeded the response of the control neuron (Figs. 1.8 and 1.9) and this demonstrates specificity of instrumental conditioning in respect to output.

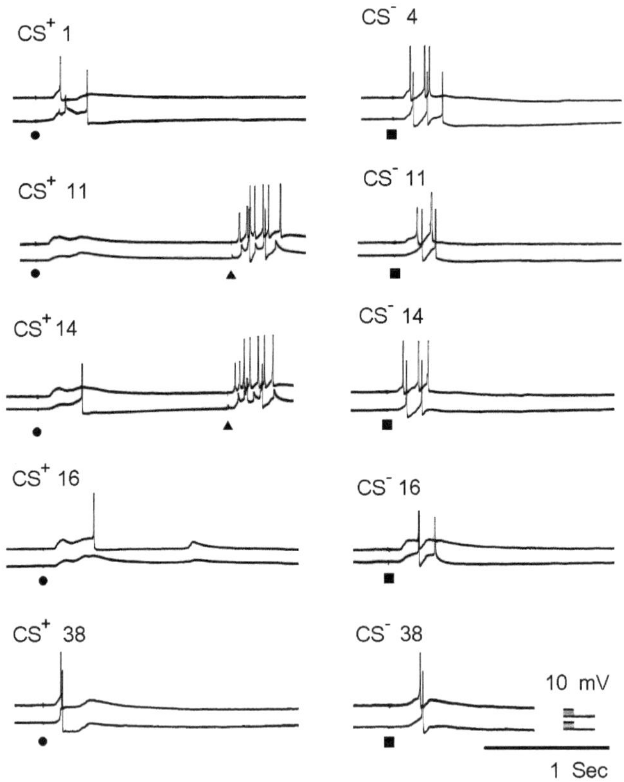

Fig. 1.7. Intracellular recordings of neuronal responses during elaboration of a local instrumental conditioned reflex with the trained neuron RPa2 (top in each frame) and the control neuron LPa3 (bottom). At the left: the number of the CS^+ is indicated for each exposure. At the right: the responses to the CS^-. Calibrations are in the figure. Stimuli 11 and 14 produced incorrect responses and painful US were presented.

Average data are presented in Figs. 1.8 and 1.9. The control neuron, as well as the trained one, exhibited a decreased response to the CS^+ at the beginning of the training. In the middle part of the learning, approximately at the same period of time when response of the trained neuron failed, response of the control neuron increased (Fig. 1.8) and only after that it decreased.

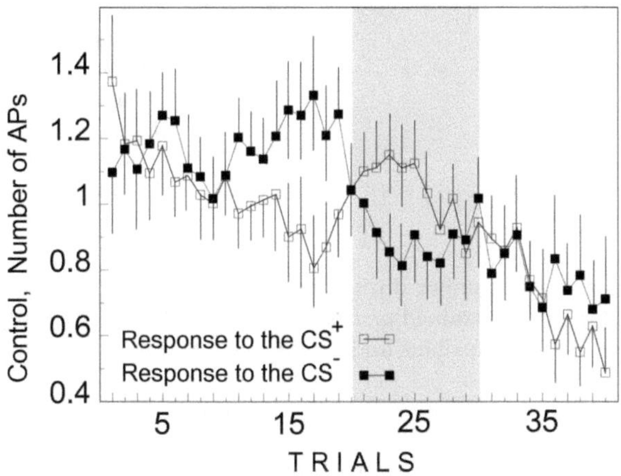

Fig. 1.8. Change in AP number during instrumental conditioning in the control neurons of mollusk Helix. Ordinate, AP number; abscissa, number of trials; medians and confidence intervals are shown. Trials 20-30 are indicated. During this interval of training, control neuron response to the CS^+ overcomes the local maximum. Symbols are at the Figure.

Comparing Figs.1.6 and 1.9 we see in both cases that the responses to the CS^+ exceed those to the CS^- and at first glance these responses seem similar. A selective increase in the conditioned response compared with responses to the CS^- during training is well known. Nevertheless, the animal somehow distinguishes the logical difference between classical and instrumental conditioning. It "fears" the CS^+ during classical conditioning and an increased response anticipates an emergence of the painful US. During instrumental conditioning, it learns to generate response to the CS^+ in order to prevent an appearance the US [1270]. Therefore, the neuronal activity observed in our experiments may be considered as representative instrumental action of the entire mollusk.

Decision-making is not a continuous process, but it consist of, at least, three levels (but usually this is a more complex multi-step process). Firstly, there is choice of the dominant demand, secondly, there is choice of the channel of action in order to satisfy this demand, and thirdly, this is the method of action in the chosen channel. An animal evidently acts as if it hopes to achieve a specific result when it generates an instrumental reaction. This means that

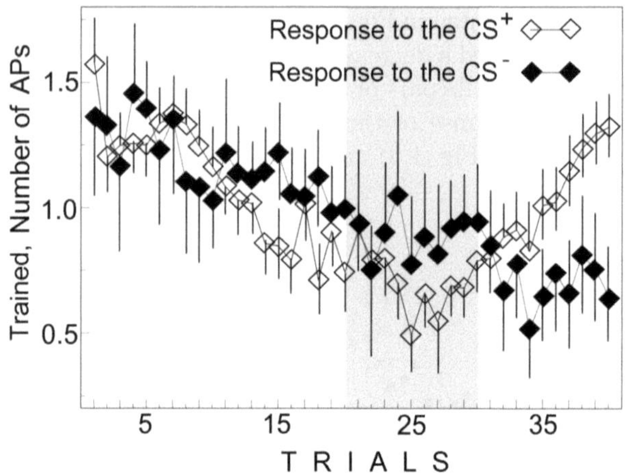

Fig. 1.9. Change in AP number during instrumental conditioning in the response to the CS^+ and CS^- in the trained neurons of mollusk Helix. Ordinate, AP number; abscissa, number of trials; medians and confidence intervals are shown. Symbols are at the Figure.

past experience allows the animal to predict probable future events in cases where properties of the environment are steady. Comparison of the cases in which the US appeared either after generation or after failure of the AP in the given neuron provides the essential criterion for this neuron's participation in the instrumental reaction: absence of a difference in a neuron's response to the CS^+ preceding a US means that the neuron does not participate in an instrumental paradigm. An absence of the difference throughout the brain means that the paradigm is not instrumental at all.

Throughout a learning session, the neural system consecutively acquired information as to which kind of learning was presented, whether a reaction of the neural system must be generated or inhibited and which instrumental reaction is correct. This process follows a multistep course and may occur at the single cell level. It may be possible for neurons to evaluate the significance of the difference between the appearance of simple events, such as numbers of trials ended or not ended by punishment, etc. During habituation and classical conditioning, neurons can perform such an evaluation by a selective modulation of their excitability [1258, 1259]. Nevertheless, we do not know whether each neuron evaluates the learning process separately, on the basis of its synaptic influx and transient regulation of the AP threshold, or whether this requires more neurons. As a first approximation, we may proceed from the hypothesis that one neuron is sufficient. However, one cannot discard the fact that neurons are also subjected to learning states of the whole neural system by all their chemical and electrical connections.

1.2.2 *How does a neuron reveal, which type of learning tasks it has encountered?

At the very beginning of training, neurons demonstrated similar dynamics of cellular responses for both forms of conditioning, classical and instrumental: an absence of learning-related changes in the responses to the CS^+. Further, the neural system discloses the particularities of the experimental procedure so that it begins generate the correct reaction. We tried to trace this process through various learning sessions (Fig. 1.10). A slow down during several trials before the start of changes in CS^+ response may be connected with recognition of the regularities of the learning process. The data presented in Fig. 1.10 describe how information related to the experimental procedure was acquired by neurons during training. The significance of the differences, as presented in Fig. 1.10, reflects only statistical evaluation of information that has been received by the neurons. We do not know what level of significance is sufficient for a neuron's decision-making. Nevertheless, these evaluations are rather essential. At the beginning of training, a whole brain and each neuron of the brain does not have at its disposal any information regarding the form of current training. This information does not exist if the neuron cannot take into consideration the researcher's intention. As far as the training develops, this information appears bit by bit and we may observe how neurons change their behavior. This process is illustrated in Fig. 1.10.

During an experimental procedure that corresponds to classical conditioning, the CS^+ precedes the US significantly more often than the CS^- precedes the US. There was a statistical significance to the difference in the numbers of times the US was given after the CS^+ or CS^- by a given stage of classical conditioning. An animal acquires knowledge about which tactile stimulus would more frequently precede the US (Fig. 1.10A, rhombi). This became significant only after 5-7 combinations of the tactile stimulus and the US. At this time, appearance of the US after the CS^+ still did not depend on generation or failure of an AP in response to the CS^+ (Fig. 1.10A, squares), corresponding to a classical procedure. Increase in the response to the CS^+ began only after around 7 combinations of CS^+ and US (Fig. 1.6).

In our experiments, participation of a trained neuron in the instrumental reaction was predetermined by the conditions of the experiment, while the statistical significance of its participation in an instrumental reaction increased with the data accumulation (Fig. 1.10B,rhombi). This significance was determined by fulfillment of two conditions: more non-delivery than delivery of US after AP generation in response to the CS^+, and more delivery than non-delivery of US after AP failure. After trials 7-10, the trained neuron acquired sufficient information for generating a conclusion about its participation in the instrumental reaction. Nevertheless, it still did not form a correct reaction (Fig. 1.9). It only began to search for an effective instrumental reaction, which may consist of the generation or failure of an AP. Note that trained and control neurons initially had a kindred function.

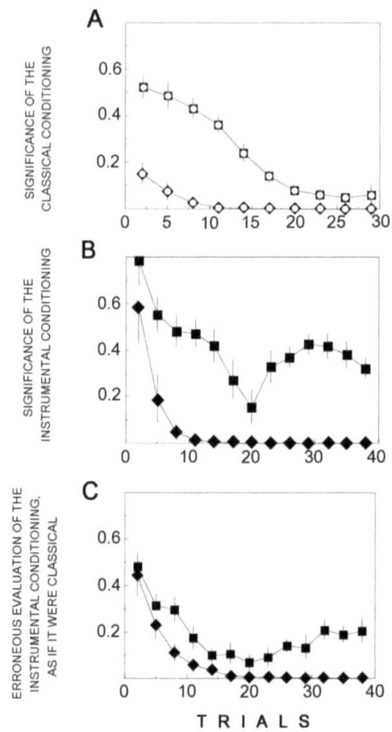

Fig. 1.10. Statistical significance of information collected by neurons during classical (A) and instrumental conditioning (B,C). The significance of a difference in the number of events corresponded to the presence or absence of the properties examined and was evaluated by the test for binary sequences. Significance of the data accumulated during the given number of trials (abscissa) was calculated for each neuron and mean values from the whole sample of neurons (banned every 3 trials) and confidence intervals ($p < 0.05$) are shown. A - squares (with circle) demonstrate the statistical significance of the conclusion that neurons generate or fail to generate an AP with different probability in response to CS^+ always followed by a US. Rhombi (with circle) designate the significance of the difference in the numbers of times the US was given in the responses to the CS^+ and CS^-. B - rhombi give the significance of trained neuron participation in generation of the instrumental reaction. This was determined by the product of two probabilities, the presence or absence of a US after an AP generation and the presence or absence of a US after an AP failure. Presence of this regularity is not sufficient, since absence of a US after an AP generation may be due to habituation, while presence of a US after an AP failure may indicate classical conditioning. Squares correspond to the significance of control neuron participation in generation of the instrumental reaction. This was determined by the product of two probabilities: the probability of trained neuron participation in the reaction and the probability that trained and control neurons either do or do not generate an AP in the same trial. C rhombi as in A. Acquisition is slower than in A, since not each CS^+ was followed by the US. This depended on the presence of the instrumental reaction generated by the trained neuron. Squares plot the significance of the conclusion that during the course of instrumental conditioning the control neuron generated or failed to generate an AP in response to the CS^+ with different probability in trials in which a US was given. This conclusion indicates that the control neuron did not respond to an instrumental procedure as if it were classical conditioning. The Fig. 1.10 was redrawn in accordance with the data [1260].

Up to trials 15-18, both the trained and the control neurons decreased their reaction to the CS^+. A temporal coincidence between output reaction and the US may be more easily interpreted by a neural system as a possible instrumental reaction than a temporal coincidence between absence of output reaction and the US. Absence of a reaction fails to provide information about which reaction must be generated in order to prevent a US appearance. In our experiments, when a CS^+ failed to generate an AP in the trained neuron and caused a US delivery, it also induced APs in many other neurons, including the control neuron and other similar neurons, as well as those producing a postsynaptic potential during AP failure in the trained neuron (Fig. 1.7). Therefore, AP failure in response to the CS^+ may be the first attempt of the neural system to counteract a US appearance after an AP generation. However, if a temporal coincidence between an AP and the US is more effective for generating the supposed instrumental reaction, then the neural system must decrease its reaction to the CS^+.

Determining which neuron should be responsible for an instrumental reaction is a further problem that a neural system needs to resolve. Although our experimental procedure was directed to the trained neuron, reactions of both trained and control neurons were weakly, but very significantly, correlated (coefficient of correlation $r_{627} = 0.26, p < 0.0001$). Therefore, control and trained neurons sometimes generated similar reactions and control neurons acquired erroneous information as if its response participated in the instrumental reaction. In order to evaluate to what extent the control neuron received information that its participation is important for the generation of the instrumental reaction, we compared the number of events during which the control and trained neurons generated similar reactions (both generated or both failed to generate an AP in responses to the CS^+ in the same trials) and the number of events when reactions of the control and trained neurons did not coincide with respect to AP generation in response to the CS^+ (Fig. 1.10B,squares). At the beginning of training and up to 15 trials, it was uncertain whether a control neuron participates in the instrumental reaction (probability of participation was around 0.5). In the middle of training, when a control neuron generated an erroneous instrumental reaction, the trained neuron decreased its participation in the reaction. At this point, control and trained neurons failed to generate APs in counterphase. However, till the very end of training the control neuron did not acquire information as to whether it participated in the instrumental reaction or not. At the same time, participation of the trained neuron was highly significant (Fig. 1.10B,rhombi).

During instrumental conditioning the appearance of a CS^+ was not always followed by a US, because this depended on generation or failure of an AP in the trained neuron's response to the CS^+. Therefore, the difference between the responses to the CS^+ and CS^- (Fig. 1.10C,rhombi) was acquired later than in the case of classical conditioning (Fig. 1.10A,rhombi). However, the control neurons preferred to respond to the CS^-, rather to the CS^+ earlier than it began to generate an erroneous instrumental reaction to the CS^+

(Fig. 1.8). As pointed out earlier, during instrumental conditioning the control neuron did not receive reliable information about its participation in the instrumental reaction (Fig. 1.10B,squares). At the beginning of training, the experimental procedure for the control neuron may look like classical conditioning with partial reinforcement. The control neuron acquired this information during trials followed by the US. Fig. 1.10C (squares) demonstrates the significance of the conclusion that the control neuron showed different probabilities of generating or failing to generate an AP in response to the CS^+ (in the trials followed by the US). This conclusion was not significant up to 10 trials and, therefore, the stimulus paradigm it received corresponded to a classical paradigm. However, the control neuron could not elaborate classical conditioning, since up to the 10th trial the contingency between CS^+ and US was still absent (Fig. 1.10C,rhombi). In the middle of training, when a control neuron generated an erroneous instrumental reaction, it received significant information that a US was present after a CS^+ when the neuron generated or failed to generate an AP with different probabilities (Fig. 1.10C,squares). This situation ceased to correspond to the classical conditioning paradigm.

During classical conditioning, a CS^+ always preceded the US, independently of generation or failure of an AP in the response to the CS^+ and, therefore, there is no difference in the number of times a US was given after AP generation in response to the CS^+ and after AP failure in response to the CS^+. During successive trials, the neurons accumulated data that a US appearance does not depend on their reaction to the CS^+. During the first half of training, the neurons showed the same probability for generating or failing to generate an AP (Fig. 1.10A,squares). Therefore, AP generation in response to the CS^+ was not the product of an instrumental reaction. At the end of classical conditioning, the neurons generated or failed to generate an AP with different probabilities (neurons mainly did generate APs). However, at this time, the nervous system was responding to the experimental procedure as if it were classical conditioning.

These observations revealed distinct differences between a classically versus an instrumentally conditioned increase in spike number. Inasmuch as these differences were observed in the same identified neurons, they strongly suggest that the mechanisms that produce the conditioned changes are also different. For example, a cellular response to a painful US increases during training only in the neurons responsible for the instrumental reaction [1260]. It might be expected that the dynamics of responses to the US reveal the greatest difference between the two forms of learning. During classical conditioning [527, 1325, 508, 401] and instrumental conditioning [553, 1245], a CS^+ paired with a harmful US reduced behavioral and neuronal responses to the harmful US and thus had a defensive function. As for an instrumental reaction, it prevents the appearance of a US.

Learning makes easy a choice of profitable behavior, because the sense of learning consists in the prediction of an environmental reaction. Our data demonstrate that even behavior controlled by a single (trained) neuron develops as if evaluation of the future result of an instrumental action restricted by a given neuron is predicted by the processes running within this neuron.

1.3 Location of functions in the brain

Brain has architecture that is specific to a given species and gradually develops during evolution and during individual existence, though correspondence within the species is not ideal. Subtle structural peculiarities of the brain can depend on heredity, but they are not decisive for brain function. On the contrary, brain functions tend to remain steady after a light or even an intermediate crash in morphological structure. Brain morphology of an adult human has the most complex organization. Therefore, the brain cannot be considered as the gland that produces behavior. Rather, neurobiologists and theorists are inclined to accept another extreme and consider the brain as a super perfect construction with an ideal structural design. Given appropriate neural centers, it would be naturally at the beginning of a book on brain to describe the functions of all brain sections. Although such a description is out of the scope of our book, we will be compelled sometimes to appeal to localization of some brain functions. Unfortunately, at the present time, we know the roles of various brain centers only approximately. Some functions of brain seem to be distributed within neural tissue and some brain centers seem to have reference to a great number of functions. For example, brain imaging studies in humans indicate that learning and memory involve many of the same regions of the cortex that process sensory information and control motor output [1287].

There are obstacles preventing comprehension of function localizations in the brain. In order to determine functions of a neural center we may observe activity in this center (electrical or chemical processes, temperature, blood flow, etc.) during functioning or evaluate the peculiarity of functioning after influence to this center (electrostimulation, pharmacological blockage, cooling, extirpation, etc.). Both methods are beneficial for simple systems but inappropriate for complex ones with relation to complex functions. We know what makes heart, kidney, chewing muscle, tear glands and so on. We can explain what functions brain performs although somewhat more vaguely and verbosely than, say, kidney. Yet, we do not know what functions are performed by the neocortex, hippocampus, nucleus caudatus or amygdala if we are not satisfied by simplistic responses like, "Neocortex is the center of high nervous functions". Which of brain's supposed functions ought we to choose: excitation or inhibition, synthesis of amino acids or endorphins, elaboration of temporary rhythms or carrying out the relay functions, unit of errors or unit of forecasts, a memory store or only the memory about something special? Artificial hindrance of performance of a function by a given neural center may not illuminate the problem, since brain possesses strong compensatory capabilities and function must be preserved at the expense of another brain area. Many functions are not, really, scattered in the brain, but they are capable of being compensated. This would not mean, however, that removed of a brain center was not essential in the normal conditions. Much more important would be those discoveries that demonstrate irreversible damage of a function after brain damage. However, a powerful mechanism of compensation prevents irreparable damage or makes it infrequent.

Recently, methods have been developed for visualization of living brain activity during normal behavior: positron emission tomography and functional magnetic resonance. These methods give admirable portraits of the working brain and bring to light sequences of activations in many brain areas during normal behavior. It is also possible now to do non-invasive irritation of deep brain areas. Although these methods are based on recording of nonspecific processes such as oxygen consumption, glucose turnover and blood flow, they have disclosed some important details of neural function.

There are many examples that point to a given brain area as being related to a given brain function, but this pertains to input and output that is sensory and motor functions. Injury of afferent or efferent pathways damages image recognition or motor control and leads to the most prominent impairment of behavior. This is a trivial result, since interruption of direct connection must result in the failure of corresponding function. For example, olfactory nerve damage inevitably leads to anosmia. In the same way, it was established that rostral regions within the neocortex control motor and executive functions, whereas caudal regions process sensory signals [1196]. In vertebrates (except fish), each motor neuron is always part of a pool of tens, hundreds, or thousands of similar neurons that serve a common purpose. This repetitive arrangement contrasts with the pattern of connections in invertebrate muscle, where one, or at most a few, identifiable motor neurons innervate each identifiable muscle fiber, forming a stereotyped circuit exclusively dedicated to a particular function [738]. Relatively strict localization of some particular brain functions does exist [58]. Certain neural functions specifically depend on physiological processes that happen in restricted areas of brain. Even individual neurons may play a role in performing cognitive functions of the brain. Many specific cognitive functions such as language disorders, face detection and mathematical disability, are carried out by groups of highly specialized neurons whose roles are genetically predetermined. A famous example is the speech center (Broca's area) in the opercular and triangular sections of the inferior frontal gyrus of the frontal lobe of the human cortex, which is involved in language processing, speech production and comprehension. A linguistic processing system located in a specific domain in the left hemisphere is specialized for language independently of the particular modalities through which language is perceived and expressed (deficit of reading, writing or spoken language). In this area, damage leads to a specific language disorder. The circadian rhythm is also a function of individual neurons and is determined by intracellular molecular processes. Their dysfunctions cannot be compensated by other elements of the nervous system or by a reconstruction of interneuron connections within networks [58].

There are specific brain areas for emotions and cognitive signals[1] and separate brain regions are involved in different aspects of emotion[2] (positive negative, fear sadness, emotion with cognitive demands and instrumental behavior).

A local brain area may be responsible for face recognition in primates. Recently it has been described that cells in the medial temporal lobe and the hippocampus of humans were responding to individual objects (famous persons such as actresses Jennifer Aniston and Julia Roberts, ex-president Bill Clinton, the basketball player Michael Jordan, the Beatles and landmark buildings or objects) [1005]. The prominent feature of these results was the consistency of responses across different images of the same object. Sometimes neurons generated specific reaction not only to the given pictures, but also even by letter strings with their names. The number of images in the given screening session was around 100. 14% of units were selective to only a small amount of images, as 3% of the projected pictures. The cells had a very low baseline activity and responded in a selective manner. Clearly, this was a kind of invariance based on learned associations, not geometric transformation of visual structure, and these cells encode memory-based concepts rather than visual appearance. Nevertheless, detection of 3 out of 100 pictures is far from ideal as face recognition. It would be impossible to find a specific neuron for a specific face, if ideal specificity exists. It is necessary to make a reservation that this criticism is not proof that "grandmother cells" (cells corresponding grandmother, my or yours) are absent in the brain, but only that a search for them is labor and time consuming [500]. However, less selective neurons that differentiated face from non-face were found and there is a direct causal link between the activity of face-selective neurons and face perception. Microstimulation of face selective sites in the inferior temporal cortex of primates, but not other sites, strongly prejudiced the monkeys' decisions towards the face category [8] and damage of this or related systems can lead to an impairment in recognizing individuals by the sight of their faces [1037]. It was also

---

[1] The medial prefrontal cortex is one of the most frequently observed areas of activation across all types of emotion stimuli [974, 485]. An area in the left-inferior prefrontal cortex has been observed to be active across a wide range of tasks requiring subjects to retrieve words, while the tasks requiring subjects to retrieve information from a specific study episode have reliably activated the right-anterior prefrontal cortex [178]. Emotion and cognition, supposedly, activate separate areas of the frontal midline [20], but this is not for a certainly [968]

[2] There is specificity of location for approach/withdrawal valence in some brain areas and, particularly, in the frontal cortex [207, 974, 1313, 485]. Brain areas responsible for dissimilar biological motivations are different, but are partially intersected [1257]. Regions of a brain that are more active during the ambiguity (vague, because of missing information) and during risk (under varying levels of probability of the result) conditions are different [565].

revealed that there exists discrimination between faces according to an emotional relationship to a given face. So, when participants alternately viewed a photograph of their beloved and a photograph of a familiar individual (the head only), activation specific to the beloved occurred in dopamine-rich areas associated with mammalian reward and motivation, namely the right ventral tegmental area and the right postero-dorsal body and medial caudate nucleus. Activation in the left ventral tegmental area was correlated with facial attractiveness scores [57]. Interestingly, local damage to a specific area in the visual association cortex in both hemispheres or in the right hemisphere can cause face agnosia, that is, face-selective neurons in the temporal cortex are located differently on areas responsible for face agnosia [58]. Nevertheless, some face-selective neurons in the inferior temporal cortex altered the relative degree to which they responded to different members of a set of novel faces over the first few presentation of the set [1037] and, hence, response to a given face is not predetermined. Activation of some parietal neurons is necessary for visual contents but is not in itself sufficient [72].

Another example of sensory specificity is "neurons of place". In the hippocampus was found the cells that are activated only when animals were settled down in the determined place [908]. Neurons of place were found also in the neocortex. Neurons of place are not something that never changes. Ensembles of place cells in the hippocampus undergo extensive “remapping” in response to changes in the sensory or motivational inputs to the hippocampus. The nature of hippocampal remapping can be predicted by ensemble dynamics in place-selective grid cells in the medial entorhinal cortex, one synapse upstream of the hippocampus [439]. Sensory maps of the neocortex are adaptively altered to reflect recent experience, pharmacological influences and learning. Plasticity occurs at multiple sites. This view contrasts with the classical model in which map plasticity reflects a small set of cortical synapses [377].

Nevertheless, participation of a given area in a given function does not exhaust all aspects of given function maintenance, especially in connection with mechanisms of memory. For instance, the basolateral complex of the amygdala is involved in modulation of memory for aversively motivated behavior. Even so, memory for this kind of training is modulated by post-training drug infusions administered to many other brain regions. Modulation within the amygdala is not sufficient to effect memory. So, inactivation of the amygdala with lidocaine prior to retention testing did not block memory of any kind of training [821]. Paradoxically, lesion of the basolateral complex of the amygdala sometimes improves learning. Insufficiency of reversal learning after orbitofrontal damage is eliminated by additional lesions of the basolateral amygdala [1167].

Beside, damage to a specific function concerning a specific area, damage to a specific area of the brain leads to partial impairment of irrelevant functions of areas distantly located from the injured place. For instance, extirpation of the visual cortex prevents elaboration of a conditioned reflex to light, but also slows down elaboration of a conditioned reflex to sound [767]. Similarly, people with comparatively restricted lesions of the ventral right hemisphere show longer reaction times in response to targets on the left side when they pertained to the right side. The dorsal parietal area was the only brain region that increased activity on the lesioned side during the acute to the chronic stages. This was accompanied by reduced activity in the non-lesioned hemisphere [263]. Moreover, even impairment of sensory systems and efferent control may be compensated after injury. For instance, damage to cortical representation of the leg leads to lack of motor control, and then gradual recovery the function. Surprisingly, after recovery, it is impossible to find new cortical representation of the same leg [1255]. Likewise, the post-lesion lack of vestibular input of central vestibular neurons may be compensated for by changes in the efficacy of the remaining sensory inputs. However, the compensational process can also rapidly become independent of these external cues. Vestibular compensation first relies on external cues and finally on changes in the intrinsic properties of the vestibular neurons themselves [1304]. Nevertheless, the damaged function is not recovered completely. So, even if damage to vision after a stroke is ameliorated, in the chronic stage patients improved in their ability to detect stimuli, while even after many years they had a large deficit in reaction time [263]. The most remarkable example of recovery pertains to sexual function. Sexual behavior is severely compromised in rats after removal of the medial preoptic area, the main sexual center. However, sexual behavior does not disappear completely: fewer than 30% of rats with medial preoptic area lesions mounted, fewer than 15% intromissed, and fewer than 3% ejaculated [758]. In humans, sexual arousal without erection includes cortical, limbic, and paralimbic areas, but only a subpart of these areas participate in the development of an articulated sexual response including full penile erection [383].

We may conclude that some general functions have specific localization in normal brain. These are particular and ancient functions, such as circadian rhythm, breathing, self-reception in determined places, simple emotions, image receptions and efferent control having inherent or even an automatic nature. In some cases, but not as a rule, damage to these areas leads to irreversible failure of the corresponding functions. Unexpectedly, contemporary functions having a voluntary nature such as control of language and face recognition also have precise localization. This is, evidently, determined by an increase in the degree both of brain differentiation and of function localization in higher animals and especially in humans. To a lesser degree, localization of functions concerns those functions acquired during learning and connected with memory. These neural functions after brain damage are in most cases only partially deteriorated and later they are recovered, although this improvement is, nevertheless, incomplete.

1.4 Location of memory in the brain

We guess the brain is connected with behavior. When we are studying brain functions, we analyze behaviors that are connected to memory, whether recent, past or innate. This involves all cases, even if we examine brain activity at rest. We cannot avoid consideration of the memory problem, even if we take an interest in any other function. It has become evident that many nervous centers are involved in the phenomenon of memory, no single and specific memory center exists, and various parts of the brain participate in the representation of any event [1255, 845]. Various centers perceive signals, create the aware image, record new and call to mind past information and generate output reactions, by which the investigator may judge that a memory system functions properly. Learning-induced alterations in neuronal electrical activity usually coincided with formation of material traces (c-Fos immunoreactivity, transcription, etc.) in the same area. The long-term changes in neuronal firing produced by instrumental learning occur in the same areas where there were activated during acquisition of the same task. The neurons participating in learning are not without fail those neurons which store the memory. Although neuronal correlates of learning are observed in countless brain neurons, recollection of information is a much more local process. Only a small number of neurons, when compared with those initially active in learning, form and retain task-specific firing patterns in well-trained animals [1203]. This corresponds to the discoveries of I.P. Pavlov [947] of phase generalization and specialization of conditional reflexes. Concordance in time of the activity in a local brain area with the memory process does not tell us a lot. On the other hand, if stimulation of a brain area causes recollection of images [958], this does not mean that memory is kept right at this place. Stimulation may excite another brain area. This is, evidently, a very complex problem, but there is one characteristic of memory, that allows one to hope that the puzzle will be solved: memory retains material traces and they can be found. Essentially, memory traces are kept in the healthy brain, in opposition to traces of trauma, insult, infection, etc.

Where does brain store its memory and how are memories are stored within brain? We do not know how images are kept in memory. Specific images may be kept in local places and then there is a one-to-one correspondence between memory elements and spatial coordinates in the brain. In such a case an image could recall when this particular place is activated. As we will demonstrate, this is not the best idea for memory mechanism, since memory is not a local process. Therefore, before discussion of "how" memory is stored in the brain, let us consider "where" the brain stores its memory. At first glance, the question "where" is much more simple than the question "how". Yet it turns out to be impossible to find memory without finding out its mechanism. Memory activates motivational and emotional centers and another question one may ask is: where do these aspects of memory lie in the brain? Also, can perception and recollection of images out of memory be indifferent? Habituation testifies against indifference. If something is not recorded, it can not be remembered. Emotional load of an image promotes its storage.

There is a distinction to be made between safety of conscious and unconscious memories. If, as was shown, reproduction of a previously elaborated system of reflexes in rats worsens with an increase in neocortex mass that has been ablated after 180-200 pairings of CS^+ and US (a 3-linked chain of conditional reflexes with an interval between CS^+ and US of a few seconds), the reflex system does not depend at all on the mass of the cortex removed after 350-400 pairings (phase of automation). Neocortex does not participate in automatic behavior [1250]. As we already have mentioned, reproduction of more complex behavior is more sensitive to deterioration. "Complex behavior" in this context means a participation of awareness in the control of behavior. Automatic behavior may also be complex (at the stage of information acquirement), but during the automation phase it is launched as a whole performance. Aware memory is, probably, more vulnerable, especially aware memory recall. It is sensitive to impairment to many places of brain but it has a tendency to recover, and recovery time depends on the severity of damage. In reality, patients with amnesia can show selective impairment in conscious memory in comparison with preserved skill learning. Enhanced sensitivity of a complex or aware behavior to damage is, evidently, a general rule. So, genetic loss in mice of the protein subunit responsible for fast inactivation of K^+ channels results in an alteration of the excitability and damage of complex learning, while simple learning is not altered [468]. Complex behaviors need, to a large extent, to exploit a diverse past experience within awareness.

Difference between localization of earlier and recent memories has been also described. Using functional magnetic resonance imaging in healthy older adults, it was shown that medial temporal lobe lesions typically produce retrograde amnesia characterized by the disproportional loss of recently acquired memories; the hippocampus participates in processes of memory lasting a few years, while entorinal cortex is associated with memory extending up to 20 years [511].

The precise role of specific brain structures in storing/retrieving remote memories currently remains unclear [657]. If centers of memory exist, the search for them is complicated because of the existence of the powerful apparatus of self-recovery. It has become apparent that no single memory center exists, and many parts of the nervous system participate in the representation of any single event. Extirpations of brain tissue or other influences to neural centers have not revealed the center of memory and give rise to the hypothesis that the ability to memorize is an inalienable property of any neuron. K. Lashley formulated the law of mass action, according to which the extent of a memory defect in rats may be correlated with the size of the cortical area removed, but not with its specific location [705]. That is to say the cortex (and may be the brain) operates as a whole. Later this result was confirmed in rats [1250], in chickens [1040] and even in the mollusk Aplysia. The gill withdrawal reflex of Aplysia is generally depicted as a simple behavior mediated by a simple neural circuit in a simple organism. However, the reflexive withdrawal of the gill and other mantle organs is anything but simple. Even here, memory may not be localized to specific loci, but rather may be an emergent property of physiological mechanisms distributed throughout the entire circuitry [769, 278].

High neural centers, such as the cortex and hippocampus, play a colossal role in mammalian behavior. They have a screen-like structure and represent the world as a point-to-point correspondence, for instance, body surface, vision field, etc. This representation is continuous and closest points of the world are represented in adjacent points of brain. Of course, this is not ideal correspondence, since a closed object in three-dimensional space (as, for instance, our own body) is impossible to represent at the open two-dimensional surface of the cortex or hippocampus. Nevertheless, this does not prevent the creation of an inner model of the outer world and additionally to correlate events with a determined point in time. After cortex removal, an animal is a severe invalid. Observing such an animal, one may detect a deficit of almost any behavioral function with the exception of the inborn behaviors and the simplest forms of learning. These invalids behave much more primitively, than a simple vertebrate, fishes or reptilians, which do not have a cortex at all. Interestingly, learning after brain injury slows down, but the vegetative component of learning almost does not differ from the norm [103]. Therefore, how may we agree that the temporal lobe or hippocampus is the centers of learning or memory? Evidently, higher neural centers in mammalians have some high functions and the absence of these functions prevents the realization of primitive functions. One such function is, evidently, the possibility to integrate past events into a whole, intact and ordered representation of the world that relates to aware perception. Can awareness be present, even in a primitive form, in an isolated part of the neural system or in one neural cell itself? We will return to this problem later. Although a memory center is absent, maybe particular elements of memory, as images, are stored locally (grandmother neurons)? This suggestion looks doubtful, too. Usually, brain injury impedes the most complex forms of learning, memory recall, and weak, recent and unessential memory. Behavior is recovered during the weeks or months after injury, so evidently, memory existed and exists in the remaining regions of the brain. Long-term memory, evidently, never disappears, only its reproduction is worsened. It is difficult to imagine that a memory initially is recorded in one place and later is over-recorded in other places, or, say, is scattered through brain tissue.

If a function is once recovered after a temporary down-turn of some brain area (for instance, cortical representation of the paw), repeated down-turn of the same or any other brain area does not hinder the same function. Therefore, even localized functions have the potential to become non-localized. As for memory, even if a special form of memory is localized in a certain part of the brain, it can be delocalized after deletion of this zone. This means that even before removal, the memory traces were spread within the brain but only a limited volume was actually responsible for this trace. Lesions of different brain areas show specificity, but shortly after injury non-specific impairments of memory also surface. Damage of various brain areas induces temporary impairment in long-term memory. This suggests absence of the localization of specific memory element in specific brain zones [1316]. It is not without possibility that temporary impairment of behavior after brain damage arises not because of breakage of the neuronal structure, but as a result of influence of the wound itself. Brain may feel its wound affecting its chemical background. Recovery from mild head injury is displayed typically 1-3 month post-injury. Temporary boundaries of available memory after traumatic amnesia are gradually narrowed. Memory is partially affected during injury of neural system. Wonderful, these harms finally become too little to be a problem.

Recordings of neuronal activity during learning have shown that a large amount of brain neurons change their behavior. The frontal cortex, hippocampal system, basal ganglia, hypothalamus and other areas, each of which has neurons whose activity undergoes systematic evolution during learning [1006, 662, 1348]. Even during simple classical conditioning of the gill-withdrawal reflex in the simple nervous system of Aplysia hundreds, if not thousands, of neurons are active during learning [476]. Nevertheless, various neurons do not promote learning in the same degree. When, for example, monkeys were trained to learn new visuomotor rotations for only one target in space and neuronal activity was recorded in the primary motor cortex before, during and after learning, specific elevations of the firing rate of neuronal activity were only observed in a subpopulation of cells with directional properties corresponding to the locally learned rotation; that is, cells only showed plasticity if their preferred direction was near the training, while non-responding cells did not participate [951].

Learning may be artificially directed to a restricted population of neurons. Capability for learning is displayed in restricted parts of the neural system, when they are artificially divided from the rest of brain: in the peripheral neural system (without head) [376, 345, 1342, 1351], in a brain slice [734, 949], in a tissue culture [670, 1121, 371] and in an isolated neuron [1157, 348, 1419, 1301]. Neuronal analogs of learning are observed after transference of CS^+, US, or instrumental reaction in the neuron's vicinity [616, 710]. These primitive forms of behavior are not concerned with awareness, or at least we do not have evidence that any degree of awareness is present, say, in isolated brain slices (an example of tissue culture). We also do not know if diverse past experience concerning different habits may be combined into whole representation within isolated parts of brain, but this question is related to the problem of awareness.

When we are speaking of memory that is spread within the brain, this means that each image is spread. A specific image is never harmed; memory damage concern to more general categories: impairment related, say, to recent or unessential events, and so on. We cannot remember the scrap of a given image. Yet, this pertains only to memory. Partial loss of function happens during perception of actual events. A patient with damage to the visual cortex may see only half of a visual scene [263]. Output of motor reactions after brain injury also may be damaged partially. Therefore, if each image is spread within the brain, how may we resolve the "binding" problem that is a necessity to unite a whole body of information relating to a given image within a single neuron? An image cannot be spread over brain as a photo. Each feature of the image must be spread and each image must be projected through a great number of neurons. This would be possible if each neuron remembers little by little about each and every event. It would be sufficient, if a neuron would remember only the necessity to generate or fail to generate a reaction in response to a given stimulus. The outer limit of images which a neuron theoretically may discriminate is 2^N, where N is the number of synapses converging to a given neuron. This colossal number to a marked degree exceeds the number of images which we meet during our life (2^{22}, if we remember an image each minute). In order to have the capability to recognize 2^{22} images, a neuron must maintain specific chemical processes in 22 different synapses and all these processes ought to have the capability for mutual interactions. However, even this biochemical system is overcomplicated. Later we will give a theoretic description for these processes.

Thus, if the center of memory exists, it possesses some very perplexing properties. When memory grows old, it is rewritten from the place to the place, while a special "library" preserves important data. Besides, skill memory is kept in files of fast access and when a species is perfected it opens new memory centers, especially exposed to spoiling. This picture looks strange and unbelievable. Additionally, memory centers seem to have a common urgency system, which temporarily turns off all memory centers when damage threatens any one of them. In other words, memory may be spread within the brain, but access to memory depends on many circumstances. These data mean that one cannot identify neural centers with the elements of learning machinery. It is impossible to hope to find centers for short and long-term memory, apparatus for recording and reproduction, timer pulsing, a catalog of addresses, a coincidence detector, etc. All these functions are concentrated in each place of the neural system. Probably a small part of neural tissue is capable to learning and memory and, maybe, a morphologic unit of brain a neuron - serves as its functional unit. It important to understand what is the upper limit of neuron function.

Nonetheless, in neurobiology there are examples of learning forms that appear to be localized in narrow areas and are incapable of compensation after removal of these areas. These are the songs of birds and classical eyeblink conditioning. The latter example has undergone extensive analysis. Classical eyeblink conditioning (about 100-200 pairing) typically involves paired presentations of a tone or light as a conditioned stimulus (CS^+) and a periorbital shock or air puff as an unconditioned stimulus (US) and represents unconscious skill learning. The essential neural circuitry for acquisition and performance of classical eyeblink conditioning is found disposed in the cerebellum and related brain stem structures (especially the interpositus nucleus). The local nature of eyeblink conditioning is displayed in morphological alterations that accompany its manifestation; the number of excitatory synapse within the interpositus nucleus increases after training by 1.6 folds [640]. At the same time as this system appears to be highly localized, many other brain areas beside cerebellum are recruited during eyeblink conditioning. Learning-related activity has been found, for instance, in the auditory cortex [1441, 1356], hippocampus [865] and in a variety of other structures including the neostriatum, thalamus and trigeminal nucleus [1175]. Comprehensive analysis reveals that even though learning-related plasticity occurs in many regions of the brain, maintenance of plasticity is critically dependent on processes that occur in the cerebellum.

Although the cerebellum appears to be absolutely necessary for establishing the basic eyeblink conditioning in all training situations, the cerebellum alone does not encode all features of eyeblink conditioning. Other brain structures and circuits are critically involved in encoding various aspects of classical eyeblink conditioning and it is likely that plasticity in at least some of these areas may be established independently of the cerebellum. While interpositus nucleus lesions abolished eyeblink conditioning established with a tone CS^+ and air puff US (a somatic response), the same lesions had no effect on conditioned changes in heart-rate (an autonomic component of the response). This means that when eyeblink conditioning cannot be recollected after lesion of the interpositus nucleus, safety of the memory is not disturbed.

Eyeblink conditioning is a simple form of automatic skill learning, characterizing by a low specificity in respect to CS^+ and CS^-, which was never paired with the US. Only after 200 presentations of CS^+ and US and 200 presentations of single CS^-, conditioned motor responses in neurons of the cerebellar interpositus nucleus began to be slightly different (CS^+ - 85%, CS^- - 60%). Neuronal response to the CS^+ exceeded response to the CS^- with latency at 200-275 ms (75-0 ms before US). Responses with shorter latencies were not different [865, 411, 1175]. Interestingly, another well known higher localized specific response, to faces of famous persons, also had long latency, between 300 and 600 mS after stimulus onset [1005]. It is necessary to denote that specific response to the CS^+, typically has short latency, from 7 to 100 mS and usually from 10 to 30 ms [1292, 1003, 1351, 352, 1037, 1356, 951, 8]. Evaluating the biological significance of the signals shows that the sensory part of responses is specifically altered. By the way, during eyeblink conditioning to a click (as a CS^+), specific response to a click in the auditory cortex of the awake cat shows latency as short as 8-12 ms [1356]. A secondary (1216 ms) temporal component of response to the click was not specific. The executive parts of responses also may change, but they are observed already after decision-making. Thus, later components are generated after completion of decision-making. Therefore, a highly localized eyeblink conditioning response in the cerebellar nucleus is, evidently, an executing component of the reflex and it, certainly, is not connected with the memory.

As in the case of eyeblink conditioning multiple brain areas are involved also in other forms of primitive behavior: in conditioning of siphon withdrawal in Aplysia [769], vestibuloocular reflex gain [429], two-neuron spinal reflex [1351], etc. Simple behavior changes do not happen due to single alterations at single areas. Not only brain, as a whole, but neural centers are multitasking structures and participate in many behaviors. On the one hand, a given behavior, under artificial conditions may be controlled by a local site of brain and on the other hand, various brain areas participate in given behavior and various brain centers participate in various behaviors. An intact brain may be important for fine coordination.

We will demonstrate further that the chemical nature of memory has a firm basis. Conversely, an importance of spatial distribution of synaptic terminals is impossible to reject. Usually neurons receive heterogeneous chemical information, but they can recognize synaptic patterns which consist of inputs transmitting the same substances. The only difference between patterns in this case is a spatial distribution of the inputs.

1.5 Memory location in the pre- and postsynaptic structures

The most effortless supposition concerning the mechanism of memory is the formation of new pathways in brain during learning. When I.P. Pavlov suggested development of a new neuronal connection between cortical representations of conditioned and unconditioned stimuli, he used an analogy of a telephone exchange. In the early twentieth century, when one user wanted to establish communication with another user, he called to a telephone operator and she connected the two users by a cable. I.P. Pavlov suggested that something similar is also happened in the brain (only, of course, without participation of an operator or homunculus).

The nature of plasticity has been linked with changes in the reserves of neurotransmitter in the presynapse, in the configuration of the synaptic space, the rate of transmitter destruction, in the density of receptors in the subsynaptic membrane and in the sensitivity of the postsynaptic receptor to the neurotransmitter. These hypotheses can be grouped together on the grounds that they all relate plasticity to changes in the efficiency of synaptic transmission. If memory traces would be localized in any location, in the pre- or in the postsynaptic structure, memory will be determined by the place, and a one-to-one correspondence will have to exist between the memory elements and the points of the brain. Another group of hypotheses relates plasticity to changes in the efficiency of excitation of neurons as a result of a change in the level of their membrane polarization or the threshold of AP generation. Clearly, the various hypotheses differ only in the localization of the plastic component while a firm fixation of memory trace in brain is supposed in any case. The problem of plasticity itself is reduced to a change in the efficiency of spread of the signal along the nerve net. Memory traces have been identified with stable changes in the conduction of excitation through a neuron or synapse. This hypothesis became extremely popular during tens years and almost forced out an alternative chemical hypothesis. Really, change in synaptic response during learning was established reliably. All regularities of behavioral learning were reflected in changes in synaptic efficacy. Particularly, when conditioned response increased, corresponding EPSPs in many neurons of brain also increased. The utmost form of synaptic hypothesis is postulation of the presynaptic nature of synaptic plasticity. No attention has been paid under these circumstances to the role of intracellular biochemical reactions. However, location of a memory trace in the postsynapse admits perception of all current synaptic influences as a whole (by chemical means). Certainly, this is not a trivial task, but it at least has a decision.

New neuronal connections might also be developed. Death of neurons or the birth of new neurons may presumably serve as the basis for long-term memory. The formation of long-term declarative memory has been tried to explain by means of the involvement of constantly appearing new neurons in specific areas of the brain, which differentiate from stem cells. They develop stimulus-selective (grandmother) and behavioral selective (which encode particular behavioral acts) long-term neurons. Final differentiation of long-term memory neurons may be triggered by the 'novelty signal" from newly formed neurons of hippocampus. Such newly formed neurons included in a new environment form new synaptic contacts that then convert these neurons in "gnostic units" [1158]. For example, hypothalamic neurons in canary responsible for new song are anew born neurons [943]. Naturally occurring synapse elimination and permanent loss of axonal input has also been supposed as the attractive mechanism for information storage [738]. Instead of a new pathway corresponding to a new memory, it has also been supposed the formation of new systems of mutually activating neurons [43]. Evidently, whether a new pathway arises after learning or a preexisting path changes its efficacy, or a new combination of neurons arises, we are within the scope of the same structural hypothesis.

Intracellular electrical activity is almost impossible to record from a presynaptic area, but it is accessible from the body of postsynaptic neuron. Indirect proofs of change in the presynapse were found 30-40 years ago. It was established that during learning the membrane potential, input resistance, threshold level, membrane capacity, membrane time constant and other characteristics of cellular membrane do not change [217, 1442, 376]. Sensitivity of postsynaptic neurons to certain neurotransmitters, in many cases, also does not change. If we proceed from the principle of "on the contrary", absence of alterations in the postsynaptic site denotes the reason for the excitatory postsynaptic potential (EPSP) change in the presynaptic area. However, this principle is not appropriate in the given case. In reality, change in the response during learning concerns only a specific stimulus related to the learning procedure, such as the CS^+, while the response to the CS^- does not change. In order to examine membrane properties we need to affect neurons, for instance by a current pulse for examination of excitability or to administrate a corresponding neurotransmitter close to the cell soma in order to evaluate chemosensitivity. Such a measurement gives us insight into characteristics concerning the impact of the matter used for the measurement. For example, it shows excitability in response to current pulse and not in response to the CS^+. Whether excitability, membrane potential, etc. are unspecific parameters? We will demonstrate that this is not always correct, but now we are only pointing out: if one considers a parameter to be unspecific, this parameter cannot explain specific changes of the response and it was not necessary to examine the role of that parameter at all. Curiously, similar difficulties arise if plasticity would be located within synapses accounting for a CS^+. Specificity between changes in responses to the CS^+ and CS^- is here easily explained, but correspondence between the alterations in CS^+ and CS^- effects is modified during learning in a nontrivial way (see Sec. 1.3). Therefore, during learning, it is insufficient to change synaptic efficacy. Synaptic patterns accounting for CS^+ and CS^- signals have to be reorganized and in fact become more distinctive in the course of training. Synaptic plasticity had also been investigated by means of quantal analysis of instability of EPSP, modified during learning. Neurons, it is known, secrete neurotransmitters from synaptic terminals into portions contained in synaptic vesicles. Therefore, EPSP consists of a few portions, quanta. Alterations in synaptic strength can be encoded presynaptically as alterations in the machinery releasing the neurotransmitter, say, glutamate, or postsynaptically by changing the number or function of receptors sensing the glutamate signal. If instability of EPSP is connected with quantal fluctuations, then either the value or the number of quanta will alter after learning. A change in value of quantal size means postsynaptic localization of plasticity, while a change in the number of quanta (quantal content) means an alteration in the probability of secretion of elemental portions from the presynaptic terminal. In order to perform quanta analysis, an amplitude histogram of EPSP was approximated by the Puasson's (or in some studies by binomial) distribution [77] and experimental distribution did not reveal significant differences from the theoretical distribution. Thus, a change in EPSP amplitude was explained by the change in quantal content. However, this method is impossible to implement to a given task [1255, 890]. The same experimental distributions did not have significant differences also from Gauss's distribution (we evaluated the above mentioned data by Kolmogorov's criterion $\lambda = 0.4 - 1.1$). Quantal analysis allows evaluating the difference between the experimental and theoretical distributions, but did not give any possibility to evaluate similarity between the two distributions. Paradoxically, the smaller is the set of experimental data, the easier it is to determine an absence of significant difference between the distributions. Therefore, implementation of quantal analysis is also proof of "on the contrary". By the way, the quantal size may differ across similar synapses (with the difference more than 10 fold) [153].

The strongest evidences of existence of material traces in synapses after aversive learning come from morphological data. The number, size and vesicle complement of presynaptic sensory neuron active zones are larger in sensitized animals than controls [77]. Neural terminals are markedly expanded (Fig. 1.11). Sensory neurons from long-term sensitized Aplysia exhibit a twofold increase in the total number of synaptic varicosities, as well as an enlargement in the size of each neuron's axonal arbor. Similarly, after one-trial aversive learning in chickens, morphological traces of memory were found in their forebrain: when the chicks learned to discriminate edible grains (each chick only once pecked a bitter grain) the number of dendritic spines in the forebrain of the chick increased up to 60% [1039]. Eyeblink conditioning increases the number (1.6 fold) of excitatory synapses within the interpositus nucleus that in the brain region is essential for long-term retention of the conditioned response [640]. We are considering that if some morphological or biochemical alterations after learning are too extensive, they, probably, do not concern memory. It is enough to remember how large an amount of experience any living being collects during its life. Such massive traces cannot serve normal memory and they are more similar to the traces of injury, evoked by excessive aversive stimulation. Really, it is well known that aversive events and excessive excitation lead to cell damage and tissue growth [797, 832, 337, 1389].

Nevertheless arguments like "on the contrary" sometimes are exploited even at the present time [1017, 45, 704, 86]. The fact that the response of a neuron to frequent stimulation is attenuated, despite the fact that the same unit did not decrease in responsiveness to rare stimuli (ordinary specificity), ones earlier [217, 1442, 376, 1438] and sometimes at the present time [371, 1282] is interpreted as evidence that habituation is mostly a result of synaptic depression rather than cellular-level excitability reduction.

There are many various forms of synaptic plasticity and it is impossible to consider these phenomena here in detail. We will point out only some important features in the well-known models of synaptic plasticity. Mechanisms of plasticity for Aplysia, rat hippocampus, chick and honeybee can be slightly different, but their important features are similar [515, 1071, 86]. A logical rule for the setting up of a new pathway during learning is suggested by D.O. Hebb [530]. He postulated that changes in synaptic efficacy take place when a presynaptic cell participates in the firing of a postsynaptic cell, or, in other words, both cells generate AP, but the postsynaptic cell generates a later AP. The Hebbian scheme implies participation of both the presynaptic and postsynaptic neurons. The cellular mechanisms underlying Hebbian plasticity have been well studied, mostly using the example of long-term potentiation (LTP) in regions of hippocampus and neocortex: enhancement of neuronal responses after frequent or intensive stimulation. Low-frequency stimulation was found to induce long-term depression (LTD). Yet, firing produces LTP at low frequency Correspondingly Besides, LTP is induced by stimulation that results

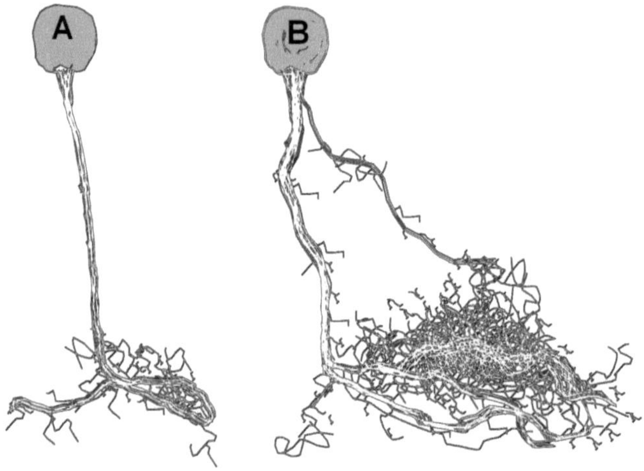

Fig. 1.11. Growth of synaptic terminals in the sensory neuron of Aplysia after sensitization. The Fig. 1.11 was redrawn in accordance with the data [77].

in strong postsynaptic activation, while LTD is induced by stimulation that results in weaker postsynaptic activation [746]. Similarly, a single tail shock in Aplysia produces short-term sensitization that lasts for minutes, whereas repeated tail shocks given at spaced intervals produce long-term sensitization that lasts for up to several weeks. Short-term behavioral sensitization lasts minutes to hours and correlates with an increase in synaptic strength between the sensory to motor neuron connection referred to as short-term facilitation.

Thousands of investigations have been undertaken in the field of synaptic plasticity. At present, this is the most broadly studied phenomenon in neurobiology. Typically, the problems discussed were: this circuit is critically involved or does not involved, presynaptic mechanism or postsynaptic, excitability changed or did not change, these protein kinases are essential and those are not essential, Ca^{2+} participates or does not participate, protein synthesis is important or ribonucleic acid (RNA) one is important, etc. [21]. Nevertheless, many fascinating details[3] of the cellular and molecular mechanisms that underlie synaptic plasticity have been established and they, mostly,

[3] When synaptic stimulation, usually glutamatergic, reaches a given threshold or is repeated a number of times, it causes a persistent increase in the level of the diffusible second messenger cyclic AMP by activating the enzyme adenylyl cyclase. It also enhances intracellular Ca^{2+} levels and leads to spike-timing-dependent plasticity. Up-regulation and trafficking of AMPA-type receptors and a change in the

relate to postsynaptic phenomena [389, 1033, 890, 731]. The contribution of postsynaptic mechanisms to learning was earlier underestimated. Really, the induction of LTP/LTD by correlated pre/post spiking is accompanied by an immediate and persistent enhancement/reduction of the intrinsic excitability of the presynaptic neuron. This was established in the first report of LTP [135]. In addition to changes in global presynaptic excitability, correlated pre- and postsynaptic activity also results in modification of the local postsynaptic excitability [290].

Cooperation of many presynaptic fibers is often needed to produce LTP, since individual unitary synapses are too weak to cause strong postsynaptic activation, whereas LTP depends on general integral power of excitation [1148]. Long-term potentiation or depression of the synapses may be dependent not only on strength and frequency of stimulation, but also on the temporal order of pre- and postsynaptic activity (spike-timing-dependent plasticity) [1279]. The timing of the postsynaptic AP relative to the EPSP determines the sign and magnitude of synaptic modification. Small differences in the timing of pre- and postsynaptic activity can determine whether synapses are strengthened or weakened. When a presynaptic AP precedes a postsynaptic AP (usually 10-25 ms), this leads to LTP. Reversed order of excitations lead to LTD. Spike-timing-dependent plasticity can be evoked in almost natural conditions. Repeatedly pairing visual stimulation and neuronal spiking induces rapid changes in the spatiotemporal receptive field of the neuron. The sign and magnitude of the receptive field modification depends on the relative timing of the pairing, in a manner consistent with spike-timing-dependent plasticity of the excitatory synapses on the recorded cortical neuron (± 25 ms) [833]. The order of signals (the US after the CS^+) is critical for behavioral plasticity, too, if one does not bear in mind that the time delay for real conditioning is in the order of seconds [740]. Additional peculiarities of the LTP phenomenon have been found. The NMDA receptors are largely blocked by Mg^{2+} at hyperpolarized membrane potentials, but the block can be relieved by depolarization, leading to the idea that this receptor can serve as the coincidence detector for pre/post activity [555, 290] and for change in synaptic efficacy. Synaptic transmission is enhanced if the NMDA receptor detects the co-activity of the presynaptic (release and binding of glutamate) and postsynaptic neuron (enough depolarization to expel Mg^{2+} from the channel pore)

---

correspondence between two types of glutamate receptors, AMPA and NMDA, also plays a critical role during long-term synaptic facilitation in hippocampal pyramidal neurons and in Aplysia [686, 734, 860, 85, 713, 1246, 1317, 1433, 86, 290, 1316]. The tetanus that elicits NMDA-dependent LTP produces LTP of the EPSP and, concomitantly, LTD of the inhibitory postsynaptic potentials (through $GABA_A$ receptors) that leads to enhancement of excitability ([1218, 766]) and depends on postsynaptic depolarization. Chemoreceptors for GABA and glutamate switch between extrasynaptic and synaptic localizations by lateral diffusion with the rates around tens of minutes [1246]. This time approximately corresponds to the speed of development of LTP.

[1316]. So, strong dependence on the temporal order of pre- and postsynaptic excitations (mechanism of Hebbs's rule) is excellently explained by the link of the voltage-dependent blockade of the NMDA receptor with Mg^{2+}.

In mammalians, excitatory synapses, the object of plastic reorganization, are often located at the dendritic spines, at a long distance from the cell soma and separated from dendrites by thin necks. The regulation of diffusional coupling across this thin neck provides a possible mechanism for determining the amplitude of postsynaptic potentials [138]. It was supposed also that back-propagating of action potentials into dendrites of hippocampal and cortical pyramidal cells provide synapses by depolarization and are responsible for Hebbian plasticity. However, this logical scheme may not be absolutely correct. A back-propagating AP invades the dendrites, but may be insufficient to relieve the Mg^{2+} block of NMDA receptors. Really, back-propagating of action potentials is not always required for LTP and a strong synaptic stimulation is able to induce LTP even when back-propagating of action potentials is blocked [751].

Our memory is constantly renewed and there is a problem to explain how modified synapses keep old information when new information comes in. Retention is an active process and memory-storing synapses must somehow retain the capacity for ongoing plasticity if old information would be preserved in the face of new learning. Electrically induced LTP in the hippocampus is rapidly reversed when animals receive novel information through a new or enriched environment [290]. New learning incorporates a partial trial of old patterns and this retards the recording of new information [5]. At least the early phase of LTP and LTD can be reversed by neuronal activity following an induction protocol: de-potentiation and de-depression. The seizure activity following tetanus and low-frequency stimuli was found to reverse LTP. Similarly, reversal of LTD can be achieved by subsequent activity, typically high-frequency stimulation. De-potentiation stimuli are milder in intensity than those used in induction of LTP and LTD [1432]. The phenomenon of depotentiation is in poor agreement with properties of memory, since memory continuously changes, but never disappears.

LTP can develop either gradually, or by the all-or-none principle, depending on the intensity of stimulation, but gradual development, if it is observed, takes tens of minutes after powerful stimulation [586, 15, 414, 1432, 124, 374, 1367]. The learning process may be prolonged and learning traces are usually exceptionally protracted, but this time is not the phase of time-consuming ripening. A necessity for time delay in order to develop an effect is not typical for learning, whose consequences are the most noticeable immediately after training. At the same time, traces of damage are usually expanded in time and appear in space gradually. An LTP is comparable with cell excitotoxic damage [1255, 785] and an LTP-related cascade of biochemical events coincides with the damage-related cascade: participation of glutamate receptors, Ca^{2+} elevation, dependence on protein synthesis, neural growth, dependence on strong excitation, blockage by GABA-inhibition or hyperpolarization, etc. We will discuss this problem further.

Thus, localization of short-term plasticity within synapses has never been demonstrated. Some indirect but important data such as spike-timing-dependent plasticity, the role of NMDA receptor as a coincidence detector, properties of dentritic spines, etc. are in accordance with supposed mechanisms of synaptic plasticity during LTP, but they are poorly timed with behavioral plasticity: Hebbian plasticity requires too short a delay in a millisecond scale between pre- and postsynaptic excitations and this contradicts the properties of behavioral plasticity. Long-term plasticity within synapses was demonstrated only on the level of morphology and only for anxious or inadequate learning, such as LTP. If current electrical activity could influence the structural organization of brain tissue, this is a very long lasting process and not compatible with the rate of behavioral plasticity. It would be possible to suppose that morphological reorganization may play a role in long-term memory storage, which is formed after memory consolidation over the course of several hours and keeps for days and years. Really, compared with non-stable biochemical processes and with fast- alternating electrical processes, morphological structure of brain, when it is represented at the histological level, looks steady, fine ordered and well suitable as a memory substratum. Nevertheless, synaptic proteins and structures are not stationary, but rather are highly dynamical. Synapses, even in adults, continuously arise and degenerate [267]. New-born neurons usually also are short lived [492, 332]. Video recordings from hippocampal neurons allow the visualization of actin dynamics in living neurons. These recordings revealed large actin-dependent changes in dendritic spine shape. Spines contain high concentrations of actin, suggesting that they might be motile. Visible changes occurred within tens of seconds, suggesting that anatomical plasticity of synapses can be sufficiently rapid for long-term memory, but are too slow for current memory [389] and it is not clear how synaptic efficacy can be precisely maintained without suffering from accumulative drift in the face of molecular turnovers of synaptic machinery [1316]. On the other hand, a large body of data has accumulated that it is in disagreement with the morphological nature of memory traces. In behaving animals, single neurons can intermittently participate in different computations by rapidly changing their coupling without associated changes in firing rate [352].

Is LTP an authentic model of learning? There are many parallels between LTP development and behavioral plasticity between synaptic plasticity and aversive or stressed (hippocampus- and amygdala-dependent) learning relative to pharmacological, morphological and physiological characteristics [1218, 86]. Rodents can readily be tested about their memories for places, objects, and odors and these studies have revealed that lesions of the hippocampus and related structures interfere with long-term storage of these kinds of memory [845]. However, nobody has demonstrated that modification of synaptic pathways during normal, non-stressed learning hold permanent features.

Synaptic reactions do change during learning, but any signal usually activates many synapses at the given neuron and we record the reaction of an entire synaptic pattern. If the neuron recognizes a synaptic pattern, all together, the synapse may have a different efficacy, when it is included in different patterns [1264, 1272, 1292, 1064]. This means that when single neuron recognizes a synaptic pattern, as a whole, synaptic memory does not exist at all. Learning may depend on rapid modulation of effective connectivity with time constants of modulation in the range of from ten to one hundredth of a millisecond. Structural or anatomical connectivity should be distinguished from functional connectivity. At present, proofs relating to a connection between a change in synaptic efficacy during LTP and memory are absent [750].

Currently, it is evident that the Hebbian process is an oversimplification [281, 377]. A coincidence of pre- and postsynaptic excitations cannot be a decisive mechanism of plasticity. If we perceive stimuli, they must induce postsynaptic AP in some neurons. This automatically satisfies the Hebbian rule. The environment sometimes may demand that this reaction fail (for example, as a result of habituation), but Hebb's rule requires an increase in this pathway. A change in synaptic efficacy often depends on its inherent properties and is not dependent upon behavioral needs. For instance, synapses with a low release probability display facilitation and augmentation, whereas the synapses with a high release probability supplied by the same axon may exhibit paired-pulse and frequency-dependent depression [1225]. On the other hand, transgenic mice (lacking a subunit of the AMPA glutamate receptor) may develop LTP and spatial learning in the absence of synaptic strengthening [1112]. Moreover, in most studies there is a specific range of repetition rates that must be used for successful induction: at frequencies below 10 Hz, LTP cannot be induced [751]. We would like also to remind that synaptic plasticity localized in determined points of the brain meets difficulties in furthering a plausible explanation of the numerous facts of memory recovery after injury. After all, individual cell participates in the storing of many patterns of activity.

1.6 Plasticity of excitable membrane

Excitability of neurons depends on the short-term history of the membrane potential (e.g. refractoriness and accommodation). However, properties of neural cells in normal, no-learning conditions (the rare presentation of indifferent stimuli) can be described by the all-or-none principle [815], which means that the threshold for AP generation does not depend on the magnitude and variety of the input signal, and after threshold attainment the neuron generates an AP of maximal amplitude. A model of a neuron possessing such properties was named the formal neuron. Nevertheless, it turns out that excitability is a changeable parameter and depends on long-term neuron history [428, 255].

During learning, the reactions of many neurons to signals change in correspondence with animal behavior. Theoretically, a neuron may change its reactions either passively or it may acquire new properties that are not consistent with the all-or-none principle. Generally though not always, responses to biologically significant stimuli (for example, conditioned ones) increase and responses to insignificant stimuli (for example, habitual ones) decrease. How does a neuron decide which stimulus is more important in the current behavior? An elementary explanation is that synaptic efficacy changes for the specific input: the synaptic input corresponding to a more important signal becomes stronger. However, there is no evidence that during normal learning, such as habituation as well as classical or instrumental conditioning, the efficacy of the synapse changes independently from the activation of other synapses.

Change in synaptic efficacy is not the exclusive manner of storing information in the brain. Control of excitability may serve, too, as the means for modulation of neuronal activity. Excitability usually increases [1258, 1355] during augmentation of the biological importance of the signal during conditioning and decreases when it falls during habituation [1266, 1267, 641, 1272] and during extinction of conditioned reflexes [1259]. During instrumental learning, excitability also changes, but in a complex manner [881, 1270, 1351]. Learning also leads to changes in a post-spike afterhyperpolarization [24, 865], the AP amplitude [723, 1266], the AP duration [641, 547, 713, 245], input resistance of neurons [348, 54] and the conduction velocity in the axon [1314, 1351]. The more important the stimulus, the more powerful an AP it evokes. Plasticity of excitable membrane demonstrates a postsynaptic mechanism of learning with its powerful biochemical machinery. Thus, a neural cell might take on a role of self-contained unit within the brain. During development of visual circuits, plasticity of excitable membrane also plays a central role [992].

In the last half of twentieth century, the paradigm of the formal neuron appeared strong. It was substantiated by physiological achievements in the investigation of giant axon of squid, which produced an analytical description of the behavior of an excitable membrane [548] that was so fine that this till this present time it is still the basic means for analysis of membrane currents. Besides that, theoretical generalizations of the neural network theory have lead to the advent of a new branch of science with its adherents, journals, conferences and accomplishments. However, investigation of excitable membrane plasticity did not develop on the back of experiments devoted to synaptic plasticity, particularly, LTP. Instead, easily reproducible phenomena that, as expected, are related to learning and memory have become the firm scientific background of synaptic plasticity. Work on excitable membrane plasticity, in contrast has been seen as sporadic attempts to explain a few data that are contradictory to conventional thinking and relate to vague and previously unexplored chemical events in neural cells. In many cases, researcher published their results dedicated to excitable membrane plasticity and never returned to the problem, without enlarging scientific knowledge, confirmation or refutation [1309, 135, 1157, 641, 752, 982, 524, 348, 547, 1419]. Excitable membrane plasticity has remained the concern of only a few researchers, who continued investigations during many years [1253, 1355, 24].

Nevertheless, leading physiologists over and over again returned to the idea of a connection of brain functions with the individual, goal-directed behavior of neural cells and with the modulation of neuronal membranes by means of chemical processes, which might or might not have electric exhibitions [1126, 18, 19, 661, 181, 43]. Now we satisfactorily understand the LTP phenomenon, but memory comprehension is till rudimentary. At present, the role of neurons in signal processing has evolved conceptually from that of a simple integrator of synaptic inputs until a threshold is reached and an output pulse is initiated, to a much more sophisticated process with mixed analog-digital logic and highly adaptive synaptic elements [647]. This enriches calculating function of a neuron, but very likely does not allows for the hypothesis of memory as a reorganization of the neural network. Versatile complexity of neuronal chemical interactions is still underestimated.

Recently a new interest in the plasticity of the excitable membrane has flourished [713, 71, 1166, 54, 1024, 291, 1278, 1339, 1367, 255]. Voltage threshold of an AP and other membrane parameters can be spontaneously changeable [71, 482]. A preliminary change in neuronal excitability by blocking of subunits of potential-dependent K^+ channels impairs complex learning, but more simple learning was found to be normal [468]. Such enhanced sensitivity of complex learning to various harmful influences, as we had described earlier, is a general property of the learning mechanism and a weak influence of any impact to simple learning does not undermine the significance of this impact. Sluggish change in electrical activity, neuronal environment and chemical disturbance lead to a homeostatic change in excitability [15, 672, 14, 1386, 296, 791], which supplements the homeostatic forms of synaptic plasticity [887, 966] is the property of living beings, which regulates its internal state in order to maintain a stability). Synaptic plasticity and plasticity of the excitable membrane can mutually assist reorganization of the neuronal reaction. Change in the ion currents of an excitable membrane of a neuron has been supposed to influence postsynaptic potentials affecting the same neuron and to sharpen the input specificity [1185, 1351, 33, 54, 1331, 997].

One consequence of experience is to alter the biophysical and integrative properties of neurons [1004, 14], which can respond to changing inputs by adjusting their firing properties, through the modification of voltage-gated channels [672]. It is important to note that besides homeostatic forms of reorganization of excitability, further development receives investigation of excitability during learning [670, 865, 713, 1100, 1004, 762] and LTP [33, 15, 54, 414, 1367]. Difference in excitability could also play a role as a means for maintenance of inborn specificity in image recognition [1308, 1339].

The threshold, as one expects and as usually observed, is independent of the input signal. Therefore threshold specificity in respect to different signals, as a rule, is not examined at all. In a majority of cases, change in excitability during learning is considered as an unspecific parameter and as a steady alteration concerning any input of the given neuron. The unselective form of excitable membrane plasticity cannot be the basis for the selectivity of neuronal reactions [48, 818] and such unselective phenomenon might influence plasticity but cannot explain the main particularities of learning. Nevertheless, sometimes the specific, in relation to the input, modifications of threshold were described [1258, 1259, 291, 414, 997, 1339], although there are different explanations for this non-standard phenomenon: the selective change in excitability looks like an unconventional result.

The various pieces of discovered data concerning plasticity of the excitable membrane sometimes seem contradictory, but they are clearly dependent on the methods of measurement of excitability. Measurement of characteristics associated with excitability is not a simple task. Spike amplitude is the easiest controlled factor, even in the course of extracellular recording, but changes in AP amplitude are usually small, with infrequent exceptions. AP duration, afterhyperpolarization and membrane potential are accessible during intracellular recording. Nevertheless AP duration and afterhyperpolarization are formed after the AP generation has been completed and, hence, role of excitability in the given spike generation is already completed, too, and may affect only the follow APs. Membrane potential is a non-specific parameter and moreover it is usually stable during learning. AP threshold is the most important factor directly connected with actual spike generation and it is necessary to discuss methods of how to find the threshold in experiments.

Threshold is sometimes an ill-defined value. There is a difference between voltage threshold (the voltage value at which the membrane is able to generate sufficient current to drive the AP without further applied current) and current threshold (the minimal amount of sustained current needed to generate a spike). A fast and powerful input signal is an appropriate condition for calculation of the voltage threshold, while a sustained and small input for current threshold. Sustained and small input signals generate a spike at a voltage that is smaller than voltage threshold [646]. Frequently used intracellular current injection [1100, 54, 14, 414, 1367] corresponds more closely to the conditions under which current threshold is observed, at the same time that synaptic current delivered to the soma has a short duration and corresponds to conditions appropriate for the voltage threshold. Therefore, measurement of excitability by the intracellular current pulse does not correspond to the thresholds that are essential for synaptic responses. Examination of the change in threshold of mollusk neurons after classical conditioning by means of an intracellular depolarizing current pulse in our experiments revealed only a weak change in excitability (Fig. 1.12).

In general, it is natural to examine a change in excitability using tests for unspecific excitability, such as by the response to the current pulse. However, in order to evaluate specific change in excitability within the given response, a threshold ought to be measured within this response. Excitability within responses has been determined by measurements of levels of membrane depolarization immediately prior to AP generation. The simplest method for

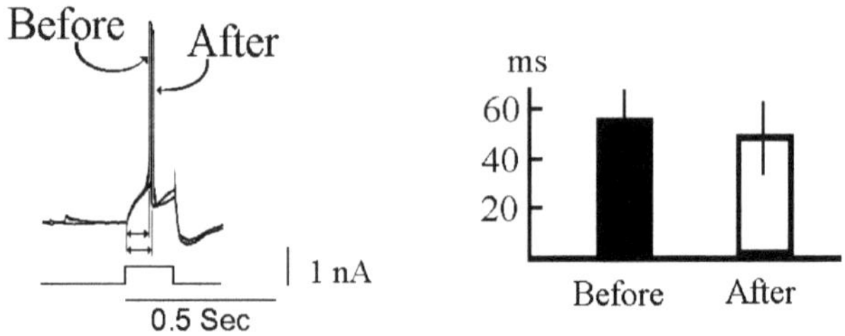

Fig. 1.12. Response of the Helix neuron LPa3 to current pulse (0.5 nA, 150 ms) before and after acquisition (at the left). Excitability within the current pulse was evaluated by means of AP latency in response to the current pulse. At the right, mean change of AP latency in response to the current pulse for 13 neurons (means and standard errors).

this purpose is evaluation of the given rate of depolarization, which one chooses to be rather high, so that it is observed only during an AP generation [71, 413, 672, 1308]. This method is acceptable, but depends on an arbitrary choice of the ultimate rate of depolarization, which itself may alter during learning and, besides, the evaluation is rather rough. The threshold also can be found as a maximum curvature at the leading front of an AP (Fig. 1.13), to be exactly the depolarization necessary to activate an action potential [1258, 1259, 54, 1339, 1367]. We determined maximum curvature (Fig. 1.13) according to the points of intersection of tangential lines at the points of inflections of the EPSP and at the leading edge of an AP at the intersection of two tangents [620]. Both methods are appropriate [1043]. An AP waveform cannot be analyzed when a neuron failed to generate an AP.

A perfect illustration of the importance of choice of the method of excitability measurement is the investigation of LTP in the rat hippocampus. When LTP was induced by pairing subthreshold synaptic stimulation with backpropagation of AP into the dendrites, current threshold in response to current injection increased, voltage threshold of spontaneous APs (measured as the voltage where the first derivative of the action potential waveform exceeded 10 V/s) remained stable, while voltage threshold of spikes arising from the potentiated EPSP was lowered up to 2 mV and voltage threshold of unpotentiated inputs was less increased [374]. Thus, we may conclude that voltage-threshold from potentiated and unpotentiated inputs differs with respect to threshold within spontaneous APs, but these data cannot be compared with the results of measurements of current threshold.

The selective form of excitable membrane plasticity is characterized by different excitability corresponding to different signals (for example, conditioned and discriminated, habitual and novel). The levels of AP generations shifted to de- or hyperpolarization side during training and t his result is not fully consistent with the all-or-none principle of neural cell function.

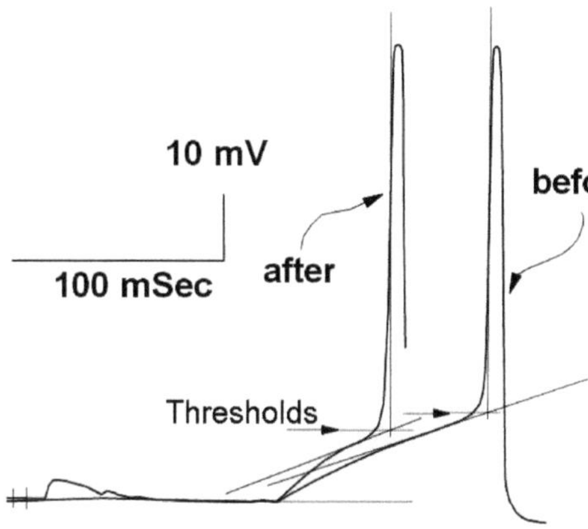

Fig. 1.13. The first action potential generated within responses to the CS^+ before and after classical conditioning in snail neuron. The method of the level of spike generation determination, as the level of intersection of two tangents is shown (explanations in the text). Threshold levels are shown by the pointers. Responses were aligned in the points of EPSP beginning. Short vertical lines at the start of frame are times of tactile stimulus presentations.

We first discovered [1253] that spike amplitudes evoked in mitral cells (pyramidal neurons) of the olfactory bulb of frog by different stimuli may be different. The effect was large enough and visible by a naked eye (Fig. 1.14), since action potential duration was 15-20 ms, ten times longer than in mammalian cells. For our knowledge, we made the first intracellular recording from neurons of the olfactory bulb of frog and firstly it was difficult to believe that so elongated large positive intracellular potentials are really spikes. Therefore, we tried to examine whether these were conventional action potentials and whether they follow the all-or-none principle. At the time of discovery, classical neuron paradigm was so strong that nobody wasted efforts to measure alteration of spike amplitude in the experiment. So, when spike amplitude turns out to be dependent on neuronal input and this had been published, certain colleagues (for example, the theorist V. Dunin-Barkovsky) considered this as a joke. Plasticity of excitable membrane seemed to contradict to the basic principles of neuroscience.

After stimulation of epithelium by the smell, the AP generated in response to olfactory nerve stimulation decreased by 10-30% and this inhibition went on for approximately 15 sec without perceptible change in the membrane potential. During this period, spontaneous spikes and spikes induced by repeated adequate stimulus have normal magnitudes. The character of spike amplitude inhibition in the responses to stimulation of olfactory nerve after adequate stimulus corresponds to the inhibition of the evoked potential recorded from the surface of olfactory bulb in the same response (Fig. 1.15). We never ob-

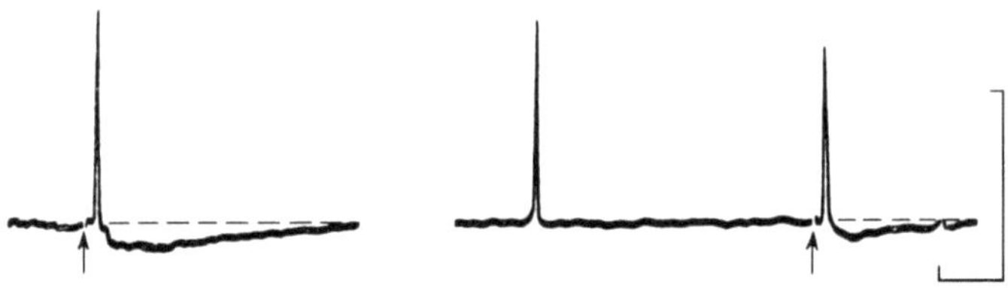

Fig. 1.14. Decrease in amplitude of spike induced by stimulation of olfactory nerve of frog after irritation of olfactory epithelium by the smell. At the left control response to the nerve stimulation (pointer). At the right - the nerve stimulation soon after presentation of smell evokes low amplitude spike (second spike), while spontaneous spike was normal (left spike). Calibration 50 mV and 100 ms.

served so large an effect in other objects (see further) and later it was always necessary to undertake labor consuming measurements.

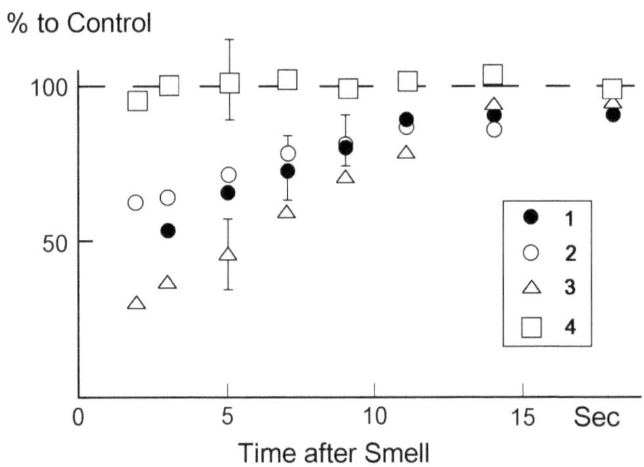

Fig. 1.15. Inhibition in the olfactory bulb of frog after adequate stimulus. 1) Amplitude of spikes in response to olfactory nerve stimulation after presentation of smell; 2) evoked potential in olfactory bulb after smell; 3) the same in the embryonic olfactory cortex of forebrain; 4) evoked potential in the olfactory cortex after stimulation of olfactory tract. Control, before adequate stimulation 100%.

A decrease of the spike amplitude probably inhibits activity of the next neuron, since it was accompanied by reduction of the evoked potential in the olfactory cortex. Probably, the small somatic spike had incomplete amplitude also in the axon and this leads to decreased postsynaptic potentials in the cortical neurons. At the same time, the evoked potential in the olfactory cortex in response to stimulation of the olfactory tract (axons of mitral cells) does not undergo inhibition (Fig. 1.15). Inhibition of the evoked potential in the cortex under these conditions was more pronounced (Fig. 1.15). Thus, neurons were inhibited exclusively for one stimulus, but not for another. This is a very parsimonious inhibition that extends the possibilities of neurons and selectively blocks the neuron from some types of activity but leaves it free for other types of activity. This parsimonious inhibition, evidently, does not require supporting stimulation, like postsynaptic inhibition, and it is difficult to explain by the classical processes on the neuron membrane. We supposed [1253] that this phenomenon relates to neuronal plasticity and some types of events in intracellular organelles are essential here.

Similar phenomenon we further discovered in the general cortex of the turtle forebrain when we investigated habituation to light flashes [1267]. During habituation, the amplitude of the AP evoked by a habitual stimulus falls when compared with spontaneous APs and an AP in the response to a rare stimulus (Fig. 1.16).

The results of these experiments suggest that plasticity is connected with a selective change in the state of the neuron, which is manifested only during the action of a particular stimulus. Habituation of turtle cortex neurons to light flashes resulted in a selective decrease in the AP amplitude measured from the point of maximal curvature at the start of the AP to the top of the AP. The result of intracellular measuring corresponded to the results obtained during extracellular recording (Fig. 1.17) when the AP amplitude can easily be measured without any element of subjectivity and without utilizing the point of maximal curvature [1266]. Thus, AP waveform selectivity for the different responses evidently is not an epiphenomenon. After training, neurons can reveal different excitability within different responses. Change in the amplitude of the first extracellular AP in the response to habitual stimulus was as small as 3-5%.

General cortex is the higher neural center of the turtle and therefore we cannot determine whether the decrease in amplitude of an AP is connected with impairment of the transmission of excitation to the next neuron. It is impossible to rule out that a slightly decreased somatic AP generates a normal reaction in the target neuron. On the other hand, the value of the EPSP and the threshold of an AP are directly connected with the output signal of the neuron. It is important to reveal the principles governing changes in EPSP and the level of an AP threshold during habituation. We have investigated this problem in neurons of the mollusk Helix.

Habituation of identified neurons for defensive closure of the pneumostome in the Helix mollusk to tactile stimuli produced decrease in amplitude of the first synaptic AP in the response to habitual stimulus [1272]. At the same time, an AP in the response to the rare stimulus (another tactile stimulus, directed to other point of the body) did not change. The response was restored after a long interruption in stimulation and after a change in the parameters of the stimulus. The rate of habituation increased with an increase in the frequency

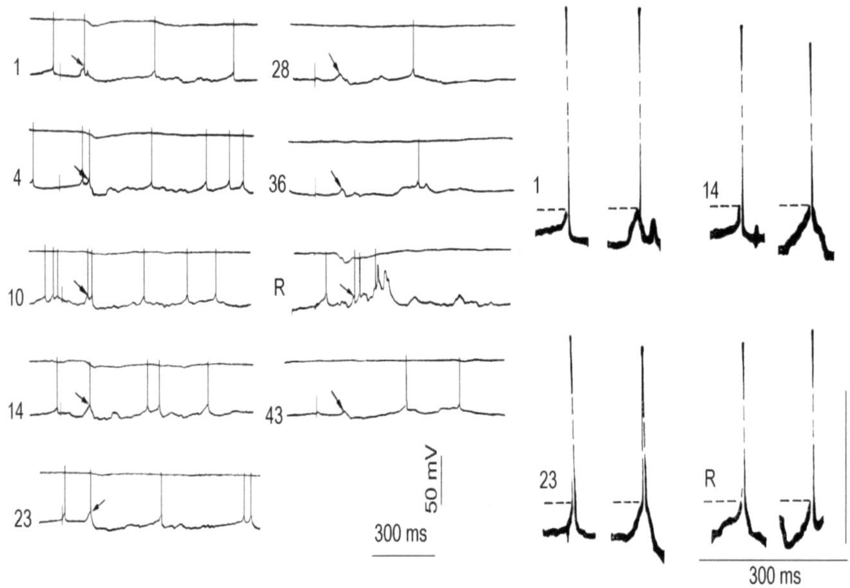

Fig. 1.16. The effect of habituation on the neuronal responses in the turtle general cortex. At the left upper beam, recording of the evoked potential from the surface of the cortex; lower beam intracellular recording from the basic cortex neuron. The number of stimuli is pointed out at the left. The moment of presentation of the habitual stimulus is pointed out by the negative-positive line (shorter than APs). R response to the new (rare) light stimulus. Pointer the place of generation of the short-latent AP and levels for its measure (maximum curvature at the AP front). At the right change in AP amplitudes; parts of the recording, presented at the left, were augmented. In each frame, at the left, first spike - spontaneous AP, second evoked short-latent AP. The Fig. 1.16 was redrawn in accordance with the data [1267]

of presentation and during a repeated series of habituation. No significant shifts of membrane potential were noted during habituation.

A decrease in AP amplitude during habituation proceeds in parallel with an increase in the threshold of the same AP [1272]. The cause of the blockade of an AP during repetitive stimulation was the combined action of two factors: a rise in threshold and a fall of the EPSP. Change in the threshold was also selective, since threshold within a response to the rare stimulus did not increase. The selective raising of the threshold relative to the ordinary stimulus indicates that the neuron can identify the stimulus and change the state of its excitable membrane rapidly.

The fall in amplitude of the AP was very small in terms of absolute magnitude. When averaged for all identified neurons it amounted to 3-5% of the

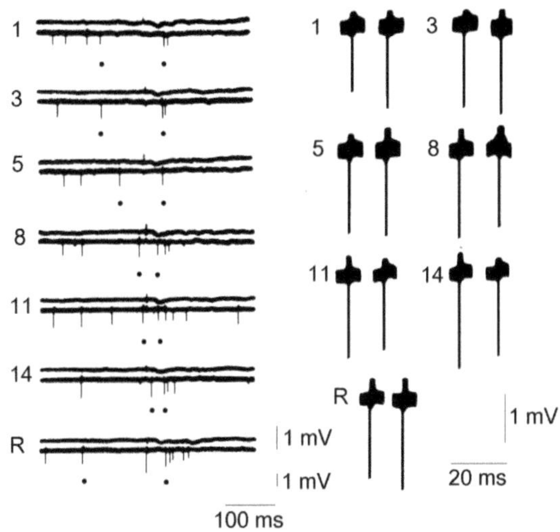

Fig. 1.17. Neuronal responses of the general cortex of turtle to repeated presentation of light flashes and rare stimuli. At the left upper, evoked potentials, lower extracellular activity. Numbers of habitual stimulus are pointed out, R response to the rare stimulus (another light stimulus). The points indicate APs, which are shown at the right: in each frame, at the left- spontaneous AP, at the right evoked AP. The Fig. 1.17 was redrawn in accordance with the data [1266].

mean AP. The threshold of generation of the first synaptic AP rose during repetition of the stimulus on an average of 20% after the habituation series.

Neuron excitability can both selectively decrease (during habituation) and selectively increase (during classical conditioning). Selective decrease in threshold we observed in land snails Helix during classical conditioning [1258, 1259]. A typical example of neuronal activity during classical conditioning is presented in Fig. 1.18. Tactile stimulation of the foot was used as the conditioned stimulus (CS^+). A similar tactile stimulus directed to another point on the foot served as the discriminated stimulus (CS^-). Cutaneous electrical stimulation of the foot caudal area served as the unconditioned stimulus.

During acquisition, the number of AP in response to the CS^+ in both recorded neurons increased from 1 to 2-3 AP (combinations of CS^+ and US number 2, 20 and 30). AP threshold in response to the CS^+ during acquisition decreased. At the same time, the response to the CS^- during acquisition revealed a small change in the number of APs as well as in the threshold value (trials 1, 21, 27). At the end of the acquisitions, the threshold in the response to the CS^- exceeded the threshold in the response to the CS^+. During an

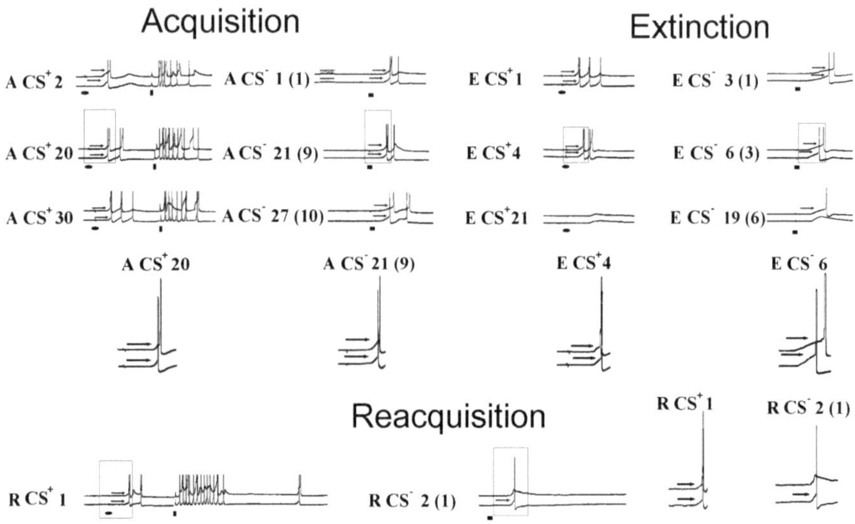

Fig. 1.18. Representative intracellular recordings of neuronal activities during acquisition (letter A), extinction (letter E) and reacquisition (letter R) of the neuronal analog of a classical conditional reflex. In each frame, at the top - activity of the neuron LPa3, at the bottom - RPa3. The arrows indicate thresholds. The threshold was measured from the membrane potential level to the point of maximal curvature at the leading front of the AP. The number of CS^+ at each exposure is indicated. For the responses to the CS^- the number of the preceding CS^+ is indicated (in the brackets - the ordinal number of the CS^-). Sections marked in A CS^+ 20, A CS^- 21, E CS^+ 4, E CS^- 6, R CS^+ 1 and R CS^- 2 were magnified (X 1.8) and are presented at the end of acquisition, extinction and reacquisition. The real amplitudes of the APs are shown only in the magnified responses. CS^+ - ellipsis, US vertical rectangles, CS^- - horizontal rectangles. Calibration, 10 mV (trial A CS^- 1). The Fig. 1.18 was redrawn in accordance with the data [1259] Methods. The training procedure consisted of the elaboration of a neuronal analog of classical conditioning. Acquisition consisted of 25-35 combinations of the CS^+ and US and 8-15 presentations of the CS^-. The interval between the CS^+ and US was 1s. The CS^- was never paired with the US and was presented every one to four combinations over the course of training. An extinction series (after a 5-10 min break) consisted of 15-20 presentations of the CS^+ and 6-12 presentations of the CS^-. Reacquisition (after a 20 min break) consisted of 5-15 combinations of the CS^+ and US and 2-6 presentations of the CS^-. Properties of our neuronal model of learning were similar to the properties of a well-known behavioral conditioned reflex of the defensive closure of the pneumostome in Helix [810].

extinction series, the number of APs in both the response to the CS^+ and the response to the CS^- decreased and the AP threshold in these responses increased. After a 20-minute break, the number of APs in the response to the CS^+ recovered (reacquisition), but did not reach the amplitude found at the end of acquisition. Response to the CS^- did not increase after the break. An enlarged version of some traces in Fig. 1.18 demonstrates selective change in AP threshold in response to the CS^+ compared with the thresholds in the response to the CS^-.

There is peculiarity of thresholds during classical conditioning. We examined whether the decrease in the first AP threshold coincided with the decrease in the second AP threshold in the same response. The above-mentioned selective change in the AP waveform which we use for determination of the threshold applied in general only to the first AP (Fig. 1.19). During the last part of acquisition, the threshold of the first AP in response to a CS^+ differed significantly from the threshold of the first AP in response to the CS^-. At the same time, the threshold of the first AP in response to the CS^+ was smaller than the threshold of the second AP in the same response ($p < 0.05$). Thresholds of the second AP in responses to the CS^+ and CS^- did not differ significantly. A change in AP threshold during habituation is also an integral factor of the first AP in the response [1272].

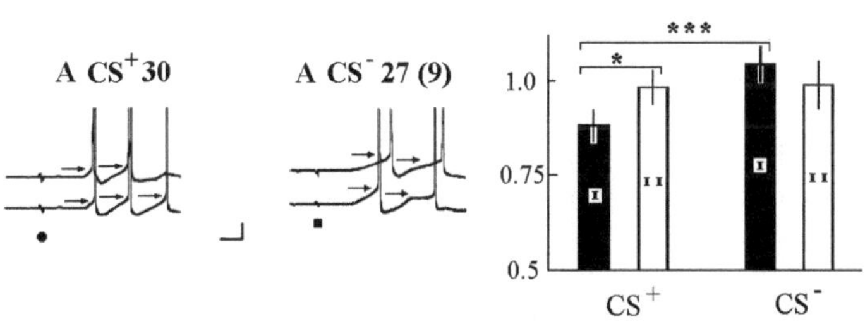

Fig. 1.19. The change in the first and second AP thresholds (are indicated by the arrows) during the last part of acquisition. Examples of response to the CS^+ (at the left) and CS^- (middle). At the right, averaged changes in the first (I) and second AP (the latency 50-200 ms, II) thresholds in the responses to the CS^+ (left) and CS^- (right) during acquisition (trials 21-30). The thresholds of the second APs were normalized according to the average threshold of the first AP in the same neuron. Asterisks indicate a significant difference (Student t-test, $*p < 0.05$, $***p < 0.001$). Calibration 10 mV, 0.1 s.

Thus, the decrease in the threshold of the first AP only poorly promotes the generation of the next AP in the response and hence, change in excitability within the response is short-term and transient. Thus, specific change in excitability concerns just the first AP in the response. This corresponds to general observations that only short latency responses selectively change during learning, compared with discriminated signals. Current changes in excitability of a neuron in time are probably linked with the degree of the real participation of given neuron in current behavior.

The AP threshold, latency, EPSP slope and amplitude (Fig. 1.20A-D) changed in a rough correspondence to the number of AP changes (Fig. 1.20E). During acquisition the number of APs in response to the CS^+ on average increased and in response to the CS^- decreased, while during extinction both responses decreased. During the reacquisition, the number of APs in the responses to the CS^+ increased rapidly.

During training, a change in spike threshold, spike latency and postsynaptic potential slope and amplitude after pairing also displayed selectivity for responses to the significant and insignificant stimuli (Fig. 1.20A-D). All these characteristics correlated with the efficacy of the corresponding stimulus for the spike generation. The properties of the AP threshold, EPSP and the number of APs in the response correspond to well-known properties of behavioral classical conditioning, but in spite of the existence of a roughly inverse relationship between AP threshold and EPSP (Fig. 1.20), there is a significant difference between the dynamics of the EPSP and the threshold at the end of acquisition and at the beginning of extinction. The cause-and-effect relationship between the strength of the presynaptic volley and the excitability was evidently weak[4].

Although it is clearly that AP amplitude is suitable for measurement, after learning it changes only for a few percentage points [1267, 670, 586, 1004, 1367] and sometimes alterations are not found at all [865, 1428]. A change in excitability after learning and LTP has been described in many reports, but the results were mostly acquired by testing with a current pulse, while the threshold within the response to current pulse may differ from the thresholds within responses to, for instance, CS^+ and CS^-. Therefore, these reports exposed unspecific change in excitability, and it would be important in these cases to examine a selective change in thresholds. Plasticity of excitability was persistent during days or weeks, but later passed away [1351, 1100], and these

[4] A correlation between the EPSP slope and the AP threshold was related, at least partly, to their common dependence on trial number, and it was possible to evaluate this correlation by means of analysis of covariance. Changes in threshold, the slope of the EPSP and AP latency during pairing are inter-dependent. The negative correlation between the AP threshold and the number of APs does not decrease when an EPSP slope was kept constant [1259]. Hence, reliability of AP generation depends on the threshold itself and not only on the slope of the EPSP. The AP latency depends on both cellular and presynaptic processes and a change in the AP latency within the response strongly correlated with a change in the threshold of the same AP and weakly but significantly correlated with the EPSP slope. The influence of threshold as a covariate on the latency of an AP was much more pronounced than the influence of EPSP slope. Threshold of AP and slope of EPSP are not connected by a one-to-one correspondence and their connection was determined by their dependence upon the trial number.

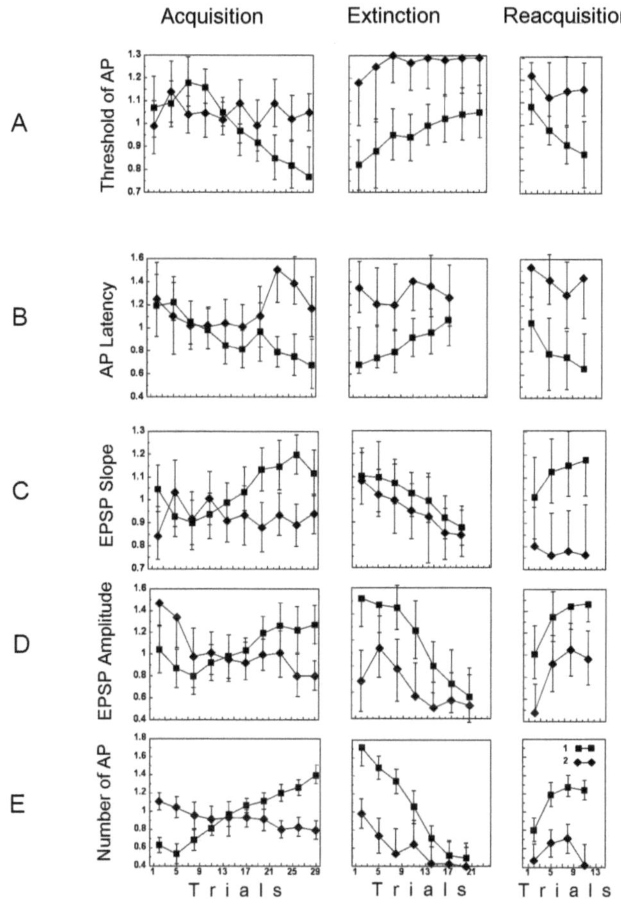

Fig. 1.20. Behavior of neuronal activity during acquisition, extinction and reacquisition. The symbols are denoted in the figure (E), 1 - responses to the CS^+, 2 - responses to the CS^- vs. the trial number. For each neuron the corresponding characteristics were normalized to a mean response - separately for acquisition, extinction and reacquisition. A- AP thresholds; B- AP latencies; C - slope of the EPSP; D - EPSP amplitudes. The values and confidence intervals ($p < 0.05$) in the plots (A-D) are calculated by means of a two-way ANOVA with interactions between the trial number and the type of stimulus - CS^+ or CS^-. E the number of APs, the time window over which the APs were counted was 0.5 s; medians and confidence intervals are shown (Mann-Whitney U test, $p < 0.05$). AP latency in response to tactile stimulus was measured from the onset of the EPSP to the onset of the AP. The slope of the EPSP was measured as the tangent of the angle of the EPSP growth. The EPSP amplitude was measured when a neuron failed to generate AP, the AP threshold and latency when a neuron generated an AP in the response, while EPSP slope and number of APs in the response were recorded for every response. The Fig. 1.20 was redrawn in accordance with the data [1259].

results are impossible to consider as convincing. Since prolonged recording in an experiment is impossible at the present time because of technical reasons, more intracellular recording is necessary in order to evaluate excitability. The data mentioned here were obtained in separate experiments at the neural tissue taken from trained and naive animals. Neurons from heavily trained animals may be more sensitive to damage (see Chapter 3) and were more injured by the electrode. Particularly, the effect may be observed because in the trained animal some neurons with a low threshold were recorded [1100].

Many reports describe a change in excitability during learning within the responses to stimuli whose significance changes during learning [1355, 641, 24, 525, 45]. However, in cited investigations did not test excitability within responses to discriminated stimuli. The selective form of excitable membrane plasticity has been described in other laboratories [1157, 982, 1419, 374, 997, 1339]. During elaboration of a conditional reflex in the Helix mollusk, an AP threshold within the response to a CS^+ decreased to a greater degree than a threshold in the response to a US, while pseudoconditioning did not change excitability within the response to a CS^+ but augmented excitability in the response to a current pulse [752]. It has been reported that the excitability of cortex neurons during conditioning increases while spontaneous activity does not change, which was explained by the specific geometry of neuronal inputs [1355, 48], but it could also be explained by selective change in excitability. Selective change in excitability was found later in other cases after LTP in the hippocampus, cortex and cerebellum [586, 291, 414, 374] and was explained too by the local modification of excitability in the restricted dendritic area, connecting this excitability to potentiated and non-potentiated inputs.

We think that this assumption of specific excitability in the restricted dendritic area cannot explain the specificity of normal learning. Certainly, if synaptic input for the CS^+ lies, for example, close to modified channels, an alteration in these channels could underlie long-lasting changes in the AP waveform just within the response to the CS^+ and may facilitate spike generation. An EPSP evoked by a CS^- might not undergo amplification because of its spatial remoteness from these modified channels. Therefore, a change in excitability will be selective but steady, not transient. In order to accept this hypothesis, we have to speculate that locations of the conditioned and discriminated excitations become different during acquisition, draw together during extinction and rapidly diverge during reacquisition. This does not look more probable than a transient alteration of excitability within the response to a given stimulus. In particular, selective change in excitable membrane properties was demonstrated in an isolated neuron when microelectrodes played a role as source of a stimuli with an invariable location [1157, 1419, 1301] and when a rare stimulus was a fragment of a habitual stimulus [1272].

An unexpected outcome of investigation into excitable membrane plasticity has come from the analysis of neuronal reaction selectivity during image recognition. Direction selectivity in simple cells of the cat primary visual cortex has usually been explained by a selectivity of inhibitory processes: in the preferred direction excitations surpass inhibitions. However, it has been proved that excitations and inhibitions in the visual cortex and retina are tuned for the same direction and therefore selectivity cannot be explained by the spatial organization of excitatory and inhibitory synaptic inputs [997, 307]. During habituation and conditioning, too, the EPSP and inhibitory postsynaptic potential changes in the same direction. Moreover, the time of the greatest inhibition was close to the time of the least excitation for the same spatial input and a difference in their latencies cannot explain the selectivity. The spike response is far more direction selective than synaptic response and its selectivity may be explained by a nonlinear decrease in spike threshold for the preferred direction.

Sensory neurons in the rat barrel cortex also demonstrated different excitability to the preferred and non-preferred orientations of the whisker angular deflection (Fig. 1.21A,B) [1339]. The difference in the threshold was 2.1 mV. The mean spike response was more sensitive to the direction of whisker deflection than the mean synaptic response. Spike threshold was measured at the maximum of curvature at the AP front. Only the first AP in the response was measured. The rate of rise of the synaptic response, EPSP slope, and threshold were connected by a negative correlation. This correlation is considered as the reason for existence of different thresholds in the same neuron. The authors explained the change in excitability as the result of an increased proportion of Na^+ channels entering the inactivated state during the slower depolarization. This explanation is plausible if depolarization is long enough or large enough. The difference in latencies between an AP for preferred and non-preferred direction was only 5 ms, while EPSP amplitude was a few mV [1339] (Fig. 1.21).

Negative correlation between EPSP slope and AP threshold was observed also in our experiments [1259], but cause-and-effect correspondence between these factors was absent (Fig. 1.22).

High thresholds, as a rule, are observed in the responses with larger AP latencies and lower EPSP slopes, but correspondence between the threshold and the slope of the EPSP was not well-defined. Thresholds may differ when EPSP slopes almost coincided and a smaller EPSP slope may correspond either to smaller or larger thresholds (Fig. 1.22,right). Fig. 1.22 (left) demonstrates a weak correspondence between AP threshold and EPSP slope for within-cell comparisons. The correlation ratio for the dependence between AP threshold and EPSP slope tended to decrease after factoring out the covariate trial number, and only for a few cells did the covariate tend to increase, when the trial number was kept constant [1259].

The examples in Fig. 1.23 demonstrate that the AP threshold, AP latency and EPSP slope are not connected by a rigorous dependence. After training, the AP latency in the response to the CS^- may be smaller (top left) or bigger (top right) than it was in the response to the CS^+. Similarly, the EPSP slope may be bigger (bottom left) or smaller (bottom right) in the response to the

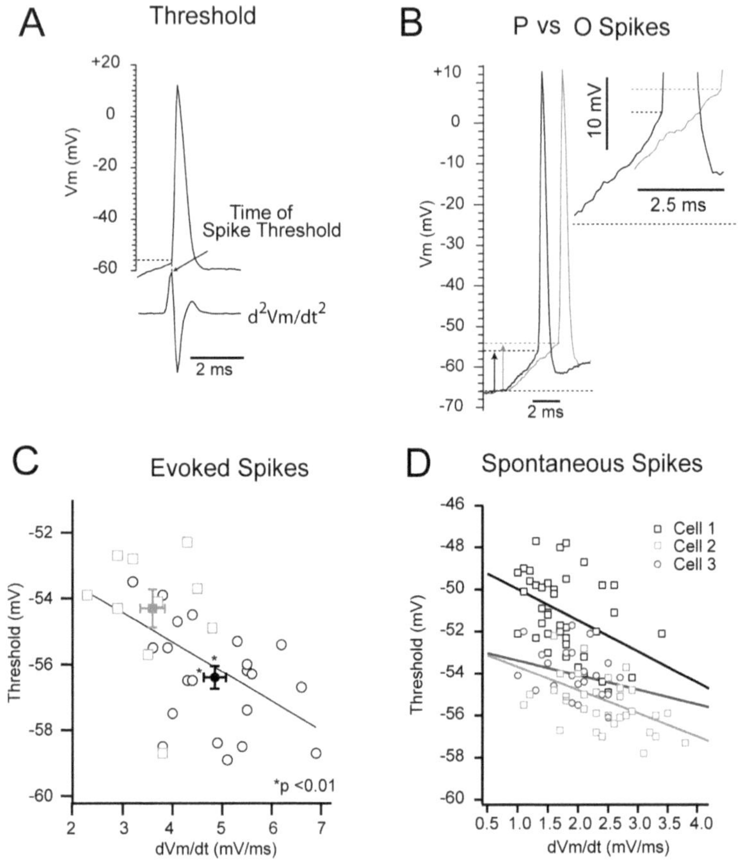

Fig. 1.21. Spike threshold and the preceding dVm/dt. A, The level of the spike threshold was determined as the voltage from the baseline to the time of the peak of the second derivative. B, A preferred direction - (black) and opposite direction - (gray) evoked spike are shown. The rate of rise preceding the preferred direction of the spike was 1 mV/ms greater than that of the opposite direction - evoked spike, and the spike threshold of the preferred direction was 2.0 mV lower than that of the opposite direction-evoked spike (vertical arrows). The inset shows the final 2.5 ms of the synaptic response leading to a spike, the time period in which dVm/dt was determined. C, Threshold and EPSP slope of all evoked spikes for opposite direction (squares) - and preferred direction (circus) - are shown. The spike threshold of the preferred direction-evoked spikes was significantly lower, and the d_m/dt was significantly higher than that of opposite direction -evoked spikes (mean SE). D, Threshold versus EPSP slope for spontaneous spikes from three sample cells, including the cell shown in B (cell 1). From [1339] with accomplishment.

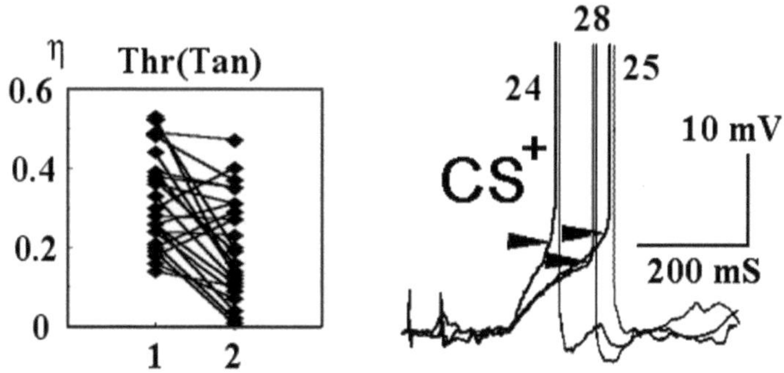

Fig. 1.22. An overview of correspondence between an AP threshold and an EPSP slope during acquisition of classical conditioning. At the left, the correlation ratio for an AP threshold and EPSP slope in the majority of the cells decreased when the trial number was kept constant. The presence of dependence (not only linear) between the variables was estimated using analysis of variance. Analysis of the influence of covariates on correlation ratio shows how a given dependence would have changed if the covariates were kept constant. The influence of a given covariate means that the covariate affects either the shape of the dependence or the spread in its value. Each diamond represents one cell. Ordinate: correlation ratio (η) of the dependence of an AP threshold (Thr) on the EPSP slope (Tan) in the same response. Abscissa: 1 main effect, correlation ratio. 2 the correlation ratio decreased when it was factored out in analysis of a covariate trial number. The relationship between an AP threshold and EPSP slope is partially explained by their common dependence on trial number. At the right, an example of the responses of the Rpa3 neuron to the CS^+ during the last part of acquisition. The oscillogramms are matched in the membrane potential levels and the onset of the EPSPs. The number of CS^+ at each exposure is indicated. Clear-cut correspondence between AP threshold and EPSP slope was evidently absent. The Fig. 1.22 was redrawn in accordance with the data [1259]

CS^- (Fig.1.23). However, in both cases the threshold in response to the CS^- exceeds the threshold at the response to the CS^+.

We evaluated correlation between the EPSP slopes and the AP thresholds in direct experiments. Change in strength of tactile stimulus lead to an increase in EPSP slope and amplitude, but does not affect AP threshold. This is just the famous all-or-none principle and it is almost always correct (Fig. 1.24).

An AP threshold decreases with the growth of the depolarization slope, but it remains almost constant with only a slight increase when the amplitude of the input signal increases, in spite of EPSP increases also in this case [1043, 1165, 413]. Therefore, selective increase in synaptic efficacy in the conditioned input cannot explain enhancement of excitability in the same input. Examination of the change in excitability in response to a stronger tactile

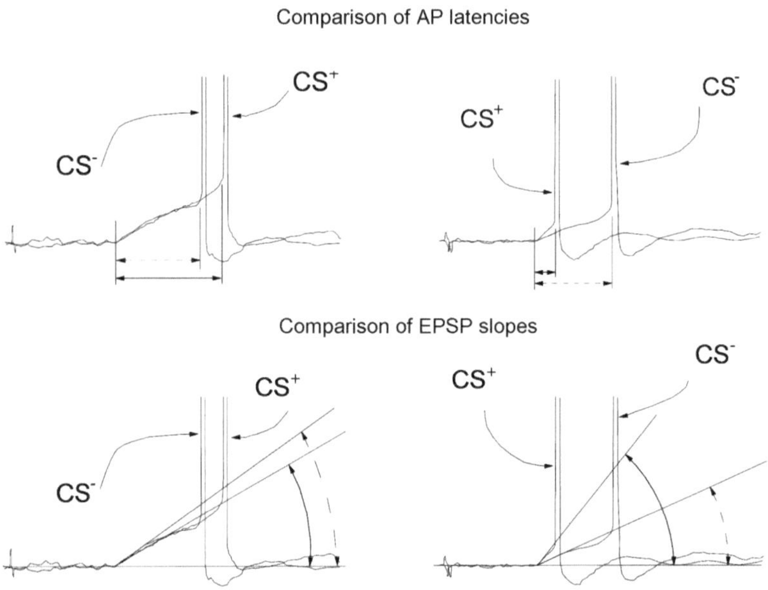

Fig. 1.23. Weak interdependence between AP thresholds, AP latencies and EPSP slopes. Responses of mollusk neuron RPa3 to the CS^- and CS^+ after training are aligned at the point of the beginning of the EPSP. Two examples are shown at the left and at the right. At the top, imprecise changes of the AP latency in the responses to the CS^- and the CS^+ are shown. At the bottom, imprecise changes in the EPSP slope are demonstrated. Methods of measurement of AP latency and EPSP slope are indicated.

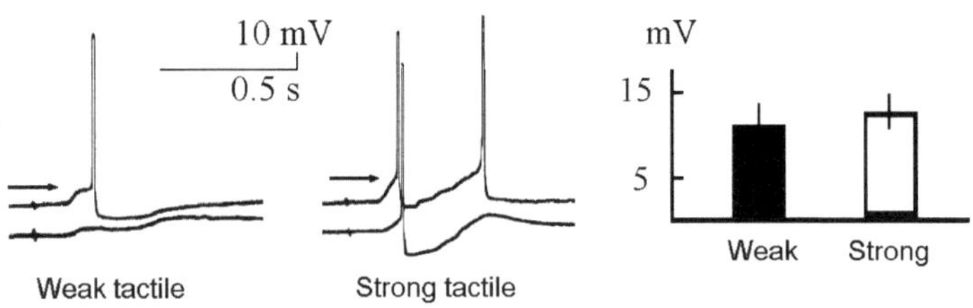

Fig. 1.24. The change in the AP threshold within responses of the Helix neurons LPa3 (top) and RPa3 (bottom) to weaker (left) and stronger (middle) tactile stimuli. At the right, mean change of AP thresholds for 18 neurons are shown. The results were obtained before acquisition. Thresholds are indicated by the arrows.

stimulus did not reveal a significant difference between excitabilities within the response to the weak tactile stimulus directed to the same point of the foot (Fig. 1.24). This inspection was not examined in rat barrel cortex [1339].

We believe that a cause-and-effect connection of the negative correlation between EPSP slope and threshold in the experiments of [1339] is not a correct conclusion. Firstly, excitability and EPSP slope were larger for the preferred direction, exactly as we have described properties of EPSP slopes and thresholds for the CS^+ and CS^- during classical conditioning; change in excitability and EPSP slope are parallel processes. Secondly, both EPSP slope and threshold are slightly increased with the strength of a stimulus that is connected with a positive correlation according to the experimental [1259] and theoretical [1043, 646] data. Thirdly, for the same value of EPSP slope, the threshold of the evoked AP was 2 mV smaller than the threshold of the spontaneous AP (Fig.1.21C, D). This difference is compatible with the whole value of the effect discovered in this investigation and, hence, a one-to-one correspondence between these characteristics was absent.

Investigation of selectivity of neuronal response sometimes reveals no specific change in excitability. In the visual cortex of cat, neurons respond selectively to the orientation and direction of movement of an object, but the threshold for the AP generation, which has been measured within the responses, did not depend on stimulus orientation, and thus could not account for the observed difference in the transfer function [1308]. However, only the threshold for the first AP in the response has to be measured, while 10 or more APs were generated and measured in each response. Possible participation of selective excitability in the recognition of images is a basic result, since it means that capability for reorganization of thresholds may be long-term. Then, when a neuron selectively reorganizes it excitability, its may maintain this property for a long time. Each time in the future when it will be necessary to recognize the signal, selective excitability has to recover it after access to long-term memory. Certainly, this suggestion needs examination.

The selective change in threshold cannot be explained by non-specific alterations in the neuronal state, such as the change in input resistance, ion concentration, the membrane potential, etc. These and any other changes in neuronal state may determine the selective change in excitability only if these alterations are transiently induced by the current stimulus. This is the reason why it is difficult to use patch recording, pharmacological blockage or other methods of channel investigations during learning. It would be convincing to compare responses from one entire experiment under normal conditions and another entire experiment under exposure to some one factor.

Neurons appear to evaluate the probable consequences of a signal not only according to its strength but also according to its significance, and then transiently change their own excitability. A neuron somehow chooses an appropriate excitability, but the mechanism of the transient reorganization of excitability is still unclear. Modifications of excitability during learning are in accordance with the assumption about rapid reorganization of Na^+ channel properties: the number of active channels in the membrane and their opened-closed characteristics. Although this scheme is not a final hypothesis, it is in accordance with all known data. Thus, during habituation, drawing of Na^+ channels into the signal-induced excitation is impeded, while after classical conditioning, CS^+ more easily draws Na^+ channels into the reaction. Consequently, during habituation, Na^+ channels are but sluggishly concerned in AP generation and a spike may not be generated at all. Correspondingly, for classical conditioning, the effects should be opposite. These conclusions are supported by the data concerning change in latency, threshold of AP and number of APs in the response to learning related signals (see Sec. 1.7).

The AP start depends mostly on the potential-dependent Na^+ channels, while other channels such as K^+ and Ca^{2+} are involved later. Na^+ channels interact with second messengers, undergo post-translational modifications and this regulation allows electrical excitability to be tuned to changing physiological needs [55, 790]. Neurons may respond to changes in their input by adding Na^+ channels that alter electrogenic properties of the cell membrane [1215]. However, expression of new channel requires few hours, whereas the phosphorylation of channels already presented in the membrane takes a fraction of second. Chemical diffusion through cell is slow process and can take 10-100 ms, but diffusion of little molecules between neighboring synapses has a shorter character time. Theoretical accounts demonstrate that the change in excitability during learning may result from a shift in the voltage dependence of Na^+ channel activation or a change in the probability of their transition into an open state [514, 1044]. Change in excitability during LTP [1367], LTD [745], stimulus detection [14], cocaine-withdrawal [1428] and conditioning [1351] depends on the function of voltage-gated Na^+ channels and on participation of the system of second messengers. There are indications that second messenger cyclic AMP participates in the adaptive reorganization of excitability. Thus, excitability of central auditory neurons is, probably, regulated by cyclic AMP pathway. Elevation of the intracellular cyclic AMP level slowly (tens of minutes) increases the spontaneous activity compared to tone evoked responses, while responses to tone and to glutamate increased similarly [1119]. However, there is a shortage of direct experimental data on the participation of Na^+ channels and cyclic AMP in the reorganization of behavior.

The same neuron or synapse may participate in the analysis of different stimuli. How quickly can modulation of Na^+ channels be expressed? This time should be shorter than the EPSP duration, since the same neuron may generate different AP waveforms in response to an EPSP evoked by different signals. Rapid conditioned alterations have been observed with a latency of from 10 ms to 30 ms, but it is unclear if they are transiently evoked by a CS^+, or preexisted without connection to the current stimulus. We have shown directly that the AP waveform changes within 50-200 mS, from the first to the second AP within a response to the CS^+. During execution of a sinusoidal flexionextension movement of the wrist in a tempo of 1 Hz, motor neurons in cortical structures and spinal motoneurones modulate their excitability with congruent time courses as the motor program [152]. Since the excitability modulation precedes the movement, one can suppose that high-speed adjustment of excitability may control a current behavior. The firing threshold seems to be a dynamic property of an excitable membrane, but it must rebuild their properties rapidly.

It is necessary to make a reservation concerning the speed of signal recognitions. This process uses only short-term memory. 10-30 ms for decision-making is related only to signals participating in the current learning, as the CS^+ and CS^-, to be exact in respect to current signal and without participation of long-term memory. What is really necessary for signal recognition is the merging of information from activated synapses and correlation of this information with the current expectation of reward. Participation of long-term memory requires at least a ten times longer time interval for recognition than, for instance, recall of famous faces [1005, 8].

Rapid transmission of intracellular signals may also be realized through the cytoskeleton, which forms continuous, dynamic connections between nearly all cellular structures, and these connections present an enormous surface area upon which proteins and other cytoplasmic components can dock [696, 230]. Transmission may have a selective direction through narrow channels in microtubules, since both the neuron soma and dendritic tree are penetrated by a system of microtubules. The cytoskeleton network not only contains many filamentous elements, but its structure also changes constantly in time [140]. The chemical and physical properties of the cytoskeleton suggest that it may have subtle and important effects on the way in which information is passed[5].

Co-existence of two thresholds in a neuron looks like the occasion to reject the all-or-none principle. Nevertheless, neurons generally either do generate an AP or do not generate it at all. Change in AP amplitude during learning is usually small. Consequently, it is not a good idea to reject this basic principle of neuron function. However, our results mean that the all-or-none principle ought to be modified. Fig. 1.25 demonstrates the results of a computational

[5] The cytoskeleton adjusts Na^+, K^+ and Ca^{2+} channels. Intracellular mechanical signals may be rapidly transmitted from cell surface receptors to distinct structures in the cell and nucleus, including ion channels, nuclear pores, chromosomes, and, maybe, individual genes, independent of ongoing chemical signaling events. This is directed propagation, where the force is intervened through a narrow channel. Whether the effects of the cytoskeleton on ion channel activity are a direct mechanical effect or an indirect effect due to such changes as regulation of kinases sequestered at the cytoskeleton remains a controversial issue [787, 583]. It was shown that after a cell surface integrins were mechanically stressed neurotransmitter release from motor nerve terminals can be detected within 1020 ms [230]. A signal caused by a conformation transition in a microtubule-associated protein may propagate over the microtubules at a speed of 8-800 m/s [696]. This speed is sufficient for considering that input information arriving at synapses generates specific second messengers at a remote place and at the almost the same time.

experiment illustrating such a modification. Suppose that one combination of excited synapses corresponds to a CS^+ and another such combination corresponds to a CS^-. The properties of excitable membranes in our model were controlled by supposed chemical reactions occurring in neural cells (for details see Sec. A.1).

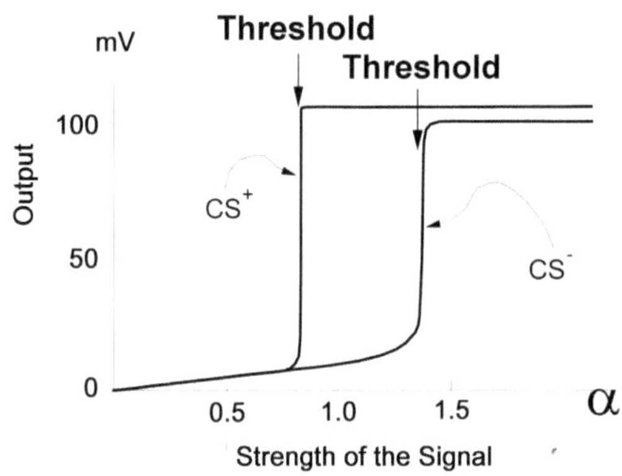

Fig. 1.25. Modified all-or-none principle. Activation functions for responses to the CS^+ and CS^- are shown for the 10-synapsed model neuron after 20 combinations of a CS^+ and an US. The CS^+ corresponded to the pattern: 111111000, CS^- to: 0000111111. During the test, after learning the CS^+ was - $\alpha\alpha\alpha\alpha\alpha\alpha0000$ and the CS^- - $0000\alpha\alpha\alpha\alpha\alpha\alpha$, where α changed in the limits $0 < \alpha < 2$. Abscissa - α value. Ordinate - maximum membrane potential deviation in response to the signal, calculated as described in the Hodgkin-Huxley model. The Fig. 1.25 was redrawn in accordance with the data [1259].

After conditioning, the thresholds within the response to a CS^+ and a CS^-become different and a neuron chooses a branch of the activation function (Fig. 1.25), which, presumably, leads to a healthy result. However, for the given combination of inputs, the model exhibits (after learning) the all-or-none principle of spike generation, when intensity of the inputs changes.

1.7 The chemical nature of memory

1.7.1 Chemical traces of memory and chemical sensitivity of memory

Many indirect observations illustrate the chemical nature of memory. For example, stability of memory after severe brain damage with catastrophic impairment of neural network structure demonstrates that a chemical hypothesis may be correct. When partial amnesic effects are ameliorated spontaneously or after remembering, this indicates to impairment of memory recall, while memory storage is preserved even during temporary amnesia. Where else, except in the chemical substratum, may temporary missed information be maintained? In addition, reorganization of neural pathways is a very long-lasting process when compared with chemical reorganization. It could be supposed that a copy of all supplies of long-term memory in each cell might be the safest way to protect behavior from brain damage. Remaining cells will possess all of the essential information required for survival. However, this guesswork is not proper.

Brain does recover its function after a long period of rehabilitation, days, weeks and months, but animals cannot to survive even a couple of weeks in the wild being completely helpless. Brain injury in ontogenesis and phylogenesis leads to weaker damage of behavior. Therefore, distributed properties of brain functions are not preferential in evolution and do not appear. In contrast, strict localization of function in brain structure increases in evolution, reaches a maximum in human and is completely absent in hydra, the most primitive multicellular animal [1255]. For example, complete extirpation of the neocortex renders a cat invalid, without any possibility for recovering, while in a turtle, extirpation of forebrain, together with the incipient cortex retains its viability. Habituation in the cold-blooded animals is observed even after decapitation. Localization of functions in the higher mammalian is not evidently an evolutional achievement. Behavior really recovers after brain damage, but rehabilitation of patients becomes possible only after the development of civilization. Capability of brain to recover its functionality is obviously explained by the ability of a brain part or even individual neurons to perform the functions of an entire brain. Rehabilitation is not, evidently, the result of successful brain construction, but rather the property of a neuron itself, which is simultaneously both the brain element and the living being. Of course, this idea may be right only if "cell consciousness" is rather primitive.

Experimental evidences exist verifying a close link between learning, memory and intracellular biochemical processes [1255, 663, 1039, 657]. In human, memory acquired in early childhood is fragile (infantile amnesia), but memory of the primitive animals much more stable. Their memory may stay intact while brain morphology is strongly reorganized in the course of individual development. During insect and amphibian metamorphosis, a subset of neurons within the central nervous system persists to participate in adult behavior, while neurons that will no longer be necessary die and some adult interneurons are born postembryonically. Hence, adult motoneurons, as well as some interneurons and modulatory neurons, are persistent cells, but neural networks undergo reorganization. In accordance with their new behavioral roles, these neurons undergo striking changes in dendritic morphology, intrinsic biophysical properties, and synaptic interactions [259]. However, a conditioned reflex acquired by a tadpole is sustained after metamorphosis [842]. Similarly, avoidance of a specific odor passes through stages of metamorphosis in fly [1276]. Conditioned odor avoidance produced in larvae still was present in adults 8 days later. Such memory through metamorphosis was specific to the temporal pairing of odor and shock. Presentations of odor alone or shock alone did not produce change. Part of larval neurons survive, some changing their peripheral targets and others innervating targets. Soma of neurons that persist during metamorphosis may serve as neuronal substrates of memory through metamorphosis. These facts are difficult to explain without an assumption of the chemical substrate of memory. After amnesia, long-term memory does not disappear: information can be recovered by means of reminder or by influence of some drugs [1255, 1039, 657]. Chemical alterations during learning have been discovered in many brain areas, but after the firm fixation of information, chemical traces of memory are concentrated in specific areas of the brain[6]. Damage of these specific brain areas together with their traces of memory does not lead to disappearance of acquired information. Evidently, memory traces in other brain areas from early stages of learning do not disappear absolutely, although they are not usually accessible.

Thus, learning leaves chemical traces in the brain. Moreover, chemical impact leads to alteration in memory. Fear conditioning is disrupted by inhibition of cyclic AMP-dependent kinase [86]. The second messenger cyclic AMP has a very fast and profound effect on many neuronal genes. Molecular genetic investigations of Drosophila olfactory learning have uncovered numerous genes whose gene products are essential for memory formation. Genes that impair olfactory memory when they are disrupted: adenylyl cyclase (that produces cyclic AMP), phosphodiesterase (that degrades cyclic AMP), protein kinase A (that leads to protein phosphorylation), etc. These genes have a prominent role in mammalian and non-mammalian organisms [298]. Reactivation of NMDA receptors is also necessary for the long-term storage of old memories in neural circuits.

Animal behavior and including learning change also after alterations in neurotransmitter systems of brain [680, 53, 481], after administration of neuropeptides [983, 314], after the addition of antibodies to many brain antigens [603, 973], after blockage of cytoplasmic flow [397] and as the result of activation of immune processes [62]. Single gene modification in Drosophila flies determines agressive behavior [1310]. Both genetic and immunologic memories depend on protein synthesis. Therefore, it was natural to suppose that neuronal memory is also associated with a similar mechanism and that acquisition

[6] The long-term changes in neuronal firing produced by instrumental learning occur in the same areas of the rat cerebral cortex [1203] and forebrain of chicks [1040], where the transcription was activated during acquisition of the same task. During learning, content of myscarinic cholinergic receptors in specific area of brain increases [1040, 53], synthesis of proteins and ribonucleic acid (RNA) increases [574, 768], expression of early genes (which ensure activation of other genes and thus regulate the order of protein synthesis) strengthens [42, 690] and the system of second messenger cyclic AMP is augmented [811, 20, 86]. Brain areas where neuronal activity is correlated with learning are narrowed after fastening of habits [1094, 489, 561].

of new memory is combined with the synthesis of new proteins. This parallel, however, is superficial, since neuronal memory is dynamic and established rapidly, when compared with genetic and immunologic memory.

There are some general molecular events related to memory storage that have been, in particular, identified in Aplysia sensory neurons, chicken forebrain and mouse pyramidal neurons of hippocampus [86]. These are: equilibrium between kinase and phosphatase activities at the synapse, retrograde transport from the synapse to the nucleus, activation of nuclear transcription factors, activity-dependent induction of gene expression and synaptic capture of newly synthesized gene products. The location of these events, supposedly, moves from the synapse to the nucleus and then back to the synapse. Following transcriptional activation, newly synthesized gene products have to be delivered specifically to the synapses whose activation originally triggered the wave of gene expression. That is, hypothetically, the products of gene expression are delivered throughout the cell but are functionally incorporated only in those synapses that have been tagged by previous synaptic activity [86]. This description looks like a version of the chemical mechanism of memory, but it cannot explain the correspondence between chemical and informational processes and factually is in accordance with the synaptic plasticity hypothesis and may plausibly explain development of new pathways in ontogenesis, during damage, etc. Participation of the discussed chemical processes may be reduced to servicing. Transport from the synapse to the nucleus is non-specific and it cannot deliver a message to precise sites in the nucleus. A specific message cannot be transmitted by means of diffusion of unspecific substance. This is possible only through specific channels. For example, newly synthesized proteins have to be selectively transported to activated synapses without altering the function of all other synapses in the activated cell (hours and days) [1316]. A cell must somehow integrate signals from a pattern of synapses and translate information into chemical language. All aforementioned facts testify to the availability of chemical alterations in the brain during learning. However, the data add little to comprehension of chemical coding of information.

There is agreement between physiologists, pharmacologists and psychologists about the existence of at least two phases of memory that proceed through temporally distinct phases: short-term (minutes - hour) and long-term (hours days years) [1039]. Disruption of one does not disrupt the other. Sometimes ones distinguish instantaneous (seconds) and intermittent (hours) phases of memory. Instantaneous memory maintains traces of current signals. Phases of memory have specific sensitivity to influence different substances. Various phases differ by their steadiness and stability to amnesic impact. Long-term memory is considered to be the most firm and is very resistant to hurt. Memory transformation into a long-term phase is called consolidation of memory. In a simple form, consolidation hypothesis postulates existence of only two sorts of memory storage, each having a specific chemical characteristic and a different dependence on pharmacological impact, short-term and long-term memory [1039]. The cell nucleus is considered to be a participant in long-term memory, while short-term memory operates without nuclear application [86].

The most important property of long-term memory consolidation is its dependence on protein synthesis and on synthesis of ribonucleic acid (RNA) [477, 397, 1040, 20, 86]. Indeed, formation of long-term memory is impossible after blockage of protein synthesis, at least for stressed learning or learning with negative reinforcement. An overwhelming majority of data was achieved for specific, negative forms of learning and very rarely normal appetitive learning may depend on new proteins. In special circumstances, it was reported, instrumental learning with feeding reinforcement, also may depend on protein synthesis. So, early consolidation of instrumental learning requires protein synthesis in the nucleus accumbens [536] and strong inhibition of protein synthesis in the motor cortex (for 4 days) by means of anisomycin retarded appetitive learning [768]. Protein synthesis in experiments is usually blocked by antibiotics, but to our knowledge, nobody has observed amnesia in humans after antibiotic injections. In animal experiments typically anisomycin or electroshock are used in order to inhibit protein synthesis. On the other hand, anisomycin, which disrupts traumatic memory consolidation, inhibits anxiety-like behavior [250]. Anisomycin can stimulate stress-activated protein kinases, and after anisomycin administration an animal behavior may look forgetful, but an animal may simply not execute a negative habit because its negativity was weaker after injection. Besides, anisomycin was described as downregulating gap-junctional intercellular communication [905] and this may prevent normal behavior. Blockage of protein synthesis leads also to enhancement of amino acid concentrations, because amino acids cannot be included in the protein structures [1040]. Many amino acids (especially glutamate and GABA) are neurotransmitters and affect the process of learning. A comprehensive review of the protein synthesis literature leads to conclusion that long-lasting memory is observed despite protein synthesis inhibition [1049].Therefore, the theory that memory consolidation is accomplished through macromolecule synthesis is premature.

Recently, a newly described phenomenon, memory reconsolidation, had been reported to occur after negative learning. During learning, punishment was presented immediately after a conditioned stimulus. Later representation of the CS^+ alone evoked fear and the reaction of avoidance was accomplished and the animal escaped punishment. The phenomenon of reconsolidation was revealed by means of post-training with an intra-amygdala infusion of the protein synthesis inhibitor anisomycin. Amygdala, supposedly, is involved in modulation of memory for aversively motivated behavior. Anisomycin produced amnesia of the past memory if administered immediately after retrieval, that is, in the time period when consolidation had already been completed [880]. After retrieval, anisomycin leads to disturbance of the habit in the next trial, that is, next CS^+ does not evoke avoidance, although without anisomycin avoidance would be still robust. This looks like forgetting the habit. The phenomenon seems to be that consolidated memory becomes labile again after retrieval and reconsolidation of the stable memory after retrieval requires protein synthesis.

Reconsolidation is an extremely firm hypothesis, since this is an energy- and substance-consuming process. Of course, each particular image recollection may correct memory, if it is enriched by new details. Fact recollection itself also may leave a trace in memory. Nevertheless, uninterrupted memory reconsolidation through an entire life, in a body, looks improbable. All the more, the aforesaid data are easy to explain without such a strong suggestion. We may suggest a more parsimonious explanation. Absence of a US leads to extinction and the next presentation of the CS^+ may decrease instrumental reaction and increase probability appearance of the US. This standard scheme of extinction works properly for an appetitive conditional reflex because the animal immediately discovers the absence of a US. During aversive conditioning, abolishment of a US will for a certain period remain unknown to the animal: it is unclear, to the animal why the US is absent, because after it has accomplished the correct instrumental reaction or because of change in environment and abolishment of punishment at all. As a result, extinction of aversive habits is delayed. When the animal perceives the CS and accomplishes avoidance reaction, it does not receive punishment, but does feel fear. He is afraid the punishment and therefore escape it. Since anisomycin inhibits anxiety-like behavior [250], animal later remember the fact receiving of the CS and the fact of absence of fear. This must lead to reconsideration of the degree of danger. In the next test, the CS^+ will be less connected with fear, animal will not perform the reaction of avoidance and this will be look like amnesia. Essentially that anisimycin-induced anxiety-like behavior tightly interacts in time with the situational reminder. Microinjection of anisomycin, administered 1 hour after the reminder, did not affect the anxiety-like behavioral response 3 days later [250]. Similarly, when electroshock, which is distressed closely to time of aversive learning trial, evokes amnesia, this may be connected with remembering stronger aversive event comparing with the weaker negative US. Therefore reconsolidation may be epiphenomenon. All the studies devoted to reconsolidation are based on punishment or aversive training and "reconsolidation" was detected only in the absence of significant extinction [341]. As we have explained, anisomycin may assist to display extinction. In particular, the dependence of "reconsolidation" on protein synthesis decreases with the age of memory [616]. Phenomena similar to "reconsolidation" were observed also after administration of drugs, which do not affect protein synthesis directly. These are opioid peptides, GABA, norepinephrine, cyclic AMP, acetylcholine and glutamate [821, 616].

The border between short-term and long-term memories is vague. The duration of memory components can vary with different drugs, tasks, reinforcement and species [316]. Spontaneous recovery or reminders-induced recovery of memory over a time course of weeks or months has been reported in animals which were initially thought to be amnesic [871, 1316]. Memory for a well-learned instrumental response does not require protein synthesis-dependent reconsolidation as a means of long-term maintenance [537]. Short intervals for consolidation also were demonstrated. During classical conditioning of negative reinforcement in earthworms (the occurrence of a shrinking response), inhibition of mRNA synthesis by actinomycin-D or protein synthesis by anisomycin blocked consolidation of the long-term memory, when either of these two compounds was injected into the body cavity of the worm within 25 min of conditioning [1324]. Amnesia usually, with the rare exceptions, more concerns to recent events. Sometimes these are years, sometimes minutes. This means that consolidation of memory either continues uncertainly long or amnesic agents could affect the recall of memory instead of memory storage.

There are many doubts that memory is coded in the protein structures. Previously, the data concerning memory transfer from one brain to another by chemical substrates was seriously discussed, but now this idea is only of historical interest. J.V. McConnell [816] described memory transfer through cannibalism in planaria. Also G. Ungar [1286] and some others reported memory transfer in rats. If memory transfer would be confirmed by subsequent investigations, the suggestion of chemical coding of memory would be irrefutable. Memory transmission through metamorphosis and through heredity also is in agreement with the reality of a chemical code for acquired memory. Existence of the brain-blood-barrier may be interpreted, as the defense of newly synthesized brain antigens from immunological attack of the organism and this indirectly testifies in favor of the chemical nature of memory. However, consensus deems it unfeasible to transfer memory acquired during learning from brain to brain [194, 1040]. The effects that had been observed appeared as the result of such accompanying circumstances, as an augmentation of motor activity, induction of stress or depression, administration of endogenous opiates, etc. This, however, does not concern inborn behavior. Some general predispositions to an inborn form of specific behavior do exist and we will discuss this problem later. The possibility of behavior transfer using chemicals is indirectly supported by the numerous data involving the induction of motivational behavior by means of chemical substances. This is related to sexual behavior, feeding, drinking, etc. [1257]. State-dependent learning is another example of selective chemical influences to memory reproduction: some pharmacological modulations (for example: cholinergic substances) act as a specific chemical environment, and memory recall is possible only on the same chemical background, which takes place during learning [53]. Nonetheless, this is unlikely to assume transfer of ordinary habits between brains. Moreover, transmission of chemical memory between different regions of the same brain is also doubtful. For example, lobotomy as a form of psychosurgery, consisting of cutting the connections to and from the prefrontal cortex, results in major personality changes and these changes are irreversible [1294, 584], although the pathway for transmission of chemical information stays intact. Similarly, when the corpus callosum connecting the two halves of the brain is severed, we have the phenomenon of the split-brain and the patient perceives separately the data acquired from the left and right sides. This impairment also does not recover.

The surgical operation to produce this condition is rarely performed, usually in the case of epilepsy, and this operation reduces the severity and violence of epileptic seizures. Hence, even if chemical memory exists it is not a material substance observable regardless of neuronal structure, as, say, a hard disc containing information independent of a computer.

1.7.2 Biological meaning during habituation is acquired or lost by chemical means

Any brain activity is based on transformation between chemical and electrical processes. Only at the final stage does a nervous process convert to mechanical movement. A decision to generate or fail to generate a spike may be related to changes in synaptic efficacy and excitability and both changes depend on chemical reactions within neurons [827, 20]. Therefore, chemical processes do participate in learning and memory. Really, supporters of the memory hypothesis as the reorganization of a neural network (and even uttermost form of this hypothesis: memory in presynapses), fairly do not consider contradictions between synaptic and chemical memory [86]. Therefore, chemical processes certainly participate in memory and even if they have ancillary, non-informational functions, they will be augmented during learning and their blockage will interrupt memory consolidation. These processes may be associated with augmentation or decline of general metabolism or else certain informational metabolic pathways. However, there is a question, to what extent these chemical reactions participate in memory. They may provide energetic processes during learning [31], morphologic reorganization [77] and homeostatic recovery [478] that is, do not participate in the control of informational significance. When chemical processes have non-informational functions, the role of chemical reactions is reduced to ensuring the presence of memory traces in the space of a neural network.

On the other hand, chemical alterations during learning could perform informational loading and there has to be a parallel between elements of information and chemical processes or substances. During learning, responses to biologically important stimuli usually increase and responses to insignificant stimuli decrease. When neurons decide which stimulus is more important in the current behavior, is the decision made by chemical means? As a maximal presumption, different perceptual signals give rise to different chemical processes in the neurons. As a minimal presumption, chemical specificity may correspond only to a difference between signals, participating in current behavior and other signals. Let us consider how neurons distinguish a biological important signal from insignificant ones.

Learning depends on various chemical processes within neurons, but it is unclear whether these chemical processes have any informational significance. We tried to examine to what extent the signals having different biological significance may have a different chemical basis [1264]. If they did, injection in a neuron vicinity of the same biologically active substance before and after learning would have different effects on the response of the neuron. As a simple example of learning, we chose habituation in a mollusk and studied whether a change in reactions of an identified mollusk neuron for defensive closure of the pneumostome during elaboration of the neuronal analog of habituation is connected with modification of the chemical processes induced in the neuron by a habitual (tactile) stimulus and whether this process differs from chemical processes induced by a new stimuli (light). Let us suppose that the chemical processes evoked within the neuron by the stimulus are modified during its repeated presentation, when the biological importance of a novelty is reduced and response to the stimulus decreases correspondingly. Then, microiontophoresis of the same substance before and after habituation would have different effects on the response of the neuron, if, of course, this substance somehow interacts with the learning-related chemical processes. As an example of a biologically active substance we used acetylcholine, which exerts modulatory action by means of muscarinic receptors [676, 1036]. Acetylcholine is readily availably to the learning process, is not a toxic agent and is naturally presented in neural tissue [183, 366, 481]. For instance, when in the rat somatosensory 'barrel' cortex the temporal frequency of whisker deflection, were modified by cellular conditioning, administration of acetylcholine during testing revealed frequency-specific changes in response that were not expressed when tested without acetylcholine [1140]. Acetylcholine also modulates an action potential waveform and thus connects synaptic and membrane processes [386].

Habituation in our experiments was accompanied by a decrease in the number of action potentials in the response and by elevation of the threshold of AP generation, but was not accompanied by any appreciable change in membrane potential. The number of APs in response to switching off the light in the experimental camera increased after habituation and the threshold of the first AP in this response fell below the control level[7] (Fig. 1.26A). This is

[7] **Methods**. The electrical activity of identified snail neurons for defensive closure of the pneumostome was recorded intracellularly. Tactile stimulation of the foot was used as the repeated stimulus. The specificity of habituation was tested by a rare stimulus, switching off the light in the camera for 2 seconds. To intervene in the neurochemical process beginning immediately after the action of the stimulus, the barreled electrode was inserted under visual control towards the soma of the neuron to be recorded. The barrels of a multiple-barreled electrode for microiontophoresis were filled with saturated solutions of acetylcholine chloride, atropine sulfate (an antagonist of muscarinic receptors) and sodium chloride. The drugs were applied by current-balanced microiontophoresis. The training procedure consisted of the elaboration of a neuronal analog of habituation. Repeated tactile stimulus was presented 30-50 times with an interval between the trials of 20-30 sec. The novel stimuli were presented before and immediately after the course of habituation. Acetylcholine was administered for 5-8 sec before the stimulus during a 10 sec period. Both the tactile and the light stimuli were tested

in agreement with previously discussed data that a neuron may have different excitability corresponding to different signals.

The modulatory effect of acetylcholine in our experiments was exhibited almost in the absence of any direct shift of membrane potential under the influence of acetylcholine, which is in accord with previously published data [272, 355]. The influence of acetylcholine to the responses was blocked by atropine (Fig. 1.26B), in agreement with the data that indirect action of acetylcholine to synaptic responses is insured by the muscarinic receptors. Although atropine prevents the modulatory action of acetylcholine, it did not exert any influence on responses evoked by the tactile stimulus and light (Fig. 1.26B), either before or after habituation. Postsynaptic sensitivity to acetylcholine did not change after habituation. Those experiments (9 of 34 experiments) where acetylcholine displayed direct excitatory action to a recorded neuron (1-4 mV depolarization), modulatory action of acetylcholine was not significant. This is in agreement with properties of neurons in the rat somatosensory system induced by a sensorysensory association [303].

The number of APs in response to a tactile stimulus slightly decreased on the background of acetylcholine before habituation, but sharply increased after inhibition of the response during habituation (Fig. 1.27). Disinhibitory action of acetylcholine to frequent stimulus was found also in the rat's hippocampus [676].

Qualitatively similar results were received in the control series when we used the tactile stimulus on the background of acetylcholine as a repeated stimulus, while the same stimulus without acetylcholine administration served as a rare stimulus (Fig. 1.28).

Changes in the number of APs, at least partially, may be explained by changes in excitability (Fig. 1.29). Acetylcholine, it is known, may play an important role in the regulation of neuronal excitability [272, 355]. Before habituation, the thresholds of AP generation in response to light and tactile stimuli changed weakly after administration of acetylcholine. After habituation, introduction of acetylcholine disinhibited response to the habitual stimulus but raised the threshold of the first AP in response to light.

The effects of acetylcholine on afferent response properties could not be predicted from its ability to excite a cell. This suggests that the mechanism

---

by the acetylcholine microiontophoresis twice during each experiment, before and after habituation. Atropine was administrated for 10-12 sec before a stimulus during a 15 sec period. We examined the action of atropine to response evoked by the tactile stimulus without and during acetylcholine microiontophoresis. In the control series of experiments the tactile stimulus on the background of the acetylcholine was used as the repeated stimulus (acetylcholine also was administered for 5-8 sec before the stimulus during a 10 sec period), while the same tactile stimulus without acetylcholine administration was used as the rare stimulus. The interval between trials in the control experiments was 0.5-1.5 min. We measured the number of APs in the response (time window 3 sec) and the threshold of the first AP in this response.

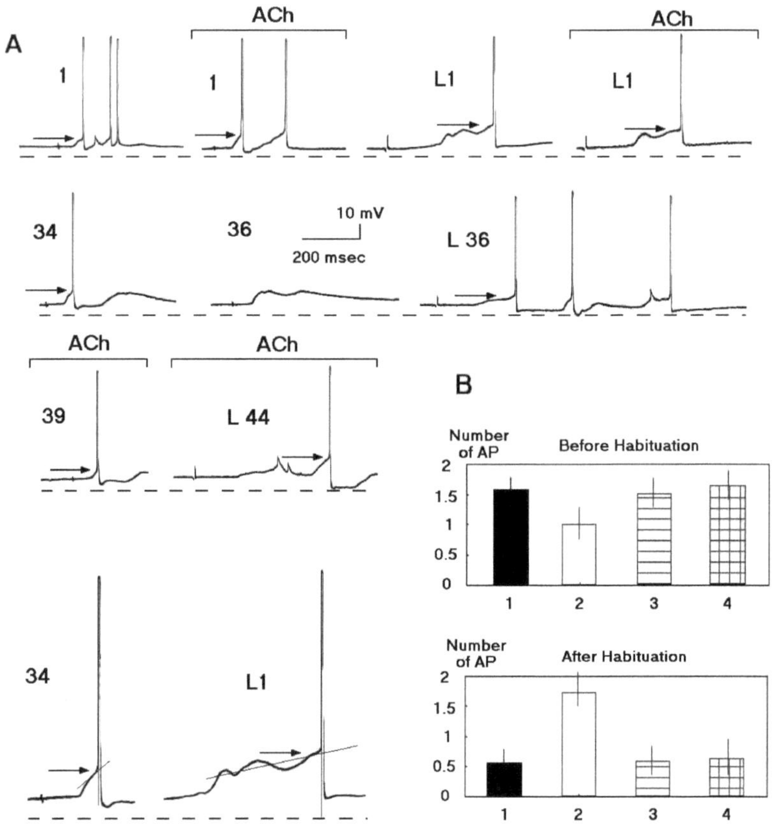

Fig. 1.26. Cholinergic influences to the neuronal activities before and after elaboration of the neuronal analog of a habituation to repeated presentation of the stimulus. A - Representative intracellular recording of the activities of the identified neuron (Lpa3) during habituation. The numbers of the habitual stimuli (tactile: 1, 34, 36, 39) and the number of those habitual stimuli, which preceded the rare stimulus (light: L1, L36, L44), are indicated. Responses on the background of acetylcholine (ACh) are indicated by the line (current - 90 nA). Time interval between the responses in the first line (initial responses) was 5 min. Mean level of the membrane potential in the neuron was -61.2 mV. Dotted line indicates the level 65 mV. The arrows indicate thresholds. Responses 34 (simple case) and L1 (complex case) are magnified (X 2) and method of the threshold determination is presented (see paragraph 1.7). For the complex case, the tangent intersects the voltage trajectory near to the proposed point of inflection by as much as 5 points. Calibration, 10 mV, 200 ms. B A muscarinic nature of the cholinergic influences to the responses before (at the top) and after habituation (at the bottom). Ordinate: number of APs in the responses. Abscissa: 1 mean response to the tactile stimulus; 2 mean response to the tactile stimulus on the background of acetylcholine; 3 - mean response to the tactile stimulus on the background of muscarinic antagonist atropine; 4 - mean response to the tactile stimulus on the background of atropine and acetylcholine. Confidence intervals are shown ($p < 0.05$; t-test).

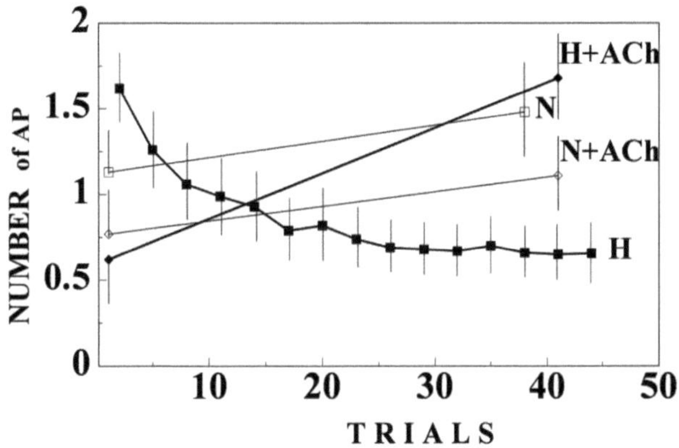

Fig. 1.27. Neuron acquires chemical specificity during habituation. The symbols are denoted in the figure. 1 - Responses to the repeated presentation of the tactile stimulus. 2 - Responses to switching off the light. 3 - Responses to the tactile stimulus on the background of acetylcholine. 4 - Responses to switching off the light in the background of acetylcholine. (C,D). Ordinate - the number of APs in the responses. Abscissa the trial number. The mean values and confidence intervals ($p < 0.05$) in the plots (A-D) are calculated by means of a two-way ANOVA with interactions between the trial number and the type of stimulus habitual or rare stimulus.

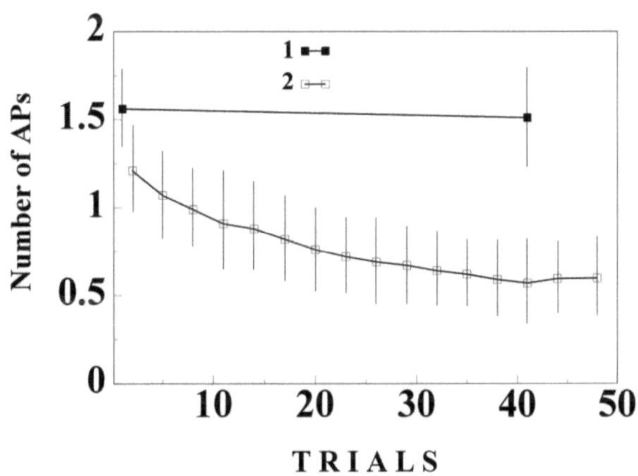

Fig. 1.28. Behavior of neuronal activity during habituation, the results of the control experiment. Repeated presentation of the tactile stimulus in the background of acetylcholine was used as the habitual stimulus, and tactile stimulus without microiontoforesis of acetylcholine was used as a rare stimulus. Abscissa the trial number. The symbols as in the figure 1-27.

by which acetylcholine modifies the effectiveness of somatic inputs is different from the mechanisms leading to excitation.

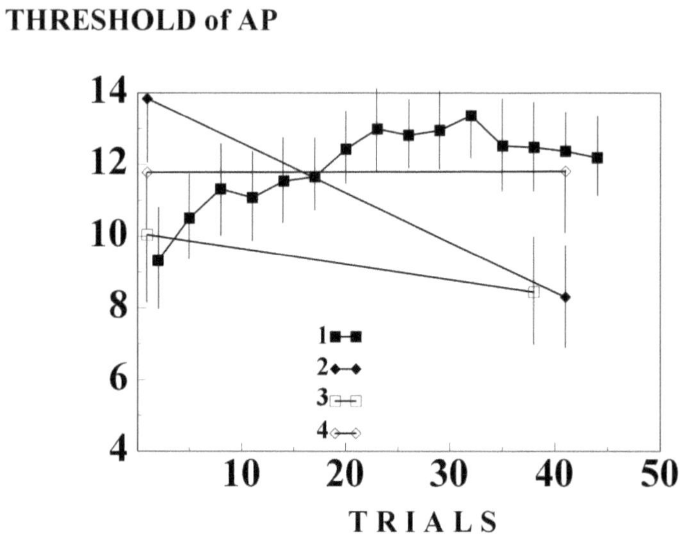

Fig. 1.29. Chemical source of change in AP threshold during habituation. Ordinate, the threshold (mV) of the first AP in the responses. The symbols is in Fig. 1.26

Thus, we have found that the same substance (acetylcholine) in the same identified mollusk neuron affected responses evoked by the same tactile stimulus differently before and after habituation. We may conclude that the chemical process evoked within a neuron by the stimulus is rebuilt after habituation. Moreover, the same substance in the same identified mollusk neuron and in the same time (after habituation) affected differently to responses evoked by habitual and rare stimuli. Evidently, the chemical processes within the neuron evoked by the habitual and novel stimuli are different. Atropine blocked this specific action of acetylcholine, in agreement with data that the muscarinic cholinoreceptors play a role in neuronal plasticity. We suppose that the chemical processes discussed are within the recorded neuron, because of relatively local action of the acetylcholine in the vicinity of the neuron soma and because a threshold is the intra-neuronal property of an excitable membrane. This experiment does not reveal the peculiarity of chemical reactions, but it does prove specific chemical changes during learning.

ACh in our experiments helped to reveal the properties of a chemical process evoked within the neuron by the given stimulus. This phenomenon may be related to another well-known property of ACh. A number of cholinergic substances are capable of producing state-dependent learning [877, 53], in which an animal retrieves the newly acquired information only if this animal is in the same physiological state as it was during the acquisition phase. In particular, habitual stimulus in a new chemical environment may be perceived as a new stimulus. It has been shown that ACh can induce state-dependent learning at the cellular level [1140]. However, state-dependent learning, which evidently might take place in our experiments, is not an alternative explanation of our results. The existence of a chemical difference between a habitual and a novel stimulus is difficult to reject. State-dependent learning in its turn may be based on the chemical specificity raised during learning.

During habituation, a neuron acquires the ability to regulate an excitability of its own membrane depending on the biological significance of stimulus novelty. In response to the habitual stimulus, the neuron displays lower excitability than in response to the novel stimulus. This regulation is based on chemical processes the nature of which is not yet known. All that can be suggested is that chemical processes arising in the neurons after the action of the stimulus changes after habituation and that the result of action of the novel stimuli differs qualitatively. Experiment allows making this conclusion without any a priori knowledge of the underlying anatomy, physiology or biochemistry. This conclusion does not depend on knowledge about neural pathways for the tactile and light stimuli, types of the neurotransmitters, etc. However, there is the question, whether the same synapses induce different chemical reactions before and after habituation, or new chemical reactions were induced by new synapses, which were activated, as the result of habituation. The second possibility looks less probable, since habituation, disinhibition and other forms of behaviorally-related plasticity may be induced by "artificial synapses": microelectrodes for application of electrical current or drugs [1157, 662, 1419, 1173]. The chemical process evoked within a neuron by a stimulus is rebuilt after habituation. The problem of finer specificity is as yet open. It is much more difficult to hope that informational significance may provide the chemical difference between memory elements (for example, different habitual or different novel stimuli), since such a supposition is too complex to be accepted unconditionally. A wide variety of memory elements implies the participation of informational macromolecules in a quantity that exceeds the genetic memory. As a lesser assumption, chemical specificity may only provide a difference in the biological significance of the reactions. At a minimum, this assumption is, evidently, correct. There are also intermediate assumptions. Chemical specificity may correspond to classes of objects (faces, foods, places, dangers, etc.). Elements of memory may also correspond to specific chemical processes in the given brain, while the same elements differ in other brains. At present we may conclude that, at least during the fresh memory (tens of minutes), some aspects of information store have a chemical nature.

1.7.3 The direction to chemical specificity of memory

If remembering of specific information is somehow reflected in origination of chemical specificity in the brain, these chemical traces must be extremely tiny and subtle in order to contain the quantity of information that we acquire during life. Such traces are difficult to find by direct experimental measurement. Nevertheless, some indirect evidences exist. One such piece of evidence is transmission of acquired behavior through metamorphosis [842, 1276]. Let us evaluate, now, if neuronal memory may be transmitted through heredity.

Some forms of relatively complex behavior are genetically predetermined. Moreover, inborn actions may become aware. For example, breathing and heart palpitation can be controlled by consciousness, although such control is impeded. Genetically predetermined actions may be not only automatic. Newborn primates snatch things by palm without any visible learning and newborn rats avoid lit places. Some forms of behavior look innate and faintly depend (if depend at all) on learning. Inborn kittens, blind and scarcely dried off, demonstrate an aggressive reaction, and hiss and grin under danger [1273]. Adult cats keep secret the location of the nest and bury their faeces. Astonishingly, cats living in domestic conditions after defecation also scratch the floor and try to bury faeces, although this is impossible to accomplish in the firm floor. When a cat fulfills the innate command "to bury" its faeces, it is calm and does not note faeces further, although they are openly present on the floor. Interestingly, this instinct does not decay and cats during the years continue these senseless actions. These observations do not prove that same habits may be absolutely innate and are displayed without any learning. So, salmon find their mother's river from the ocean, but homing is not innate behavior; salmon remember the smell of the river they left in their youth [1095]. In the human, launching, entraining and expulsion events are interpreted causally by young infants, but there is no good evidence that these representations are innate [1088]. However these facts demonstrate an inclination to some types of learning. For instance, imprinting arises rapidly, is very stable and does not decay. A chicken interprets as the mother the first moving thing that it sees and never changes this habit. Almost any signal appearing in the sensitive period of infancy acquires traits of the "key" signal for imprinting. If learning is rapid (one accidental pairing, as during imprinting), it may slip away from attention in an experiment. The principal importance of the existence or absence of inborn actions is determined by the possibility to transfer acquired information to posterity. Rates of learning in young animals and in the adult are different. They are much higher in the young, and at a venerable age the rate of acquiring new information continues to decrease [1379]. An instrumental conditioned reflex in 2-week old rats is acquired after one or, rarely, two pairings of correct movement with food reinforcement, compared with 5 pairing needed in adults [1252]. Young animals have a predisposition to rapid and robust learning. Their learning is similar to imprinting; it is rapidly automated and resistant to decay after abolition the reinforcement. Therefore, early acquired habits are easy to identify as innate behavior. Nevertheless information can probably pass through a genetic barrier. When, in infancy, an animal promptly remembers and firmly keeps in mind association between the signal and the vital event, this cannot be any unspecific signal. Only the signals that are similar to the natural signal may play a role as a key signal. For example, when 1-3-days chicks take the first moving object for their mother, this object must be by size, color, direction and speed of relocation similar to a fowl [1040]. Then they remember this object and follow it. Although this is the acquired habit, there are restrictions for a key signal, which have to correspond to some determined features.

This means that some general information is inherent and may be transmitted with the genetic material. Another remarkable example of the dependence of behavior on genetic chemical properties is seen in Drosophila flies. Males and females fight with distinctly different styles, and males but not females establish dominance relationships. As well, males (but not females) perform an elaborate and innate courtship ritual directed toward females (but not males). Male courtship requires products of the fruitless gene, which is spliced differently in males and females. When the male form of the protein is expressed in females, the females will mount and direct the courtship toward other females or toward males that have been engineered to produce a characteristic female odor [308]. Hence, peculiarities of behavior are determined by the genes.

Strictly speaking, not only macromolecules of deoxyribonucleic acid (DNA) may serve as a substrate of transmission, but also a cytoplasm of the egg. Hereditary predisposition to specific behavior is not, of course, transmitted in each individual case. This event is (if exists at all) extremely infrequent. However, if this is principally feasible, it would be a decisive argument for the chemical nature of memory. Even a weak possibility for transmission of acquired behavior to posterity is important for comprehension of the memory mechanism, which in this case has a chemical nature. If the particular unit of memory is kept in the salient feature of the neural network, a way for altered brain structures to influence the genome in absent. This way must pass through the gametes. Particularly, sperm motility varied depending on the social status of male mice and becomes enhanced due to environmental factors experienced by male mice during maturation [669]. Besides, this again brings up the question of concerning the genetic acquisition of features obtained during an individual life: the question that was brought up by J.B. Lamarck and was closed down by Ch. Darvin.

Does a possibility for memory transmission through heredity exist? Let us believe that production of counterfeits (forgery) is a normal way of technical progress. When one fabricates some item, say a calculator, in an underground factory there is a possibility unintentionally to improve this item, since sometimes a mistake in the fabrication may result in an upgrading. Further fabrication of this upgraded item may lead to a new useful mistake and so on, until the factory will produce a computer. Formation the brain in evolution by means of mutations one may compare with the fabrication of spoilt counterfeits. If this long way can be considered to be probable, then in order to inherit habits one must seat a monkey at the computer and wait till a new version of Windows is created.

It is difficult to believe that neuronal memory may arise as the result of mutation, but nobody rejects this option determinedly. Chemical transmission of memory from parents to offspring needs two factors: chemical memory must exist and the way for a spread of modified memory through heredity must exist, too. We may only conclude that some primitive forms of behavior are hereditary and now will demonstrate that influence of acquired memory to the genome at least, does not contradict experimental data. One of the possible forms of influence of acquired information to offspring (epigenesis) is transmission of factors affecting activity of preexisting genes.

One of the most elusive questions in the study of memory is the nature of the enduring molecular changes that underlay memory storage. New memory may be connected with a new macromolecule, modification of preexisting macromolecule, change in gene activity or with mixtures of relatively simple substances. The separate question is how chemical alterations may become persistent.

Proteins are the main working horse of cells and neurons exceed all other cells by their intensity of biosynthesis [927]. Biological activity of proteins is determined by their three-dimensional structure, which depends on a succession of amino acids, constituting the protein. On the other hand, a one-to-one correspondence between amino acid sequence and spatial structure is absent, and there are several stable configurations of the same protein. Actual configuration may depend on conditions of assembling, microenvironment, posphorylation, etc. Thus proteins structure is a candidate mechanism for memory storage. In particular, prion-like proteins represent auto-replicative structures that may serve as a persistent form of information. Prions (infectious proteins) exemplify a novel mechanism of biological information transfer based on self-propagating changes in protein conformation, rather than inheritance of amino acid sequence. A prion may be intrinsically pathogenic, but may function as a cytoprotective molecule [519]. It has been proposed that prions play a role in memory storage [612].

Another possibility for information storage is spatial travel of preexisting proteins. Trafficking of chemoreceptors and potential-dependent channels plays a critical role during long-term plasticity, as we have described for LTP. However, supposed variants of proteins in memory coding (with the exception of synthesis of macromolecules, when information can be coded in their amino acid sequence) do not explain the variety of memory elements. Although a protein is a very complex molecule, a restricted number of its stable conformations exist. And although protein modifications sometimes may be persistent, the variety of possible memory elements is a decisive factor. As an example, scars also may be persistent, but only at a stretch of the imagination a scar may be considered as a memory, and we will argue further that LTP is something like a scar in the brain. Such chemical processes ensure only spatial modification of the network or morphological modification of the neuron itself. Especially intracellular integration of synaptic influences (the binding problem) still waits an explanation. It was supposed also that memory is stored in astrocytes by organizing the activity of ion channels, as a closed, high resistance, state of the gap junctions, and is not associated with a strict physical location [220]. This function may be served by astrocyte gap junctions and suggests that agents that selectively block these gap junctions should disrupt memory. Indeed, general anesthetics disrupt both gap junctions and consciousness and during loss of consciousness, memory does not function. Nevertheless, it is incompletely clear how memory may be recovered after one regains consciousness. Besides, not every impact, that decreases awareness, blocks also gap junctions.

Any long-term trace of memory must be embodied in some steadfast change in the brain, whether we are speaking about restructuring of brain construction or chemical reorganization. Nevertheless, long-term memory cannot be considered as utterably stable and, once being formed, kept intact for an entire life. A short (few minutes) disruption of the blood circulation evokes deep amnesia, more serious than mechanical brain damage. Morphology of synaptic connections usually does not change after short-term impacts, and neurons also survive an accident overload. Life is more resistant to failure of oxidative metabolism than memory. The neural network is too well-built in order to be the source of memory. On the other hand, a memory trace cannot be as firm as, for example, the DNA structure, which is not destroyed after a short arrest of blood circulation. Memory, evidently, needs energy and has a dynamic nature. Nevertheless, complete blockage of brain electrical processes during hibernation, or, contrarily, complete distortion of electrical activity during an epileptic seizure disrupts consciousness, but leaves memory intact. This means that a dynamic process connected with memory has a chemical rather than an electrical character. These chemical traces concern, evidently, memory retrieval, since amnesia may sometimes be reversible even after a severe crash, as after the vegetative states of consciousness. What constitutes long-term memory itself is still unclear at the present time.

Neurons, like other cells, are controlled by the portions of DNA in the cell nucleus that are expressed. Proteins governing the structure and function of a neuron depend upon which RNA chains are produced by DNA in the nucleus – which, in turn, depends upon the regulatory proteins bound to the DNA. RNA and DNA codes consist of nucleotide sequences (bases, which are slightly different for DNA and RNA, but we will not go deeper into details). One amino acid in a protein is coded by a sequence of the three bases.

An expression of the protein in the cell is very stable. So, one minute exposure to nerve growth factor in early developmental converts a cell into a neuron and to acquisition of membrane excitability through a signaling pathway requiring immediate-early genes for induction of sodium channels. With few exceptions, protein have a relatively short half-life (hours or few days) compared with the duration of memory (days, weeks or even years).

RNA is even less stable than proteins. DNA is a stable molecule. Although DNA is persistently attacked by nucleases and mutagens (in each human cell, about 500 bases suffer oxidative damage per day [1132], DNA structure is constantly repaired. The normal direction for informational flow in a cell is: DNA RNA protein substrate. The process of making RNA based on DNA information is called transcription, while the process by which RNA produces proteins is called translation. The reverse route (DNA RNA protein substrate) is fraught with difficulties that are insurmountable, since the reasonable way for influence of the substrate to the consequence of amino acids in the protein is unknown, doubtful and may be absent.

1.7.4 Nontemplate RNA synthesis?

It is known that basic RNA synthesis is accomplished by copying DNA with DNA- dependent RNA polymerases participating, and triphosphates of ribonucleosides are used for the synthesis. In addition to DNA-dependent RNA synthesis, there are a number of pathways of nontemplate RNA synthesis that are catalyzed by various enzymes [990, 265]. In this case ribonucleoside triphosphates are used in the process of synthesis, and the enzyme that polymerizes ribonucleoside diphosphates is known (polynucleotide phosphorylase). There is almost nothing we can say about the biological function of most of these enzymes. Nevertheless, if learning induced nontemplate synthesis or modification of RNA in neurons is possible, in spite of a rapid destruction of these RNAs, acquired information may be stored in DNA, since reverse transcription of RNA DNA is a carefully established fact. Reverse transcription has in fact been shown to be activated after learning [1070]. If RNAs participate in the process of memorization (as a template for synthesis of specific proteins or peptides of memory or in some other fashion), their synthesis, modification or degradation must take place during active learning.

The order in which nucleotides are joined in a polynucleotide chain in systems *in vitro* depends on the specificity of the given enzyme and on the concentration of substrates; the order of joining of nucleotides *in vivo* is not known for most enzymes. The order of nucleotides might depend on properties of the substrate, which absorbs nucleotides and which is altered as the result of learning. In accord with the principle of cross-stereocomplementarity [825] there is a correspondence between spatial structures of a protein and coding its DNA. At least in some cases this principle works well. Correspondingly, if the cellular antigen, as, for instance, modified protein, which was developed during learning, will absorb ribonucleoside diphosphates and polynucleotide phosphorylase will combine these bases into a chain, the protein created by this chain will be complementary to the antigen and thus will serve as the learning-related trace. We do not know if this principle holds true. It is necessary also to consider that thus far we do not know any template-dependent RNA polymerase that utilizes nucleoside disphosphates and it is difficult to express a specific conception as to how nontemplated synthesis can participate in the formation of RNA of a concrete structure.

As is known, the signal reaching the neuron leads to chemical or conformation changes in the membrane. With a change in the biological significance of the stimulus as a result of learning, the chemical process that this stimulus gives rise to in the neuron changes (see previous paragraph), specifically, the state of the excitable membrane. Considering that learning is accompanied by modification of membrane proteins [356], we may hypothesize that specific determinants formed during learning selectively adsorb nucleotides that develop a chain localized near the nontemplate polymerases. Information on the nucleotide sequence carries a specific determinant that develops during learning and serves as a kind of template, such as, for example, the enzyme complex in bacteria that catalyzes nonribosomal (independent of the RNA template) synthesis of peptide antibiotics [490, 1187]. The ability of nontemplate polymerase to carry out the reaction that is the reverse of synthesis pyrophosphorylation of RNA with the formation of nucleoside diphosphates [990] ensures a rapid destruction of these RNAs after their translation and preparation of substrates of nontemplated RNA synthesis for receiving new bits of information. The existence of such a reverse reaction could explain why learning is made easier not only due to RNA of a brain, but also due to RNA of the liver or of yeasts [899]. Probably, the effect is linked to the enrichment of the supply of nucleotides in the brain. We examined the influence of ribonucleoside diphosphates to the process of the simple learning, habituation [1263].

Experiments were carried out on basically identified neurons LPa3 and PPa3 of the snail connected with the reactions of breathing and avoidance. Adaptation to repeated tactile stimulation of the mantle at intervals of 5-7 sec was developed. Specificity of adaptation was controlled by presentation of a rare stimulus (turning off the light) 1-2 min before and 5-7 sec after training. Activity of the neurons was recorded by intracellular glass microelectrodes filled with a solution of potassium citrate in experiments with the pure control (learning without introducing the substances) and with solutions of corresponding nucleoside diphosphates in the other cases. We tried to determine if learning would be disturbed if the concentration of one of the substrates of template (triphospates) or nontemplate (diphospates) RNA synthesis were changed. For this purpose, we introduced cytidine triphospate (CTP), cytidine diphospate (CDP), or uridine diphospate (UDP) into the neuron. In subsequent learning, this could lead to synthesis of the RNAs with an excessive content of cytidine or uridine nucleotides. As a control, we introduced cytidine monophosphate (CMP) into the neuron, which is not capable of polymerization but can participate in other metabolic processes proper to cytidine nucleotides. Another control was a mixture of approximately equimolar concentrations of CDP, UDP, adenine diphosphate (ADP), and guanine diphosphate (GDP). If the proportion of concentrations of ribonucleoside diphosphates in the neuron was approximately the same, its uniform change would not substantially distort the structure of the RNA synthesized during learning.

After the substances were introduced, the character of spontaneous activity did not change. Habituation training in the pure control was accompanied by a decrease in AP in response to repeated stimuli, and an increase in the threshold of generation and latency of the first synaptic AP elicited by this stimulus. Habituation training did not result in an increased AP threshold elicited by the rare stimulus. The membrane potential remained almost unchanged regardless of training.

The effect of nucleotides on learning was studied soon (in 3-12 min) after their introduction because after a certain time metabolic processes of the neuron would have had a substantial effect on the concentration of the substances introduced. Experiments indicated that the dynamics of decreasing the AP during habituation did not differ after introducing CMP, CDP and CTP (Fig. 1.30).

We must consider, however, that the nucleotides were introduced into a single neuron, but habituation was developed by the whole nervous system of the mollusk. The AP number in the response is heavily dependent on the power of the stimulation reaching the neuron, which is superthreshold in AP generation, and reflects primarily the passive participation of the neuron in brain activity. For this reason, we had to isolate the individual neuron's contribution to the learning process, which is determined by a local change in excitability of the recorded neuron.

Nucleotides mostly elicited an initial increase in the AP threshold in response to the first stimulus of the habituation series. A comparison of all curves in Fig. 1.31, however, shows that the initial increase in the threshold cannot be the reason for the blocking of its further increase with administration of CDP and UDP. Thus, the initial threshold was approximately identical after the mixture of nucleotides and CDP was introduced, but habituation was accompanied by an increase of the threshold only in the first case.

Changing the threshold of AP generation accounts for the individual contribution of the recorded neuron to the process of habituation since the threshold is the property of the excitable membrane and, under normal physiological conditions; its value is practically independent of stimulus. After administration of CMP, habituation was accompanied by a selective increase in the AP threshold elicited by the repeated stimulus (Fig. 1.31,A 2). Such changes in excitability during development of habituation were also observed when no substances were introduced (Fig. 1.31,A in the frame). CDP and UDP completely prevented an increase in the threshold in response to repeated stimuli (Fig. 1.31,A 1 and B 4). Habituation had become nonspecific. A change in thresholds after administration of CTP had a qualitatively normal character although it was somewhat lower (Fig. 1.31,A 3). This slight drop may be linked to the fact that the CTP preparation contained up to 2-3% CDP formed in the CTP after purification and during storage and use. The CDP

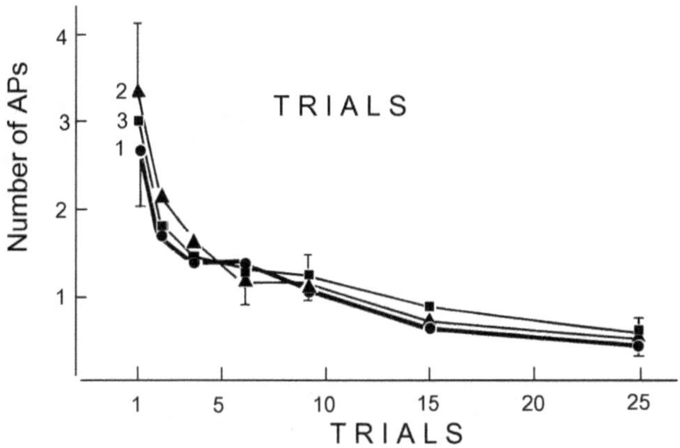

Fig. 1.30. Decrease in AP number in neuronal response according to degree of habituation after introducing CMP, CDP or CTP. Abscissa : Number of presentations of the tactile stimulus. Ordinate: AP number in the response. 1) Development of habituation with CDP; 2) with CMP; 3) with CTP. Standard errors are indicated. Methods. The substances were introduced into the neuron by intracellular microionophoreses (current, 0.5-1.0 nA; time, 30-120 sec). One of the substances (or the mixture) was introduced into each recorded neuron in low or high concentration. After the first dose, a short training series (seven stimuli) was administered. After the subsequent second dose, a complete habituation series was administered (30 stimuli). There was a 10-12 min interval between the short and long training series. The amount of substance introduced into the neuron depended on the charge passing through the microelectrode. The first dose was $5 \cdot 10^{-9}$ C in all cases; the second doses: CMP - $2.5 \cdot 10^{-8}$; CDP - $3 \cdot 10^{-8}$; CTP - $4 \cdot 10^{-8}$; UDP - $2.5 \cdot 10^{-8}$ C; mixture of nucleotides, 10^{-7} C. The difference in amount of electricity passing through the electrode for the cytidine nucleotide can be explained by the fact that the molecular charge differs: CMP < CDP < CTP. The mixture of four nucleotides was introduced in doses four times greater than in experiments with a separate administration of nucleoside diphosphates. The approximate amount of substances introduced into the neuron was 10^{-13} mole; concentration of the substances in the neuron was brought to approximately $0.5 \cdot 10^{-5}$ M. Nucleoside phosphates pass through the membrane poorly. For this reason the chemical action was limited by the recorded neurons.

+ UDP + ADP + GDP mixture did not prevent an increase in the threshold during training, although the increase was retarded (Fig. 1.31,B 5).

Assessment of threshold is a labor-consuming operation, while evaluation of AP latency does not have any of the subjective factors that appear when it is necessary to evaluate the threshold in a response to a stimulus connected with learning. For this reason, we studied the effect of nucleotides on the dynamics of change in latency of the first synaptic AP elicited by a repeated stimulus. After CMP, CTP and the nucleotide mixture were administered (Fig. 1.32), habituation was accompanied by the usual increase in AP latency (control:

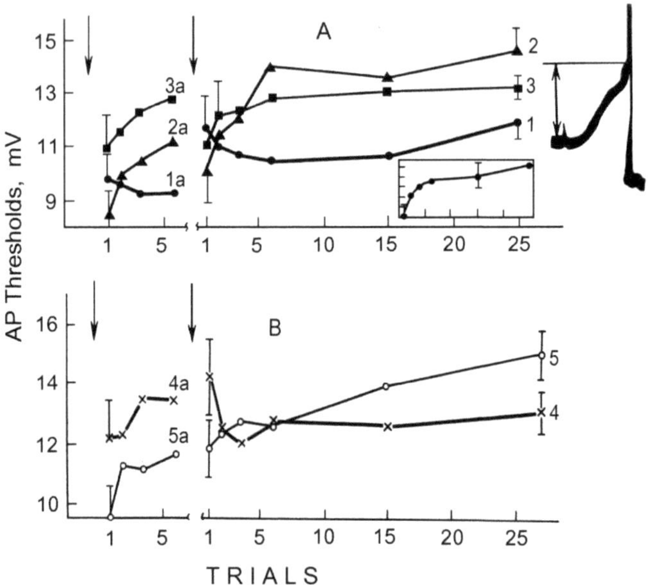

Fig. 1.31. Change in neuron excitability during habituation after administration of nucleotides. Ordinate: threshold of AP generation, mV. The threshold was measured from the level of membrane potential to the point of greatest curvature on the leading edge of the first synaptic AP in response to the stimulus (inset, on the right). Fine arrow, administration of a small amount of the substance: 1a-5a: short habituation series after this administration. Thick arrow, introduction of larger amounts of the substances; curves 1-5: long habituation series. A) In the frame: curve of change in excitability during habituation without administration of substances, in the same coordinates. B) 4a and 4: Habituation after administration of UDP; 5a and 5: after administration of the mixture (CDP + UDP + ADP + GDP). Other notations as in Fig. 1.30.

straight line in the frame, obtained in experiments without administration of the substances). After introducing CDP and UDP, however, the coefficient of regression dropped sharply although not to zero (Fig. 1.32). Incomplete blocking of increase in AP latency during the process of habituation after administration of CDP and UDP is evidently coupled with the absence of suppression of excitation in the presynaptic pathways that are not affected by administration of the substances.

Latency of the AP depends on the threshold of its generation and on the characteristics of the stimulation that reaches the neuron. These results mean that biochemical processes of memory are reflected in the changes of thresholds and latency of action potentials in responses of neurons to a stimulus, which biological significance of novelty changes as a result of habituation. This regulation is based on chemical processes the nature of which is not yet

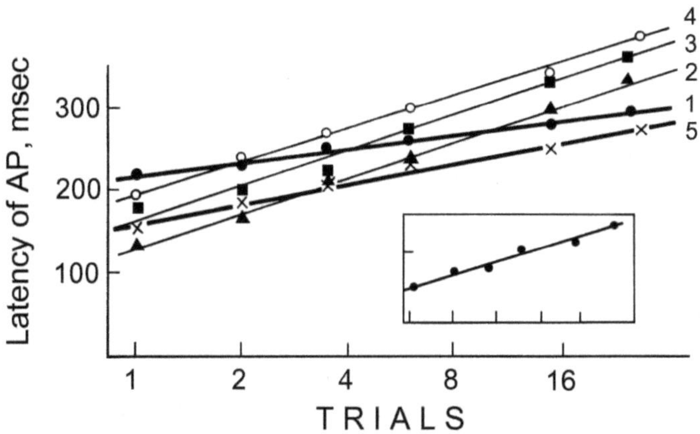

Fig. 1.32. Change in AP latency during habituation to tactile stimulus presentations after administrati on of ribonucleoside phosphates. Abscissa: number of tactile stimulus (logarithmic coordinates); ordinate: latency, ms. Notation as in Fig. 1.30. Lines of regression constructed. In the frame, change in the latency during habituation without administration of substances in the same coordinates.

known. Our data indicate that local chemical action on the recorded neuron can lead to a change in the learning process only in that neuron and change in excitability is its contribution in a whole response alteration. Actually the procedures we used did not affect the dynamics of decrease in AP value elicited by the repeated stimulus. Evidently this is linked to the fact that the response of the neuron to superthreshold effects is only slightly dependent on its own excitability. This means that in studying the chemical mechanisms of plasticity by means of action on a single neuron it is scarcely possible to use such parameters of electrical activity as the AP value in the response of the neuron or the frequency of spontaneous activity. The most sensitive parameter is the threshold of AP generation, and its latency is also acceptable. It must be noted that apart from complete objectivity of measuring latency, this parameter can be used not only in intracellular, but also in extracellular recording of neuron activity. Intracellular administration is weakly reflected in the frequency characteristics of peak activity, since the process of learning was hindered in only one neuron.

The evidence gathered in this study indirectly shows that learning is linked to synthesis of RNA from nucleoside diphosphates and is apparently not dependent on a template. Actually, a significant increase in concentration of one of the substrates of nontemplated RNA synthesis completely blocked the learning process. Blocking was rapid and practically complete, but for template synthesis we might expect an incomplete inhibition because of the great similarity of the complementary template nucleotides. The substrate of template synthesis (CTP) affected learning much less than CDP and UDP. CDP and UDP are not among the agents that have generally toxic effects. The similar effect of CDP and UDP is also evidence of this. The weak effect on learning of the mixture of four nucleoside diphosphates indicates that the participation of all four ribonucleotides is necessary for normal memorization. The only known reaction of this type is the synthesis of recognized RNA sequences. These data are in agreement with the hypothesis that the learning process is linked to nontemplate synthesis, but does not prove this assumption definitively. Besides, we have investigated only short-term effects. If long-term effects of ribonucleosides do exist (we know nothing about it), the same mechanism may work with the participation of reverse transcription.

1.8 Preparing of 'a whole' out of mutual interactions

We had also another reason to assume an affect of ribonucleosides to the process of learning. Chemical specificity after learning may arise without acceptance of the complex mechanism of non-template protein synthesis. As components of proteins, peptides and many amino acids are neurotransmitters and realize intracellular communications, while some components of RNA chains, bases, participate in intracellular communications and are second messengers. Our data are in accord with the role of the chemical interactions of second messengers in a learning, which may integrate synaptic activity in the whole image. A chemical substrate of memory can be satisfactorily described as the mixture of simple substances within neurons. Functional interaction between chemoreceptors is an ancient mechanism. Even assembles of bacterial chemoreceptors that control chemotaxis work in a highly cooperative manner [1163]. Although networks of biochemical pathways have often been thought of as being only neuromodulatory, these pathways can also integrate and transfer information themselves. The temporal dynamics of chemical signals are at least as fundamental as changes in electrical processes. Interacting chemical pathways could form both positive and negative feedback loops and exhibit emergent properties such as bi-stability, chaotic behavior and periodicity. The efficiency of a synapse probably depends on the combination of simultaneously activated synapses to which it belongs [1264, 1272, 1292, 1064]. This is shown, particularly, by the possibility of disinhibition with a reduction in stimulus strength and by means of intracellular microiontophoresis. If part of a population of initially activated excitatory synapses evokes a larger EPSP than all the population, EPSP cannot be represented as the sum of the potentials generated by each synapse separately, for the whole must be greater than its parts. Chemical processes initiated in the synapses may perhaps interact in the neuron.

Simple enzymes 'read' the concentration of their substrate, produce a corresponding level of product and generate a monotonic relationship between input and output, which saturates as the concentration rises. The activity of protein can be altered by enzyme-catalyzed modification. Besides reaction to substrates, some proteins respond specifically to light, temperature, mechanical forces, voltage, pH, etc. Input-output relationships are often extremely rapid, less than 50 μs. However, the 'wiring' between molecules depends on the diffusion coefficient and is rather slow. Biochemical signals might be generated and act locally in the cellular compartments and these processes are sometimes as slow as seconds or minutes [611]. For small molecules, second messengers and ions (as cyclic AMP) diffusion through a cell is relatively rapid, 100 ms [162] and diffusion between neighboring synapses may be as prompt as 1 ms. Considerations of memorizing by a mixture of substances are rather indeterminate. We tried to put these ideas into the model of neuronal memory and it was necessary to modify the classical model of a neuron [815, 548].

Considering that information arrives at a neuron through chemoreceptors and transmits to other cells through potential-sensitive channels, it would appear reasonable that chemical processes involved in learning and memory consolidation begin at the chemoreceptors and terminate on the excitable membrane channels. We can suppose that the chemical singularity during learning is generated as result of interactions of second messengers, which are specific to corresponding synapses. The simplest assumption is that a memory mechanism includes paired interactions of the excited synapses. This presumption we put as the basis of model of our neuronal chemical memory. The most frequently meeting pairs of simultaneously excited synapses are bound to their second messengers by chemical bonds. Thus, we have a second order model relative to the space of the neuron inputs. Synaptic influences in neurons are summed up in non-linear fashion. So, subthreshold synaptic inputs are modulated by voltage-gated channels. Blockage of transient potassium or potential-dependent sodium channels in neurons linearizes the summation [1289, 1166].

We have considered the model of single neuron learning. The starting point for this neuronal model is that the properties of excitable membranes are controlled by biochemical reactions occurring in the nerve cells, which has N inputs and one output. The set of excited and non-excited synapses is regarded as the input signal. Besides these N inputs, we introduce a special reinforcement input (this is the input that receives an US). The reinforcement from the environment arrives at the entire neural network as a response to the action of the whole organism, usually causes a generalized reaction of the neural system and, therefore, reaches almost each neuron. The introduction of the reinforcement input is supported by the experimental data that the analogs of associative learning may be executed at the neuronal level and, hence, conditioned and unconditioned stimuli converge in the neuron (see Sec. 1.6). Let us suppose that in the case of synapse excitation the specific second messengers x_i are generated in the cell, where i is the synapse number. In the case of reinforcement input excitation - the second messenger R is generated. The substances xi interact with each other in pairs and generate the instantaneous memory information molecules w_{ij} about the current signal, namely second-order components, which partially merge information entering to activated synapses. Our model functions on the basis of probable intracellular chemical reactions. The set of chemical reactions corresponds to the set of differential equations. During learning the products of reactions are accumulated and then used for signal recognition, prediction of its probable consequences and decision-making related to the reconstruction of a neuron's excitability. The neuron has N inputs, one output and a special input for the reward.

It is well known that during AP generation high-threshold potential-sensitive calcium channels are opened and the intracellular concentration of Ca^{2+} ions rapidly increases. Thus, the availability of calcium ions in the cell immediately after AP generation serves as an indicator of the neuronal response to the input signal. Intracellular Ca^{2+} concentration may serve steady sign of firing activity [1135]. Let the molecules of the instantaneous memory w_{ij}, the result of simultaneous activation of synapses number i and j, interact with the second messengers R and Ca^{2+} ions. The molecules of the second messenger R are then transformed to an inactive form. As a result, they form the short-term memory molecules W_{ij}^{+} and W_{ij}^{-}. The availability of the W_{ij}^{+} molecules facilitates AP generation to obtain reinforcement (US) from the environment and the W_{ij}^{-} molecules prevent AP generation. The process of short-term memory formation is competitive. The reactions involved in this process cover all possible combinations of the neuronal response to the input signal and the subsequent reaction of the environment that serves to develop instrumental neuron learning.

We can assume that chemical singularity during learning is generated by interactions of inner messengers specific for the corresponding excited synapses and for the rewarding input. We may imagine a process of recognition in a neuron as the interaction of second messengers x_i, x_j, produced in the synapses i, j (activated by a signal that it is necessary to recognize), with the molecules of short-term memory W_{ij}. There is a need in the search to establish a correlation between the appearances of this signal in the past, corresponding actions and the consequent appearance of rewards or punishments. In any computer, a search for the required signal in a memory is the most time and energy consuming operation, with the impossibility to exclude exhaustion among its components. In agreement with the idea of E.A. Liberman [735], this operation in the "molecular computer" is rather cheap; it happens at a cost of heat movements, which dictate the conditions for a combination of complementary molecules. Complementary molecules meet by chance, but their interactions are nonreversible. This is a stable state, since at this state the system is at the bottom of a local minimum of energy, such that small variations from this minimum increase the local free energy in the system. Coupling in the form of complementarity is the form of natural selection at the biochemical level [326].

In the succession of random collisions, molecular forces choose those pairs of x_i, x_j, which correspond to the indexes of memory W_{ij}. Therefore, the input signal chooses the subset of memory molecules that corresponds to the set of pairs of activated synapses and information concerning lucky W_{ij}^{+} and unlucky W_{ij}^{-} events as that information becomes available. The set of the chemical equations is described by the set of first order differential equations according to the laws of chemical kinetics (see Appendix A.1).

We considered the effect of regulation of the properties of sodium channels. This allows simulation of the changes in the neuron's electrical activity parameters occurring during learning, associated with its excitability (amplitude, duration and generation threshold). A neuron must decide, to generate or not to generate an AP. Neuron electrogenesis is determined by the properties of the membrane receptors and ion channels, which are controlled by metabolic processes occurring in the cell.

The neuronal model exhibits different excitability after the learning procedure relative to the different input signals (Fig. 1.33).

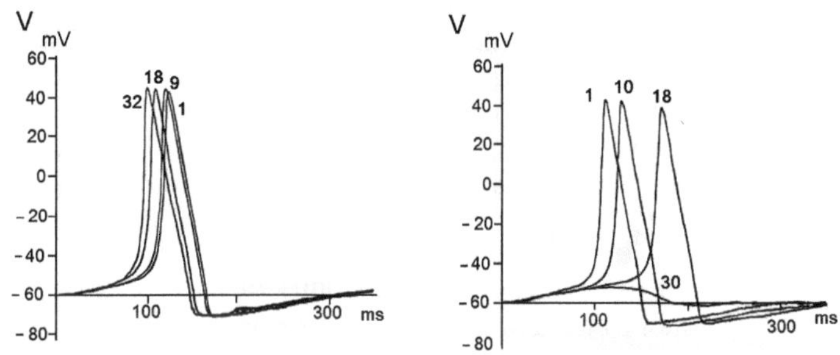

Fig. 1.33. Change in the spike responses of a 10-synapses model neuron during successive application of two different signals. At the left a neuron receives a reward for the AP generated in response to an input signal 1111110000. At the right the reward came following the AP absence in a response to the input signal 0000111111. The signals were presented in turn. Numbers designates the count of learning cycles. Ordinate membrane potential, mV. Abscissa time after the stimulus presentation, ms.

The neuron became more excitable within the response to a stimulus when AP generation lead to receipt of the reward. In this case, threshold and latency of the AP in the response to the CS^{+} decreased to the 32nd combination of the CS^{+} and reward (Fig. 1.33, left). When, oppositely, AP failure in the response to the signal was rewarded, excitability within this response decreased and, finally, the neuron fails to generate an AP in response to stimulus 30 (Fig. 1.33, right). This corresponds to the experimental data. A neuron executes a complex task. It classifies the stimuli not only according to their strength but also according to their biological values and transiently changes its own excitability. Fig. 1.34 demonstrates selective change in the threshold within response to the CS^{+} during classical conditioning. In the same period of training neuron displayed greater threshold within a responses to the signal CS^-, which differed on the reinforced signal CS^+. The more difference was the more differed threshold within the signal.

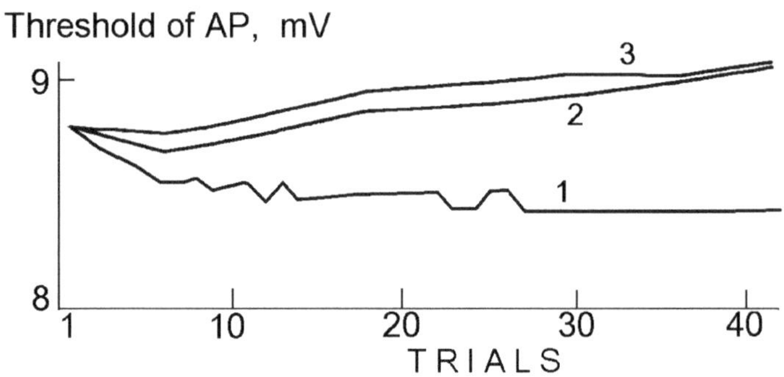

Fig. 1.34. Dependence of the AP threshold (mV) within the response on the number of a learning cycle during classical conditioning. A 10-synapsed model neuron received a reward, when it generated an AP in the response to the signal 1111110000 (curve 1). Signals 0111111000 (curve 2) and 0011111100 (curve 3) were applied following every fifth applications of signal 1 and did not coincide with a reward.

Our model demonstrates the principle possibility of the existence of chemical intra-neuronal memory based on a mixture of relatively simple substances. It does not need the construction of very complex macromolecules. Extension of the model from interaction of two to three second messengers makes capacity of neuronal memory more massive than it is known from experiments: a thousand variants for three-fold interactions in a 10 synapse (10 second messengers) neuron. This exceeds the specificity found in experiments with face recognition. Cooperative collision of several molecules has a low probability and may be accomplished by sequence collision and combining of two-three and afterward the next molecule [362]. Nevertheless, participation of several molecules in the same chemical act is sometimes possible: simultaneous collision of 3-4 Ca^{2+} ions with the Ca^{2+} sensor is possible and necessary for triggering transmitter exocytosis (Thompson 2000). Composition of the mixture may be stored in the DNA structure for producing simple peptides consisting of 2-3 amino acids. After second messenger interactions, molecules of instantaneous memory w_{ij} pick out molecules of short-term memory W^+ij and W^-_{ij} for choice of the output reaction. These processes must be rather rapid. Which combinations of memory elements will be activated in the current behavior is determined by the environment and current motivation [1141]. Maintenance of the existing combination of molecules W^+_{ij} and W^-_{ij}, which determines participation in the current behavior, is a slower process and needs a search of the location in the linear structure of DNA that must be faster than in a cellular volume. Choice of the current combination of short-term memory molecules W_{ij}^{+} and W_{ij}^{-} has to serve as an apparatus of long-term memory.

The order of the model is restricted by the low probability of chemical reactions requiring a simultaneous collision of three or more substances. The model allows for pair interactions between excited synapses and correctly classifies the number of binary input signals of the order of N^2, where N is the number of neuron inputs. The number of second messengers participating in a pair (or three-fold) of interactions is critical. A large variety of receptors is available in neurons, but simultaneous or even the successive interaction of several substances is an impracticable requirement. For this reason, the number of interacting substances should be less and the informational power of a neuron should be less, too, and a neuron perhaps only roughly evaluates signals. One second messenger may maintain a change of non-specific chemosensitivity or excitability through regulation of its absolute concentration. Two second messengers may maintain non-specific plasticity by means of regulation of their relative concentrations. Three second messengers may support rough specificity. Such a model describes the way for merging of input information and thus solves the 'binding problem', which never has been solved before. This is an important theoretical conclusion, but we cannot be sure of the physiological relevancy of the model. A detailed description of the model is given in Appendix A.1 [1056, 1055, 1044], but a reader may miss it without a loss.

Investigation of long-term memory is now in its infancy. In a majority of researches experimenters use a stationary environment when the existence of short-term memory is sufficient for proper functioning. A non-stationary environment requires a long time for investigating the reorganization of learning after a change in an environmental condition, for example, acquisition of one habit after preliminary acquisition of another or sudden replacement of rewards. Nevertheless such attempts were undertaken [1291, 352, 299, 21, 166, 944], but body of data till is too small in order to make any conclusion. Properties of complementary molecular interactions might be the basis for changing the quasi steady combination of short-term memory molecules to another quasi steady combination under control of long-term memory (about which we know almost nothing). Molecular interactions depend on context, since molecules will have diverse properties dependent upon microenvironment and the nature of the other molecules to which they are coupled [326]. DNA may participate in the reorganization of this context, but cannot preserve the entire long-term memory directly. This becomes obvious, if we take into consideration that a general volume of information in long-term memory to a marked degree exceeds the amount of genetic information. For instance, at least some persons can recite texts, the complexity of which surpasses the sequence of bases in their DNA. The arrangement of attributes in DNA cannot increase a possible capacity of storage, since previous events in memory must be arranged in time. Naturally, we are discussing here a suggested role of DNA only as an example of ordered structure. This reasoning is concerned with any construction ordered in space. Therefore, firstly, memory is scattered amongst neurons, secondly, memory constituents are not allocated to neurons and, thirdly, a neuron is not able to store a whole memory. If neurons store a coarse version of considerable volumes of memory, this, at least, does not contradict data known at the present time.

1.9 *Forward propagation of prediction

We have demonstrated experimental and theoretical data indicating that a single neuron may be not only a minimal structural unit, but also a minimal functional unit of a neural network. The ability of a neuron to learn is based on the property of selective plasticity of its excitable membrane. In brains, the neurons are integrated into complex networks by means of synaptic connections. As we have demonstrated, reorganizations of both synaptic processes and excitability are essential for learning and it is important to evaluate the correspondence between plasticity of excitable membrane and synaptic plasticity. Both forms of plasticity may recognize an input signal as a whole. A chemical input signal fires modification of potential-gated channels and excites them, while excitable membrane in its turn affects its own synaptic input. As a result, the waveform of a generated AP depends upon the already completed chemical calculations. Do these calculations have an effect on the target neuron, too?

This cannot be accepted with definiteness, but some data support such a notion. First of all, selective decrease of AP amplitude in neurons of the olfactory bulb of the frog [1253] affect the olfactory cortex of forebrain. Reduced somatic AP somehow exerts an effect on the axon and monosynaptic output reaction. This effect is also selective: after adequate stimulation of olfactory epithelium, direct stimulation of the olfactory nerve produces a smaller monosynaptic reaction in the olfactory bulb neurons and the target neurons in the forebrain, while direct stimulation of the olfactory tract (axons of the same neurons) does not exert a standard reaction in the forebrain. In addition, high-frequency stimulation of the mossy fiber axons in the cerebellum [54] or Schaffer collateral pathways in hippocampus [1222, 586, 446] induced an increase in excitability through a monosynaptic connection in the target neurons. Sometimes one has doubts as to whether the AP waveform determines the postsynaptic reaction evoked by the AP, since mismatch may exist between the AP broadening and the increase in the monosynaptic EPSP slope [547]. However, this result was not unexpected, since the broadening of the AP occurs after the time moment when the EPSP slope has been already formed (line in the Fig. 1.35).

Synaptic delay is rather short and only a rapid alteration in the leading edge of an AP can ensure the participation of the excitable membrane in the selective change in the next EPSP evoked by this AP.

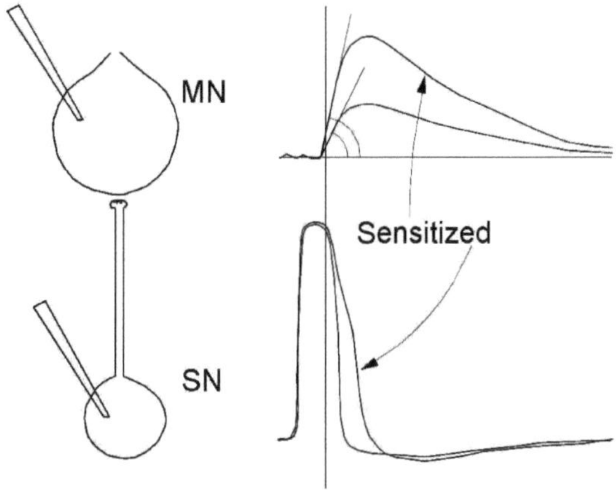

Fig. 1.35. Correspondence between AP duration in the presynaptic neuron (SN) and EPSP in the postsynaptic target neuron (MN) (at the left). At the right, sensitization evokes AP broadening in the presynapse (bottom) and increase in EPSP slope in the postsynapse (top). Synaptic time delay is smaller than the AP duration and therefore a cause-and-effect connection between the back front of the AP and the EPSP slope and amplitude cannot be direct.

Change in excitability of the postsynaptic neuron may also exert the reverse influences to activity of its own synaptic inputs and thus regulate the synaptic potentials, which it accepts. Neuronal excitability is dependent on the distribution and properties of ion channels in a neuronal membrane. And activation or inhibition of potential dependent channels may compliantly augment or diminish the amplitude of postsynaptic potentials, both the EPSP and inhibitory postsynaptic potential, converging to a given neuron. Specific effect of tetanic stimulation depends upon a complex interaction of the stimulated synapse with all of the other synapses onto the common target cell and it appears that the postsynaptic cell somehow regulates its presynaptic inputs [1359]. Supposedly, Na^+ channels not only amplify the generator potential in the subthreshold voltage zone before each action potential, they also depolarize cellular membranes, unblock NMDA receptors and increase EPSP [319]. So, a specific change in excitability may control modification of responses to a specific synaptic pattern, but this control concerns only the later part of the synaptic potentials, since the slope is produced before activation of voltage-gated channels in the same neuron, analogously, as we have described for influence in the target neuron (Fig. 1.35).

A parallel potentiation in synaptic activity and neuronal excitability is enhanced input-output coupling in a cortical neuron [670, 291, 997]. The time course of EPSP potentiation development and change in excitability may differ [54, 1367]. Correspondence between processes in presynaptic and target neurons might depend on involvement of the soma of the presynaptic neuron in the process, as happens in the natural condition. For example, during habituation in mollusks, different branches of presynaptic neurons produce synchronous fluctuations of unitary EPSPs in different postsynaptic neurons [760]. However, correlation between change in presynaptic excitability during learning and modification of reaction in the postsynaptic neuron is not a simple one-to one correspondence.

It was established [1258, 1270] that the primary appearance of APs during pairing in response to an initially ineffective stimulus does not correlate with the decrease in AP threshold, but is determined by the increase in EPSP. Prolongation of pairing after the rise of an AP results in a decrease in the threshold in that response. The presence of an AP in the current response may favor reorganization of thresholds in the following responses. This means that the primary AP generation in response to an earlier ineffective stimulus is unlikely to be due to the local reorganization of excitability in the given neuron but is determined by synaptic plasticity and, maybe, by reorganization of excitability in the presynaptic neuron. After a primary appearance during conditioning, the AP was generated in a sporadic fashion and the growth of excitability could promote the reliability of this AP generation. This is probably the reason why the dynamics of the number of APs were similar to the dynamics of the AP threshold at the end of acquisition, at the beginning of extinction and during reacquisition, when the probability of AP failure was minimal. At the end of extinction, when this agreement was the worst, the AP failure was maximal (Fig. 1.20).

Although initial appearance of the APs during training is determined by EPSP augmentation and reorganization of excitability participates only at the latest stage of training, it is necessary to emphasize that while a signal induces an EPSP in recording neurons, the presynaptic neurons generate the APs in response to the same signal. When an AP is already generated, change in excitability developes rather rapidly. Therefore, EPSP plasticity in the postsynaptic neuron may be accompanied by a preliminary change in excitability in the presynaptic neuron [1253, 641, 446].

The kinetics of presynaptic action potentials is modified after learning [245, 1314, 1351, 71, 687, 1359] and an axon may generate APs of different waveforms [826, 1314, 1360]. The possibility that a change in the conduction velocity of an axon during learning may be the result of a change in cellular excitability also cannot be excluded [1351]. This alteration in the axons may ensure change in synaptic efficacy in the postsynaptic cell. Hence, a modification of the excitability in a presynaptic neuron, at least in some cases, is accompanied by changes of the monosynaptic responses in the target neurons; the existence of a cause-and-effect correspondence between these two events has to be the subject of further investigations. Na^+ channels in the presynaptic terminals show faster inactivation kinetics than somatic channels [362]. Therefore, subtle alteration of the AP amplitude in neuronal soma [1266, 1272] may augment transformation of the Na^+ channels, the axonal spike and Ca^{2+} inflow in the presynaptic boutons, thus affecting synaptic transmission in the target neuron. Therefore, in comparison with the small somatic change in excitability, modification of voltage dependent channels in presynaptic terminals must be more efficient for control transmitter secretion.

We may conclude that a neuron generates a more powerful AP if a stimulus with greater biological significance acts. An action potential is an output signal. A more powerful AP is a more potent output signal and it may generate stronger postsynaptic potential in the postsynaptic target neurons. An output reaction is a gradual function of the biological significance of an input signal, when this signal is a superthreshold. Biological significance of the signal is analyzable by calculation of the prediction P (see Appendix A.1), as a fraction of rewards or punishments $|P|$, which was accepted by the animal in the past after AP generation in this neuron. Punishment is considered to be a negative reward. The negative prediction P means that, in the past, an AP generation lead to a punishment or to the absence of the reward. This has to lead, in the future, to an increase in the threshold and to a decrease of AP generation probability. If, in this case, an AP for all that will be generated, its amplitude will be decreased and efficacy of the next neuron activation will, too, decrease. The same magnitude of prediction P controls threshold, output reaction and has an effect on response of the target neurons. Therefore, prediction P is formed in the neuron, propagated in the target cells (Fig. 1.36) and related to the statistical significance of the reward expectation. We may express the state of activation function of a neuron as follows:

$$y = (1 - P)\psi\left(\sum x_i - \theta_0(1 + P)\right), \tag{1.1}$$

where P is the prediction of an expected reward and ψ is a binary threshold function. If the postsynaptic response $\sum x_i$ exceeds the threshold, $\psi = 1$; otherwise $\psi = 0$. When a memory of the neuron is absent, threshold $\theta_0 = const$. The function $\psi(V)$ was computed by using the Hodgkin-Huxley equation. In a simplest case, it is a binary threshold function. Prediction, computed by a neuron, modulates its threshold and magnitude of output signal and thus is transmitted to the postsynaptic target neuron (Fig. 1.36), which will change its input signal in correspondence with the prediction P.

We may consider the input signal of a neural network as being analogous to the conditioned stimulus, or discriminated stimulus, and the difference between the desired and the actual output of a network as being analogous to the unconditioned stimulus.

Fig. 1.36 demonstrates a neuron that decreases its threshold if a significant stimulus acts. The change in the excitability in presynaptic neuron is accompanied by changes of the output reaction in its axon. A more powerful spike produces a more potent output signal and generates a stronger postsynaptic potential in the postsynaptic target neurons. Thus, the output reaction of neurons ensures transmission of the prediction of the reward.

Neural networks consisting of neurons based on twin interactions are second order networks. Such networks are able to solve much more complicated

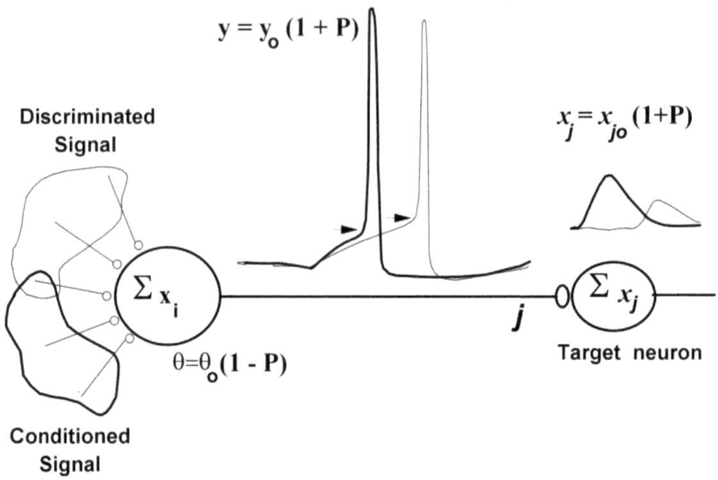

Fig. 1.36. Forward propagation of a prediction. Threshold θ_0 and output signal y0 correspond to absence of memory of the presynaptic (left) neuron. x_i $(i = 1, ...N)$ are inputs of this neuron, which in general are gradual. $x_j = x_{j0}(1+P)$ is a component of the input signals of the corresponding target neuron and x_{j0} is this input signal, when prediction occurs within the presynaptic neuron $P = 0$. Neurons use a prediction P in order to tune their own excitability and modulate their own output signal.

problems than those solved by neural networks of the first order. In particular, second order networks may be invariant to translation, rotation, and scaling of the input signal [965].

We have constructed a network algorithm (forward propagation of prediction) according to the following rules: 1) learning is a fairly local process within individual neurons (chemical model); 2) network error is generally a common factor for all neurons and for each one it is proportional to the number of connections with the network output; 3) change of excitability in presynaptic neurons is accompanied by changes of the monosynaptic responses in the postsynaptic neurons [1262].

Our algorithm does not use chained derivatives and the expression of correction values of the network weights include only neuron inputs and global network error. Prediction is transmitted from one neuron to another, but each neuron corrects the prediction according to its own experience. Global error acts equally on all nodes, making the learning scheme biologically plausible. This is rather suitable for large applications. However, the algorithm is generally not fully independent of the network structure. Nevertheless this dependence is much weaker than that of back-propagation. It reflects not the structure of the connection between nodes, but only the number of paths existing from a node to a network output. For large networks, this scheme gives an acceptable solution within less time and computational cost (Fig. 1.37).

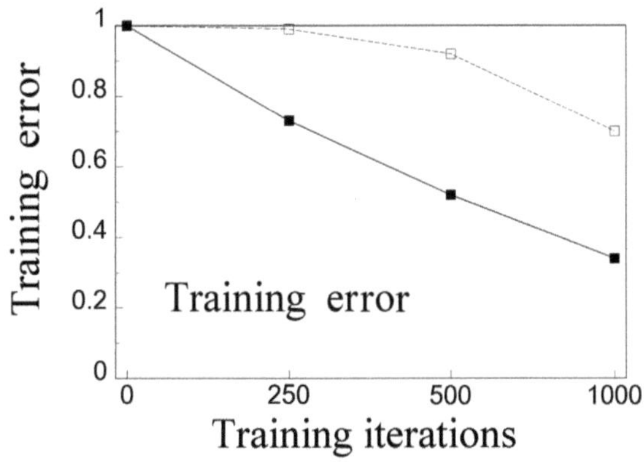

Fig. 1.37. The plot of the averaged training errors against the training iterations for error back-propagation (dotted line) and forward propagation of prediction (solid line). The network consisted of five (three hidden) layers. The connections linked the nodes from one layer to the nodes in the next layer. The number of nodes in each layer was 20.

Forward propagation of the prediction algorithm was carried out in a problem of letter and figure recognition in an the example of a two-layer neural network having 20 neurons in the first layer and a single neuron in the second layer [1262]. Every image was presented by a 20x20-matrix with binary elements. The algorithm found the precise solution after some hundreds of presentations of the learning sequence. The influence of patterns of distortion on the accuracy of recognition was also investigated. The distortion was performed by inverting every pixel of the pattern with some given probability. When the inversion probability was equal to 0.14 then the probability of true recognition of the distorted patterns was up to 0.99. An algorithm of forward propagation of prediction is simple in computer realization, weakly depends on the structure of a neural network and is closer to neurobiological analogues than previously known algorithms. The main feature of this algorithm is that the adjusting of a particular neuron in the network is fulfilled according to the input signal and to the overall error of the network.

Properties of such a model neuron resemble the properties of a neural cell. An increase in complexity of the neuronal model is more than compensated for by simplification of neural network tuning. Tuning of each neuron in the network requires no other information than its inputs (which are modulated by means of predictions in the presynaptic neurons) and network error, which is the same in all neurons. A neuron transmits the prediction of a reward, that is a partially processed signal, into neuronal output. Such neural networks only weakly depend on network structure and a large neural network may be easily constructed and extended. At last, it is tempting to suggest that real brain uses forward propagation of prediction or some similar algorithm in its operations.

2

The verve of injured neurons (a single neuron tries to survive)

The neural system somehow gives rise to a sense of being and Self. Does this immense phenomenon result from the particular activity of specific cells interconnected in certain circuitry in the innermost place of a brain? If this is factual, our task is to discover the shape of this activity, the structure of a neuronal circuit, characteristics of corresponding neurons and, may be, other non-neuronal cells and their location in the brain. This, by the way, is just the problem that neuroscience really tries to solve. Now, let us this problem is already solved. And our response is this: the sense of being arises, when at least 14 glutamatergic neurons located in the 5th layer of the medial prefrontal cortex, connected by means of electrical synapses in a circuit together with 21 astrocytes and generates synchronous activity in a γ-band. Might we, after all, say that now the sense of being is clear for us? Definitely not. We hoped to realize why our Self possesses a mysterious free will, how it is continued in time, restricted in space and located in our body. And, mainly, how one guesses that one is alive and why this is so valuable? In the end, any living cell possesses aspiration to life. At least it behaves as if it does aspire. Desire to live is the core sense and it somehow connects with the sense of being.

2.1 Neurons and glia operate together

The death and birth of neurons is a normal process in brain. The nervous system is a central regulator of an organismal life span, and brain performance is accompanied by neuronal birth and death [157, 1350]. During ontogenesis, axons emit signals that play a role in target field development and these signals regulate target cell proliferation, differentiation and survival. Target-derived molecules, such as neurotrophins, transport also in a retrograde fashion along axons to bring neurotrophin molecules from postsynaptic cell populations (both neurons and glia) to presynaptic cell bodies [687]. Cells in the brain of adults die and are born, too. These processes are non-uniform: neurons that are in close contact may elicit markedly different responses to increasing age that can result in the loss of specific populations of neurons [1379]. For instance, the rat supraoptic nucleus does not loose neurons in old age [396].

In the adult mammalian brain, neurons and glial cells are continuously passed away and generated in retina, olfactory bulb, inferior olive, cerebellum, thalamus, locus coeruleus, hippocampus and neocortex [492, 1216, 332] and, possibly, in other areas. The number of new generated cells (glia, and, in a larger proportion, inhibitory interneurons releasing GABA) decline during several weeks, but a smaller percentage of them continue to survive. In olfactory bulb of adults, approximately 1% of the total interneurons die and are added each day [332, 759]. Therefore, cellular birth and death may somehow participate in cerebral function. The number of neurons in a nervous system depends on environmental circumstances [943, 797]. In addition, the location of neurons that are particularly vulnerable to injury suggests that damage to neurons may participate in the control of behavior. Emotional, homeostatic, perceptive and motor centers such as the cortex, hippocampus, specific hypothalamic nuclei, amygdala, medulla, and cerebellum are more sensitive than other brain area to metabolic stress [1228]. It is tempting to speculate that the sensitivity of these nervous centers to injurious factors is not only an aggravating opportunity, but is essential for the fulfillment of higher neural functions.

2.2 Death through necrosis (murder of cells) and apoptosis (suicide of a cell)

Cells can die via two manners, necrosis (death from injury incompatible with life) or apoptosis (programmed cell death or cellular suicide). Necrosis and apoptosis occur by a mechanism that has been at least partially conserved through animal evolution [1176]. The superficial difference between these two forms of death, swelling during necrosis and shrinking during apoptosis, is based on a difference in ionic processes. Necrosis is a pathological form of cell death caused by the direct result of external insults such as physical injury, energy depletion, toxic insults, acidic or osmotic damage, hypoxia and ischemia, leading to cell swelling and lysis with release of intracellular material and with the disruption of external and internal membranes [1176, 1198]. Apoptosis, in contrast, is an inborn cellular program of cell death associated with activation of a genetic plan [83, 351, 390, 1389, 942, 1198]. Apoptosis is often connected with the development and reorganization of brain, when a cell has two fortunes: to die or proliferate. The default apoptotic pathway is shutdown when development is complete, and growth factors are no longer required to prevent death, but most animal cells in adults retain the ability to activate their suicide program when they have lost their function during natural development or because of serious injury. During apoptosis, the nucleus and the cytoplasm are condensed and fragmented, and the cell shrinks. A curable necrotic damage that does not lead to death also revives developmental mechanisms that are connected with growth but not with proliferation. Apoptotic cells die being sufficiently healthy for their utilization during phagocytosis by macrophages or neighboring glial cells. Cells that have undergone necrosis usually are not phagocyted by macrophages.

Local injurious influences that are incompatible with life can, in the same tissue, evoke both necrosis and apoptosis. The ischemic center is characterized by cell necrosis, while apoptotic cells are located in the zone around the ischemic focus, the ischemic penumbra [532]. Hypoxia [1054] and ischemia [1318, 748] induce necrotic and apoptotic cell damage in a close relationship between them. After ischemia, neurons are the main cell type that dies [260], because of their high sensitivity to energetic metabolism. There are procedures that ameliorate cell damage and prevent cell death and these protective procedures reduce both necrotic and apoptotic cell death [59].

The healthy cell responds to impairment by a defensive homeostatic reaction and its reaction displays necrotic syndromes. If deviation from the norm is recoverable, the cell recuperates and remains intact, but if the severity of damage exceeds the possibilities of homeostatic compensation, the cell dies through necrosis. The necrotic cell dies in a struggle, the apoptotic cell passes away without struggle and, moreover, undertakes efforts in order to die when its environment is hopelessly altered, such as when a cell is deprived of growth hormone during individual development [227]. A short and strong impact usually evokes necrosis, while relatively small and protracted influences cause apoptosis. For example, calcium ionophores (which increase permeation for Ca^{2+}) can induce either apoptosis or necrosis in cultured cortical neurons, dependent on ionophore concentration: a large concentration produces necrosis, a small and prolonged concentration provokes apoptosis [507]. Similarly, the length of time a neuron spends in a depolarized state following hypoxic depolarization is a critical determinant of the extent of reversible necrotic damage or irreversible apoptotic cell death [233]. Increased duration of acidosis to 60 minutes induced irreversible damage and neuronal loss in the hippocampus, while 30 minutes damage exerts reversible necrotic alterations [1366]. Tissue injury is related to the total duration and intensity of tissue depolarization and not to the frequency of depolarization [1027].

Strength and stretch of harm determine further the route of damage, but the quickness of each of the routes to death is also different. Necrotic and apoptotic deaths proceed in different time scales. Glutamate can induce either early necrosis or delayed apoptosis in cerebellum granule cells [41]. Apoptosis is a slow form of cell death, lasting for many hours or days, [83, 351, 390, 942]. In contrast, necrosis is a form of cell death that occurs rapidly (minutes or tens of minutes) in response to severe insult [546]. For instance, development of necrotic symptoms, such as a rapid loss of spine and dendrite structure [1426], change in spontaneous action potentials (APs) [836] and intracellular Na^{+} increase [416], is observed during the first few minutes after severe ischemic damage, anoxia or mechanical injury. Within two minutes, protective mechanisms are activated, create an elevation in intracellular Ca^{2+}, develop of Ca^{2+} wave and quickly returns Ca^{2+} to background levels [642]. Protective events within a neuron can prolong necrosis for several hours or days. For instance, approximately 24 h after axotomy, sensory neurons of Aplysia display neuritic outgrowth, enhanced cyclic AMP level and augmented excitability [1195].

A necrotic cell tries to survive by means of activation of recoverable homeostasis, but an apoptotic cell reverses its resources into destruction and does not turn on homeostatic mechanisms. Many molecular, biochemical and genetic events occur within cells undergoing apoptosis. Some of these events are incompatible with cell survival because they have irreversible, catastrophic consequences. The onset of such changes marks the point of no return, an event termed 'the commitment-to-die' [227]. The way to the point of no return may be dependent on cell line and initial conditions, but the execution phase of apoptosis is evolutionary conserved and occurs with amazing temporal and morphological uniformity in most if not all cell types regardless of the circumstance and impacts used to induce death. Once this phase of apoptosis begins, death is inevitable [843]. Apoptosis proceeds with central control and cell dies conventionally after receiving of the signal to die and passing over the point of no return. After passing over this point, homeostasis does not work for cell recovery and this, evidently, is not associated with the exhausting of compensatory capabilities. Although harmful influences are slowly accumulated before a cell reaches the point of no return, there is a special signal to die and this signal may be outer or inner. A cell, for example may receive the inflammatory factor interleukin 1 (a member of the signaling family for the induction of apoptosis), or when cells lose unmanageable amounts of intracellular K^+, cells have to initiate and execute the K^+-sensitive suicide cascade [1386].

Oppositely, a cell may be subjected to necrotic damage only partially, depending on environment, on localization of damage, on impairment of the specific metabolic path and in accordance with experience. Protection at the early stage of necrosis also may be specific (see further) and lead the cell into an unusual stable state. During early development, the necrotic process is reversible [532]. For instance, alterations evoked by ischemia [1426], hypoxia [836, 233] and acidic stress [1366] may be recovered after cessation of injuring factor. Besides, there are many treatments that not only prevent development of necrotic damage (apoptosis also is possible to prevent), but treatments that reverse necrosis that already has begun.

If neuronal damage and death participate in an on-line control of animal cerebral activity, the mechanism can only be necrotic, which is relatively rapid, reversible in its early stages and may be specific in respect to prehistory. Further, we will pay attention, before all, to necrotic damage, but relative to apoptosis, we have to make one important remark. The commitment-to-die is the important point, connected with the switching off homeostasis. If a cell's aspiration to live lies in the basis of the beginnings of psychics, then a solution of this paramount brain mystery is very important to understand: how does this distinctive property of the living cell disappear when a cell passes through the point of no return? At present, this problem remains a subject of considerable debate and controversy and details of this phenomenon are still necessary to determine. Nevertheless, even the existence of a point of no return is important: its existence means that a state of aspiration to life is a qualitatively special state. A given cell may or may not possess an aspiration to life, but it is impossible to live or to feel half-alive.

Necrotic damage is an extremely complex phenomenon. To be more accurate, the homeostatic compensation of necrotic damage is the most complex. Different forms of compensation are categorized by the means, times, pathways of recovery and even by a final point, which, generally speaking, does not coincide with the initial state. Compensation recovers the life of a cell and, therefore, is almost as complex as a life itself. We will describe only those attribute of damage that may concern mechanisms of brain behavior.

Neuronal damage and death have been extensively investigated [343, 797, 921, 1188, 796]. Mediators of the processes related to neuronal damage are excitatory amino acids (glutamate, NMDA, kainite, AMPA, etc.) retrograde transmitters (arachidonic acid, NO), neuropeptides and neurohormones, cytokines (interleukin 1, 6, 8, neurotensin, thyrotropin-releasing hormone etc.) and corticotrophin-releasing factor and these mediators are interdependent. For example, NMDA receptor activity leads to enhancement of intracellular Ca^{2+}, release of nitric oxide (NO) and increases the formation of toxic hydroxyl radicals.

Swelling of cells begins from Na^+ influx through a hole in the cellular membrane after physical damage or through potential-dependent channels and glutamate-gated channels, because of physiological activity. Nevertheless, Na^+ influx is not the primary reason of necrosis or the single reason of damage. The increase in intracellular Ca^{2+} is an important reason for neuronal injury, and damage is suppressed by inhibition of endogenous Ca^{2+}. The rise in intracellular calcium may result also in the activation of Ca^{2+}-dependent proteins and further increase the rate of membrane depolarization that leads to uncontrolled cellular swelling and necrosis [158]. A large intracellular concentration of Ca^{2+} in natural conditions usually follows excitation. Glutamate receptors (AMPA, NMDA, kainate and metabotropic receptors) are the main channels for transmission of excitation in the central nervous system and this excitation sometimes is excessive. Detrimental action of glutamate is, in particular, connected with a high permeability of different types of glutamate receptors to positive ions, rapid loss of membrane potential and extensive excitation.

The rapid component of most excitatory postsynaptic potentials (EPSP) could be accounted for by activation of AMPA glutamate receptors, which perform 'fast' synaptic transmission in the central nervous system and might be crucial contributors to injury [555, 585, 694, 671]. Depolarization also favors chloride ions to enter into neurons and creates an osmotic disequilibrium, driving water into neurons and producing cell edema [1045]. Ionotropic glutamate NMDA receptors have at least an equal, if not higher, permeability to K^+ as to Na^+, though their permeability to Ca^{2+} is several folds higher than to Na^+ or K^+ and activation of NMDA receptors potently induce cell damage [555, 1386]. Metabotropic glutamate receptors are coupled with G proteins, consisting of, at least, eight subtypes. Each subtype controls a specific metabolic path and has a unique role [743]. Activation of G proteins may potentiate NMDA and AMPA receptors [1290] and control cell behaviors through regulation of metabolic pathways right up to involving the genetic information [315, 1333]. This connection is essential, since it ensures change in connectivity of gap junctions, in production of the second and retrograde messengers, inositol trisphosphate turnover, intracellular Ca^{2+} and protein kinases. G proteins thus connect extra- and intracellular environments and can reorganize ion homeostasis of neurons. They can exert either stimulatory (G_s) or inhibitory (G_i) action on metabolic pathways and so support equilibrium around an optimum [504, 520, 462, 200, 443]. A powerful apparatus of G proteins promotes cell survival and determines mental health. Dysfunction of multiple neurotransmitter and neuropeptide G protein-coupled receptors in frontal cortex and limbic-related regions, such as the hippocampus, hypothalamus and brainstem, likely underlies the complex clinical picture of psychiatric illnesses that includes cognitive, perceptual and affective symptoms. Thus, the same biochemic machine regulates cell damage and psychics.

Because excessive excitation usually causes cell damage, escaping of excitation typically protects neurons from neuronal death. Damage can be prevented by inhibitory neurotransmitters, by antagonists to various excitatory amino acids, by hyperpolarization, by K^+ channel openers and by Ca^{2+} or Na^+ channel blockers [295, 573, 1085, 585, 783, 671].

In contrast to the ionic mechanism of necrosis that involves Ca^{2+} influx and intracellular Ca^{2+} accumulation, excessive K^+ efflux and intracellular K^+ depletion are key early steps in apoptosis [1386], although necrotic cells also lost their Ca^{2+}. High concentration of intracellular K^+ is a distinctive feature of any cell. Physiological concentration of intracellular K^+ acts as a repressor of apoptotic effectors, at the same time as a massive loss of cellular K^+ may serve as a disaster signal allowing the execution of the suicide program. Cellular K^+ loss stimulates post-translational maturation and release of the inflammatory factor interleukin 1. However, although cellular K^+ depletion is important in apoptosis, in many cases particular apoptotic activating signals act together with low K^+ to initiate the death program [1386]. Astrocytes play an essential role in the maintenance and protection of the brain and can sense the extracellular K^+ concentration. During injury, neurons lack intracellular K^+ and its concentration in the extracellular space increase, but the recovery of extracellular K^+ is retarded because of glial toxins [733].

2.3 Neural and glial cells assist in survival

Neuronal life and death is controlled by glial cells [885, 332]. Nevertheless, when we speak about a brain, we imply neurons. Really, that's the truth, but not the whole truth. Brain consists of neurons and glial cells: astrocytes, oligodendrocytes and microglia. It had been thought that glial cells only physically support neural networks, ensure a reinforcing cage for proper functioning of the neuronal scheme and do not themselves participate in the mental function. As a simplification, this was close to reality, but at the present, many data are not kept within so straightforward a representation. Brain activities are a coexistence of the several cell's populations. Neurons are born, live and die within the brain. Glial cells also are born, live, die and interact with the neurons. This interaction sometimes looks like collaboration and sometimes like a fight. Neurons are also not homogenous and differ by their locations, chemical distinctions, morphology and functions. Different strains of neurons collaborate or battle as well. Some neurons excite their target cells, first of all by glutamate, and some neurons inhibit them, first of all by GABA. Correspondingly, some neurons harm their target cells and some protect them. However, this simple rule does not operate in all cases. Cells in many cases require extracellular factors that are produced by other cells in order to stay alive [1176]. Moreover, the same impact in some cases may protect cells, but in other circumstances may harm them, such as NO production [1439, 360] and microglia activation [848, 695], or the same impact influences differently on different (even close) neurons [1427]. Soon after damage, inhibitory processes, which could protect a cell, are usually declined, but this is not connected with the loss of GABA interneurons [403]. In the same way that an excitation does not always damage cells, cell protection may proceed without inhibition. What is more, excitation may sometimes protect cells.

Damage and death of brain cells provokes an immunological reaction, which is ensured by glial cells. Immune functions as phagocytosis and cytokine production have emerged 700 million years ago in starfishes and sponges [971]. Ancient immune defense mechanisms were very effective, but lacking memory function. 500 million years ago, the basic system for creating the genetic material for recombination and mutation was developed to establish variability and diversity of proteins such as immunoglobulins. NO is a factor of innate immunity in early phylogenesis, while opioid peptides - in the late phase of phylogenesis. NO was present in phagocytic cells and guaranteed an effective defense. For those ancient mechanisms, the trigger to activation is stress that in invertebrates is channeled through macrophages, while stress response in vertebrates induces activation of the hypothalamus, adrenal gland, pituitary gland, etc. More recent, opioid system is much more specific in respect to harmful impact [910, 141]. Both astrocytes and microglia are cells that provide immune surveillance in the central nervous system [824]. The role of microglia sharply differs from the role of astrocytes. Microglia comprises several percentages of cells in the brain. Under pathological conditions microglia become active and surround damaged and dead cells, like phagocytic macrophages of the immune system [384]. Microglia are activated during various neurological diseases and this has earned them a reputation as endogenous malefactors, but the activities of these cells are for the most part beneficial. becoming destructive only when they escape from the strict control normally imposed on them, that is after severe damage [1109]. The executive functions of microglia can change not only in magnitude but also in quality. Depending on the magnitude of microglial reaction, on the type of stimulus and on the concurrence of other local factors, microglia can contribute to defense and repair. When damage is harsh and cellular homeostasis disrupts proper functioning, mechanisms of defense fail and microglia are involved in the establishment of brain damage [848]. However, although cells of microglia are intimately connected with inflammation, neuronal damage, infection, etc., cells of microglia, probably, also directly participate in mental activities and participation in cerebral function is determined just by their servicing neuronal damage and protection.

On the microglia membrane, a special type of Na^+ channels and several K^+ channels are present and they are important for microglial activation. Many neuronal signals, such as ATP, NO, substance P, excitatory amino acids, or pro-inflammatory cytokines may provide the stimuli for microglial activation [305]. Particularly, as a main instrument of neuronal excitability, tetrodotoxin-sensitive Na^+ channels participate in activation and phagocytosis of microglia [274]. Ca^{2+}-dependent glutamate signaling between astrocytes and neurons is potently amplified in the presence of inflammatory microglia [119]. Astrocyres are the most abundant non-neuronal cells in mammalian brain and in humans constitute 50% of the total brain volume [260]. In the human cortex, there are 3.4 astrocytes for every neuron [885]. The processes of one astrocyte contact tens of thousands of synapses. For instance, one hippocampal astrocyte makes contact with 100,000 synapses and astrocytes themselves are functionally compartmentalized [526]. Therefore, individual astrocytes can integrate signals from numerous synapses, and signal back to multiple synapses. Glial cells regulate basic neurotransmitters, glutamate, GABA and taurine, which ensure the large share of excitatory and inhibitory transmissions and which are important for neuronal damage-protection. Cycling of neurotransmitters, ions and intercellular messengers are the basic means for interaction of glutamatergic neurons, GABAergic neurons and glia. Glia cells capture glutamate from the synaptic cleft, convert it into glutamine, which is transported into the neuron and is reconverted into the neurotransmitter molecules [1139]. Astrocytes are very sensitive to glutamate released from synaptic terminals [209]. Intracellular Ca^{2+} rises in glial cells and triggers glutamate release that modulates neuronal function [119]. The uptake of glutamate into astroglia is the predominant mechanism to terminate glutamatergic neurotransmission and to prevent neurotoxic extracellular glutamate concentrations [387]. Disruption of astrocytic support leads to excitotoxic damage [512]. The assistance of glia to metabolic processes of neurons is evident, but in addition, glial cells contribute to information exchange in the brain. Certain astrocytes receive direct glutamatergic and GABAergic synaptic innervation from neurons and respond to synaptic activity by releasing transmitters that modulate synaptic activity [675, 538, 27].

In some cases, astrocytes can enhance excitatory synaptic transmission. Thus, astrocyte-derived glutamate preferentially acts on extrasynaptic receptors, augments neuronal excitability, supports neuronal synchrony and can influence damage by release of glutamate. In other situations, astrocytes release molecules of the ATP that suppress synaptic transmission. Many other receptor types are also expressed on astrocytes, including opioid, dopamine, acetylcholine, glycine, serotonin, β-adrenergic, and purinergic receptors; astrocytes express also ion channels, both ligand-gated and voltage-dependent [675]. In astrocytic membranes, K^+ channels predominate over other ion channels and this prevents their excitability [885]. Astrocytes participate in many physiological functions of neurons. They have been implicated in dynamic regulation of neuron production, synaptic network formation, neuron electrical activity, neuronal assembles and specific neurological diseases [1012]. Glia displays the remarkable capacity to discriminate between different levels and patterns of synaptic activity [209]. Astrocytes and microglia intimately participate in neuronal sensitization not only via the release of glutamate, but by evoking changes in synaptic ion homeostasis, too [305]. In response to neuronal signals, astrocytes can signal back to neurons by releasing various neuronally active compounds, such as the excitatory neurotransmitter glutamate. This bidirectional communication system between neurons and astrocytes may lead to profound changes in neuronal excitability and synaptic transmission [209].

Cooperation of brain cells is brightly displayed during spreading depression that is characterized by rapid depolarization of both neurons and glia. The extracellular ionic shifts during spreading depression include an increase in K^+, and decreases in Na^+, Cl^-, and Ca^{2+}, but intracellular Ca^{2+} in astrocytes increases in response to membrane depolarization. Astrocytes normally extrude calcium during spreading depression, resulting in rapid recovery of the levels of extracellular Ca^{2+} [1362]. Astrocytes play a role in brain homeostasis, regulating the local concentrations of ions, amino-acids and other neuroactive substances [964, 733]. Microglia also can sense homeostatic disturbances. Neurons and glial cells recover homeostasis that was disturbed during functioning and ensure joint regulation of damage-protection. At the same time, they, probably together, control behavior. And all this coexistence is somehow converted into our identity. Processing of information is important, but not the single attribute of brain. Existence of neurons and glia between life and death turns their indifferent "calculations" into cooperative being. Therefore, interaction between physiological activities of neurons and glia is a properly established fact, but the role of astrocytes in behavior is less clear. For instance, functionally syncytial (that is electrically connected) glial cells are depolarized by elevated potassium to generate slow potential shifts that are quantitatively related to arousal levels of motivation and accompany learning

[697]. It was established also that inhibition of astrocyte-specific metabolic pathways rapidly impairs vision and learning [538].

Nevertheless, physiological processes in glial cells are slower than in neurons and therefore they cannot participate in fast behavioral activity. Astrocytes react to neural influences slowly, during a few seconds [964] and this is not compatible with the time scope of current behavioral actions. Ca^{2+} concentration in astrocytes may also alter in response to neuronal activity during seconds, because of Ca^{2+} outflow from intracellular stores [964, 538]. However, direct influence from an astrocyte to a neuron's electrical activity may be rather rapid, since the rise time of the astrocyte-evoked slow glutamate-mediated inward current in neurons is 60 - 200 ms, and the current can be extremely large in magnitude because vesicles within astrocytes are large [526]. Responses of astrocytes to signals from the environment or blockage of metabolic pathways takes minutes [538, 27]. Development of neuronal damage after different impacts is also slow, minutes or tens of minutes [416, 836, 1366, 1426]. Within 2 minutes of stroke onset, neurons and glia undergo a sudden and parallel loss of membrane potential caused by failure of the Na^+, K^+-ATPase pump and disturbance of neuronal homeostasis[36]. The reaction to acute neural injury is a massive expansion of the microglial cell population, which peaks a few days following injury. At the same time, the initial stage of microglia activation from minutes to a few hours following injury is related to neuronal compensation, rather than to injury [695]. A characteristic time for initial microglia reactions is minutes [742, 36, 384]. This does not allow one to exclude microglial cells from consideration of their possible participation in the regulation of behavior. Whereas microglial cell bodies and main branches stay stable for hours, their evenly distributed and highly ramified processes in the brains of anesthetized animals are remarkably motile (1.5 mkm/min) [384]. Astrocytic processes also dynamically alter their apposition to synapses in response to environmental cues, although transmitters do not change their own membrane potential [27].

Current processing of information in a brain takes a fraction of second to complete. These fast-acting episodes concern the reception and recognition of signals, decision-making and actions of previously alerted brain. Reaction to unexpected signals and recollection take extended periods of time, right up to seconds and more, while reorganization of brain states, such as sleep-awareness, hunger-satiation, memory consolidation, etc. proceeds in a middle-time scale and may take minutes or more. Rapid processes in a brain are, evidently, executed by neurons, while in the slow reorganization of brain conditions, both neurons and glial cells may, in principle, participate together. Obviously, service of glial cells to neurons is not restricted only to support of neuronal electrical activity. Glia is, doubtless, capable of making a larger contribution. It is impossible to exclude the option that glial cells not only serve brain, but also participate in the slow gradual tuning of cerebral function. For example, they could participate in the sluggish recording of memory and in its rapid reproduction, all the more that astrocytes are richly equipped with the tools for reorganization of intercellular connections: gap junctions. Taking into consideration that astrocytes slowly react to neuronal activity but promptly exert an effect on neurons, they could rapidly modulate behavior in accordance with previously prepared situation. Thus, neural and glial cells agreeably interact in support of the physiological function of brain. They communicate between outer and inner environments by means of second and retrograde messengers and an exchange of ions. Let us now consider processes of survival. We will argue further its importance for mental function.

2.4 Spread of damage within a tissue

Interconnection between cells, evidently, has arisen in evolution together with the origin of cells themselves. It seems likely that an early step in the evolution, a multicellular organism represented the association of unicellular organisms that formed colonies. The single-cellular animals when food supplies are exhausted form colonies and show social behavior in a primitive form. Cytokines, cyclic AMP, NO and arachidonic acid regulate colony formation. Injury of a neural system returns inner physiological processes to path of growth and development and activates the same factors. Increased neurotransmitter release (glutamate; substance P and ATP) activate both second order neurons and surrounding glial cells, which produce cytokines and other inflammatory agents: tumor necrosis factor, NO, ATP, etc. [824]. If these ancient processes participate in the cerebral function, they probably play a role in establishment of the background states of a neural system, depending on its current needs, and do not use past experience for upgrading of their activity. For instance, shortage of nutrients consistently causes hunger in spite of when one earlier already experiences this sense. However, more recent defense systems, neuropeptides and, in particular, the opioid system, may more closely be connected with acquired behavior.

Cells in multicellular organisms are incessantly interconnected. During injury, interaction between cells is reorganized and cells begin to express ancient properties. A dynamic process of damage eventually propagates with time from the center of damage to neighboring tissue [532]. Any local injury tends to spread to distant parts of the nervous and non-nervous systems. The extent of severe damage is determined by an outflow of the deleterious contents of dying cells, such as free radicals, and by mediators of cell death such us the inflammatory managers cytokines, retrograde transmitters, etc. These substances are produced by various cells, but especially by glia, for instance, by microglial cells. Compressive injury induces changes in Ca^{2+} levels in glial cells that can spread through glial networks along the white matter [742]. Experimentally evoked elevation of intracellular Ca^{2+} in astrocytes evokes elevations in the internal Ca^{2+} of adjacent neurons [526] and this can aggravate damage. Astrocytes can communicate with each other through Ca^{2+} waves, providing a potential mechanism for the propagation of information over large distances [27, 526]. Similarly, in response to neuronal activity, elevated Ca^{2+} in astrocytes leads to glutamate-dependent Ca^{2+} elevations in neurons and can induce membrane depolarization that eventually can trigger action potential discharges [964]. The spread of cellular damage in the brain increases with the augmentation of intercellular coupling through gap junctions, which provides a fine regulation of surrounding neuronal activity [311, 844, 1011].

Opening of a gap junction can exacerbate cell damage, since when coupled cells have undergone to injury, destructive molecules spread from more to less injured cells helping the former and harming the latter. Health promoting molecules such as necessary metabolites spread from less injured to more injured cells, harming the former and helping the latter [260] and this ameliorates cell injury. In addition, since a gap junction behaves as a resistance connecting two cells, when current flowing from a more injured cell depolarizes a more negative cell, the same current makes the first cell less depolarized [109]. Therefore, when two cells interact through gap junction the damage to the injured cell decreases at the same time as a healthy cell experiences injury. Gap junction, certainly participate in a cell damage, but this participation is ambiguously. Although no direct studies of cognitive or affective effects of coupling have yet been published [257], there are much indirect evidences. By means of gap junctions, cells endure troubles together.

2.5 Cell coupling through gap junctions

Gap junctions unite cells into a tightly interacting group. Neurons in such a group exchange simple substances and ions. The neurons in the group are electrically coupled and, may be informationally integrated. At least, flows of spike activities that are usually observed in such a group are highly correlated and this indicates that the members of a group participate in a current behavior in a similar manner. Coupling between adjacent cells occurs as hemichannel proteins from one cell dock with hemichannels (connexins) from a bordering cell and form gap junctions. Connexins are large protein family with a number of isoforms [1066]. The large internal diameter (1.2 nm) of many gap junction channels allows not only a flow of electric current, largely carried by K^+ ions, but also an exchange of nutrients, metabolites, second messengers, cations and anions such as cyclic AMP, cyclic GMP, inositol trisphosphate, glucose, and Ca^{2+} [109, 257, 1155]. Hemichannel openings can be revealed by a large linear current and flux across the membrane of small fluorescent molecules and even single-hemichannel openings may be observed [1227]. At least some hemichannels, which are half gap junction channels not connected to bordering cells, can open to allow passage of molecules between the cytoplasm and extracellular space [260].

In particular, energy-rich molecules of ATP releases through connexin hemichannels [160]. There are several connexin proteins, connecting adjacent cells. The connexin36 was found in oligodendrocytes, both amoeboid and ramified microglial cells and in neurons. Astrocytes showed no detectable expression of the connexin36; their gap junctions are formed by connexin43 and some other connexins [941]. Microglia exhibit little coupling [800], but their coupling may increase after activation of microglia. Intercellular coupling between neuronal and microglial populations occurs through connexin36 gap junctions [329]. Electrical coupling in microglia, as in neural tissue, is a dynamic process. Gap junctional communication between microglia can be regulated oppositely by calcium- and protein kinase C-dependent pathways [800]. Gap junctions are also found between many types of neurons, between astrocytes, between astrocytes and oligodendrocytes, and in a few somewhat controversial instances between astrocytes and neurons [260, 329, 1227].

Moreover, a group of three cells, neuron, astrocyte and microglia, may function as a unit, since Ca^{2+}-dependent glutamate signaling between astrocytes and neurons is potently amplified in the presence of inflammatory microglia [119]. The gap junction protein connexin36 is expressed widely along the mammalian brain, implying that there are undiscovered electrical synapses throughout the central nervous system [257, 926]. In many regions of the mammalian central nervous system mixed synapses are also found, allowing chemical and electrical transmission in a single synaptic contact [260, 874, 1155]. Coupling is the attribute of healthy tissue and functioning of gap junctions changes during injury. During prolonged ischemia and hypobaric hypoxia, the pattern of cell coupling is reorganized, although, as a rule, electrical uncoupling is observed [587].

Cell coupling is, typically, low in the injured tissue and closure of gap junctions during damage may have a protective character. For instance, strong ischemia evokes myocardial infarction and closes gap junction channels [523], although sometimes gap junction augmentations during damage may have a direct protective effect. For myocardium tissue, a coupling is the necessary condition for its rhythmic contractions. Usually, damage reduces gap junctions between cells of the same type. Many chemical substances affecting damage, although not every one, influence gap junctions too. Some substances affect damage independently on their influence to coupling. For example, Ca^{2+} participates in many cell functions and may serve as a mediator of damage, but it evokes uncoupling only in pathologically high concentrations [257]. Coupling is connected with the spread of damage and the spread of Ca^{2+} also depends on the value of coupling. Evidently, in this case gap junctions close as the result of Ca^{2+} evoked damage and not because of a Ca^{2+} enhancement itself.

On the other hand, metabotropic glutamate receptor agonists and low pH directly reduce electrical coupling and aggravate cell damage in many and various tissues [598, 1366, 257, 1015, 701]. Potentiation of gap junctions depends on postsynaptic receptors. Neurotransmitters, neuropeptides, neurohormones and cytokines, affecting behavior, modulate coupling in different tissues [697, 1371, 1103, 834, 1192]. Arachidonic acid is also involved in gap junction inhibitions [1048] and volatile anesthetics, affecting consciousness, reduce gap junction conductances to zero during a second [1030, 109, 257] and this indicates that modulation of coupling may participate in aware behavior. A classical activator of gap junctional communications is cyclic AMP and it plays a role in the regulation of intercellular communications. Elevation of cyclic AMP increases gap junctional conductance within minutes [1059, 1030, 1223, 1072, 1160]. On the other hand, cell damage, if the cells stay alive, activates the cyclic AMP system [76], but this activation is not the attribute of damage and has a protective effect [1421, 229, 652, 856]. Astrocytic gap junctions may play a neuroprotective role, since they are under control of the cyclic AMP system, which is tightly connected with defense from damage. Likewise, nitric oxide (through cyclic GMP and stimulatory or inhibitory G_s/G_i proteins) can decrease or increase junctional conductance [109]. Nitric oxide has protective properties at low concentrations and detrimental properties at high concentrations [1439, 360]. In physiologically normal concentrations, NO reduces coupling between neurons, but gap junctions formed by different connexins can be affected oppositely by NO [260].

One amazing property of gap junctions is to connect mainly inhibitory neurons. Electrical synapses occur between excitatory neurons of the inferior olivary nucleus, but, more frequently, gap junctions connect inhibitory GABAergic neurons. Dendrodendritic and dendrosomatic gap junctions usually connect GABA-containing interneurons, in distinction to excitatory principal neurons. Gap junctions were found between inhibitory interneurons of the neocortex, hippocampus, ventral tegmental area, midbrain, thalamus and olfactory bulb [260, 1182, 240, 521, 541, 1155, 1302]. So, in contrast to inhibitory cells, which are often coupled, excitatory cells in the cerebral cortex are rarely coupled [257]. Electrical coupling evoked more precise and strong cortical inhibition [192] and promote oscillatory rhythmic activity [541]. Uncoupling of gap junctions lead to desynchronization of γ-band oscillatory activities in neuronal populations. It is important to note that such high-frequency oscillations are, supposedly, intimately connected with the state of awareness. This electrically and chemically multiple system accepts multifaceted information through excitatory inputs and exerts an inhibitory controlling action, which may be protective against excitotoxic damage, since excitation and inhibition are main factors controlling cell existence. Besides, this is a unique demonstration that excitatory and inhibitory processes are non-symmetrical and they are different not only by a sign of their action to neuronal activity. Correspondingly, damage and protection are also non-symmetrical and are not opposites, since small damage excites cells, while large damage leads to both inhibition and impairment of function. Protection either inhibits cells or excites (depending on the scale of escalation of damage), while overprotection only inhibits function for a long time or even leads to death [469].

2.6 Multiple pathways for cell survival

2.6.1 Damage through excitation and the paradoxical properties of an injured neuron

Usually, although not in every case, damage, evoked by various factors, injures cells through excessive excitation [797, 642, 742]. Necrotic neuronal damage often is a fee paid for excitation. Mild or strong excitations are at the core of injury evoked by various insults. Neuroexcitatory mechanisms are involved in ischemic neuronal injury, in damage due to neurotoxin injections, in mechanical damage, in damage as a consequence of oxygen or glucose deprivation, etc. [331, 337, 799, 832, 1389]. Mediators of the processes related to neuronal damage are excitatory amino acids, especially glutamate, retrograde transmitters, free radicals, intracellular Ca^{2+}, impairment of energy metabolism, intracellular acidification, etc. Intensive influx within the cell of positive ions, Ca^{2+} and Na^{+}, depolarizes the cell and hurts it. On the other hand, strong depolarization is characterized by a disturbance of homeostasis of intracellular Ca^{2+} and Na^{+}-influx [437]. Although suppression of Na^{+}-channels exerts neuroprotection [627], Na^{+} ensures immediate neuronal reactions and does not control long-term cellular homeostasis. However, Na^{+} metabolism is closely connected with the maintenance of excitability and may serve by a satisfactory indicator of a neuron's conditions on the scale of damage-protection.

Not only excitation causes damage, but also damage produces changes of excitability. Excitability of neurons decreases during apoptotic degeneration, but membrane potential and AP amplitude remain stable [546]. During necrotic damage, excitability changes in a rather complex way. Nerve fibers develop anomalous excitability at or near the spot of nerve injury. The mechanisms include unusual distributions of Na^{+} channels, and an increase in intracellular Na^{+} as well as a deficit in central inhibitory mechanisms following nerve injury [416, 1195, 1439]. Particularly, peripheral axotomy induces long-term hyperexcitability of centrally located neuron soma in diverse species including the Aplysia that serves as the main object for investigation of behavioral plasticity. Long-term hyperexcitability of soma is selectively expressed in the sensory neurons of Aplysia having axons in cut or crushed nerves rather than nearby, uninjured nerves [453]. Hyperexcitablity of axotomized dorsal root ganglion neurons also depends on up-regulation of tetrodotoxin-sensitive sodium channels, which are a basic instrument of excitability [327, 626]. After spinal cord injury, an increase in excitability of neurons may be connected with the shift of the expression of Na^{+} channels from a high-threshold tetrodotoxin-resistant type to a low-threshold tetrodotoxin-sensitive type [1382]. Sodium up-regulation is usually observed in such cases.

Damage distorts membrane properties. Both injury and excitation evoke depolarization and the growth of excitability in neurons. In dying neurons, a phase of enhanced excitability before death is inevitably transformed to an unexcitable state. Such a property of neural tissue, to be activated under the influence of a harmful impact before death, condemns an injured neuron to display paradoxical properties. An injurious agent (asphyxia, etc.) usually evokes short-term depolarization and excitation, which, after a few minutes, is replaced by a slow compensatory hyperpolarization. This is followed by a persistent depolarization until the membrane potential reaches zero and excitability decreases, even if the injurious agent is removed [81, 797, 1213]. In such cases neurons inevitably die. In certain circumstances, any one of these phases may be absent. For instance, vulnerable neostriatal neurons respond to ischemia with a membrane depolarization and resistant neurons with a hyperpolarization [221].

The same succession of phases may be observed for behavior. For example, morphine injection induces short-term hypophagia, which is followed by a period of hyperphagia and then by persistent hypophagia [144]. The nociceptive responses to the formalin test also may be divided in three distinct phases: acute pain, inhibitory (interphase: pain decreases) and tonic pain [454]. Similarly, the ventilatory response to acute hypoxia in mammalian species is biphasic, an initial hyperventilatory response is followed by a reduction in ventilation within 23 minutes below the peak level [558].

During necrotic damage, properties of excitable membrane undergo reorganization because of disturbed ionic equilibrium. Enhanced survival may be sharply, but in a rather complex way, related to membrane potential [405]. Correspondence between excitability of an injured cell and characteristics of its membrane also becomes paradoxical. Spontaneous discharge may be inhibited, when input resistance and the level of membrane potential are not affected [836]. Sometimes, when there are no changes in membrane potential of neurons, input resistance and action potential amplitude of certain identified neurons are increased [13]. Many pharmacological substances affecting the state of neuronal damage-protection also do not alter membrane potential but change the threshold of AP generation or refractoriness [1200, 502]. Damage may selectively affect K^+, Ca^{2+} or Na^+ currents [836, 433] and may depend, or not on NO production [551]. These paradoxical properties may be easily explained by non-monotonous changes of excitability during dying.

The process of cell recovery from damage disturbs the coordination between membrane potential and excitability so that a change in excitability after damage is not the direct consequence of a shift in the membrane potential [1213, 1423, 81, 604]. The relationship between threshold and other membrane factors is distorted when measured during ischemia and after its release. During ischemia, there is disproportionately greater refractoriness and, after release of ischemia, supernormality develops and excitability increases. The former could be due to interference with the recovery from inactivation of Na^+ channels evoked by some intracellular metabolic disturbance created by ischemia. After recovery, the properties of excitable membrane return to the norm. Processes of neuronal damage and death are illustrated in Fig. 2.1.

Normally hyperpolarization removes the membrane potential from the threshold level of AP and decreases excitability, while depolarization increases

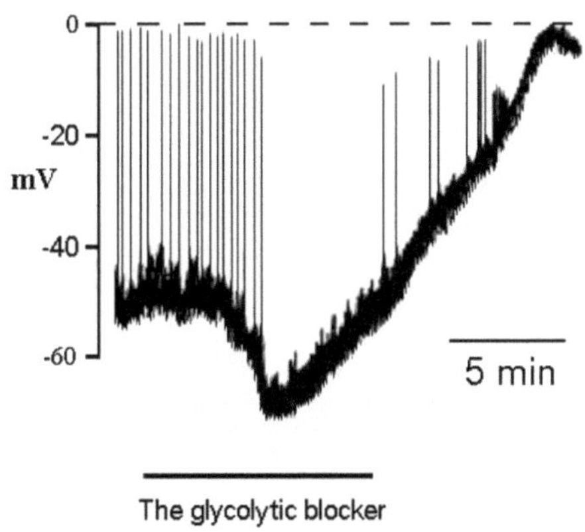

Fig. 2.1. Metabolic disturbance changes electrical activity in an inspiratory neuron in the brainstem-spinal cord preparation of rat. The glycolytic blocker iodoacetate irreversibly abolishes respiratory rhythm and elicits a hyperpolarization, followed by a prominent depolarization. The alterations depend on Ca^{2+} entry, K^+ -channels, Na^+/Ca^{2+} exchanger functioning, etc. Dotted line indicates zero level of membrane potential, when the cell is near the death. Spike amplitude decrease irreversibly. Calibration in the figure. The Fig.2.1 was redrawn in accordance with the data [81]

it. Specifically, when the membrane potential in a healthy neuron to some extent transitorily decreases, AP height, EPSP amplitude and the threshold of AP generation decline, whereas excitability, refractoriness, AP duration and spontaneous activity enlarge. However, when membrane depolarization is large enough or dragged out, the neuronal state worsens and the above-mentioned correspondences are disturbed. After the occurrence of persistent depolarization when cells are near death, excitability decreases, the AP threshold increases and the Na^+-dependent AP decreases and disappears (see Fig. 2.1,bottom). These disturbances are typical not only, because of excessive excitation, but are observed after any harmful impact, although particular details may differ. Mild damage usually increases excitability, while severe damage decreases it and thresholds may increase during depolarization evoked by injury. Paradoxical properties of electrical activity of injured neurons are illustrated in Fig. 2.2.

Strong depolarization increased activity and then evoked inactivation, excitability declined and the neuron failed to generate APs. Short-term decrease in the depolarizing current diminished inactivation and AP generation recovered. This phenomenon was discovered by N.E. Vvedensky, more than hundred yeas ago [1311]. Injury of neuromuscular junction leads to development paradoxical state of the tissue. While responses of tissue to stimulus after injury decrease, the higher irritation the smaller becomes the response. Thus, small

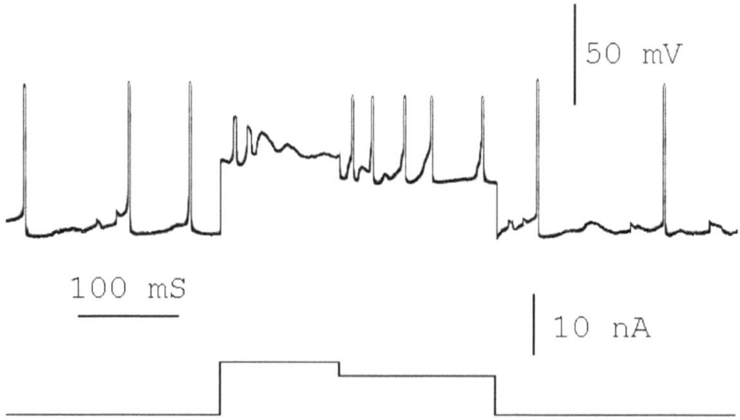

Fig. 2.2. Paradoxical effect of decrease in excitatory current to intracellular activity of neuron of the parietal ganglion of Helix. The neuron was hyperdepolarized by the direct excitatory current and failed to generate spikes. Decrease the degree of the hyperdepolarization improved neuron's state and recovered its possibility to generate spikes. At the top, spontaneous neuronal activity and response to the direct current injection; at the bottom, value of depolarizing current. Calibrations in the figure.

damage excites, whereas large damage inhibits and when damage is large, responses increase, if strength of stimulus decreases.

In Fig. 2.3 is schematically presented some changes in neuronal excitability at the scale of damage-protection. An increase in depolarization brings a membrane potential closer to its threshold so that the neuron generates spontaneous spikes and coupling, as we already pointed out, increases. We will demonstrate further that such enhanced activity may serve as the basis of trial-and-error when the brain searches for a way to avoid excitotoxic damage. Additional depolarization, nevertheless, evokes excitotoxic damage and decreases of excitability. Frequency of spike generation increases, while each spike becomes smaller, smaller and eventually is totally blocked. That is, further depolarization injures neurons and leads to a paradoxical phase, in which decreasing the excitation results in an increase in excitability. In such cases, hyperpolarization protects neurons and paradoxically recovers excitability. Thus, compensational recovery of membrane potential leads to generation of action. Compensatory hyperpolarization is in many cases connected with the activation of K^+ current. As an example, infusion of K^+ channel blocker in the hippocampus induces release of glutamate from nerve endings that results in neurodegeneration. Under these conditions, $GABA_A$- mediated transmission may paradoxically increase (!) neuronal excitation [956]. This is explained by the property of GABA receptors to protect cells, not only to inhibit them. Similarly, anoxia (2-5 min) hyperpolarized dopamine-containing

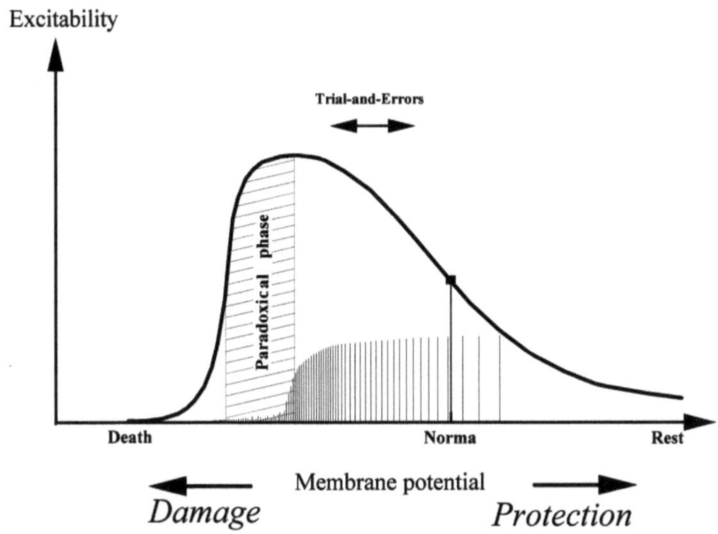

Fig. 2.3. A schematic illustration of the dependence of neuronal excitability on excitotoxic influences. Abscissa the shift to damage or protection (or in the first approximation a membrane potential). Ordinate the change in excitability. A paradoxical phase borders the interval in which hyperpolarization or decrease in the level of damage leads to reaction augmentation. The changes in amplitude and frequency of spikes are shown along the abscissa axe.

neurons through K^+ current and inhibited spontaneous AP. Block of potassium current lead to anoxia-induced hyperdepolarization and AP inhibition, but neuronal activity recovered after decrease in the hyperdepolarization (one minute after end of anoxia) [836].

Protective hyperpolarization may be also connected with the outflow of K^+ from a cell. Two-pore-domain K^+ channels are a diverse and highly regulated superfamily of channels that are thought to provide baseline regulation of membrane excitability. Moreover, opening of these channels protects neurons against both ischemic and epileptic insults [407]. Another reason for compensatory hyperpolarization may be an augmentation of the electrogenic Na^+, K^+ pump, such as following release of ischemia. Heightened activity of the electrogenic Na^+, K^+ pump leads to hyperpolarization [502]. As we already have pointed out, excitatory amino acids are the main mediators of cell damage. Activation of glutamate receptors depolarizes neural cells, ensures enhancement of intracellular Ca^{2+}, and activates G proteins that reorganize cellular metabolism. Besides excitatory amino acids, many other neurotransmitters also affect cell damage and, in the first turn, these are the inhibitory amino acids GABA [295, 593] and taurine [573, 697, 1085, 1090, 1387], which are the most abundant inhibitory neurotransmitters in the brain and they act against damage and protect neurons. Some $GABA_A$-agonists provide complete histological protection against hypoxic-ischemic injury [470]. In addition to its direct inhibitory action through the ionotropic $GABA_A$ receptor, GABA exerts a variety of modulatory actions through the metabotropic $GABA_B$ type of receptor by means of G proteins [1024].

GABAergic interneurons in the hippocampus and neocortex contain the calcium-binding proteins parvalbumin and calbindin, preventing damage of these interneurons, when most other neurons eventually cease to live as a result of ischemia [409, 465] or excessive excitation [588]. Excitatory and inhibitory spinal pathways showed different susceptibility to ischemia. Interneurons of excitatory segmental pathways are less sensitive to ischemia than motoneurons and motoneurons were less sensitive to ischemia than interneurons of inhibitory pathways [655]. This may be determined by the presence of calcium binding proteins. In parvalbumin knockout mice, paired-pulse depression in synapses between interneurons and Purkinje cells of cerebellum converts to paired-pulse facilitation [199], that is, sensitivity to excessive excitation is higher in the absence of calcium-binding proteins. Nevertheless, during regular functioning, GABAergic interneurons undergo intensive damage and death, since they receive strong excitatory input from pyramidal cells and they are intensively replaced by newborn neurons. Pyramidal cells, receiving inhibitory influence from these interneurons are thus protected and more stable. One must recall also that inhibitory GABAergic interneurons are frequently electrically coupled by means of gap junctions.

A cell uses many synaptic pathways for chemical control of damage, not only glutamatergic and GABAergic synapses. Non-amino-acid neurotransmitters also influence cell damage-protection. Acetylcholine can induce neuronal damage [929] and, on the other hand, damage leads to a reduction in acetylcholine receptors, while anticholinergic actions produce protection, evoked by toxicity [210], infection [1204], ischemia [12] and extensive intracellular Ca^{2+} following some insults [994]. Dopamine [789, 1107] protects cells against damage; dopamine-containing neurons are very stable [803, 1380] and dopamine-receptors are connected with the inhibitory G_i protein [1422]. Serotonin ameliorates cellular damage [1370, 1107], plays a pivotal role in the homeostasis of neural tissue [69] and stimulates cell proliferation. Peptide neurotransmitter substance P also protects neurons [866, 364]. Motivationally relevant substances, neuropeptides and neurohormones also affect the processes of neuronal damage, when they act on behavior[1]. On the other hand, damage increases expression of neuropeptides.

[1] For example, cholecystokinin, leptin [175, 176, 1031], estradiol, angiotensin II [1051, 495], neuropeptide Y [1, 232], vasopressin [1136], androgens [1388, 11, 978, 934], estrogens [1327, 1346] and prolactin [392] each exerts a strong effect on the damage-protection processes and they can either ameliorate or exacerbate cell death, depending upon the concentration, level of motivation and location of the tissue.

Besides numerous neurotransmitters, neuropeptides and neurohormones, various second and retrograde messengers, also interact with cell damage and protection. Let us consider the roles of these substances in neuronal function, damage, and behavior. However, keep in mind that any (or almost any) factor affecting neuronal behavior, affects neuronal survival, too.

2.6.2 Second messengers and cell survival

Second messengers, in any cell, evidently, execute general basic functions, but particular functions may appear and disappear in specific cells. Cellular damage and damage-related processes are closely connected with the systems of second messengers, Ca^{2+} and cyclic AMP, which participate in signal transduction within a cell and control a variety of processes that are important for survival. Injury increases cyclic AMP and Ca^{2+} concentrations within a cell. First of all, the effects of cell damage are strictly related to Ca^{2+} homeostasis [1236]. Damage of different origins injures cells through Ca^{2+} influx, while inhibition of endogenous Ca^{2+} suppresses damage [642], Ermak et al. 2002, Linda et al. 2004). Some chemical factors, known by their detrimental action, also affect damage through stimulation of intracellular Ca^{2+}. NMDA enlarges damage through enhancement of Ca^{2+} [1128, 10]. Retrograde messengers, arachidonic acid, NO and their metabolites, may affect damage by means of distortion of Ca^{2+} homeostasis [23, 388, 1236]. Augmentation of intracellular Ca^{2+} may also be evoked by cytokines [339, 798]. On the other hand, Ca^{2+} elevation is not an obligatory attribute of damage. For instance, neuronal swelling may be independent of the elevation of intracellular Ca^{2+} [540]. Usually Ca^{2+} penetrates a cell as the result of activation of excitable membrane, but intracellular Ca^{2+} may mediate damage independently of change in excitability [123]. The spread of Ca^{2+} in tissue depends on neural-glia interactions. Ca^{2+} diffuses through gap junctions to evoke Ca^{2+} signals in neighboring unstimulated astrocytes, elevates Ca^{2+} in adjacent neurons and thus aggravates damage [381, 763, 526].

Ca^{2+} concentration in astrocytes may be quickly elevated because of Ca^{2+} outflow from intracellular stores [538], and treatment of astrocytes with inositol trisphosphate, controlling Ca^{2+} influx from intracellular stores, protects astrocytes against oxidative stress [1362]. Recently a new metabolic path for protective activity of neurons through σ_1 receptors was discovered. These receptors are intracellular proteins on the membrane of the endoplasmatic reticulum of neurons and they modulate Ca^{2+} mobilization from inositol trisphosphate-gated intracellular pools. The remarkable ability of σ_1 receptor ligands is to block or delay the deep depolarization that accurs immediately after stroke, although they have excitatory effects at low concentrations. Such a deep depolarization, in the absence of σ_1 receptor agonists, inevitably leads to cell death [36]. This collapse in membrane function results from failure of the Na^+, K^+ pump and, in the given case, protection of neurons independent of glutamate receptor mediation, since extracellular glutamate begins to accumulate only after occurrence of the depolarization [808]. Gap junctions in astrocytes are involved in mediating intercellular Ca^{2+} signaling throughout the glial syncytium, but they are controlled by neurons [878, 1047]. In response to neuronal activity, Ca^{2+} concentration in astrocytes rises and can lead to glutamate-dependent Ca^{2+} elevations in adjacent neurons and, hence, although Ca^{2+} ion is a typical second and not a retrograde messenger, these ions may affect remote cells [964]. Depending upon local Ca^{2+} concentration, astrocyte Ca^{2+} waves can either increase or diminish nearby neuronal activity [885] and thus align a level the difference between the states of near neurons. Rapid Ca^{2+} elevation within astrocytes may, probably, protect them, but injure adjacent cells. Nevertheless, the velocity of Ca^{2+} signal propagation is in the order of tens of micrometers per second, i.e. two to three orders of magnitude slower than AP propagation [160]. Therefore, regulation of intracellular Ca^{2+} cannot participate in the on-line control of performance, but may exert a slower influence on behavior.

For example, disturbance of intracellular Ca^{2+} may be important in affective states [1190] and play a crucial role in the pathogenesis of Alzheimer's disease [1144]. In bipolar disorder patients, abnormalities in Ca^{2+} homeostasis were found and these abnormalities may be linked to disturbances in the function of G proteins that mediate cyclic AMP signaling [358]. Modulation of intracellular Ca^{2+} through σ_1 receptors is a necessary component of cocaine and ethanol-induced motivational effects, and above all, activation of σ_1 receptors decreases anxiety. They act as antidepressants and are involved in learning, response to stress, addiction and pain perception [808]. The acute and repeated stimulations of a σ_1 receptor subtype improve behavioral depression [1283]. Thus, Ca^{2+} homeostasis directly or through activation of σ_1 receptors participates in fine-tuning of attributes of behavior, and may by likened to fine-tuning the regulation of the mood. At the same time, impairment of second messenger systems does not lead to immediate death, but provokes delicate changes in behavior.

A second messenger cyclic AMP such as Ca^{2+} ions, is tightly connected with processes of damage-protection. However, Ca^{2+} is an important modulator of damage, while cyclic AMP mediates protective effects in cells. Cyclic AMP, as Ca^{2+} ions, affect various intracellular processes[2] and can penetrate

---

[2] Cyclic AMP is synthesized from ATP by adenylate cyclase through activation of the G protein, whereas cyclic AMP decomposition is catalyzed by the enzyme phosphodiesterase. The simplified sequence of cyclic AMP- dependent events is as follows [1352]: a signal arrives at the corresponding receptor on the surface of a cell, which leads to the production of a second messenger cyclic AMP (or in other cases ions of Ca^{2+}), which in turn activates protein kinases. For example, protein kinase A is normally inactive. Cyclic AMP binds to specific locations on the regulatory units of the protein kinase and causes dissociation between the regulatory and catalytic subunits, thus activating the catalytic units and enabling them to phosphorylate substrate proteins. This protein kinase moves to the cell nucleus, where it activates a cyclic AMP-response element-binding

through gap junctions, too, but a wave of cyclic AMP cannot spread very quickly, since intracellular stores for cyclic AMP are absent. Cyclic GMP and cyclic AMP pass equally well through various hemichannels, whereas other nucleotides including AMP, ADP, ATP, cyclic TMP, and cyclic CMP are not effective [483]. The cyclic AMP role in cell protection was investigated especially carefully. Cyclic AMP molecules are important in many biological processes and in any one of cell types. It is used for intracellular signal transduction, such as, say, transferring the effects of hormones that cannot move through the cell membrane. Cyclic AMP connects events in outer and in inner environments and with cyclic AMP cells are capable of distinguishing different outer factors (for instance, stress and mutagens) [812], activating protein kinases, increasing expression of a large number of genes and controlling functions of ion channels. Augmentation of the second messenger cyclic AMP converts a cell to activation of developmental plasticity and causes, for instance, the formation of new synapses between rat hippocampal neurons in primary culture [402] and induces morphological changes within Aplysia sensory neurons [1195], dependly on protein synthesis [909]. Thus, cyclic AMP metabolic pathways regulate cellular homeostasis by means of activation or repression of transcription in response to extracellular signals [510].

In bacteria, cyclic AMP is a chemoattractant and controls mating. In multicellular animals, cyclic AMP is very important for neural function, but, for sure, in this case it is not concerned with mating only, although cyclic AMP is activated by steroid sex hormones [1146] and cyclic AMP pathways determine sexual differentiation of the brain. An example of this divergence is the response of cyclic AMP system to GABA. In the male rat brain, GABA action leads to increased phosphorylation of cyclic AMP response element binding protein, whereas GABA action in the female brain leads to a decrease in phosphorylation of this protein [68]. The general function of cyclic AMP is to support cellular activity during stressful or injuring impacts. In neurons, cyclic AMP increases excitability and allows intensive functioning when it is necessary. Factually, in bacteria, cyclic AMP plays a similar role, since it increases expression of enzymes that can supply energy independently of glucose, when the glucose-dependent pathway of ATP synthesis is exhausted. According with the "traditional" role of cyclic AMP, as the substance that counteracts stressful conditions, injury of any kind evokes a prominent enhancement of intracellular cyclic AMP in neurons and this is intimately associated with neuronal defense[3]. In spite of this protective action, cyclic AMP increases excitability

protein, or other extracellularly regulated transcriptional factors. The activated cyclic AMP-response element-binding molecule is bound by a binding protein, which coactivates it, allowing it to switch certain genes on or off. Besides, cyclic AMP can regulate synthesis at the posttranslational level, for example, affecting the expression of NO [1352].

[3] Levels of the cyclic nucleotides cyclic AMP (and cyclic GMP, too) are positively correlated with the severity of acute trauma in humans [76]. So, after an hypoxic-ischemic insult cyclic AMP-dependent phosphorylation can be persis-

of neurons without increasing symptoms of excitotoxicity. In most cases, the treatments that excite neurons lead to their damage, while the treatments that inhibit neural cells protect them. This is not the case for treatments affecting the cyclic AMP system. Axonal injury elevates cyclic AMP and increases excitability of sensory neurons of Aplysia, in the hippocampus and other brain areas [1195, 1159]. In its turn, activation of the regulatory cyclic AMP transduction cascade produces acute sensitization in rat sensory neurons [147], increases potential-dependent Na^+ current and excitability with a latency of some minutes after acute cyclic AMP activation [330, 1377]. Activation of cyclic AMP pathways potentiates not only excitability but also augments postsynaptic potentials in neurons of various brain areas [1159]. Beside enhancement of gap junctions, excitability and synaptic efficacy, cyclic AMP activates other neuronal functions and favored expression of a mature ion channel pattern in astrocytes. Nonetheless, knockout mice with deletion of binding protein of cyclic AMP-responsive element performed normally in the Morris-water maze and fear conditioning [1316].

Nevertheless, even after chronic enhancement of the cyclic AMP activation (up to 7 day), the ability of sensory neurons to be sensitized is maintained and does not appear to be down-regulated [147]. Cyclic AMP-evoked increase in excitability allows intensive functioning of neurons [812] and does not lead to cell death. Considering that excitotoxicity is an important reason for cell damage, cyclic AMP excitation itself is not the reason for neuronal protection. A pharmacologically reinforced cyclic AMP signaling in rat glial cell cultures depresses oxygen radical formation in microglia and the release of cytokines which mediate oxidative damage and secondary astrocyte activation [1096]. Evidently, turning on the cyclic AMP cascade ensures that the neural system can compensate for detrimental consequences of injury, but the resulting steady state is not the component of previous steady state recovery in a neural system and rather accompanies only a transfer to a new stable state. Cursory evaluation of complexity of these biochemical events (which, in reality, are much more complicated) shows that cyclic AMP cannot participate in rapid on-line processes, but may be important during reorganization of cell states during homeostatic survival.

---

tently activated in surviving neurons and this phenomenon is closely associated with protection of these neurons [1315, 1214]. Artificial augmentation of cyclic AMP pathways exerts neuronal protection and promotes cell survival after different harmful clashes [613, 843, 953, 1421, 229, 652, 856]. Likewise, insulin-mediated protection is dependent on cyclic AMP accumulation [761]. Inhibition of cyclic AMP degradation increases cyclic AMP concentration within cells and protects against oxidative stress in rats [463]. Probably, transcription factor cyclic AMP-responsive element binding protein is required for the survival of neurons [747]. Besides that, cyclic AMP strongly stimulates gap junctions, which decrease in number during serious damage and thus ameliorate injury. Cyclic AMP prevents also apoptotic cell death [629].

Different types of adenylyl cyclases and protein kinases may have specific (even opposite) effect on cell functions during acute and chronic action [889] and, hence, the same molecules of cyclic AMP may lead to different outcomes and this is important for the organization of behavior. Nevertheless, when evaluating a possible role of the cyclic AMP system in behavior, one must realize that the protective result of cyclic AMP influences is too prolonged in time. The question is how this activity participates in a mental function. The target gene activation in response to cyclic AMP is typically observed after several minutes. The protection begins slowly (at least in minutes and usually much longer), it needs continuous metabolic and energetic support, and evidently, weakly depends (if at all) on real-time activity and learning, and some more memory. It had been supposed, that cyclic AMP pathways were critically involved in memory formation (LTP is implied) [20, 298]. The cyclic AMP system and intracellular Ca^{2+}, as with other pathways connected with damage-protection may execute its function only as a slow or middle-time regulator of behavior. For example, a homeostatic control of neural activity is considered to be monitored through activity-dependent changes in second messenger AMP [296]. Also, rest in Drosophila, which has common features with mammalian sleep such as prolonged immobility and decreased sensory sensitivity, is connected with cyclic AMP signaling. Activation of cyclic AMP pathways is inversely related to the duration of rest, thus demonstrating a cyclic AMP role in waking and rest homeostasis [534]. Some forms of behavior, which are connected with chronic damage and the depolarization of neurons such as pathologic drug-dependence, are also determined by the cyclic AMP system. Opioid [1362] and cocaine [850] dependencies develop through activation of the cyclic AMP pathway, impairment of Na^{+}, K^{+}-ATPase activity and depolarization. Cyclic AMP pathways may be also concerned in recovery of functions [953].

2.6.3 Intercellular protection by retrograde messengers and cytokines

Besides second messengers, retrograde messengers also exert a direct influence to intracellular metabolism and are serve by the means for coordination of adjacent neuron activities and neuro-glial interactions [1128]. These tools provide intercellular transmission. Small molecules of retrograde messengers penetrate the cellular membrane and exert their influence apart from a synaptic transmission. Then the systems of second and retrograde messengers interact. For instance, cyclic GMP mediates the modulation of transmitter release by NO [991]. Some conventional neurotransmitters, such as GABA, dopamine and neuropeptides may also reach the intercellular space, but they interact with chemoreceptors. Retrograde messengers are released from regions on the postsynaptic cell that do not display typical morphological specializations for secretion (e.g. vesicles, active zones, chemoreceptors). In principle, the messenger can act on any part of the presynaptic cell: its terminals, axon, dendrites or soma. Retrograde messengers, endogenous cannabinoids, lipids (arachidonic acid) and the gases NO and CO are membrane-permeable and may penetrate the outside of a mother cell and within adjacent cells in any location, but usually cannabinoids act at the presynaptic axon terminal, while the gases NO and CO act globally.

A system of retrograde messengers is involved in the regulation of vital brain functions and they can, depending on conditions, cause an inhibition of cyclic AMP or a sharp rise in its concentration [402]. Potent chemical control of protection is accomplished by cannabinoids, which are the primary psychoactive component of the cannabis plant (marijuana), but they are naturally produced in the bodies of animals and their receptors, coupled with the G proteins, are widely presented in the brain [23, 402, 1138, 1285]. The cannabinoid system is currently thoroughly studied retrograde signal system in the brain [23]. Natural cannabinoids are derived from arachidonic acid and their amount increases during damage. They are synthesized for actual needs and are not stored for later use. Activation of cannabinoid receptors temporarily reduces the amount of conventional neurotransmitter released and this permits the postsynaptic cell to control its own synaptic inputs.

The effect of the cannabinoid secretion depends on the nature of the conventional transmitter that it controls, since it may reduce both excitatory and inhibitory synaptic inputs. Cannabinoids cause an inhibition of cyclic AMP, but concurrent stimulation of cannabinoids together with dopamine (as evidently happens during behavioral reinforcement) leads to a sharp rise in cyclic AMP concentration, reduces the secretion of NO and the inflammatory cytokine from microglia, and facilitates neurogenesis [402]. Canabinoids are potent regulator of mood and cognition. The sensitization of cannabinoid receptor-mediated G protein signaling in the prefrontal cortex is one of the factors in the pathophysiology of suicide and post-mortem studies have shown a higher density of the cannabinoid receptor in the prefrontal cortex, striatum and anterior cingulate cortex of schizophrenics [1305]. Canabinoids, as other substances affecting damage-protection, regulate many life-important processes: anxiety, pain, epilepsia, insomnia, depression, enhance appetite and produce hypothermia, which reduces brain damage [610]. Arachidonic acid, NO and their metabolites have a number of common targets on which they may exert similar or opposite actions, and have a crucial role in the regulation of inflammation, immune responses and cell viability [23, 1431]. Arachidonic acid is found to be cytotoxic at concentrations that overlap physiological ones and evoked necrosis [985]. At present, good evidence exists for inhibition of voltage-gated Ca^{2+} channels, a direct action on the synaptic release machinery at the presynaptic terminal and to the opening of K^{+} channels on dendrites The retrograde messenger of a global action, free radical NO is a small molecule and it can relatively quickly (milliseconds to seconds) spread to adjacent cells. Therefore, there is short-term activity-dependent retrograde signaling [23] and this may be a reason for parallel modulation of synaptic efficacy and excitability during learning. A release of NO affects numerous other cells in the environment without regard to their functions and synaptic relationship to the releasing cell [23]. The balance of NO levels in the tissue may be crucial for orienting microglial reactions towards neuroprotection or neurotoxicity and it is likely that a low NO promotes protection, while a high NO results into neurodegeneration, since enhancement of NO concentrations is toxic to surrounding neurons [848, 749, 360][4]. Evidently, during the reversible stage of damage, when NO concentration is small enough this retrograde messenger protects adjacent cells, but after damage, NO concentration increases and kills neighboring cells. By the way, taking into account that NO is a retrograde messenger, it is natural to suppose that the neuron emitting NO is injured itself, but protects (even if at the beginning) neighbor cells, since NO concentrations reaching these cells have to be smaller.

Besides the above-mentioned direct pathways affecting damage, there are substances that modulate damage-protection. Powerful participants in the processes of damage-protection are cytokines. Cytokines are peptides that are produced by every cell type in the body: neurons, glia, astrocytes, etc. [120], but predominantly by activated immune cells such as microglia and are involved in the amplification or reduction of inflammatory reactions[5]. However, this enhanced expression is not the cause for neuronal damage. Defiantly, mice, lacking receptors of tumor necrosis factor, demonstrated increased neuronal degeneration [190]. The the colony stimulating factor cytokine also demonstrates "protective" reaction [120]. As part of the scope of our book, it is important to consider manners of cytokine influences, since the same substances modulate slow modifications of behavior. So, although maximal expression of cytokine interleukin-1 occurs five days after experimental cerebral ischemia [467] and this time robustly exceeds the time that is necessary for completion of mental functions, cytokines may participate in cerebral activity. In low doses, (picomolar range) interleukin-1 elicits a slow-wave sleep [1046]. At the

[4] In particular, high NO concentrations inhibit neuronal Na^+, K^+-ATPase, while low concentrations activate K^+ channels and Na^+, K^+-ATPase [677]. Some reactions to excitatory amino-acids are accomplished with NO participation, as, for instance, excitatory amino-acid NMDA augments damage through enhancement of NO [1128, 10, 388], while inhibition of AMPA receptors depend on NO production, and this could offer protection [585]. Because of NO influence on neighboring cells, it may selectively affect background neuronal activity without influencing excitability [551].

[5] The cytokine family includes interferons, interleukins, colony-stimulating factors and various growth factors. In humans, more than 100 cytokine genes are expressed. They can cross the bloodbrain barrier and function as a microglia activator [404]. Lesion in the nervous system is followed by an inflammatory process that induces rapid activation of glial cells. Inflammation can cause neuronal damage, but may exert neuroprotection at some stages, like many other active treatments. During protective reactions, inflammatory cells generate cytokines, activate growth signaling and stimulate oxidative reactions [214]. For another cytokine, the tumor necrosis factor, expression is highly upregulated early after injury and is often implicated in promoting neuronal degeneration.

same time, it may contribute to healing before reaching its maximal concentration. Some cytokines even more potently affect a mental activity, in particular the cytokines neurotensin and thyrotropin-releasing hormone. Although neurotensin is a proinflammatory neuropeptide [215] and it is elevated in plasma during damage [340], it stimulates intestinal wound healing [172], minimizes tissue damage [179], reduces oxidative damage [63] and counteracts ischemic damage [1239]. Thyrotropin-releasing hormone also exerts protection against damage [150, 600, 1302] and, moreover, it has been shown to be able to reverse damage to glia and neurons [955]. Both neurotensin and thyrotropin-releasing hormone protect cells against damage in physiological doses. Nevertheless, these cytokines inversely affect awareness: neurotensin decreases arousal and calm animals without sedating them [155], while thyrotropin-releasing hormone wakes animals from anesthesia [989]. Bath application of thyrotropin-releasing hormone resulted in a transient cessation of spindle waves, which is prominent during slow-wave sleep and rhythmic burst firing [168], and maybe this is one of the causes of its awakening action. Evidently, thyrotropin releasing hormone strains defensive functions[6], while neurotensin is probably connected with the reinforcement system of brain[7]. Neurotensin and thyrotropin-releasing hormone, evidently, protect cells by different means and, in particular, they differently affect neuronal activity. Neurotensin, at the first decreases in membrane resistance by the activation of a cationic conductance, depolarizes and activates neurons (100 nM - 1 μM) and after that evokes inhibition, at time scales of a few minutes [104, 231, 905]. Primary depolarization depends on suppression of K^+ conductance, but did not depend on Na^+ potential-sensitive channels, GABA receptors and ionotropic and metabotropic glutamate receptors [231, 905]. The first AP in the response usually has a larger amplitude, than following APs, as usually happened after depolarization-evoked excitation under normal conditions, when excitatory effects were compensated

[6] Thyrotropin-releasing hormone has anti-epileptic effects, decreases energy expenditure, regulates sleep, cognition, locomotion and mood [310, 1321]. Vegetative reactions and arousal level are augmented by thyrotropin-releasing hormone, which elevates blood pressure and heart rate, and is responsible for an increase in the respiration rate [447], antagonizes hypothermia and respiratory depression [122]. However, thyrotropin releasing hormone inhibits positive motivations: decreases the desire for food [239, 1321], inhibits penile erection (100 and 500 pmol) [554], alcohol intake [685] and development of physical dependence on opiates, suppresses the abstinence opiate syndrome but not the analgesia induced by opiates [122].

[7] Neurotensin is localized in regions containing either cell bodies or terminals of dopamine neurons [266, 155]. Dependending on the doses, it augments or diminishes nociception. Low doses of neurotensin (0.03 pmol0.03 nmol) facilitate nociception, whereas greater doses (1.330 nmol) are more potently antinociceptive than morphine, but in an opiate-independent manner [179, 164]. It is an anorectic peptide and restricts food intake [1060, 1025, 155], changes sensitivity to drugs of abuse, but differently for different rat strains [89] and evokes hypothermia [1239, 164].

by manually hyperpolarizing the cell to control values [104]. In addition, neurotensin increased the frequency of both excitatory and inhibitory spontaneous synaptic currents and this effect depended on presynaptic potential-sensitive Na^+ channels [905].

Thus, during damage, neurotensin evidently augments homeostatic activity of cells. It evokes strong, but short-lived excitation and prolonged inhibition. Neurotensin probably stresses the compensation and affect of neuronal activity in a conventional way. That is, neurotensine directly counteracts injurious factors to recover an initial state and, probably, counteracts a basic component of damage, high activity. Damage-protection is similarly dependent on the influence of neurotensin and on a shift of the membrane potential if one considers only weak linear deviations, when compensation recovers the lost equilibrium. High neurotensin concentration decreases Na^+, K^--ATPase activity [917], suppresses the hyperpolarizing of potassium currents, activates a cationic conductance [104, 231] and stimulates inositol-trisphosphate production that provokes Ca^{2+} release from intracellular stores [763]. This cannot be the primary cause of protection, and, oppositely, can promote development of excitotoxicity, but prolonged action is inhibitory concerning both neuronal activity and common physiological functions.

Thyrotropin-releasing hormone elicited a transient hyperpolarization (of a few seconds) of the cell membrane and an increase in the membrane conductance to K^+, followed by an enhancement of the generation of action potentials due to Ca^{2+} entry from the extracellular space. In addition, the input resistance of the cell membrane increases during the facilitation (few minutes) [928, 339]. Facilitation stimulates the release of prolactin and growth hormone, which usually protect cells. The hyperpolarizing response was transient despite the continuous leakage of thyrotropin-releasing hormone from the pipette. The enhancement of the spike generation was not due to membrane depolarization, since thyrotropin-releasing hormone causes a burst of action potentials without any detectable change in the resting membrane potential and the first AP after hyperpolarization was small, like during the damage depicted in Fig. 2.1. Hyperpolarization also tended to be gradually reduced by repeated administration of thyrotropin-releasing hormone, and the second application of thyrotropin-releasing hormone immediately enhanced the spike generation without eliciting the hyperpolarization. Action of thyrotropin-releasing hormone to an inhibitory neuron differed from action to excitatory ones. Thyrotropin-releasing hormone application to the GABAergic thalamocortical or hippocampal neurons resulted in depolarization, and increased excitability as well as increase the AP firing frequency and membrane resistance. This effect is mediated by a decrease in K^+ conductance [168, 310]. The mechanism of the facilitatory action of thyrotropin-releasing hormone is different from the mechanisms of conventional excitatory neurotransmitters or from shifts of membrane potential [928]. However, thyrotropin-releasing hormone exerts a classical excitatory influence, compatible with the effects of change in membrane potential to inhibitory GABAergic neurons and thus increases a direct inhibitory protection. However, such classical influences are distorted during interaction with the excitatory neurons, and the correspondence between excitability, membrane potential and spike amplitude testifies more to a complex non-linear effect of thyrotropin-releasing hormone than to compensatory processes. It produces transient inhibition and protracted excitation. Protection evoked by the thyrotropin-releasing hormone may be understood as a distortion of excitable membrane properties and reminds one of alterations that are observed during damage-related activation of the cyclic AMP system and other protective excitatory mechanisms. APs arise without depolarization during an increase in membrane resistance and, at the same time, amplitudes of spikes decrease. This ought to prevent the development of excitotoxicity. In addition, thyrotropin-releasing hormone triggers intracellular Ca^{2+} release, which opens Ca^{2+}-activated K channels, evokes transient hyperpolarization and does not directly modulate Ca^{2+} channel activity [339]. Nevertheless, while such alterations delay the harmful consequence of damage and extend the amount of time required to search for a new homeostatic equilibrium, these alterations do not make the tissue healthy.

2.6.4 Protection through a detoured route

The first acute reaction of a neuron to damage is usually an excitation. The "reaction to damage" by its meaning ought to be clearly defensive and an excitation, as its primary meaning, is a protective action. Of course, not every excitation in an organism has an adaptive character, and it may acquire pathological character. As we already have described, strong excitation leads to damage and further excitation. This excessive excitation is a symptom of cell damage and factors that powerfully excite neurons lead to damage, too. Nevertheless, activation of the cyclic AMP system induces excitation and generates a protective response. The cyclic AMP cascade is not the only example of protective activity that is accompanied by an excitatory influence. There are also a few other factors that protect brain through excitation. Some cytokines, for example, interleukin-1 [467], granulocytemacrophage colony stimulating factor [120], thyrotropin-releasing hormone [928, 310] and anesthetic fentanyl (which decreases injury despite it activates epileptoid activity) [862] also excite and protect neurons. Evidently, protective processes in the given cases are not reduced to counteraction between inhibition and excessive excitation. We may suppose that compensation of damage through augmentation of excitation is connected with a search for a new point of equilibrium that does not coincide with initial normal conditions. The general sense of acute excitatory reaction of an organism or its cells may be not only a defense, but may constitute an attack, too. The efforts that are directed to compensation of injury are targeted not only to the outer environment, but also to a homeostatic recovery of one of the possible stable points of the inner state. Therefore, one may say figuratively that protection through excitation is more reminiscent of aggression, while defense may be correlated with recovery of the initial state.

Evidently, the role of any neurotransmitter is not limited by its influence to membrane potential. Synaptic influences also induce alteration in concentration of intracellular Ca^{2+} and other second and retrograde messengers. As a result, they affect intimate properties of brain cells, damage, protection, homeostasis and proliferation. The possibility of so versatile a control of damage-protection using synaptic process suggests that such alterations of the neuronal state somehow depend on physiological activity of brain and may perform mental functions not connected with unequivocal pathology. These intimate properties of cells somehow participate in information processing.

Flexible dependence of damage from heterogeneous parameters (or, more exactly, the absence of a general rule for a correspondence between the state of damage and the specific magnitude of particular parameters) may be demonstrated by the example of the injurious influences of volatile anesthetics. The most voltage-gated ion channels, glutamate receptors and many of second messenger systems are relatively insensitive to volatile anesthetics in clinically relevant concentrations, however, general anesthetics potentiate the GABA_A and glycine receptors [394, 408, 1050], hyperpolarize neurons and decrease in excitability [52]. Therefore, anesthetics do not evoke damage, and, oppositely, protect cells. Nevertheless, some gaseous anesthetics halothane, desflurane and isoflurane in anesthetic doses cause cytotoxic effects, contribute to tissue injury and reduce antioxidant defense mechanisms in cells of various lines [693, 1147]. Besides, preconditioning with the volatile anesthetics, when they were administered before severe damage, reduces following neuronal injury caused by overstimulation of glutamate receptors and by ischemic events [126, 302, 130, 354, 6] that are typical for preliminary actions of deleterious agents (see further). Certainly, volatile anesthetics reduce gap junctions [109, 257, 1332] and this may be the reason for their deleterious affects. By the way, connexin hemichannels are blocked by hyperpolarization of the plasma membrane [260], although hyperpolarization itself protects neurons. Does this mean that gap junction conductivity is the most important of parameters? This is not true, since in other cases other parameters are vital. In addition, the state of gap junctions affects the spread of damage in tissue, rather than damage itself.

Our description demonstrates that any known particular characteristics of an injured cell such as augmentation of intracellular Ca^{2+}, increase in excitability, cell volume, free radical enhancement, etc. cannot be considered as the features of damage and the opposite changes cannot be considered as the sign of protection. Damage comes when the possibility for compensation is exhausted. Augmentation of the cyclic AMP system in tissue is the evidence of compensational processes participating in the modification of many other metabolic pathways: an increase in gap junctional coupling, O_2 uptake potentiation [716], gene expression involved in the regulation of Ca^{2+} homeostasis [44], etc. In particular, the cyclic AMP system regulates the basic instrument of ion homeostasis of cells, Na^+,K^+-ATPase. Homeostasis of the sodium pump can be enhanced [1143] or inhibited [691, 1364] by the cyclic AMP system and the effect may be different in different areas of brain [105]. Amplification of cyclic AMP does not routinely convert a sick cell into healthy one. This second messenger works intensively for repair. So, acute enhancement of cyclic AMP (20 min) by exposure of cyclic AMP agonist produced a sensitization of rat sensory neurons and a release of immunoreactive substance P, which (such as the cyclic AMP) produces protection [668, 866, 364] and excites neural cells. At the same time, long-term exposure of the agonist does not appear to downregulate its ability to augment substance P release, to increase cyclic AMP production or to be sensitized [147]. That is, despite chronic treatment, neurons are capable of undergoing a long-term sensitization that does not downregulate, but requires the continual presence of a sensitizing agent. Production of cyclic AMP is not evidently the result of damage or protection. This is an expendable substance needed for functioning of the compensatory process. Yet, it is necessary to remark that the cyclic AMP system is not unique in organization of protection. Other examples are some neurotransmitters, retrograde messengers and cytokines. The cyclic AMP system is rather the best investigated.

2.7 Nonlinear dependencies of doses, time and reciprocal interactions

When harmful factors directly affect a neural tissue, parameters of tissue change. In the first approximation, this change is in accordance with a linear law: the stronger the impact, the stronger the damage. Injury is determined by outer environment and usually one may see some damage characteristics immediately, for instance, an increase in excitability. Certainly, living systems are non-linear and effects of harmful factors may be accumulated as far as they are built up. For example, an injurious factor activates NMDA receptors and extracellular Ca^{2+} ions penetrate into a cell. A rise in intracellular Ca^{2+} can affect the second messenger inositol trisphosphate, switch the effect of Ca^{2+} from inhibitory to stimulatory [831], and in such a way cause damage [1208] and involve neighboring cells through gap junctions. Then characteristics of cell damage need an extended time in order to be exhibited as an impairment of physiological functions. Symptoms of damage may be displayed in a catastrophic fashion and their traces may appear suddenly or after delay. However, there is another reason for delay in expression of damage: development of compensational processes that counteract damage.

Tissue reacts to injury with a homeostatic compensation of damage. A state of neural tissue on the scale of damage-protection is determined by a mutual degree of damage and compensation, namely, by the strength of the harmful influence, by the power of protective compensation and by the amount of time elapsed after injury. When the injury is small, homeostasis can compensate damage, but if the damage is strong, it may provoke breakage of the compensatory mechanism itself, the damage becomes unmanageable and cells die. Compensation of severe damage may proceed through a long circuitous route, but even after a strong harmful impact, at the early stage of a desperate injury, protective mechanisms, nevertheless, are activated and the damage develops gradually. A qualitative distinction between mild and of severe injury is the inability of homeostasis to overcome damage, while very mild damage may be overcompensated. For example, small concentrations of poisons often display beneficial effects[8]. Opposing influences of low and high doses may be observed also at the level of behavior: low doses of cocaine and other narcotic drugs facilitate learning and memory performance, while high doses disrupt performance [1153]. Various factors interact with a tissue in a rather complex way, but some general properties may be described. Large and small doses of an agent or stimulus commonly evoke opposite effects (this was perceived in antiquity, in the times of Hippocrates and Paracelsus) [328, 1066]. Medications after all represent extremely reduced concentrations of poisons, while the substances that are obligatory for normal performance, such that their absence is hurtful, become cytotoxic at high concentration[9].

Nevertheless, an opposite is also held true, when physiologically relevant concentrations of some neuropeptides are inhibitory and protective, while lower doses have an opposite, excitatory (although not always harmful) effect. So, opiate receptors can couple with the stimulatory (G_s) or inhibitory G_i proteins [520, 569, 666], and ultra-low concentrations (1 pmol) increase the action

[8] This, in particular is seen in low doses of NO [991, 360], arachidonic acid [985], reactive oxygen species [79], Ca^{2+} [126], metabotropic glutamate receptor agonists [555], estradiol [1346], etc. Infectious proteins, prions, are not necessarily toxic; they can have a protective effect [519]. Similarly, a mild microglial response is able to protect, but when mechanisms of defense fail, microglia is involved in the expansion of brain damage [848, 1109].

[9] For example, although dopamine usually protects neurons, during severe but transient ischemia and other harmful treatments, dopamine levels are 300 500 times above baseline. Such a high dopamine concentration may exacerbate tissue damage. As well, Cl^- entry via GABA-receptors inhibits neurons and protects them [593], while accumulation of intracellular Cl^- decreases the pH and damages the cells [797]. Also, normally Ca^{2+} performs a variety of important functions, and absence of inositol-dependent Ca^{2+} elevation in inositol-receptor knock-out mice makes the animals especially sensitive to injury, but high intracellular Ca^{2+} injures cells [763, 526, 1000]. As well, K^+ may also play an ambivalent role in cell damage [836, 885, 733, 407]. Normally, the K^+ current hyperpolarizes neurons, decreases excitability and protects cells from damage. However, large or prolonged K^+ leakage leads to the loss of intracellular K^+, to decreases in membrane potential and to damage of neurons. On the other hand, a large increase in extracellular K^+ may also protect neurons.

potential duration and are thus excitatory. This involves not only opiates, but also cytokines, substance P and motivationally-relevant substances[10].

Excitatory actions of small doses of opiates and some other substances may also have a physiologically relevant effect, when compensatory regulation is directed against an over-protective state of cells. Thus, electrical responses of neurons, in many cases, also non-monotonously depend on the strength of the attack and many biologically active substances, affecting damage-protection processes, change the sign of their influence to neuronal activity, depending upon concentration. Exactly as a weak harm leads to development of compensation and inhibition, while a strong injury directs to damage, vulnerable neurons in the same neostriatal tissue respond to ischemia with a membrane depolarization and resistant neurons with a hyperpolarization [221]. Different influences of the week and strong stimuli may be connected with activation of different receptors, different metabolic pathways, or with a different thresholds for the emergence of detrimental and protective reactions. For example, NO through protein phosphorylation or S-nitrosylation can affect electrical activity of the neurons in an oppose direction, and the choice of the pathway depends on NO concentration [1439, 10, 360]. As a result, low NO concentrations are protective, while high concentrations injures cells. At the same time, the glutamate that usually intensely excites neurons sometimes decreases neuronal responses, when it activates specific group II III metabotropic receptors [1290]. Turning on different metabolic pathways, probably, decides the final fate of a cell. Thus, high neurotensin concentration decreases Na^+, K^+-ATPase activity [917] and may lead to a cell death, while a small concentration can activate homeostatic protection and prevent damage.

Phases of damage and protection after substance administration [104, 231, 905] or harmful stimulus presentation [797, 81] may be sequentially developed in time in the same cell. Phases of excitation and inhibition in the same neuron may also arise in succession after a shock and the effect is sometimes confined to the given neuron, without participation of the entire neural network.

---

[10] For example, at low concentrations, μ, δ or κ opioid peptides as well as morphine and other opioid alkaloids elicit dose-dependent excitatory prolongation of the calcium-dependent component of the action potential duration of sensory dorsal root ganglion neurons, whereas application of the same opioids at higher concentrations results in inhibitory shortening of the AP duration [1123]. Higher concentrations (nmol or micromol) decrease the AP duration, evoke antinociception and are, therefore, physiologically relevant. An overdose of opiates is lethal. Likewise, ultra-low doses of neurotensin facilitate nociception whereas greater doses are antinociceptive [179]. Similarly, neuropeptide Y [617, 621], substance P [1002, 1001], galanin [1338], and oxytocin [869] protect neurons on the whole at higher concentrations. In addition to previously mentioned substances, the neurons change the sign of their responses to cocaine [1153], galanin [1338], angiotensin II [1212], estradiol [1346] and cholecystokinin [294, 1193, 531], depending upon concentrations.

Therefore, sequential actions of two different factors on the state of cell damage-protection are, as a rule, also not additive. Administration of the first biologically active substance changes background state of the neuron at the time of administration of the second substance. And the same applies also to sequential actions of damage and substances or to injurious shocks [1091, 1221, 73, 774, 979, 1431]. Non-linearity of effects is connected with counteraction of the protection developed by a first factor with the damage, evoked by a second factor. Acute and chronic influences, too, give an example of compensatory processes depending on previous experience [889, 126, 127, 1104]. Mutual actions of different detrimental factors on a cell state, as well, are frequently not additive [917]. Nevertheless all these examples relate only to short-term experience after preliminary damage and do not exceed a few hours. It is unknown precisely whether residual traces of damage may be specific and kept for a long period. However, there are indirect indications to learning-related changes of homeostasis. Such a change in a neuron's state in time after a single stressful event shows that susceptibility of a neuron to injury is dependant on previous experience, at least on a short-term scale.

All these mechanisms seem to be included in the brain 'hardwire'. Another cause of singular dependencies of damage on concentrations of detrimental agents may be compensational homeostatic processes and this machinery may somehow participate in behavior plasticity.

2.8 Homeostasis as a resetting and reorganization

2.8.1 Homeostasis against death

Homeostasis is a key idea in biology. By mean of restorative, compensational mechanisms, homeostasis maintains stability of physiological systems and holds the parameters of an organism's internal milieu (or correspondence between these parameters) within limits that allow for survival [203]. We understand homeostasis as a life-supporting mechanism incorporated into a cell. The main property of homeostasis known to date is its existence in an organism and in any cell. The details of its functioning are destined to be elucidated in the future. We usually see real results of its efforts instead of homeostasis itself, but in some cases some elements of its performance become visible. Homeostasis is a material object allowing the explaining of brain/mind relations without attracting the specter of the "Homunculus".

In spite of the paramount importance of heredity for the existence of life, an individual being does not connect with heredity. If somebody has been already born, he will grow up, learn, eat, love, create and die. However, failure of homeostasis immediately breaks down life. For example, crash of the energy-dependent protein Na^+, K^+ pump, which normally removes Na^+ from a cell and transfers K^+ into the cell, is followed by a catastrophic loss of membrane potential and an uncontrolled influx of Ca^{2+} [158]. So, within only 2 minutes of stroke onset, neurons and glia undergo depolarization caused by failure of the Na^+, K^+ pump [36], while recovery to baseline Na^+ and K^+ levels, mediated by Na^+, K^+-ATPase, requires no more than tens of seconds. Loss of the Na^+, K^+ pump homeostasis is not some exceptional example of homeostasis importance. Death always comes, ultimately, through failure of some kind of homeostasis. In its turn, damage stimulates a protective homeostatic mechanism, in particular synthesis of molecules with protective properties [1198].

Homeostasis usually cares for each parameter in the bounds of its optimum, when this is possible. When this is unattainable for a given particular parameter, and in given circumstances, homeostasis tries to safe life by another route. It is forced to support a correspondence between vital parameters and does not maintain each of the parameters. Homeostasis reorganizes the manner of action in order to recover the injured function with an end-point goal to maintain life. How does the organism and its cells know that just this or the other specific combination of inner constants is optimal, must be maintained and should preserve life? If an organism directly controls function, instead of concrete optimal parameters, what criterion does it use?

Just as the very life of an organism depends on homeostasis, i.e. maintenance of an optimum performance, so a neuron's behavior depends on neuronal (i.e. ion, water, protein, etc.) homeostasis. A neuron, like a whole organism, preserves optimal levels of its variables in order to maintain functions of different levels of complexity: membrane potential, cell volume, pH of inner environment, etc. Whether homeostasis of an organism is based on homeostasis of its cells is not established definitively, but in the scope of our discussion, this idea looks plausible. In any case, considering the homeostasis of a whole organism, we come to a cell, particularly to neuron homeostasis.

As we have already indicated, there are two ways for homeostatic compensation to occur: upkeep of initial conditions and recovery of each parameter or partial reorganization of the system in order to recover a function [656]. Homeostasis regulates a number of physiological parameters and it can support any factor at a value that differs from its predetermined standard due to the compensatory changes of other parameters. Non-recovererable alteration in one parameter leads to the modification of a whole set of optimal values. At the same time, after severe damage homeostasis can recover a function at the cost of reorganization of the system itself. Transfer of homeostasis from maintenance of simple function to a more complex one is distinguished not only by a change of controlling parameters, but it acquires a new quality.

For example, damage leads to recuperation of the developmental plasticity; it often triggers an inflammatory response and produces cytokines. After nerve transection, regenerative nerve sprouts growing into all directions [1439]. Following injury, the central nervous system initiates a transient increase in growth-related proteins and generates new cells at the same time as injured cells produce processes [359]. Growth of new dendrites, axon branches and synapses after neural system injury is an example of the creation of new structures. For sure, in this case the neural system does not recover its preliminary existing structure. Therefore, protection against damage may proceed either by means of compensation of corrupted parameters or by the reorganization of neural courses of action. During the first way, the neural system tries to correct the parameters that deviated from optimal value because of regular performance or weak damage. The second way is implemented when optimal the value of a parameter is impossible to recover because of harsh injury or continuous action of the harmful factor. In this case, the previous optimum is not recovered; the neural system produces a new optimum and in this way compensates the detrimental function. Simultaneously, optimal levels of other functions, which initially were normal also change and after reorganization, the system maintains a new optimal set of parameters. The first and direct way of compensation is rapid and may participate in a current neural activity. Development of new or modification of previous function is, as a rule, a more protracted process and it may be essential only during long-term reorganization of behavior, as, for example, after stable change in environment, memory recovery after amnesia or recapture of paralyzed extremities after insult.

Homeostatic protection is easily observed at the cellular level and can be local. For instance, suppression activity in the single hippocampal neuron by overexpressing an inward-rectifier potassium channel results in an increase in synaptic input, which restores the activity to control level [185]. After a weak injury, endogenous protective mechanisms are stressed to compensate cellular damage and to recover the initial state. Weak deviation is directly recovered by homeostasis. For the example considered in Fig. 2.1, the neuron generates hyperpolarization in order to compensate excitotoxic damage. Such occurrences are ordinary in the neural system. Because excessive excitation usually causes cell damage, stopping of excitation typically protects neurons from neuronal death. Damage can be prevented by inhibitory neurotransmitters, by antagonists to various excitatory amino acids, by hyperpolarization, by K^+ channel stimulation and by Ca^{2+} or Na^+ channel blockers [295, 573, 1085, 585, 783, 671]. Several lines of evidence show an important link between cytoskeletal integrity and cell viability [583] and the importance of G protein signaling pathways [1333]. Homeostasis may be directed not anly against damage but against excessive protection, too. This is natural, since in the opposite case optimal parameters would be unstable. Therefore, over-protection is also compensated for by a homeostasis and in this case endogenous damage decreases over-protection. This is observed, in particular, during drug dependencies.

Besides a direct regulation of excitability, inhibition of electrical activity and any other kind of counteraction with excitotoxic influences, a neuron possesses even more diverse and indirect methods to avoid damage. A specific organization of neuronal parameters is directed to maintenance of the simplest neuronal functions, such as a membrane potential and a level of excitability, which depends on correspondence between Na^+ and K^+ ion concentrations and some other ions. However, neurons with similar membrane properties may have different proportions of conductances. As a result, widely different combinations of currents can produce similar firing properties and, consequently, the number of one kind of channels can slowly fluctuate depending on the numbers of each of the other kinds of channels in the neuron[11]. It is supposed, therefore, that the final level of excitability of a neuron is tightly controlled, rather than the number of each kind of ion channel individually [792]. At present, it is impossible to pick out which cellular mechanism is more essential for ion homeostasis that supports membrane potential and excitability, but one may conclude that a neuron has numerous means to preserve the ion equilibrium. Level of electrical activity of neurons is a more general parameter, than a level of excitability and it depends on the correspondence between synaptic strength and cellular excitability. Alterations of synaptic influx cause an analogous stabilization of neuronal activity[12]. Similarly, steady alterations of neuronal excitability lead to a corresponding tuning of synaptic influx, so that the resulting neuronal activity is unwavering. Change in potential-dependent channels can affect the properties of chemosensitive channels[13]. In

[11] For example, when cultured cortical neurons were incubated for several days with tetrodotoxin, blocking Na^+ channels, this prevented spike generations. However, subsequently, when the tetrodotoxin was washed out, the neurons were more excitable than before the tetrodotoxin treatment. An increased in Na^+ channel densities and a decrease in K^+ channel densities caused this enhancement of excitability [312]. An obvious example is adjustment of membrane potential and excitability through ion equilibrium regulation, particularly, through maintenance of the correspondence between inner and outer concentrations of K^+ and Na^+ by means of Na^+, K^+-ATPase [158]. Loss of intracellular K^+ during ordinary excitation or after injury may be recovered by the Na^+, K^+-pump. A mechanism of ion compensation also may be developed on a basis of the calcium-activated [334], voltage-independent K^+ channels [165] or ATP-dependent K^+ channels [684], which provide a connection between cellular metabolism and membrane potential. Depolarizing agents induce also intracellular acidification and attenuate neuronal excitability, while activity deprivation increases the amplitude of Na^+ current [312] and thus augments excitability. Compensation may also be based on the homeostatic balance between producing free radicals and free radical scavengers [337].

[12] Loss of $GABA_A$ receptors and following enhancement of excitability in the cerebellar granule cells leads to a change in the magnitude of a voltage-independent K^+ conductance that maintains normal neuronal behavior [165]. Another option for the development of compensation is the protective action of taurine release in response to an excess of excitatory amino acids [1085]. A balance between synaptic and intrinsic properties is also maintained, in part, via regulation of voltage-gated Na^+ currents, resulting in a stable neuronal input-output function [992].

[13] For example, the excitatory synaptic inputs increase and inhibitory inputs decrease when the neuron is deprived of activity [312]. Likewise, chronic blockage of Na^+ channels in rat visual cortex increases postsynaptic accumulation of AMPA receptors and amplifies the postsynaptic response of Na^+ channels [1337]. Some authors assign a steady maintenance of the level of neuronal activity as plasticity [1277]. Homeostatic forms of plasticity might provide the global negative feedback that is necessary to maintain synaptic strength and plasticity within

respect to 'baseline' levels of neuronal activity, homeostatic compensation recovers the previous level of this characteristic. Similar activity patterns can be produced by different underlying mechanisms [791]. However, in respect to particular characteristics such as excitability and chemosensitivity, a neural system does not recover a previous level of their state and compensate their deviation by means of regulation of a more general function: level of activity. At the same time, an individual neuron remains within a given level of excitability (or chemosensitivity), although values of a given conductance may be significantly altered [792]. Adequate network performance might entail a yet unknown universal sensor of network performance.

These data show that a neuron can maintain a stable level of its activity, although it is not evident that a neuron supports this level all the time. Really, a prolonged attack can change activity level on a steady basis. Potential-dependent channel density depends on energetic resources that are necessary for proper functioning of the channels. If excitable cells cannot continue to furnish the energy demands of active ion transporting systems, this leads to catastrophic membrane failure, causes a decrease in channel density and reorganizes the level of neuronal activity [158]. On the other hand, intensively of spike production can change the energetic balance of a cell. In such a way, it is possible to trace the hierarchy of functions. The quality of relatively simple functions may be evaluated by numerical parameters, millivolts for membrane potential, frequency of spikes for neuronal activity, ATP concentration for energetic resource, etc. However, the manner of evaluation of end-point function, cell survival is not clear. There are multiple pathways leading to neuronal death and even more numerous mechanisms protecting a cell from death, but any one of them is not obligatory. Damage is not just a distortion in membrane potential, intracellular Ca^{2+}, free radicals, cell volume, sodium pump, gap junctions or something else. A neuron may be injured if the homeostatic control of any parameter that participates in neuronal function is disturbed, but on the other hand, each of the parameters may decline from the homeo-

---

a functional dynamic range by scaling the strength of all synaptic inputs up or down [312, 398, 966]. Development of homeostatic plasticity demonstrates compensation of disturbed function, when recovering of previous optimal parameters of activity is impossible. This is an important example of coexistence of two forms of homeostatic compensation. It was recently found that a homeostatic form of plasticity tends to restore neuronal activity to 'baseline' levels. Depending on the level of activity, homeostatic plasticity modulates miniature excitatory and inhibitory postsynaptic currents, their excitability, the number of synaptic receptors and their location [1278].

static equilibrium without symptoms of damage of electrogenesis[14]. These all may proceed due to compensatory changes in other parameters [792].

Intracellular sensors detect activity levels and cause modification of ion channels. The change in channel density should be occurring on a time scale of minutes or hours rather than the milliseconds or seconds involved in neuronal signaling. Neurons that are constantly "self-tuning" may have quite different conductance densities at different times. The state of neurons may change rapidly, but background state of neurons is usually steady. At any time, different neurons are in different steady states. As the simple example, some neurons are active, while other neurons are silent. This means that certain neurons generate a rhythmic activity, certain others generate occasional APs and certain ones do not generate spike activity at all. Neurons can sometimes transform their activity from one type to another spontaneously or after direct action to the recorded neuron [1265, 1291, 352].

Processes of damage and protection defy theoretical descriptions. We may conclude that on the one hand there are a number of factors that are sustained at the optimal value. On the other hand, if it is necessary, when compensation of particular factor is impossible, an organism is competent to reorganize its state and activity so that it preserves its viability without restoration of initial optimal conditions. The precise modelling of such processes will meet enormous difficulties because of complexity of regularities, abundance of variables and uncertainty of initial conditions, even if we would be acquainted with all details of damage development. Nevertheless, the system is a subject to analysis, if general rules of regulation are clear. In our case, they are satisfactorily described by the example of regulation of interaction of excitability and damage. These rules we can express as activation of the protective mechanism at the early stage of damage, development of compensatory influences, inhibition of functioning during severe damage and development of a paradoxical phase, when a counteraction with a pathological excitation leads to function recuperation. Of course, this is a crude simplification and we still cannot explain how an organism searches for the effective way for survival.

2.8.2 Sensors

The normal state of a cell is not a relaxation. Just as a cell executes its physiological functions, its state deviates from a norm and recovers anew. For example, generation of each nervous impulse leads to a loss of a part of intracellular K^+ and the cell must spend energy in order for homeostasis to return

[14] For example, an excitotoxic mechanism cannot be the sole reason for cell death [221]. Neuronal damage can proceed without acidosis [1241, 1319], hypoxia may injure neurons in the absence of hypoxic depolarization [233], neuronal swelling regulation may be independent of the elevation of intracellular Ca^{2+} [540] and excitotoxicity [1062]. Damage to cells may occur without an increase in free radicals [337], and taurine increases neuron viability after oxidative stress without a depression of free radicals [146].

the ion equilibrium to the norm. During damage, these mechanisms continue to function. However, running of homeostasis in the norm may be considered, as running repairs when current working damage must be in a state of recovery. Homeostatic systems contain special proteins, sensors, providing information concerning deviation from the set point. A sensor to a given physical value is the part of the protein containing the sensitive chemical group to the given influence. In this way, homeostasis controls excitability of a neuron [296]. Neurons use the activity of sensors to regulate the number, kind and affinity of receptors and ion channels in their membrane. The balance of membrane proteins controls excitability in order to maintain a steady configuration of activity [792].

Each sensor is responsible for regulation of the given mother-value. For example, intracellular Ca^{2+} sensors detect a change in calcium level and trigger changes in the number, distribution or properties of calcium channels. The intracellular receptor for Ca^{2+} is calmodulin, which is a very sensitive sensor of Ca^{2+} concentration [40]. Similarly, other sensors regulate intracellular iron: the intracellular iron level affects the rate of synthesis of regulatory iron proteins [912]. Special sensors regulate energy-rich molecule concentrations, temperature, oxygen consumption, osmolarity, etc. The membranes of animal cells are highly permeable to water. Cell volume is an integral element within the cellular machinery regulating cellular performance [702]. Potential-dependent channels play roles of voltage sensors. Cells posses sensors for control of Na^+, K^+, Cl^-, Mg^{2+}, cyclic AMP, NMDA concentrations, intercellular signaling, glia, etc [296]. The whole system is as if controlled by a pool of tiny chiefs. Moreover, who is the superior, who regulates the makings, is it, indeed Homunculus? Evidently, cells have at their disposal the sensors for functions, not only for the variables, and there is a strict hierarchy of functions. Certainly, this problem deserves careful attention.

There is a working hypothesis that a molecular complementarity buffers the homeostatic system against changes [326]. Molecular complementarity may ensure the state, which is resistant to little deviations and thus provide the basis for homeostasis. Here homeostasis is like elastic forces and some similarities to elastic forces do exist. Homeostasis refers to the tendency of a system to return to a local free energy minimum after being shifted away from it. Homeostatic control is not something that depends on the number of their components and linkages. It is inherent in them. The example of homeostatic regulation is translocation of ions through a membrane, contraction of muscle, pH regulation, etc. So, an ionic pump is complementary to the membrane and it alters intracellular and extracellular concentration of the specific ion. If the pump were not embedded in the membrane, the pump would move the ions around within the cell, but not change its concentration. Binding of the pump and membrane is central to the production of a gradient vector [326].

Nevertheless, this mechanism does not disclose the mystery of homeostasis. Firstly, molecular complementarity may ensure stability only in the narrow vicinity of a set-point. Secondly, although the optimal value in these circumstances can be adjusted by supplementary molecular interactions, such regulation does not explain the wide spectrum of optima that is observed in a living system for the same factor at special conditions. Thirdly, it is not clearly, how molecular complementarities may explain the regulation of a set of the parameters and, last, it is not clear how a united regulation of parameters can guarantee the maintenance of special functions.

A more secure equilibrium can be ensured if at least two opposite forces control a set-point. The local minimum in such systems is separately regulated by each force and their mutual regulation allows for supporting equilibrium at the narrow interval of magnitudes. Of course, such a system is more complicated than a one-dimensional linear system. It may have several local maxima, singular points and its behavior is difficult to predict. An adjustment of such a system is in a fine dependence on a set of parameters. Above and beyond, there are number of interacting variable, that need to be adjusted. Nevertheless, homeostasis in living cells is an exceptionally robust system. The first significant example of homeostatic regulation by means of opposite forces ensures existence of excitability based on two potential-dependent channels, Na^+ and K^+. Hodjkin-Huxly equations, ascribing the equilibrium of ion currents, are very sensitive to alteration of any parameter, if other parameters are constant. Generation of two opposite inward and outward currents in response to depolarization, Na^+ influx and K^+ efflux, arranges nervous impulses, which are studied by amplitude and time for the reaction of excitation. Each type of ion channels supports a specific equilibrium potential and together they form a stable spike. Homeostatic compensation, as a rule, is bidirectional and is able to neutralize perturbations that either increase or decrease a corresponding parameter, say, excitability. Dual sensors ensure feed-forward and feedback control. The dominant mechanism connecting the outer and inner environments of cells is the G protein family. G proteins exert an effect on a large variety of cellular processes and the effect of some G proteins is stimulatory, while the effect of other G proteins is inhibitory [504, 520, 462, 200, 443]. G proteins can ensure a dual control, for example, control of the signaling system of cyclic AMP. So, adenosine receptors of different types are coupled to the G_s or G_i proteins and their activation decreases, or increases intracellular cyclic AMP levels [242]. This ensures proper support of the homeostatic equilibrium near an optimum.

Two, three or more factors usually control equilibrium of any intracellular parameter in a neuron. Various regulators slowly (days) control presynaptic transmitter release, the number of chemoreceptors and potential-dependent channels [296], etc. So, two regulatory proteins regulate metabotropic glutamate receptors and intracellular iron levels [912, 237]. Steady estimation of intracellular free Ca^{2+} concentration is reached by control of two conductances: leak conductance, connected with the threshold needed to fire and calcium-dependent potassium conductance, which modulates afterhyperpolarization [1135]. Several proteins regulate an oxygen sensor in brain neurons, including the Na^+, K^+ pump, ATP-sensitive K^+ channels, Ca^{2+}-dependent phospholipases and proteases and a heme-contained protein embodied within a multi-subunit assembly [158]. Around ten proteins regulate cyclic AMP metabolism in a cell [812]. Although intracellular factors are mutually connected, cellular homeostasis is so complex that a variety of local minima allow the alteration of some factors to compensate by means of a change in other factors. Equilibrium in the vicinity of a local minimum is resistant to small deviations. Particular examples of sensors in some cases have been investigated in fine details, but their consideration is out of the scope of our book.

Theoretically, all self-ordering systems must be internally coupled. The entropy production of systems at equilibrium is zero, but entropy production in a stationary state away from equilibrium has the minimum value [998]. Thus, stability at the cellular level can be achieved by a balance of opposing forces, such as is observed at the organismic level [113]. A similar mechanism is described for the heat shock response, thirst regulation, chemotaxis, etc. A jump in a chemotactic attractant concentration leads to a conformational change in the receptor that ultimately causes decreased tumbling and thus the chaotic component of single cell movement fails. An integrative sensor may be established through mutually dependent production of two signaling protein molecules, but the set point itself is under regulation, when other systems are reorganized [296]. Therefore, the separate regulation of individual characteristics does not explain a general control of cell activity.

By the way, not every property of a cell is under homeostatic regulation. For example, a set of actively expressed genes is out of homeostatic control. These features of a cell are neither steady nor homeostatically regulated. Cells of different breeds have a specific set of optimal characteristics: size, form, membrane potential, excitability, chemosensitivity and many others. Neurons also have properties that are the characteristics of their groups and these characteristics never change. If, say, the neuron is a pyramidal cell, it cannot be converted into in interneuron, and, equally, the GABAergic neuron cannot be converted into the cholinergic one, at least during the observable future. On the other hand, neurons displaying similar behavior may support these behaviors by means of different cellular means. That is, homeostasis in different cells of the same organism maintains a specific set of characteristics. We might suppose that the system of active genes not only determines the properties of a particular cell, but also tunes its homeostatic apparatus to support these particular properties, that is homeostasis itself is under the control of active genes. However, this is not obvious and, on the contrary, a homeostatic system may exist out of a genetic control. One may ask, how in such case to explain so fine and a specific tuning of different sets of parameters in different cells? It may be explained, as a neuron regulates an entire combination of optimal values, depending upon circumstances. Homeostasis, at the end, perhaps does not regulate the sets of specifics parameters, but first does regulate some other property that is universal and shared in different cells. Homeostasis somehow regulates a certain general quality of any cell. For example, a cell of every kind is alive and must stay alive. A neuron must somehow sense levels of damage itself and an acceptable degree of cellular damage may serve as a set-point for the homeostatic mechanism. The level of damage may serve as the mother-value of all values [321]. Then a neuron needs criteria for choice of the preferable action. Knowledge of membrane potential or excitability can, in the first approximation, serve for evaluation of cellular damage (see Fig. 2.3), although the phenomenon of cell damage is much more complicated. In any case, if such a general parameter as "damage" exists it could control a favorable combination of particular parameters. Moreover, this parameter (damage) might produce the general evaluation of quality of cell being, which factually is the same as elemental subjective appraisal.

2.8.3 A bit of injury is sometimes even beneficial

Compensation of injury is a much more complex process than impairment and requires resources for development of protection. Until compensation does arise, a cell can survive at the expense of its inner resources, but this time usually does not exceed a minute after severe damage. Protective mechanisms overcome injury or, at least, delay damage. Recovery is determined by inner factors and needs time. Correspondingly, a weak, manageable damage rouses protective mechanisms and may exert positive consequences. We have shortly considered this problem when discussing non-linear dependence of detrimental symptoms on strength of detrimental actions. Since olden times, physicians knew that a harmful impact might exert beneficial influence to organism. Short and abundant bleeding, stress, electroshock, etc. are long ago used as therapeutic treatments. The same phenomenon is observed at the cellular level. Organisms and all their cells undergo various deleterious tensions and must resist these influences in order to avoid damage and death. New injury is often superimposed to traces of damage and protection of previous injury.

Practically any harmful stimulus which is capable of causing injury to a tissue can induce tolerance against another harmful stimulus, if it is smaller and applied before the second stressor [328]. Thus, weak injury attenuates the impact of a following, more severe injury. At present, the beneficial roles of a harmful pretreatment have been demonstrated for a number of impacts and for various severe damages. Preconditioning by a weak or middle impairment makes an outcome of the next harsh damage less dangerous. Detrimental preconditioning may be the same or of a different type, as is the following severe insult. Weak or moderate harmful influences to auditory neurons [682], respiratory neurons [78] and cortical neurons with enhanced excitability [75] can, as an after-effect, protect against a subsequent severe impact of the same insult. Besides, an attack to one physiological system may also increase the viability of other systems. Preconditioning with electrical stimulation prevents the formation of secondary lesions after spinal cord injury [435], mild water deprivation of rats protects their retinas against damaging visible light [922] and chronic depolarization of many types of neurons enhances their survival after chemical stress [405]. Shortage of energy may also serve as a preconditioning and inhibit cell death evoked by a subsequent damage. For instance, actions of preliminary moderate hypoxia lead to significant reductions in structural damage to neurons after severe hypoxia [1054] and, similarly, a first sublethal ischemia increases the resistance of neurons to a subsequent severe ischemia [1221, 73]. Even pretreatment by reactive oxygen metabolites (the classical cause of cell damage, ageing and death) may serve as preliminary protection [222].

It is perplexing, but traumatic brain injury may ameliorate subsequent brain damage (authors hope that the readers will not implement this method for warding off consequences of car accidents). It was found that unilateral brain damage increases survival of hippocampal neurons in the injured hemisphere when compared with the contralateral hemisphere and sham-injured animals in the response to ischemic damage [924]. Likewise, damage to the ventromedial prefrontal cortex or the amygdala reduces post-traumatic stress disorder [649]. Brain damage can sensitize, or prepare, remaining neural tissue through growth of promoting and pruning-like mechanisms so that in the presence of appropriate behavioral pressure, the tissue can be drastically altered to compensate for lost function [1091]. Neuronal injury induces synthesis of molecules with protective properties including heat shock proteins, inhibitor of apoptosis proteins, uncoupling proteins, etc. [1198]. Short-term or moderate increase in intracellular Ca^{2+}, an important component of cell damage, is too a powerful tool of preconditioning. When 30 minutes before oxygen and glucose deprivation, intracellular Ca^{2+} in the hippocampal neurons was increased with selective Ca^{2+}ionophore, this greatly reduced cell death [126]. Intracellular Ca^{2+} increase is observed almost immediately after trauma and, possibly, can augment compensatory processes. Indeed, cells that survive the initial trauma subsequently (during a minute) display an enhanced capacity for calcium homeostasis and enhanced survival. suggesting that mechanical trauma can up-regulate calcium in glial cell populations [742].

One of the possible targets of protection by preliminary injury is cell coupling. Ischemic preconditioning may influence gap junction-mediated intercellular communication by activation of different kinases [449]. Gap junction uncoupling may reduces necrosis [448] and ischemic preconditioning decreases both damage and coupling [523, 853]. The closure of gap junctions plays a trigger role in the protection by preconditioning [853, 938], but acute uncoupling of gap junctions by means of down-regulation of connexin proteins prevents the beneficial effect of ischemic preconditioning [1099, 523]. Nevertheless, preconditioning might not involve cell-to-cell coupling at all. A significant reduction in infarct size was noted also in animals pre-treated by a mediator of inflammatory responses, bradykinin, prior to ischemia [979] and by nicotine exposure prior neuronal overexcitation [995]. As well, despite the evident potential for pro-inflammatory actions of agonists of adenosine receptor, preconditioning by these agonists protects against necrosis evoked by ischemia [529]. The mechanism of such defensive reaction is, supposedly, activation of specific K^+ channels, which partially compensate for the loss in membrane potential [683]. Likewise, volatile anesthetics may contribute to tissue injury, but they exert protective effects in various tissues when administered before ischemia or during early reperfusion [130, 354, 1431, 6]. Volatile anesthetics can also additionally enhance the protective effect exerted by ischemic preconditioning [302] and reduce neuronal injury caused by overstimulation of glutamate receptors [1431].

Cells defend themselves after preliminary damage through various endogenous protective mechanisms, Na^+, K^+-ATPase, second and retrograde messengers, inhibitory G_i proteins, etc.[15] These protective means are called up by the first non-fatal injury.

Pretreatment with retrograde messenger NO before ischemia can act as a powerful preconditioning mimetic (through subsequent production of cyclic GMP) [250], leads to dramatically decreased NMDA-evoked intracellular Ca^{2+} rises and was protective against NMDA-induced cell death, while, in contrast, NO and NMDA co-treatment caused a potentiation of Ca^{2+} rises and increased damage [1104]. Metabotropic glutamate receptor agonists [555], which are in large doses lethal, may in moderate doses induce immediate or long-term tolerance of ischemia or other stresses. Although normally lost of intracellular K^+ is injurious, continuous exposure of many types of neurons in cell culture to elevated concentrations of extracellular K^+ greatly enhances their survival [405] and this was supposed to be mediated by a cytoplasmic Ca^{2+} concentration penetrating in neurons through voltage-gated Ca^{2+} channels activated by K^+-induced depolarization. Interesting, the opposite also is true. When acute preconditioning by protective substances evokes overprotection, a homeostatic compensation counteracts with excessive protection, the condition of the cell shifts toward damage and the cell becomes vulnerable to injury [1257]. For example, pharmacological activation of metabotropic glutamate receptors by systemic injection of the agonist (15 min before the onset of anoxia) substantially protected retinas against anoxia-induced cell death [110]. In contrast, systemic injection of the metabotropic glutamate receptor antagonist (15 min before the onset of anoxia), significantly amplified cell death, that is, preconditioning by a protective action augments damage evoked by a following insult. This demonstrates that moderate damage and

[15] For example, sometimes ischemic preconditioning prevents impairment of Na^+, K^+-ATPase activity and this improves ion homeostasis [579]. Protection due to preconditioning may be based on arachidonic acid [137], on up-regulation of antioxidant enzyme activity (in the background) [59], on protein kinase C pathway [1431] or inositol trisphosphate receptor [1000]. Stimulation of numerous synaptic chemoreceptors, which are usually coupled to inhibitory G_i proteins, also has been targeted as a possible trigger for preconditioning: adenosine, adrenergic, muscarinic, bradykinin, opioid and angiotensin receptors [432]. Besides, overexpression of protective proteins increases resistance to damage. They bind and protect other proteins against damage, assist in dissolving aggregated proteins, in sequestering damaged proteins into aggregates and in targeting damaged proteins for degradation [60].

protection are symmetric processes, a phenomenon that is not true for severe damage and protection (see explanation in Fig. 2.3). Evidently, as a result of compensation, the influences that are usually harmful or are usually positive may change the direction of their action. Detrimental effects of protective action may be displayed also as a postconditioning[16]. These data demonstrate that tissue responds to insult by irregular in time modulation in the scale of damage-protection. Status of some tissue may be modified in a wave-like fashion. After injury there is protection, later protection is surpassed by damage and then again protection. The chemical influences that evoke or augment tissue damage exert protective effects when they are administrated in weak or moderate doses as a preconditioning before severe damage. Therefore, the phenomenon of preconditioning may be connected with the opposite effects of low and high doses of some substances. We described this phenomenon for the agonists and antagonists of glutamate receptors, Ca^{2+} potent substances, etc. Evidently, the protective effect of detrimental preconditioning is not specific and is connected rather with augmentation of general compensational possibilities of cells. This is corroborated by a cross-interaction of mild and severe damages of different categories and by an augmentation of background compensation. We had pointed out that preliminary moderate damage of one function might lead to following counteraction against the damage of another function. This is correct also for protection of one metabolic system by means of preliminary detrimental influence to another metabolic system. Existence of unspecific protection may be important for organization of compensation of the particular variable by a regulation of other variables, as this has been observed during homeostatic plasticity.

Now we may return to the problem of nonlinear dependencies of consequences of different influences on the strength of these influences. Let us consider a simple form of compensation, when homeostasis recovers distorted equilibrium and does not search for a detour. If harmful influence has a low intensity, a compensatory mechanism develops protection, compensates the

[16] For example, although preconditioning of hippocampal slices by NMDA can improve recovery following acute insults [1384], blockade of NMDA receptors has a protective effect when introduced immediately after brain injury. Nevertheless, activation of NMDA receptors at 24 h after brain injury enhances recovery of function, whereas NMDA receptor blockade is detrimental [127]. Similar results were obtained in the experiments with a rather different design [1032]. The risk of ischemic stroke is greater in diabetics. 24 hours after middle cerebral artery occlusion, the apoptosis in cells of the cortex and hippocampus located outside the primary zone was increased in diabetics rats (in primary zone necrosis). If after 2 hours of occlusion blood flow was restored, then in control rats the lesion was decreased, if it had been revealed after 24 hours. However, in the diabetic rats the lesion was exacerbated under the same conditions. The damage in penumbral areas revealed an extinguishment of spontaneous or evoked electrical activity but allowed for the maintenance of membrane potential and cellular ion homeostasis [1032].

damage and can even evoke overprotection. Damage comes into being as the result of external influences, while the mechanism of homeostatic compensation expands defense as the response to damage and compensatory protection is not an immediate reaction to injury. Therefore, damage-evoked compensation is backward damage. On the other hand, the magnitude of protection can exceed the level of the damage that has caused the compensation. This, at least, may be connected with the absence of immediate control of protection from the injury. Such a property may easy explain the protective effect of little doses of poisons. Excess of the damage-induced protection magnitude over magnitude of the damage may easily explain also the protective influence of a preliminary weak injury to following strong stress: overprotection after the first weak stress may defend against the next strong stress. Such logic is in agreement with the facts that have been under discussion. However, this approach cannot explain the phenomenon of death that may happen after a very strong attack. Really, if the magnitude of protection exceeds the value of damage, why should damage sometimes turns out to be irreversible? The response may be trivial: severe damage destroys the mechanism of compensation itself and the potency of homeostasis become insufficient for compensation.

Protective characteristics of preliminary damage are the general property of any mild attacks and they are observed during diverse circumstances, various preliminary impacts and in a variety of severe impacts. Therefore, the attribute of a given influence to operate, as a protective preconditioning, may serve as the criterion of deleterious nature of this influence. As we have pointed out, during development of damage produced by various ways, one can observe a change in different factors. Some factors are connected with cell injury, but others are the signs of a parallel enlargement of protection. For instance, damage is accompanied by activation of second messenger cyclic AMP, but this is the evidence of an augmentation of compensation. To our knowledge, nobody has implemented the augmentation of the cyclic AMP system as a preconditioning for amelioration of future damage. However, the phosphodiesterase inhibitor, resulting in an increase in cyclic GMP (Viagra), is the potent treatment for a preconditioning against ischemic injury [683]. At the same time, preconditionings with agonists to glutamate receptors, intracellular Ca^{2+} enhancement, elevated concentrations of extracellular K^+, bradykinin, agonists of adenosine receptors, some volatile anesthetics, arachidonic acid and NO are rather effective and evidently connected with the damage itself or with a spread of damage and not with compensatory processes.

Homeostatic compensation somehow triggers general protective mechanisms, which are not connected directly with the salient features attributed to the preliminary damage and are, frequently, non-specific. So, intermittent hypoxia facilitates the proliferation of neural stem cells in the subventricular zone and dentate gyrus [1435]. Gap junctions have a protective role in preconditioning against myocardial infarction [853] and a single injection of cocaine induced a time-dependent decrease in the number of an active form of the neuronal cell adhesion molecule in the dentate gyrus [774]. Excitotoxic preconditioning with electrical stimulation, too, prevents the formation of secondary lesions after spinal cord injury by means of activation of reactive astrocytes, attenuation of the edema and progressive necrosis [435]. We may conclude that moderate attacks of different kinds and, may be, any kind of attack may ameliorate or even prevent tissue damage from subsequent severe injury. Hence, injury itself, and not some mysterious attributes of particular damage, actuate a preliminary protection. On the other hand, preconditioning can counteract with any second severe attack, independently of its nature. Evidently, not only preliminary injury may be unspecific, but also its protective actions are unspecific. Consequently, preliminary damage unspecifically actuates unspecific protection, as this is observed during stress. Here is displayed an important characteristic of homeostatic compensation: compensation does not obligatorily recover just those parameters that were distorted.

Lastly, when we consider damage itself and not particular attributes of injury, we imply that these specific attributes really exist and may be somehow evaluated, as a general property, which may serve as a criterion of positive or negative states. Existence of unspecific homeostatic compensation indicates its relationship with an unspecific activator of actions that are not directed to a definite goal or agency, will, desire or something related. Therefore, when we study the phenomenon of detrimental preconditioning, we elucidate how cells regulate a hierarchy of setpoints.

2.8.4 Can homeostasis be perfected with experience?

Homeostasis, as is known, has predictive qualities and it can display anticipatory responses to an expected change in the environment in the future. Examples of such predictive quality of homeostasis are winter hibernation and seasonal migration. One important question is whether homeostasis can acquire or change such properties during individual experience. Development of damage in a tissue depends on the tissue prehistory, but this phenomenon is too different from learning. Moreover, the damage process does not need any sort of exercises or learning. The necessity to learn to die is absent. A compensatory system is changeable and dependent on experience, and we will demonstrate this further.

Protective preconditioning by means of a weak injury impels neurons to become more resistant to a subsequent damage and this resembles learning. Certainly, compensation creates a preliminary protection, which increases efficiency of subsequent compensation and this has adaptational meaning. Pretreatment is much more effective than post-treatment [130, 6, 1142] and this corresponds to the necessity of the previous presentation of a conditioned signal before an unconditional one. These attributes, obviously, are related to learning and cause the impression that the neural system learns to overcome damage. The question is: whether homeostatic compensation can be exercised, and implicit memory associated with the first damage can ameliorate protection against the next damage. Superficial observations seem to evince a positive answer to this question. Homeostasis can recover the specific optimal value or specific function, but subsequent augmentation of homeostasis seems to be unspecific.

May we consider that brain learns to compensate damage? If this holds true, do homeostatic compensational processes participate in neuronal plasticity? A positive response to this question is not automatic and may not be accepted automatically. However, protective preconditioning and learning are, at least, related phenomena. Nevertheless, the most important problem is as follows: whether compensation is perfected after reappearance, anticipates damage and beforehand develops protection against future damage. At present, only indirect data indicate that such properties of compensation are real. but this problem is very important for the discussion of the role of homeostatic compensation in behavior. Moreover, such hypothetical properties would be so effective for survival that they ought to have arisen in evolution, if the existence of anticipatory compensation is not connected with some theoretical prohibition.

As is known, learning is completed by a memorizing of information and nothing is stronger than a memory in the period of safe maintenance, resistance to extinction and to brain injury. However, this is not the case for the protective potency of a harmful preconditioning. Learning and protective preconditioning operate on different time scales. The time delay between the first middle damage and consequent strong damage cannot be too long. Injuries are factors that are most productive to promote healing when weak or middling hurt was delivered before or, rarer, immediately after severe damage. For instance, pretreatment with general anesthetics (barbiturates and propofol), used for i/v injections, offers a protective effect following ischemia, but no desired outcome has been reported by general anesthetic treatments after ischemic events [6]. A preliminary attack as short as 15 minutes prior to ischemia [979] or anoxia [110] substantially protects against cell death. Hypoxia 30 minutes prior to the ischemic injury also has a protective effect, as manifested in functional recovery (level of consciousness, motorsensory function, and simple behavioral tests) [459]. A moderate increase in the time interval between first and second damage does not prevent a protective effect of the first damage. So, hypoxic preconditioning 24 h prior to oxygenglucose deprivation (but not 4 h after) reduced both the necrotic and apoptotic cell death by an up-regulation of the background antioxidant enzyme activity [59]. Most stressors induce both rapid (within minutes) and delayed (over hours and days) protections [328]. Pretreatment with arachidonic acid leads to protection against a severe insult. After pretreatment with arachidonic acid, the best protective interval between the pretreatment and an ischemic or epileptic insult was 3 days and 1 additional day of delay prevents the tolerance [137]. In some cases, the preconditioning time may be as prolonged as several days or even two weeks [126, 1376, 560, 774]. This continual tolerance to damage relates to, for instance, the spreading depression elicited in a normal brain, which itself induces strong and long-lasting (couple of weeks) brain protection

[1376, 560]. A pretreatment with cocaine has an after-effect that lasted for at least 6 days [774] and Ca^{2+} preconditioning greatly reduced cell death during the following 7 days [126], but usually a protective period after pretreatment persists for a shorter time. However, even if we accept a two-week period as a time of preservation of the beneficial after-effect of harmful preconditioning, this cannot be compared with the strength of memory traces. And although sometimes a preliminary harmful influence may induce strong and relatively sustained tolerance to global or local cerebral ischemia, a protective pretreatment cannot become persistent. This, certainly, contradicts the suggestion that the homeostatic system can learn on a long-term basis.

Besides, independently of the preconditioning problem, processes of damage-protection are too slowly developed, in comparison with the prompt memorizing of events by explicit memory. Therefore, harmful preconditioning exerts only after-effect that cannot be considered as an adequate memory. Nevertheless, the possibility of preconditioning participation in a current behavior does not deserve to be rejected. In addition, protective preconditioning by a preliminary attack is not the only example of plasticity of homeostatic compensation.

Can we exclude or contend that the means related to homeostatic reorganization of function may be used by a neural system during learning-evoked modification of behavior, when a neural system learns to generate new activity? Homeostatic reshuffle of the system of optimal parameters side by side with protective preconditioning, is also related (at least, in outward appearance) to learning and serves, evidently, as a reason for negative results of experiments, when some important proteins were knocked-out or deleted. At the same time, the function that presumably should have been spoiled proved to be almost normal. For example, when numerous ion channels have been "knocked-out" or deleted, in many cases the phenotype of the knock-outs is less injured than would have been expected from pharmacological blockades of the same ion channel, chemoreceptor or kinase[17]. Thus, on the one hand, beneficial effects of harmful preconditioning are kept only several weeks, or shorter. But on the other hand, homeostasis can reorganize the set of parame-

[17] Many examples of mutant strains express virtually identical phenotypes as their wild-type [720].This is consistent with the interpretation of homeostatic plasticity that the absence of a gene for an ion channel can often be compensated for, as neurons self-tune to similar activity patterns with a different mix of ion channels [792]. Alteration of any parameter can be compensated for by regulations of other parameter regulations and therefore the knockout and normal neurons showed similar behavior, in contrast to acute pharmacological experiments [791]. Similar results were received also when animals with the genetically damaged cyclic AMP system had completely normal LTP and performed normally in the Morris-water maze and fear conditioning [1316], in spite of the important role of these metabolic pathways in normal animals. In the same way, in mutant mouse that were deficient in neuronally expressed connexins, a few abnormalities were found although electrical synapses might have a role in cognitive functions [1155].

ters on a permanent basis. This is closer to the period of memory self-recovery after brain trauma. These results remind us of the comparable results of experiments in search of memory localization, when extirpation of various brain areas did not lead to loss of memory function. Therefore it is possible that the phenomenon of memory recovery after brain damage is somehow related to a homeostatic compensation. However, this would signify that homeostasis sometimes receives access to memory stores.

Damage to brain cells is intimately connected with immunological reactions. Particularly, mediators of immunological response such as cytokineses, cyclic AMP, NO and arachidonic acid tightly participate in the processes of damage-protection. The same substances regulate amalgamation of cells into a colony. Does a bridge exist between immune reactions and homeostatic reactions and between homeostasis and behavioral reactions? Ancient immune defenses connected with these mediators do not connect with memory. Both protection and ancient defense from damage are non-specific. More perfect, immunological memory is based on immunoglobulins and we do not find a similarity between these two kinds of memory. However, the opioid system, participating in the perfect immune reaction, may be specific in respect to a form of damage [910]. For example, chronic treatment with the μ-opioid receptor antagonist naloxone leads to the supersensitivity to activate G protein by μ-opioid receptor agonists, together with an increase in μ-opioid receptors in membranes of the spinal cord [879]. It is not excluded that opiates and other neuropeptides may contribute and complement features acquired during life in compensatory responses. Consistent and almost incurable properties of drug-addiction demonstrate the stability of such alterations. Besides, drug addicts usually prefer their habitual treatment, although this preference may be transformed from feeble to harsh means.

Nevertheless, compensational processes, directed against both damage and overprotection, can modulate neuronal excitability and may participate in the plasticity of behavior on a long-term basis. An acute action of a factor usually has opposite results to its chronic action; such as, for example, substance P affects the excitability of neurons [804], angiotensin II acts on gap junction conductivity [317], hypo-osmolarity influences the size of neurons in the centers of thirst [1307] and noxious stimulation affects nociception [592, 840]. Homeostasis can selectively reorganize only one specific function that underwent detrimental influences. For instance, repeated striatal infusion of amphetamine resulted in an increase of excitability, in comparison with the reaction of control animals to acute amphetamine, but spontaneous firing did not distinguish between control and amphetamine-pretreated animals [450]. This demonstrates specific compensation-evoked change in excitability just within the response to substance producing sensitization.

Learning-related influences of sensory deprivation to behavior during development also indicate a connection between damage and behavior. So, ocular dominance plasticity, induced by monocular deprivation earlier in life, causes the adult visual cortex to be highly susceptible to subsequent monocular deprivation many weeks later [550]. Irrespective of whether the first monocular deprivation was experienced during the critical period (around postnatal day 28 in cats) or in adulthood, ocular dominance shifts induced by a second monocular deprivation were faster, more persistent and specific to repeated deprivation of the same eye. The capacity for plasticity in the mammalian cortex can therefore be conditioned by an experience [550]. Accelerating memory recovery after repetitive brain damage demonstrates like phenomenon. Long-term homeostasis-related changes of cell states are observed also during long-term potentiation. We will demonstrate that this process is connected with cell damage rather than with memory. However, importantly, steady traces of tetanic stimulation in some cases are specific in respect to an activated input.

Control of ion homeostasis itself has been shown to undergo behavior-related plastic changes such as habituation [1423], sensitization [1065] and conditioning [370]. Hence, homeostasis may be directed toward anticipatory compensation of the factors that lead to a disturbance of homeostasis, but distinctive features of this process need to be discovered.

2.9 Long-term potentiation as a form of cell damage

Long-term potentiation and related forms of synaptic plasticity, sensitization and postsynaptic depression are often used for investigation of learning and memory. The advantage of LTP over conventional forms of learning (habituation, classical and instrumental conditioning) is the possibility of inducing long-term and strong alterations in the neural system by means of a short treatment. Persistent synaptic enhancement is easily induced in several structures within the brain by a brief afferent tetanus or by an electric or chemical sensitizing stimulus. When the journal "Behavioral and Neural Biology" altered its title to "Neurobiology of Learning and Memory", it presented at once two reviews devoted to LTP [581, 793] in the first issue. This phenomena actually leaves traces in the neural system and have left a trace in the neurosciences. However, it is not clear whether it can serve as a model of even primitive learning. There is a certain skepticism in evaluating an accuracy of a parallel between LTP and memory, and evidence of LTP involvement in normal brain function is considered to be inconclusive [1255, 273, 632, 817, 785]. Damage also leaves traces, but a trace not necessarily represents the memory. There are many similarities between synaptic plasticity and excitotoxic damage, especially since they both are a consequence of strong excitation. On the other hand, an LTP is related to developmental plasticity. Investigations of LTP have given a basic system of proofs to relative synaptic plasticity as a key mechanism of memory. Somewhat parallel between the stressed memory and LTP does exist. This parallel is determined by their common dependence on hurtful impacts [660]. Let us demonstrate that LTP is a pathologic state of neural tissue and cannot be considered as an example of memory trace.

2.9.1 Is LTP something like an excitotoxicity?

Excessive excitation produces both excitotoxicity and LTP. An attack must generate postsynaptic firing. In other cases, LTP cannot arise. Initiating excitations may be of various kinds. Postsynaptic firing, which must accompany an initiating excitation, may be evoked by current injection, by extracellular stimulation of enough amplitude [1148], or by antidromic stimulation of soma [586, 414]. Besides a classical combination of presynaptic glutamatergic excitation and strong activation of postsynaptic membrane [54, 1148], long-lasting potentiation may be produced by intensification of the cyclic AMP pathway and consequent upregulation of AMPA receptors by phosphorylation [1159], while inhibition of the cyclic AMP pathway blocks long-term facilitation [77]. Enhancement of intracellular Na^+ is the first consequence of the excitatory current and postsynaptic Na^+ is known to induce cell damage [313] and LTP [894, 374]. Augmentation of inhibition or hindrance of excitation prevents LTP development and counteracts damage, while restriction of inhibition shifts cellular balance to excitation and thus promotes LTP and damage [632, 54, 1148]. For instance, augmentation of synaptic inhibition and membrane hyperpolarization exerts strong preventative action on cell LTP [54, 1148]. In the same way, after local blockage of inhibitory receptors in piriform cortex, LTP was increased [601]. Although LTP resembles the consequence of damage, this phenomenon is not similar to death of neurons. The origin of LTP corresponds to activation of compensatory protective processes. Nevertheless, potentiation of compensation may turn out to be insufficient and then neurons after LTP induction really die. So, slow-onset of potentiation in the hippocampus is directly associated with cell death in the CA1 region *in vivo* [785]. Long-term depression is induced by smaller and protracted excitations [746], which usually initiate a compensatory mechanism and produce cell protection. This completely corresponds to conditions of the origin of protective preconditioning by means of a preliminary weak or moderate attack. Prolonged harmful influence allows, probably, for the development of compensatory protective processes, which restrict neuronal excitation by homeostasis and overcompensation may lead to inhibition and protection. Long-term depression, like LTP, requires Ca^{2+} entry through the NMDA receptor and this suggests that long-term depression is a reversal of LTP, and vice versa [94].

2.9.2 Parallelism between damage-protection and LTP

There is almost complete parallelism between characteristics of neuronal damage and LTP. Both LTP and damage are induced by a strong excitation [746] and in both cases an intrinsic excitability is reorganized together with a synaptic efficacy [135]. Both necrotic damage and the LTP are persistent, but reversible phenomena [290]. Long-term potentiation is presented in the hippocampus, cortex and cerebellum [291], areas the most sensitive to damage. After a strong impact LTP is developed slowly [1432, 124], as the necrotic damage is developed, in the cases where compensation counteracts the damage. Growth of dendritic branches and synaptic terminals is also observed in both cases [1195].

Various pathological conditions, including hypoxia and ischemia, leading to cell damage also establish the LTP or long-term depression [1369, 1429]. Glutamatergic synaptic transmission in hippocampal CA1 areas is facilitated after traumatic brain injury [205]. The same impact (halothane anesthesia) producing severe neuronal damage can simultaneously induce in the same brain area (CA1 region of hippocampus) a synaptic potentiation and synaptic depression of nearby neurons [1369].

Damage and LTP are provoked by a similar means. Besides the chief role of strong excitation, they can be produced by the same pharmacological treatments. The substances that exacerbate neuronal damage, as a rule, also increase susceptibility to LTP and on the other hand, the treatments that protect neurons, such as glutamate receptor antagonists and $GABA_A$ agonists [1367] retard the onset of LTP [581, 793]. Facilitation, such as damage, is greatly reduced by postsynaptic injection of a rapid Ca^{2+} chelator or by postsynaptic hyperpolarization during tetanic stimulation [85]. Dopamine-containing neurons are very stable [1380] and accordingly, the same tetanus in the presence of dopamine leads to long-term depression, but not to more LTP in cortex slices. The metabolic pathways responsible for LTP and neuronal damage-protection are almost completely coincidental. Long-term potentiation, as damage, depends on Ca^{2+} influx [107, 1316] and on activation of NMDA [581, 793, 107, 751, 1316], metabotropic [555, 1290] and AMPA [731, 966] glutamate receptors. At the same time, application of an NMDA receptor antagonist blocks induction of LTP [632] and prevents cell injury. Moreover, early phases of both LTP [581, 374] and neurotoxicity [993] depend on NMDA, while their late stages depend on AMPA receptors. Intracellular acidosis, being protracted in time, is a harmful influence and leads to an irreversible depression of synaptic responses in the hippocampus, but if acidosis was shorter by 30 minutes, depression would become reversible [1366], as is observed at the early stages of necrotic damage. LTP and damage induce also a similar hormonal response and increase the synthesis of brain-derived neurotrophic factor and nerve growth factor.

Correspondingly, $GABA_A$ activation results in a decrease in LTP [593, 54, 756] and neuronal damage, at the same time as attenuation of inhibitions by $GABA_A$ receptor antagonists facilitates LTP induction [632, 581, 793, 1383] and exacerbates damage. Sometimes tetanic stimulation failed to elicit LTP unless a $GABA_A$ receptor antagonist was applied to the slices [1007, 632]. The maintenance of LTP in the hippocampus is accompanied by impairment of $GABA_A$ receptor function [1177].

Enhancement of levels of second messengers, intracellular Ca^{2+} and cyclic AMP, is also an important attribute of LTP [606, 713, 290, 1159] and damage. Both postsynaptic intracellular [581, 793, 925, 1367] and presynaptic [894] Ca^{2+} enhancement is an important condition for LTP and neuronal damage.

Retrograde messengers, the endogenous cannabinoids, the gases NO and CO, as well as arachidonic acid participate in development of both the LTP and neuronal injury [111, 925, 23]. Hippocampal and amygdala neuronal damage and LTP similarly depend on cytokines [659, 948]. Intracellular processes activated within Aplysia sensory neurons by injury, and those activated during long-term behavioral sensitization, also overlap significantly [1195]. Cell death [623] and LTP [39] may depend on protein kinase C translocation towards the membrane.

Systems of retrograde and second messengers are involved in processes of damage-protection and LTP in a similar way. Particularly, as a consequence of repeated stimulation, retrograde messenger NO modulates cellular function that leads to LTP, and NO effects depend on excitation power in the same way that excitotoxic damage depends on excitation power [991]. Similarly, activation of the second messenger cyclic AMP participate in the LTP induction similarly in fine details [811], as is observed during cell injury. High postsynaptic Ca^{2+} levels produce LTP and moderate levels induce LTD [124] and at the same time, a high level of intracellular Ca^{2+} is lethal and a moderate level is beneficial and induces tolerance to consequent damage [126]. Damage as well as LTP [581, 811, 374, 1367] depend on protein synthesis. Nevertheless, chemical similarity between cellular processes does not mean, generally speaking, a correspondence between the mechanisms of functioning. If, say, proteins participate in both a muscle contraction and thermoregulation, mechanisms of these forms of activity are not obliged to be similar. To say that something depends on Ca^{2+}, K^{+}, G proteins, cyclic AMP, ATP, and so on means to say almost nothing, because it is difficult to find some process in the living tissue without their participation. Of course, coincidence in multiple points is conceivable and participation of the same specific proteins in damage and LTP in a similar way is essential and this includes, for instance nitric oxide synthase, adenylate cyclase, protein kinases A, protein kinase C, phosphatidyl inositol kinase or haem oxygenase-2. A whole complex of biochemical and physiological processes that proceed during injury and during LTP is almost coincidental. Correspondingly, the main basis of proofs for a role of persistent synaptic plasticity in memory is defeated.

There are similarities between stressful conditions and mechanisms that occur in long-term potentiation. In both cases, neurons become hyperexcitable and the cyclic AMP system is activated [280], while Na^{+}, K^{+}-ATPase activity decreases [1365]. Synaptic change resembling LTP has been observed in the thalamus-amygdala pathway after naturally occurring fear conditioning [1035]. One-trial inhibitory avoidance produces the same changes in hippocampal glutamate receptors as induction of LTP and is associated with the delivery of AMPA receptors to synapses [1335], but unlike normal memory, the changes in AMPA receptors are not detectable for longer than 1 hour after training. Besides, chronic exposure to numerous types of drugs of abuse induces a long-term potentiation-like state in ventral tegmental area dopamine neurons, mediated via increases in AMPA glutamate receptor responsiveness and cyclic AMP response element binding protein [887]. Damage-protection and LTP demonstrate also many other common features. For instance, during induction of LTP, correspondence between neuronal excitability, input resistance and membrane potential is disturbed [374], as is usually observed during cell damage. As in the case of damage, LTP does not always depend on the presence of all its features for normal development and may proceed, for example, without participation of NMDA receptors and postsynaptic Ca^{2+}.

Signaling between astrocytes and neurons also participates in LTP [27] and damage of neural tissue. So, tetanic stimulation of corticothalamic fibers caused a long-lasting reduction in electrical coupling strength. Metabotropic glutamate receptor agonist reduces electrical coupling to a degree comparable with the effect of tetanus. Thus, metabotropic glutamate receptor may play a role in regulating the spatial and temporal coordination of inhibition to the dorsal thalamus [701]. LTP in one cell could decrease spike latency in a nearby cell by pairing low-frequency stimulation with postsynaptic depolarization. The enhancement in the neighboring cell, resulting from action of a diffusible messenger, may be blocked by loading the cell with a Ca^{2+} chelator and hyperpolarizing the cell during the time when LTP has being induced in the first cell [94]. Properties of sensitization in Aplysia also have symptoms of damage. Sensorimotor synapses in Aplysia, the basic object for investigation of sensitization, are glutamatergic [47]. Both cellular damage and synaptic facilitation in Aplysia depend on release of Ca^{2+} from postsynaptic intracellular stores, postsynaptic exocytosis, and modulation of postsynaptic AMPA receptor efficacy [731]. Growth of processes during sensitization and dependence upon the cyclic AMP system [77, 811] are also indicators of early stage of protection after damage.

However, the role of some biochemical pathways needs clarification. For example, participation of nitric oxide in LTP, in memory formation and in cell damage is not completely clear. In addition, strong hypoxia or ischemia-evoked LTP may be more sensitive to a blockade of NMDA or metabotropic glutamate receptors, than long-term depression, but non-symmetry between long-term potentiation and depression corresponds to a non-symmetry between strong damage and protection, since strong damage lead to inhibition of homeostasis, particularly to inhibition of Na^+, K^+ -pump activity [1022]. The role of protein phosphorylation also may be different in the induction of long-term potentiation and depression [632].This is determined by a parallel development of damage and compensation. Besides, homeostasis is responsible for general cell survival and is not obliged to support each particular parameter in a cell at the stable level.

2.9.3 Development and LTP

During early development, neurons exhibit intensive growth. Developmental overproduction has been observed in many parts of the central nervous system. Some of the neurons at this stage die through apoptosis. Parts of neurons are apparently more sensitive to damage, especially in the hippocampus. The injured adult central nervous system exhibits features that are also observed during development: the same properties that central neurons display during regeneration following damage to peripheral nervous system [858, 359]. This growth is accompanied by increased excitability. High threshold neurons in the leech respond by sprouting at the lesion site in the nerve root and by sprouting additional processes from the axon hillock region, i.e. from the region of highest excitability.

Damage induces neuronal alterations, which are very similar to the properties of sensitized neurons and in the artificially sensitized neurons, LTP is easier to induce. For example, it is difficult to induce LTP in chronic preparations of the neocortex. However, LTP can be reliably induced in neocortical slices or in acute preparations [1007]. The damage is magnified under in vitro conditions and this may be the reason for slice susceptibility to LTP. As well, the treatments which induce abnormally high excitability in mammals and invertebrates evoke growth of axon branches and synaptic connections. Strong neuronal activation leads to synthesis of vesicular and other synaptic components. Neural tissue displays also developmental-like properties during LTP [1255]. Both developmental plasticity and LTP require the participation of postsynaptic targets [476]. It is directly established that LTP of excitatory synaptic transmission can regulate the rate of neurogenesis in the adult rat dentate gyrus *in vivo* [171]. Therefore, LTP and sensitization may be the consequences of developmental plasticity induced by transient or persistent excitotoxic damage.

Long-term potentiation and depression are easier to induce in young animals and embryonic cultures [1432, 1159], during developmental instability of the neuronal structure. Crair and Malenka [273] have examined LTP in the thalamocortical synapses that form whisker barrels in the somatosensory cortex. The period during which LTP can be induced closely matches the critical period during which the barrels can be modified by sensory perturbations. Susceptibility to LTP in the visual cortex also coincides with the critical period of naturally occurring experience-dependent synaptic modification and, like the critical period, susceptibility to LTP can be prolonged by rearing animals in the dark [632]. Therefore, properties of LTP correspond to a normal mechanism of experience-dependent synaptic modification in the developing mammalian brain. Such a critical period of neuronal growth and death depends on the environment and this therefore reminds the memory phenomenon, but does not coincide with it, at least with the aware, declarative memory which is frequently identified with the LTP [20, 845, 86]. Yet, these findings support the hypothesis that LTP reflects a mechanism of experience-dependent synaptic modification in the developing or injured neural tissue, but there is too large a distance between the developmental plasticity and normal learning.

2.9.4 Temporal scopes of damage, LTP and learning

Long-term memory is remotely similar to LTP and the important likeness is that both phenomena are protracted and may exceed 24 hours [660]. Longevity of long-term memory may be almost infinite, although this is known only from behavioral experiments. Nevertheless, learning does not need a large time in order to remember new knowledge. Information is rapidly recorded in a memory and is stored for a long time. Declarative memory can be recollected immediately after acquiring new information. Procedural memory needs repetition of learning procedures, but this is necessary only in order to reveal regularities in the environment. Brain does not spend much time in remembering; this will be evident if one takes into account probing movements during an instrumental conditioning, when an animal examines a working hypothesis relative to the method of the achievement of a useful result. Long-term memory requires hours for consolidation, but this phenomenon is related to preservation of memory and its resistance to deleterious influences. Rapidity of remembering for the motor memory is also rather high. Of course, memorizing cannot be instantaneous, but the time needed does not exceed seconds. As we already have indicated, development of necrotic damage takes minutes or even tens of minutes. Compensatory processes can extend this time for hours and days, until healing or death will occur. Certainly, the first excitation in a response to injury may arise rapidly, but it is still a reaction of healthy tissue. So, anoxia-induced depolarization started at about 100 s, but the presence of a little O_2 (5%) lead to significant delays in depolarization responses [688]. Synaptic activity regulates the surface distribution of neurotransmitter receptors for glycine, glutamate and GABA_A and these processes are rather extended (minutes days) [633]. Advance of morphological symptoms of necrotic damage is even more protracted. During a later phase of LTP, the newly synthesized proteins have to be selectively transported to activated synapses and this takes many hours or days and evokes growth of cells [1316]. Such a time-consuming route cannot quantitatively account for fast memory formation, but it is in accord with the development of cellular protection and slow change in states of neural system, such as motivational states, cycle of sleep-awareness, etc. As for apoptotic damage, it develops too slowly for considering its role in the current behavior.

Origin of long-term potentiation and depression need a short and strong clash. However, exhibition of potentiation requires minutes or tens of minutes, before potentiation or depression reaches a steady level [586, 15, 414, 1432, 124, 374, 1367]. These time scopes are closer to organization of compensation during advance of damage than they are to formation of memory. So, ripening of LTP in pyramidal cells of hippocampus takes more than 10 and sometimes 20-30 minutes after induction, in order to be displayed [414, 374]. After potentiation, neurons in the cerebellar deep nuclei develop LTP gradually, during 10-20 minutes [15]. Some forms of potentiation develop even more slowly. Pairing low-frequency orthodromic stimulation with high-frequency antidromic conditioning of pyramidal cells in area CAI of the rat hippocampus resulted in long-lasting potentiation, but the amplitude of such potentiation took up to 60 minutes to reach its peak, much longer than standard synaptic LTP [586]. Moreover, after stimulating hippocampal CA1 pyramidal neurons with synaptic inputs correlating with postsynaptic neuronal spikes, evoked by current, the threshold decreases to minimum during 100-120 minutes [1367]. Induction of long-term depression is usually slower than induction of LTP. Thus, temporal scopes for establishment of memory and LTP are different, but LTP and response to injury are developed approximately with equal rate. As for the time required for information storage in memory, there are distinctions from LTP. The early phase of LTP is not long-term. Phosphorylation-dependent modification of synaptic potentiation is capable of supporting LTP only for 13 h. Further LTP is expired spontaneously. This extinction cannot be identified with a short-term memory, which is not terminated spontaneously, but is only susceptible to outer influences. Therefore, development in time memory and LTP contradict the hypothesis about their tight relationship, while temporal similarity between formation of LTP and development of damage and protection evidences their common mechanism.

2.9.5 Depotentiation and protection

We have already mentioned that post-tetanic change in synaptic efficacy and excitability may be reversed at the early phase of potentiation and depression. For example, the intracellular application of an inhibitor of the inositol pathway eliminated the LTP in pyramidal cells 30 minutes after paired pre- and postsynaptic activation [100]. The forces ensuring de-potentiation and de-depression do not create a new point of equilibrium depending on their mutual power: properties of neurons are returned to an initial state. In particular, a depression can erase potentiation [1022]. Inhibitors of the enzyme haem oxygenase-2, which catalyses the production of carbon monoxide, prevent the induction of LTP in CA1 pyramidal cells. Furthermore, they can erase LTP that is already established and the percentage decrease of response size closely equaled the percentage increase produced earlier by tetanus [1181]. This corresponds to a homeostatic return to the norm.

Potentiation is energy and substance consuming process and it does not have a quit status, suitable for the parsimonious storage of experience. Thus, blockage of CO production can erase LTP and for LTP exhibition, it is necessary continuously to support production of CO [1181]. Likewise, rat sensory neurons are capable of undergoing a long-term sensitization that does not down-regulate, but requires the continual presence of a sensitizing agent, specifically, cyclic AMP and therefore the ability of sensory neurons to be sensitized has to be maintained [147]. Augmentation of the cyclic AMP system means intensification of the compensational capability of cells. At the same time, the treatments that generate LTP or long-term depression may be short and their presence is not necessary for maintenance of alterations, which already have arisen. This concerns activity of NMDA and metabotropic glutamate receptors. Posttetanic long-term alterations of responses are insensitive to blockade of glutamate receptors [1022]. On the contrary, activation of NMDA and metabotropic glutamate receptors has been shown to be required for depotentiation and these general features of depotentiation were found in animals of all ages [1432]. This supports the notion that LTP is the trial of the neural system to intensify protection and recover detrimental function, particularly by means of the cyclic AMP system. Contrarily, it is impossible to conceive of the memory system needed for energetic replenishment.

2.9.6 Preconditioning of LTP and compensation of damage

We have demonstrated that weak injurious preconditioning ameliorates consequences of later more severe injury, but severe preconditioning augments damage. Evidently, a weak preliminary attack augments a protection, while a strong preliminary attack leaves mainly traces of damage.

Correspondingly, it would be unsurprising if harmful preconditioning would affect the subsequent LTP. If we consider an LTP, as a protective reaction of neural tissue, it would be natural to expect that weak preconditioning would augment LTP. Really, a brief acute swim stress experience enhances LTP [660]. As well, a combination of short episodes (2 min) of hypoxia with tetanus produces LTP significantly larger than potentiation alone and, consequently, potentiation induced by a weak hypoxia is additive with tetanus-LTP [771]. Obviously, here the compensations are added and not the damages themselves. Experiments with the stronger preconditioning (with high frequency stimulation that induced LTP) have demonstrated a reduction of the next LTP [474]. Correspondingly, low acute doses of cocaine augmented the induction (10 min before LTP induction) of LTP at the excitatory synapses of the hippocampus, but once LTP had been established, cocaine had no effect on the potentiated response. At the same time, high doses of cocaine block LTP induction [1153].

Preconditioning of LTP with any detrimental preconditioning has a definite time window, requiring over 2 minutes to develop, being very effective at 1020 minutes post preconditioning, and then ceasing to be operative at 45 minutes after the preconditioning stimulation [474]; however, precise parameters of time depend on conditions of experiments. This delay approximately corresponds to the time window for beneficial effects of weak harmful preconditioning.

The protective results of strong preconditioning sometimes were ambiguous. Severe preconditioning of hippocampal slices by NMDA (chemical insult) improves recovery following acute insults and increased cell survival. but has deleterious effects on LTP [1384]. An effect of strong preconditioning may exert a non-linear effect to LTP. Severe attack may induce stronger protection and this will exhaust the cell possibility for compensation. As a result, tetanization cannot increase this compensation additionally. In this case, cells will survive better, but LTP will not correspond to the real value of compensation. In other cases, severe preconditioning may hurt tissue, impair the protective mechanism itself and, hence, decrease LTP. Lastly, tissue may fall into a paradoxical phase, when additional inhibitory protection may animate and excite a neuron. For example, protracted intermittent hypoxia (during 3 days) decreased the LTP in rat hippocampal slices. The diminishment was biphasic, i.e. more pronounced after a 3-day intermittent hypoxia than a 7-day intermittent hypoxia. Influence of cocaine withdrawal to the magnitude of LTP also may be dependent on longevity of the withdrawal period. After 3-day cocaine withdrawal the magnitude of LTP was increased. A positive correlation exists among the 3-day cocaine withdrawal group between the amounts of cocaine ingested vs. the magnitude of the LTP observed in slices obtained from these rats. After the 30-day cocaine withdrawal. LTP magnitudes were similar to LTP magnitudes after saline administration and after the 100-day cocaine withdrawal. LTP decreased [1226]. Memory traces do not allow such an easy manipulation with their intensity. Typically, a trace of memory is either absent or present forever.

2.9.7 Specificity of LTP

Is distinctiveness of damage specific in respect to a specific insult? Naturally, this question has sense only for an early stage of damage, when neurons still stay alive. Apoptotic damage, certainly, cannot be specific, since it is under central control from the nucleus. Grow of necrotic tissue certainly can be local and specificity may be determined by the place of damage or by the manner of hurtful action, particularly by the variables that have been distorted. Pathways to death also may be different after different destructive impacts. Thus, anoxia and glucose depletion induce a notable acidosis in rat dorsal vagal neurons while metabolic arrest does not affect intracellular pH [1015]. It is unclear if such specificity is essential and has an informational component. However, the most fundamental problem is existence (or absence) of specificity of compensation or homeostasis and, consequently, protection. Compensation is, evidently, directed to recovery of loose function and its contribution to a current neuronal behavior is almost completely predetermined. However, if compensation could interact with the reasons of damage, that is, to be forestalled in respect to proposed damage, its efficacy would be strongly augmented.

The example of specific damage-protection events provides LTP or long-term depression. The induction of LTP in the hippocampus and visual cortex may concern only those synapses that were activated during conditional stimulation [1235, 632, 923, 414, 290], although such a specificity is not ideal and modifications in one input may concern changes in other inputs. It is possible to receive different LTPs of two independent Schaffer collateral pathways converging to the same pyramidal cell in the hippocampus [923]. Activity-induced synaptic modifications in neurons of hippocampus are often input-specific, i.e, only the synapses experiencing repetitive synaptic activation are modified. However, LTP induced at one Schaffer collateral input to the pyramidal cell by high-frequency stimulation could spread to other synapses on the same pyramidal cell when the unstimulated inputs were within 70 μm from the site of LTP induction. To be exact, specificity is limited by the distance and potentiation/depression is spread to adjacent synapses [290]. Specific changes in potentiation and depression only for part of synaptic input has, usually a simple reason: this is determined by a spatial remoteness of the groups of synapses. So, different synaptic groups may be located at different dendritic branches of hippocampal neurons [414]. It is supposed that there are larger changes in dendritic channels of excitable membrane than in somatic channels [1367]. Long-term depression also can be specific. So, conjunctive irritation of Purkinje cells by pulses of metabotropic receptor agonist and direct cell depolarization induces LTD and input specificity is retained in this reduced system [744]. It is possible even to evoke LTP in specific synapses of one neuron and long-term depression in other specific synapses of the same neuron [1235, 632]. The depotentiation also can be input-specific or may not be [1432].

In all these examples, what is principally important is an availability of specificity, even if not in all cases. Absence of specificity is a trivial result. After all, even learning is unspecific at the beginning of training (phase of generalization). The generation of long-term potentiation and depression depends upon associative interactions between synapses that converge on individual dendrites. The distance over which these associative interactions occur and consequently, the degree of specificity is limited by synaptic GABAergic inhibition [1235]. $GABA_A$ receptors establish a shunt between inner and outer environments and decrease the interaction of remote synapses and their correlation. Hyperpolarization, similarly to synaptic inhibition, also prevents LTP induction [1148]. When during LTP, the probability of firing of a postsynaptic neuron to a given EPSP is enhanced and, hence, excitability increases, this effect is partially determined by $GABA_A$ receptors (60% of this effect). 40% of the effect does not depend on $GABA_A$, but depends on NMDA and Ca^{2+} and this component is input specific; that is, modification of excitability could be restricted to a local dendritic area. Similarly, synaptic depression is NMDA and Ca^{2+} dependent, partially (40%) $GABA_A$ independent and this part is input specific [291, 414]. These effects are evidently determined by the protective action of $GABA_A$ receptors and the detrimental actions of NMDA receptors and Ca^{2+} increase. This rather reminds one of local damage of potentiated dendrites. Specificity of LTP is also well displayed in the example of change in intrinsic excitability, which is induced by relatively weaker inputs than potentiation of synaptic efficacy [54, 414, 1367]. Such a change in excitability during LTP may be selective, as was found in the hippocampus, cerebellum and neocortex [586, 291, 414, 374]. Particularly, such specificity can be high and voltage threshold of spikes arising from the potentiated EPSP can be lowered up to 2 mV, compared with the voltage threshold of spontaneous APs, which remained stable [374]. Evidently, during LTP, specificity of effects is relatively low and confined only by a spatial localization of the effects. Such methods of coding already have been rejected when we considered possible mechanisms of memory in Chapter 1. If the specificity of homeostatic protection is also restricted by a spatial location of protective events, homeostasis cannot anticipate salient features of damage and serve as an instrument of forestalling protection.

Selective changes of excitability during learning, which we have considered earlier, perhaps also conformed to a specific shift in the scale of damage-protection. For example, irresponsiveness of a neuron to habitual stimulus is a specific breakage of excitability to the given stimulus and this breakage proceeds without vague effects of excitotoxicity. A specific change in excitability is, probably, observed during damage, too (particularly, during LTP), although during learning the degree of specificity is higher. In both cases, during damage and during learning, an excitable membrane acquires paradoxical properties, when change in a membrane potential does not correspond to a change in excitability and this indirectly indicates a possible likeness of phenomena. Our discussion reveals that LTP and neuronal damage-protection have many features in common. LTP is, probably, the consequence of a development-related plasticity induced by the excitotoxic damage of neurons. The temporal scope of both damage and LTP coinside, but differ from the temporal scope of memory. Both phenomena are easily induced by a strong excitation, are mediated by glutamate receptors and depend on intracellular Ca^{2+} accumulation, retrograde messengers, cyclic AMP and protein synthesis. Both LTP and neuronal damage become weaker following treatments with GABA$_A$ receptor agonists and NMDA receptor antagonists. During development, neuronal damage and death are enhanced, as also is susceptibility to LTP. Susceptibility to LTP increases during tissue damage (for example, after preparing slices), but preconditioning with weak injury lessens both damage and LTP. Neuronal growth is observed during damage and development and during LTP. The hippocampus is especially susceptible to LTP and is the most vulnerable to damage. Similar influences protect against cellular damage and evoke depotentiation of LTP. The properties of LTP remind one of the properties of behavioral learning only if one considers high-stressed avoidance behavior, which induces transient damage of specific brain neurons. Maintenance of LTP requires the continous availability of sensitizing agent, while memory store does not need a support. Therefore, LTP and neuronal damage have similar reasons and similar mechanisms. Nevertheless, LTP is not a dying. This is, rather, intensification of compensation or nervous tension directed against damage. Long-term depression is, probably, not always opposite to LTP. Sometimes it may be overcompensation of weak damage and a shift to inhibitory protection (as during weak acidosis), but sometimes long-term depression may be converted into dying (as during strong acidosis) [1366].

3

Subjective nature of motivation (a single neuron can want)

3.1 Motivation as the simplest tool for investigation of the objective roots of a subjective life

3.1.1 The way a question is formulated

A brain, like any physical system, dissipates energy and moves towards equilibrium. Nevertheless, there is a difference between a contracted spring and a panther preparing to jump. A brain creates motivation and impels the organism to reach those environmental fluctuations (reward), which may prevent dissipation of its energy. For example, after exhausting inner resources of energy, hunger compels an organism to search for food, consume foodstuff and receive the possibility to replenish its energy stores. Motivation is the most ambivalent attribute of the neuroscience. A motivational concept is applied to both the brain and behavior as a correspondence of objective and subjective aspects of motivation. Motivations are not necessarily a subjective event. Being a subjective mood state, it is induced by a physiological cause and it modulates the readiness to obtain a reward. For the outside observer, a motivation is a phenomenon consisting of the modulation of sensorimotor relations that leads to the generation of actions in response to a previously ineffective stimulus until a certain optimal state of the organism is attained. This definition is similar to that of R.A. Wise [1344], who considers motivation in terms of matter. This aids scientific description of a motivation, but misses the most enigmatic property of motivation, for it is a subjective sense, too. We never resolve the problem of motivation if we consider only one side of the conundrum. Instead, one needs to reveal the physiological processes engendering this sense. One may say that generation of an action depends on an animal's attitude to the anticipated result. Therefore, we understand a motivation as an organism's subjective attitude to its current or future physiological state, which somehow modulates generation of actions until an optimal state is attained. How does this subjective attitude arise and how does it modulate generation of action? Motivation itself is modulated by an internal state of a subject, dependent on experience and on the incentive/hedonic value of its reward [195, 196].

If action is predetermined by the environment and by the state of the brain, goal-directed behavior can be treated as an ordinary reflex, i.e. as a reaction to stimuli (inner or outer), based on memory and heredity. This predetermination would preclude motivation as an exclusive object of physiological description. On the contrary, reality of a motivation is equivalent to the existence of free will, whose existence is frequently discussed and which is regarded as a valid object of philosophical inquiry and psychological analysis [1328], but has not yet been addressed in depth by neurobiologists. The problem remaining to be solved in regard to motivation is how behavior can be both unpredictable and goal-directed.

The free will problem does not have a purely academic interest. For instance, we are of firm belief that computer does not possess motivation and free will, since we know in fine detail the whole sequence of calculations that lead to the solution of a task. However, only comprehension of motivation will allow the creation of an artificial object, behaving as a living creature. It is impossible to develop artificial brain on the basis of conventional computers. Usually the choice of actions directed towards the satisfaction of an artificial motivation, is embedded by theorists in the construction of the system or is determined by memory. Therefore, "voluntary" actions are, factually, predetermined. An external programmer must constantly update the system about the proper strategies needed to overcome newly-encountered perturbations. However, it is unrealistic to predict all possible perturbations and disturbances in an environmental state. Hence, it is unrealistic to hope to develop reliable software for a robot control system. Investigation of motivation should help to design self-contained control systems which can function in an intricate environment, have their own goal and therefore be independent. Meanwhile, living systems demonstrate motivational behavior without any visible participation of a programmer. The 'ideal' software, which is embedded into brain construction, does not regiment behavior in detail. When the organism attains its goal it is in an optimal state, and no further actions are generated. A deviation from the optimum will result in a change of activity that leads to a return to the optimum. We argue in Chapter 4 that assuming the existence of non-predetermined actions is a valid theme in neurobiology. It is internally consistent and does not contradict experimental data and rigorous logic. Here we will consider the essence of motivation.

3.1.2 Motivation as a homeostatic recovery

Animals are capable of overcoming obstacles in order to attain a goal and may work hard to obtain food, water, sexual satisfaction, etc. Any motivation produces behavior directed to satisfaction of the motivation, if we discard those unreal wishes and dreams which may never be satisfied because of insurmountable obstacles, but, fortunately, this does not threaten existence. Motivation is satisfied by reward or by avoiding punishment. Any motivation includes an appetitive stage, a goal-directed search for reinforcement and a consummatory stage, the receipt of reward. At the cellular level, motivation reveals itself, as excitation of many brain neurons. For a detailed discussion of motivations, see reviews [201, 206, 297, 370, 656, 689, 799, 887, 892, 996, 1191, 1354, 1115].

The primary goal of living creatures is maintenance of their general physical integrity. An organism is a complex system. It is described by a majority of variables and it generates numerous motivations in order to correct existing deviations from various optima. Sometimes the subject is embarrassed to determine its innermost goal. A. and B. Strugatsky, in their fantastic novel "Stalker", described a unique, but almost inaccessible place. Anyone reaching this place will attain his/her main innermost desire. In the Strugatsky novel, certain youthful poet dreamed of creating a great poem. He overcame a variety of obstacles and endured awful dangers. In the end he has reached this unique place and got rich, but did not write his great poem.

First of all, let us simplify the problem and consider a one-dimensional case in which an elemental motivation depends on an imbalance of only one factor in an inner environment, which evokes the state that is somehow expressed as favorable or harmful for the subject. There are a limited number of such elemental biological motivations: feeding, drinking, respiration, temperature regulation, sexual motivation, avoidance of danger, the need to sleep, and artificial motivations (self-stimulation and drug-dependence). Motivation is undoubtedly associated with the need for stability of the internal milieu or homeostasis [203]. Standard functioning of the organism, a simple continuation of its being, is connected with expenses of energy and substances and is the origin of current requirements that must be satisfied for prolongation of life.

As we already know, homeostasis maintains stability of physiological systems and holds the parameters of an organism's internal milieu (or correspondence between these parameters) within limits that allow for survival. Yet deviation from the homeostatic equilibrium is not the sole cause for motivation and we will consider as an exception from this rule the examples of sexual and defensive motivations. The starting point of sexual motivation is not determined by metabolic needs and the starting point for defensive motivation lies in the outer environment. Nevertheless, independently from the starting point, the basis of sexual motivation, and may be as the basis of any simple and complex motivation (how to verify?), lies a deviation from the physiological equilibrium. Therefore, phenomenon of motivation is related to the phenomenon of homeostasis. Sometimes, motivation does not reach the threshold of awareness, for instance, the motivation to breathe. In such cases, we may describe the event without attracting the term 'motivation' and use as a background only the term 'homeostasis'. Aware motivation may be included in the term of homeostasis, if we accept that homeostasis can become aware. Previous considerations have shown that, at least in some circumstances, homeostatic disturbances may be connected with completing of high neural functions. We soon will return to this problem and demonstrate the tight relationship between the concepts of homeostasis and motivation.

Apparently, homeostatic motivation might actually reflect a variety of physiological settling points rather than one true set-point, arising as a balance among opposing systems [114]. Besides, there are convoluted motivations that depend on a combination of variables and which defy experimental analysis. Nevertheless, even in these cases, set-points for the homeostatic mechanism may exist according to a theory of a common currency for different motivations [197]. According to this theory, any motivation is directed to receiving a pleasure. There must be, then, a general set-point for homeostatic regulation, when a change in this particular set-point cannot be compensated. We argue that in such cases homeostasis seems to regulate some critical quality of existence, as, for example, the level of damage.

External influences, which are not coupled with the internal state, can affect behavior that does not directly promote homeostasis. Motivation can be associated with learning and memory and plays a leading role in the adaptation of animals to a changing environment. The animal that has already experienced satisfying its demands performs similar actions and satisfies the demand again in the presence of a new motivation. Moreover, a necessity to solve the task can turn into the demand and thus motivation may be complicated [148]. Motivations depending on memory may be extraordinarily specific, allowing us to get information on the essence of motivation from an analysis of the mechanisms of learning, especially instrumental learning. Nevertheless, dependence on memory does not necessarily capture the essence of motivation, for animals can demonstrate goal-directed behavior in early life, without any personal experience. Therefore, it is not necessary to touch 'memory' for description of simplified motivation. During instrumental learning, the motivation is created, affects behavior and is satisfied.

Thus, an organism generates action to support the stability of its inner milieu. If the organism has information that this particular combination of the inner constants is optimal, must be maintained, but is disturbed because of intervention of the environment or because of exhaustion of inner resources, its actions will direct it to search for the possibility to satisfy its needs. Whether homeostasis of an organism is entirely based on homeostasis of its cells is unknown, but at the least, cellular homeostasis exists and can affect behavior.

3.2 Chemical nature of motivations

3.2.1 Control of motivations by means of motivationally-relevant substances

When one experiences a motivation, some specific areas in one's brain are activated, indicating a connection of the specific motivation with a specific brain area. Moreover, electrical activation of such an area causes the same (or almost the same) motivation, and connection between an origin of the motivation and excitation of specific neurons is a robust conclusion. Nevertheless, neural excitation is usually a secondary consequence of a homeostatic disbalance, which affects some metabolic pathways. We would comprehend the sense of motivation more profoundly, if from a correlation between the behavioral event and its location in the brain will proceed an analysis of the material essence of motivation, which somehow actuates a behavior.

While it does not appear to be possible to transfer by memory-relevant substances, elemental motivations are easily induced, more or less selectively, by motivationally-relevant substances, which both affect motivation and change the state of a neural tissue. They usually do not imitate primary motivational events, do not provoke and only modulate an existing motivation. However, sometimes the effect may be so robust that a satiated animal may restart eating. On the one hand, the presence of a motivation correlates with production of motivationally-relevant substances, and, on the other hand, an administration of exogenous substances can specifically alter behaviors and modulate existing motivation. A salient feature of motivationally-relevant substance actions is their U-shaped dependence on dosage with very small effective concentrations (nmols-pmols). Several tens of motivationally-relevant substances are known [685, 266, 119, 101, 131, 200, 293, 382, 431, 475, 528, 609, 679, 910, 975, 805, 1018, 1083, 1199, 1209, 1307, 174, 703, 1172], which are usually neuropeptides or neurosteroids. Examples of motivationally-relevant substances influencing motivational behavior include the decrease of feeding after an increase in intracerebral concentration of corticotrophin releasing factor, cholecystokinin, leptin, bombesin, gastrin-releasing peptide, melanocortins, neurotensin, thyrotropin-releasing hormone, etc. In contrast, an increase in feeding is induced by neuropeptide Y, somatostatin, agouti-related protein, orexin-A, ghrelin, galanin, etc. while starvation leads to a change in the concentrations of corresponding substances in the brain. Aversive motivation and stress decrease with opioid use. The action of substance P is often opposed to that of opioids. Angiotensin II, oxytocin and vasopressin increase the motivation for drinking whereas the neuropeptide K and galanin decrease it. Oxytocin modulates Na^+ appetite, while vasopressin manages diuresis. Sexual motivation is regulated by steroid hormones and also by oxytocin and vasopressin. Estrogens and androgens act rapidly on sexual behavior in nonhormonal ways, without the participation of the genome, since they transmit a signal via membrane receptors coupled with G-proteins. Androgens appear to be the keystone of sexual arousal in both males and females. Thus, motivations are easily influenced by chemical regulation.

Yet, the possibility of chemical control of motivation does not prove its chemical nature, since chemical means may enable the destruction or excitation of a non-chemical mechanism. For example, feeding is easy to provoke or to inhibit by means of electrical stimulation of hypothalamic sites, introduction of neurotransmitters, gastric load, and other means. Nevertheless, the extraordinarily low doses of motivationally-relevant substances that can affect motivations do suggest that motivations are chemical states. Some substances of the sexual and defensive systems affect living tissue as well as behavior at doses of 0.1 - 1 fmol [436, 754, 837, 1123]. This is a tiny number of molecules, perhaps only a few molecules per cell. Hence, the effect must be magnified by a chemical cascade. However, it is not known whether these chemical cascades are different for different motivations. It is interesting to note that the most sensitive to chemical influences are just those motivations which are initially provoked by outer, in respect to brain, signals: sexual and defensive motivations. They do not depend on metabolic needs and in these cases the motivational system is finely tuned to these very small concentrations of key substances.

Motivationally-relevant substances were shown to modulate the processes of learning and memory approximately in the same effective concentrations [315, 823]. It is not clear whether the reason for their behavioral effects during learning is an interaction with the memory mechanism or modulation of motivation. However, their influence to learning depends on previous experience [108, 776, 823]. Therefore neuropeptides play a role of chemical signals rather than of instruments for intervention in intimate chemical processes of memory.

Action of motivationally-relevant substances to behavior includes participation in the same metabolic pathways as does corresponding natural behavior. These are, certainly, secondary metabolic pathways, with the exclusion of primary sensory effect in the motivational centers, which differs for different motivations. Motivationally-relevant substances exert their influences via G proteins and cyclic AMP pathways[1]. Other instruments for control of homeostasis through Ca^{2+} and K^{+} channels, intracellular Ca^{2+} mobilization and arachidonic acid also promote influence of these substances on behavior.

Evidently, motivation requires participation of the same or almost the same metabolic instruments, as in other cases of serious reorganization of neural tissue status. Time scopes for development of motivation confirm this similarity. Natural motivations develop slowly and satisfactions of those motivations, which are directly connected with homeostatic disturbance, take time. Relatively rapid satisfaction of motivations is possible for the cases of defensive, sexual and high order motivations that are aroused with participation of mental activity. Artificial change of the course of existing motivation by neuropeptides is also slow. Peptide secretion can be modulated on a millisec-

[1] For the example of feeding behavior, satiated peptide leptin affects energy homeostasis through reduction in cyclic AMP levels in the hypothalamus [1061], while an increase in cyclic AMP levels reverses the established effects of leptin on food intake and body weight [1430]. Neurotensin, which is also an anorectic peptide, is involved in mediating leptin's satiated action on feeding [1060]. Neuropeptide Y treatment and food deprivation stimulate the cyclic AMP system [1124, 224] and the effect is dependent upon the activation of the NMDA glutamate receptors [717] in the hypothalamus [898, 174, 717]. Cyclic AMP participations were also found for galanin, somatostatin, steroid sex hormones [1146], insulin [761], etc.

ond time scale [1334], but direct effect of influence of motivationally-relevant substances on neurons may be observed only after tens of seconds or minutes. Slow development of damage and protection are correlated with a sluggish influence of motivationally-relevant substances on motivation. Chemically induced motivations [285, 528, 609, 1168] are also displayed rather slowly (tens of minutes) and for a limited time period. Such a slow induction of motivations indicates a reorganization of the biochemical properties in neural tissue.

As we indicated in Chapter 2, on the one hand, damage increases expression of motivationally-relevant substances, but on the other hand, these substances reorganize properties of neural tissue and affect the processes of neuronal damage and protection. Such a connection with damage-protection was found for galanin [1437, 357] cholecystokinin, leptin [175, 176, 1031], estradiol, angiotensin II [1051, 495], neuropeptide Y [1, 232, 506], vasopressin [1136], androgens [1388, 11, 978, 934], estrogens [1327, 1346], neurotensin [172, 506], thyrotropin-releasing hormone [955, 1302] and prolactin [392]. Neuropeptides and neurohormones act on motivation and exert a strong effect on the processes of neuronal damage in the same tiny concentrations and with similar time courses. We may suppose that when these substances affect motivation, they change the level of cell damage. They can either ameliorate or exacerbate cell death. Nevertheless, because of complex non-linear dependence of behavioral and pharmacological effects on concentration and time on the level of motivation, location of the tissue in the brain, and shortage of experimental data we cannot trace a clear correlation between the influence of a given substance to motivation (increase-decrease) and its effect on damage or protection. However, the prevailing preliminary conclusion is that there an increase in motivation during cell damage and satisfaction regarding motivation when cell conditions are recovered.

3.2.2 Chemical specificity of motivations is not absolute

Motivation creates a readiness to actions, directed to satisfaction of this motivation. However, a motivation increases also a readiness to non-related actions, though to a lesser degree. Chemical and spatial selectivity of motivations is a properly established fact, supported by a limited location of active zones in brain, ultra low effective concentrations of motivationally-relevant substances and high affinity of their interaction with chemoreceptors. Motivationally-relevant substances applied to specific brain areas selectively influence related motivations. However, sometimes a motivationally-relevant substance does not strictly correspond to a definite motivation and there are side effects, which can reveal themselves without overdose even in the same minute concentrations.

A given motivation is more related to definite chemical events in a specific brain area, but the same chemical substance can affect another motivational behavior in another specific brain area[2]. Although some brain areas appear to be specific for a given motivation, the same area may be related to various motivations, depending upon the level of current motivation, the inner milieu, learning, type of activated chemoreceptors, etc. A specific metabolic disturbance may sometimes induce non-related behavior. This is displayed in the possibility of interaction between different motivations. One aware motivation is usually dominant, but a competitive motivation may augment a dominated one, or distort any of them[3]. One of the reasons of such side effects is that specific neurons in motivational centers also have non-relevant specific receptors.

A strict correspondence between motivationally-relevant substances and their mother motivations is absent. Several substances may evoke a similar behavioral outcome. Therefore, the shortage of one such regulatory substance can be compensated for by a detour to another metabolic path. We already described such property for other cases: compensation of lost functions after trauma, self-recovery of memory after amnesia and homeostatic compensation during stagnation tensions. For instance, neuropeptide Y seems to be a necessary link for feeding, but feeding behavior results from a complex interaction between neuropeptides presented in the hypothalamus. And the phenotype

---

[2] For instance, neuropeptide Y does not induce side effects after local injection into specific brain zones for feeding [131, 528], but in other areas it decreases anxiety [604], produces anxiety-like behavior [903] and is implicated in sexual behavior [780], while the satiated neuropeptide cholecystokinin also provokes aversive reactions and produces anxiety-like behavior, participates in temperature regulation, sexual behavior and analgesia [101, 293, 1207, 1338, 903, 37]. As well, somatostatin in the specific zones of feeding motivation increase in feeding, but in the locus coeruleus may contribute to facilitation of paradoxical sleep, while the effect of intracerebroventricular injection of somatostatin was less effective [1238]. Similarly, although substance P is, mainly, connected with pain, unilateral administration of substance P into the area of the nucleus basalis magnocellular was found to be reinforcing [145]. In the ventrolateral preoptic area (which is important in the modulation of sleep) substance P promotes sleep [1424], while within the neurons of the sexual center it regulates male copulatory behavior [288] and in newborn rats it increases in respiratory frequency [1001].

[3] Glucose-sensitive neurons of the feeding center in the lateral hypothalamus also respond to various noxious stimuli including cold, heat, pinching the tail and a change in CO_2 concentrations [1341]. Similarly, receptors for galanin were found in various brain areas. As an example, galanin is expressed in the same brain areas that are known to regulate both sleep and the waking state, preoptic area, hypothalamus, and brainstem nuclei and it increases paradoxical sleep [1238], while, microinjection of galanin into the medial preoptic nucleus, which is implicated in sexual behavior, does not facilitate feeding and paradoxical sleep, but does stimulate sexual behavior [136]. Therefore, the neuropeptide galanin is thought to regulate numerous physiological actions in the adult mammalian nervous system, including feeding, arousal/sleep regulation, energy and osmotic homeostasis, reproduction, nociception and cognition [703].

that relates to the deletion of gene coding for neuropeptide Y rarely includes changes in food intake. Therefore, chemical exposure to motivation is not ideally selective. The interaction among chemical components may be sufficient to confer stability without a separate controlling mechanism for each component [119, 177, 1341]. However, other reasons also promote an absence of ideal selectivity. This reason is that similar mechanisms are operational in different motivations and the same impact may give rise to various effects. Another example is orexins that regulate sleep, but participate also in other regulating systems[4]. A substance may also demonstrate side effects, if currently dominated motivation does not correspond to the basic effect of this substance[5]. Evidently, because of the same reason, weak motivations may augment a strong one[6].

This means that motivationally-relevant substances may exert unrelated action in the unrelated brain areas. Probably, even if density of chemoreceptors for a given substance in an unrelated place is low, after activation of these receptors in the favorable circumstances further biochemical events lead to the proper change of state of the specific cell. If this cell is responsible for

---

4 A lack of orexins or their receptors in knockout mice causes excessive daytime sleepiness, and leads to numerous unintended transitions between vigilant states and to the intrusion of fragments of paradoxical sleep into wakefulness [1115]. This indicates a decisive role of orexins in sleeping, since blockage of orexins can not be compensated by a neural system (as the neuropeptide Y knockout mice are compensated). However, in spite of so vigorous a connection with sleep, orexins, beside the regulation of vigilance and feeding behavior, are involved in anxiety-like behavior, [1202], control of emotion, reward and energy homeostasis [241, 1063], drug-seeking behavior and metabolic rate [1089] and core body temperature [857]. In addition, orexin produces an analgesic effect [1374], while orexin levels in the brain of suicidal patients decreases [173].

5 Angiotensin II receptors play diverse physiological roles in the nervous system, but first of all they stimulate drinking [25]. However, administration of angiotensin II to the subfornical area of ethanol-dependent animals provokes feeding, sexual reactions or arousal and does not cause them to drink water [1191]. Neurons of the lateral hypothalamus and zona inserta, which respond to the sight of food in hungry animals and to the sight of water in water-deprived animals, begin to respond to the sight of water in hungry animals after intraventricular injection of hypertonic saline or angiotensin II [868].

6 In satiated snails, food induces avoidance behavior [471]. In other cases, sexual motivation [1440] and pain [910] enhance feeding behavior. It is known that brain self-stimulation is connected with feeding-related zones, but thresholds for intracranial self-stimulation reward, it was shown, were significantly elevated during ethanol witrhdrawal and thus pathologically strong alcohol motivation was substituted for self-stimulation [1106]. Likewise, administration of the stimulant drug amphetamine in the lateral hypothalamus augments the self-stimulation paradigm and, interestingly, food restriction further enhances self-stimulation [516]. Also, after injury of peripheral nerves, neuropeptide Y contributes to peripheral hyperalgesia [1240].

improper, in relation to given substance, motivation, then provoked behavior will be also improper. Evidently, motivationally-relevant substances affect some common and important cellular properties. Their effects may directly relate to a shift of the scale of damage-protection, as we will demonstrate afterward.

3.2.3 Motivation reorganizes brain temperature and energy metabolism

According to our discussion, the side effects of various substances to inappropriate motivations may be apparently due to common mechanisms for different motivations. It looks convincing that even if various motivations are initiated by different reason and directed to receive different result, their mechanisms are related and, possibly, there exists a general mechanism of motivations. In particular, motivation to sleep, feeding motivation, thermoregulation, breathing and some other drives are connected with the necessity to recover an energy homeostasis. Appropriately, neuropeptides affecting feeding have an effect also on the sleep-wake cycle and energy metabolism: cholecystokinin, orexin, somatostatin, insulin, leptin, ghrelin, neuropeptide Y, bombesin, and interleukin-8 [131, 1207, 575]. Nevertheless, the relation of motivations to energy expenditure is not just this mechanism that joins all motivations to the phenomenon. In particular, defensive motivation is not connected with the energetic balance.

Goal-directed behavior requires a large energetic expenditure and, as a rule, motivational excitations lead to brain hyperthermia, while rewards lead to hypothermia [635]. Brain temperature rises during the sexual act [852, 1093], drug self-administration [635], drinking and feeding [4]. Cholecystokinin, which decreases feeding via A-receptors, [861] evokes hypothermia via the same receptors [1207]. Intravenous self-administration of heroin or cocaine in trained rats is accompanied by a rise in brain temperature, which begins to increase before the first instrumental reaction directed towards obtaining heroin [639]. Specifically, brain temperatures transiently increased from 3 to 4 min before the voluntary drug self-injection, decreased after drug injection (1 min), then increased, and were lowered again by the next drug infusion. After a last injection in a session of drug self-administration, temperature decreases for 10 minutes and 30-40 minutes after passive injection, while afterwards it returns to the norm [635, 636]. Heroin self-injections cease after naloxon administration, and at the same time brain temperature abruptly returns to the baseline. Thus, satisfaction removes the reason for enhanced hyperthermia during motivational states in a brain.

A rise in temperature during motivation points to metabolic activation and may potentiate damage. Indeed, the severity of brain injury is positively correlated with an increase in brain (but not body) temperature and is negatively correlated with the behavioral outcome [262]. Stimulant drugs induce both hyperthermia and pathological metabolic neural activation [170].

Methamphetamine produced dose-dependent hyperthermia and pathological metabolic activation within brain structures (especially the nucleus accumbens) showing a more rapid and pronounced temperature increase than in the muscle. In correspondence with the properties of cell damage, at low doses, temperature elevation is shorter, and overcompensation leads to hypothermia (2-3 h). At the highest dose, brain and body temperatures increased 3.5-4.0 C above basal levels and remained elevated for 3-5 hours. Stress and other high-emotional situations such as interaction with a conspecific female also induce a significant hyperthermic response in the rat [170]. As well, in the medial preoptic area, a steady rise in hypothalamus temperature is observed from the beginning of a mating bout until ejaculation, which was followed by a rapid cooling. This is in agreement with an increase in metabolic activity during copulation and a rapid fall of activity after consuming the reward [802].

Inhibitory $GABA_A$ receptors participate in the temperature homeostasis. Perfusion of $GABA_A$ agonist into the preoptic area and anterior hypothalamus increased body temperature. $GABA_A$ antagonist bicuculine does not alter normal body temperature, but under cold conditions it induced hypothermia in spite of hyperactivity. Cold exposure increased extracellular GABA and thus homeostatically reduces neuronal activity [580]. Robust brain hyperthermia is supposed [134, 852] to be a driving force underlying motivated behavior in animals. At least, we may conclude that augmentation of cellular metabolism during motivation, which is the basis of hyperthermia, side by side with the augmentation of neuronal electric activity, concerns motivational behavior. It would be interesting to examine in the experiment, if a weak artificial heating of specific motivational areas augments motivation and if cooling has a rewarding effect. Such experiments were accomplished in a thermoregulatory center and the local warming lowered body temperature, while cooling stimulated a rise in body temperature [498], and this is in correspondence with the action of $GABA_A$ agonists on the thermoregulatory center [580]. We may predict that analogous local cooling and warming in the centers of other motivations will, firstly, affect corresponding motivation and not body temperature and, secondly, cooling and warming cannot result in a symmetric behavioral output and will depend on current levels of the corresponding motivation.

3.2.4 Localization of metabolic aims of goal-directed behavior in the brain

We have demonstrated in Chapter 1 that correspondence between memory elements and locations of neurons, neuronal pathways, synapses and other objects in a brain is not a relevant principle of information storage. The memory, corresponding to a particular event, is not connected with the trace in some particular place in the brain. Nevertheless, memory is not smeared uniformly within the brain or within some neural center. In respect to memory, different brain areas are not completely equivalent.

As for motivation, its localization in the brain is based on a dissimilar principle. Sensitive zones for various motivations are localized in specific areas of the brain and these zones include primary receptive areas and high motivational centers. Different areas and neurons are responsible for the origin of different motivations. However, just as memory is not ideally spread in the brain, motivations are not localized strictly. That is, a memory is not ideally dispersed, while motivations are not ideally confined to small areas. Regions responsible for motivations are crossed and the same neurons may sometimes participate in different motivations. Some brain areas are activated across all types of emotional (that is, concerned or exceptional) stimuli [974, 485], but there are other brain regions that are specifically involved in different aspects of emotion [207, 974, 1313, 485]. Moreover, distinct classes of positive affective states are present in the brain, with separate but overlapping neuroanatomical substrates [184]. There is also a close relationship between emotion and homeostatic brain areas [289]. Nonetheless, neuroimaging studies reveal no support for the hypothesis of overall right-lateralization of emotional function [1313].

Localization of motivational zones in the brain is relatively delicate. For a given motivation, different neurons fulfill specific functions. Activations of different neurons must correspond to specific feelings. Really, if the same neuron or the same combination of neuron would be responsible, say, for hunger and for thirst, who chooses the final decision, if not Homunculus? Besides, a neuron is sufficiently complex in order to evaluate the importance of a particular current signal and to prefer to react or fail to react to this signal, while in the next moment it may evaluate another signal and again decide to react, or not. However, a neuron is, certainly, too simple in order to compare two events simultaneously and, more than that, choose for each of them the specific reactions. Thus, a neuron, on the one hand, is specific only by its location and its distinctive feature (chemical sensitivity, morphology, etc.), but on the other hand it may be specific according to the sense of the current signal. Motivational centers give an important example of such neuronal organization. Unlike organization of memory, the significance of motivational signals is given in advance.

Primary motivational centers are tuned to reveal a subtle disturbance of specific metabolic parameters and there are key sensory cells that reveal the specific important needs of an organism. Key neurons are the most specialized for detection of specific metabolic flaws, for example, change in salinity, temperature or oxygen concentration. However each such key neuron, as well as any neuron, keeps also its capability to detect any metabolic alteration, which is necessary to compensate for danger to survival. Although any neuron detects these subsidiary flaws less efficiently than specialized neurons, the neuronal substrates of various motivations are interconnected. There are metabolic aims of specific goal-directed behaviors. Detection of these metabolic flaws ensures materialization of actual motivations.

3.3 Elemental motivations emerge in a result of transient cell damage

At present no specific investigations are examining how specific brain neurons are damaged during homeostatic disturbance throughout motivational behavior. Such damage may consist of changes in a cell's morphology, ion metabolism, electrical activity, and other factors that prevent normal functioning of the cell, and these may eventually lead to cell death. Direct evidence of damage, such as cellular swelling, loss of integrity of the membrane, etc. is difficult to reveal *in vivo*. Nevertheless, the correlation between an outcome of motivational excitations and cellular damage is sometimes so obvious that indirect evidence is presented in many studies. On the one hand, there are the evidences of cellular damage during motivational behavior, and on the other hand, artificially induced damage provokes or augments motivation. As well, the same alterations of cellular homeostasis lead to damage and to a rise in motivations. Accordingly, excitation, the basic property of nerve action (and apparently the basis of motivation), causes damage. Specific examples are given below. We tried to detect primary signals, that initiate motivations and specific metabolic pathways that are typical for development of motivations.

3.3.1 Defensive motivations

The origin of defensive motivations is associated with a pain, fear, anxiety, aversion, disgust, and other signals of menacing danger. Traditionally, the defensive motivation is considered as negative. Defensive or negative motivations may be displayed in the active or passive form: one may passively avoid or actively liquidate the danger. In fact, any motivation is negative and needs satisfaction, which always has a positive meaning.

Attentive studying of neuronal activity during aversive and defensive reactions has revealed a wide distribution of excitations in the brain, but specific aspects of defensive behavior correspond to activity of relatively local areas within the brain. Defensive motivations are first of all connected with the amygdala [297, 556]. Different types of fear-conditioned behavior are mediated by separate nuclei within the amygdala: the central nucleus mediates suppression of behavior elicited by a conditioned fear stimulus, while the basolateral amygdala mediates avoidance of conditioned aversive stimulus [624, 556]. Some features of defensive behavior are also connected with other areas of brain. For instance, neurons in the rostral ventromedial medulla exert a facilitatory influence on spinal nociception "on-cells" and secondary hyperalgesia in acute inflammation [1368], neurons in the spinal cord contribute to chronic injury-induced neuropathic pain [1161], activation of neurons within the periaqueductal gray in the midbrain may be induced by a predator [7] and direct activation of these neurons leads to the fear reaction [838]. In addition, transmission and sensation of primary events for the development of avoidance and their effects are broadly scattered within high neural centers of the brain, particularly in the cortex, hippocampus and cerebellum [117, 840, 556, 628]. Different areas of brain are activated during early and late phases of acquisition of fear conditioning, during extinction and recall and, in particular, the ventral hippocampus may play a role in resolving conflicts, and specifically may be relevant to "safety signaling" [556]. Thus, the neurons that feel danger or pain, the neurons controlling action directed to avoidance and the neurons that feel the coming of safety, these are commonly different cells.

Aversive stimuli are inalienable attributes of defensive motivation. A distinctive feature of aversive stimuli is their resistance to habituation and responses to such stimuli even increase after repetition [248]. The origin of defensive motivation is directly connected with the danger of injury. Pain is important, although not an obligatory component of defensive motivation. It is induced by tissue-damaging stimuli and they evoke injury in the central neural system itself [840], that is, defensive motivation is initiated by means of damage-related events in the brain, not only in the body[7].

Glutamate receptor activation is an essential component of both the fear reaction and hyperalgesia[8]. Such an impact to alterations of cellular metabolism inevitably leads to deviation of the ion equilibrium from the optimum to excitotoxicity. Moreover, defensive motivation and negative experience left material traces in the brain, which are observed after cell injury[9].

Secondary effects of aversive stimuli increase the number of gap junctions, enhance cyclic AMP concentration [297, 373, 844, 1293] and produce the inflammatory mediators, such as prostaglandins, adenosine, etc. Cyclic AMP pathways are activated and also augmented during neuropathic pain [1161] and by a predator [7]. Inhibition of the cyclic AMP signaling pathway in-

---

[7] Aversion exerts strong excitation through the excitatory amino acid glutamate and retrograde messengers, causes depolarization of amygdala neurons, and enables Ca^{2+} entry into these cells [297, 373, 844, 1293]. Transmission of nociceptive action potentials (APs) is facilitated by modification of the voltage threshold of several ion channels, including the tetrodotoxin-resistant sodium channels [117, 840, 556, 628].

[8] Infusion of NMDA-receptor antagonist into the rostral ventromedial medulla prevents hyperalgesia [838, 1368]. The stress produced by noxious cutaneous stimulation increases glutamate concentration in hemolymph of mollusk and it remained elevated for at least hours [678]. One-trial passive avoidance training in a mammal also depends on activity of excitatory glutamatergic neurons [538].

[9] Negative motivations evoked by stress [400], such as isolated footshock, as well as inhibitory avoidance training [1365] cause direct morphological changes: atrophy of dendrites of hippocampal pyramidal neurons. Chronic intermittent restraint stress (for 6 h per day for 14 days) suppressed the adult neurogenesis, while chronic antidepressant treatment has been shown to increase adult neurogenesis in the rat hippocampus [1042]. It is well known that the hippocampus is especially sensitive to damage and this is among the first responses to deleterious influences during the injury. For instance, neurons in the hippocampus display symptoms of damage in response to a change in osmolarity [567], after inhibitory avoidance [1365], during sexual motivation [697] and in drug abusers [1034].

creases anxiety-like behaviors [936, 1296, 87], while augmentation of the cyclic AMP system produces hyperalgesia [1150] and antagonizes the analgesic response to morphine [121]. Fear reaction and inhibitory avoidance, such as aversive stimuli, are connected also with activation of synthesis of nitric oxide [838] and decrease in Na^+, K^+-ATPase activity [1365]. For instance, spread of the pathophysiology of neuropathic pain includes upregulation of nitric oxide pathways in axotomized neurons [1439, 1381]. These secondary consequences of aversive stimuli are usually observed during intensification of homeostatic compensation.

Not only pain, but also the state of anxiety is related to cell damage. In depressive patients, the prefrontal cortex loses neurons, while the number of cells in the hypothalamus increases [1009]. Growth of neurons as well as their death can be a direct result of damage to them. Anxiety-relevant substances, serotonin, GABA, corticotrophin-releasing hormone, interferon γ, and neural adhesion molecules play important roles in neuronal development, proliferation, cell-to-cell communication and intracellular signaling [1353]. Thus, defensive motivations are directly connected with cell damage, while avoidance of defensive motivation protects brain cells. Moreover, not only does negative motivation induce neuronal damage, but damage induced by the administration of cytokines [343] and excitatory amino acids [906] also increases defensive withdrawal. Neuronal-glial interactions, resulting in an immediate release of glutamate and discussed above as damaging, are broadly involved in the creation of pathological pain and passive avoidance learning [697, 1326]. Glial cells participate in defensive behavior. Both microglia (in the initial phases) and astrocytes (in the later phase) are important in neuropathic pain. Brain injury and some pain states leads to hypertrophy of glial cells. Concomitantly, production of a variety of chemokines, and other pain-producing substances increases. Their expression sensitizes primary afferents and dorsal horn neurons and therefore may contribute to neuropathic pain after peripheral nerve injury. Prevention of glial activation by pharmacological means produces an antinociceptive effect, while spinal administration of the pain mediator substance P produces robust activation of microglia [824]. Cells of microglia sense homeostatic disturbances and even relatively minor deviations from normal neuronal activity [1109]. The microglial response to danger might result from membrane breakdown products, the extracellular presence of cytosolic compounds, abnormally processed or aggregated proteins, or an abnormal abundance of the excitatory transmitter, glutamate. The executive functions of microglia can change not only in magnitude but also in quality and variability of microglial activity. Their activity is not a reflection of stimulus strength or persistence; rather, it is determined largely by the nature and context of the stimuli and the intracellular signal transduction pathways that they activate [1109].

Brain possesses special means that counteracts these negative senses in the system of endogenous opioids. Danger in the environment compels a living being to active actions, but strong negative feelings may prevent defensive behavior. Although pain or fear signal about a threat, when the danger appears in the surroundings, these feeling must not preclude an avoidance of danger. Therefore, after revelation of a threat, temporal relief allows the brain to work against damage. The opioid system performs this function well [910, 1293, 141]. Endogenous opioids work both for pain relief and to ameliorate negative feeling.

3.3.2 Respiratory motivation

Breathing is usually an unaware action. It continues during sleep and during a fall in consciousness, but breathing during alert behavior is, as a rule also unconscious and is controlled by an autonomous central pattern generator of rhythm. Respiratory motivation sharply increases and becomes conscious after a disturbance of gas exchanges, oxygen and carbon dioxide. Thus, respiration exhibits an example of parallel occurrence of two or more motivations, at least till they become conscious. Breathing is necessary for oxidative phosphorylation that uses energy released by the oxidation of nutrients in order to produce molecules of ATP that provide a highly efficient way of storing energy, used for metabolic needs. The substrate for respiratory motivation lies in the pre-Botzinger complex, a cluster of interneurons in the ventrolateral medulla and in the brainstem medulla. Neurons in the ventral medulla contain chemoreceptors, regulating breathing,, while the central command mechanism for breathing rhythm lies in the hypothalamus [378]. Preinspiratory neurons in the medulla are opioid-insensitive, while neurons in the pre-Botzinger complex are opioid-sensitive.

Respiratory motivation is connected to a decrease in the pH of the brainstem and an increase in gap junctions [1220, 125, 81, 187, 926]. Both CO_2 and pH sensors regulate breathing and both intra- and extracellular pH appear to serve as the proximate stimulus for central chemoreception. pH is sensed by many proteins which are located in low resistance gap junctions. Locations of the central chemoreceptors only partially coincide with location of the central pattern generator for breathing [378]. Neurons controlling inspiration and respiration are different cells, and the sensory neurons for respiratory motivation are, evidently, not the same cells that produce breathing.

In the basal state, all types of respiratory neurons in the ventral respiratory group of rats receive GABAergic and glutamatergic inputs. These neurons display periodic waves of inhibitory and excitatory postsynaptic potentials [907]. Although both glutamatergic and GABAergic mechanisms are involved in setting the resting respiratory rhythm, excitatory amino acids play a leading role in respiration [438]. During a shortage of oxygen, glutamatergic neurons of the ventral medulla are vigorously activated by CO_2 and by acidification, whereas serotonergic neurons are not [867]. Inspiratory bursts in the pre-Botzinger complex in mice depend on a Ca^{2+}-activated cation current linked to glutamate receptors [930]. During a powerful rise in respiratory motivation, such as in the acute ventilatory response to hypoxia, Ca^{2+} channels are activated, intracellular Ca^{2+} is accumulated, taurine falls and glutamate release is enhanced in the medulla as well as at the other sites associated with respiration [558]. We may conclude that shortage of oxygen leads to disturbance of the energetic balance in specific cells and, as a consequence, to Ca^{2+} dependent damage of these neurons. In correspondence with the phased change of neuronal activity after damage, the ventilatory response to acute hypoxia in a mammal is biphasic, an initial hyperventilatory response is followed by a reduction in ventilation within 23 min below the peak level [558]. In particular, agonists to excitatory amino acid glutamate in the doses in which it facilitated respiration, destroy the respiration-related neurons in the nucleus parabrachialis [542].

Experimentally evoked hypoxia is the main method of cell death investigation. Direct indications of excitotoxic damage are observed during the respiratory drive [125, 187, 849]. Opioids, which protect neurons against damage, suppress respiration, while substance P stimulates respiratory frequency [81, 960]. Metabolic coupling between the glia and neurons is necessary for maintaining rhythmic respiratory activity in the medulla [154, 189, 557, 622]. Dynamic change in baroreceptor-sympathetic coupling is observed in the central pattern generator during the respiratory cycle [455]. Probably, not only in the case of pathologically augmented ventilatory response, but during normal respiration, inspiratory and expiratory phases are connected with the cyclic changes of injurious and protective influences.

3.3.3 Temperature regulation

Warm-blooded beings maintain their body temperature within certain limits, that is, decrease or increase body heat. In both cold and hot environments, a thermal homeostasis spends effort for keeping temperature constant. The temperature of a body may be regulated not only by a change of metabolic rate, but by a choice of the optimal behavior: animals search for a warm or cool place, decrease or increase a motion and irrigate or sun themselves. A behavioral means instead of homeostatic regulation is used also by cold-blooded animals.

Precise thermoregulation is mediated by the preoptic and anterior hypothalamic areas. Local warming and cooling of this zone leads to lowering or raising body temperature. There are cold-sensitive effector neurons for heat production and warm-sensitive effector neurons for heat loss. Warm sensitivity results from inherent changes in cellular activity, while cold sensitivity is primarily due to inhibitory synaptic input from nearby warm sensitive neurons [498]. There are thermoresponsive Na^{+} conductances in warm sensitive neurons and some thermosensitive channels are also responsive to low pH. Several types of channels are tuned to various temperature setpoints. Since some neurons are sensitive to warming, while others to cooling, and optimal temperature may vary for different neurons, initiation of a motivation may be a united activity of a group of cells. In particular, temperature regulation is controlled by a system of neurons and not by a dispersed assemblage of autonomous regulators.

Temperature regulation is dependent on cellular events connected with damage. Cytokines cause a rise in body temperature and a clear correlation exists between thermosensitivity in the ventromedial preoptic area and neural responses to some cytokines [498]. Angiotensin II, a motivationally-relevant substance, connected with water metabolism also activates preoptic neurons [1243]. These neurons, which sensitive to temperature and osmolarity, are located side by side. Termoregulation, which is related to liquid homeostasis, also alters membrane properties. Warming decreased the amplitudes of the action potentials in the central termosensitive neurons and affected Na^+ channels [498]. Changes in the local environment of the hypothalamus (i.e. osmolality, glucose concentration, or the presence of reproductive steroids), have the greatest affect on the activity of warm sensitive neurons [498]. Probably, neurons that are sensitive to primary signals of homeostatic disturbance are susceptible to transient damage and this may be a cause of interaction between different motivations.

3.3.4 Drinking motivation

Thirst is one of the most important biological motivations. Nevertheless, investigations of drinking motivation are still almost at the beginning of their efforts. In particular, for our knowledge, nobody has compared central drinking mechanisms for animals living in water and in an air environment. This comparison could clarify how a morphologic substrate of motivation depends on the accessibility of reward, but this is a task for future investigations.

The need for liquid evokes dehydration and induces thirst. Circumventricular organs have a leading role in the control of body fluid regulation, and have significant involvement also in temperature control, feeding behavior and reproduction [268]. Initiation of drinking depends on the hypothalamic subregion of circumventricular organs, the subfornical organ, the paraventrical nucleus, the median preoptic nucleus and the arcuate nucleus [430]. It is considered highly unlikely that separate subpopulations of neurons exist in each of the circumventricular organs which subserve the control of each physiological system. Rather, it would seem likely that neurons in these circumventricular organs contribute to integrated control of complex interwoven physiological systems [268]. The neurons here are sensitive to angiotensin II and apelin, motivationally-relevant substances that stimulate drinking motivation and have effects on fluid and electrolyte homeostasis, osmotic pressure, body fluid homeostasis and they lose water during thirst [732, 1243, 25, 1018, 430]. Salt loading and dehydration increase the level of galanin [703]. Central osmoreceptors are located in the magnocellular neurons of the circumventrisular organ and they are sensitive to glutamate and Na^+ [573]. During dehydration, astrocytes change their coupling with neurons in the hypothalamus [697].

Change in osmolarity affects also neurons out of the thirst center, for instance in the hippocampus [567], but magnitude of the response is usually low.

Neurons in the thirst centers are subjected to deleterious influences during hyperosmotic conditions. They are activated due to a decrease in the hyperpolarizing influences of taurine, which participates in the compensation of cellular injury in neurons and the activation of glutamate afferents (from sensory osmoreceptors). Taurine efflux may be triggered as a response to damage-induced cell swelling [1085] and participates in the controls of osmotic balance, heart rhythm, respiration, blood pressure, body temperature, motor behavior, feeding behavior, alcohol addiction, and learning [29, 154, 189, 619, 1090]. Unfortunately, study of the physiological role of taurine is now in its infancy.

Besides a fall in taurine, Na^+ channels augment damage, and mechanoreceptors, which are activated during cell swelling, transmite osmotic perturbations into electrical signals [573, 697, 1307]. Ca^{2+} channels play important roles in the angiotensin II-induced drinking behavior [1434], but large doses of angiotensin II (¿100 nM) protect dopaminergic neurons against neurotoxicity [495]. Thirst centers support an optimal level of osmolarity and both hyperosmotic and hypoosmotic conditions increased the cytosolic Ca^{2+} level [88]. In rats that feel chronic thirst, an excitability of magnocellular neurosecretory cells in the supraoptic nucleus of the hypothalamus was enhanced [1215]. In such rats were found long-term changes in Na^+ channel expression and an increase in transient Na^+ current [498]. Thus, during drinking motivation, neurons in the thirst centers display alterations of cell metabolism which change their liquid and further ionic homeostasis. This alteration is, apparently, typical for cell injury.

3.3.5 Feeding motivation

Performance of physiological functions needs energy. Plants receive energy from the sun. Daily appearance of the sun does not depend on plants, but sometimes even plants actively search for energy, as does the sunflower. Animals obtain energy from food and have to search for the source of food. Therefore, the search for food is an example of intended actions. Food consumption recovers exhausted energetic homeostasis. During voluntary feeding, intertrial intervals, which are an indicator of the level of hunger, are directly related to homeostatic needs (dependence on levels of glucose and insulin, etc.) of the animal to consume food [50].

Thus, for feeding motivation, the metabolic signal is the energetic misbalance of the organism. Energy metabolism [285, 892] and glucose, the main source for ATP production [201], decreases spontaneously for several minutes preceding the onset of a meal. Fasting suppresses oxygen consumption and, in the short-term, feeding increases it [499]. Neurons in the motivation center of feeding use glucose in their own metabolism [201, 1191] and glucose deficiency in the plasma stimulates eating and damages neurons [1341].

Metabolic sensing neurons control the energy homeostasis and integrate a variety of metabolic, humoral and neural inputs from the periphery. Such neurons, originally called "glucosensing", also respond to fatty acids, hormones and metabolites from the periphery [726].

Feeding motivation is mainly connected with the lateral hypothalamus [1341], but other brain areas are also involved in feeding behavior, particularly, the orbitofrontal cortex, insular cortex and amygdala. Nevertheless, participation of the lateral hypothalamus is the most prominent. At present, more than ten motivationally-relevant substances are known that regulate feeding motivation and satiation [174]. For example, neuropeptide Y and insuline-like receptor signaling systems are essential for the dynamic regulation of (even) noxious food intake [1363]. Insulin receptors are ultimately expressed in the arcuate nucleus [131] and food restriction increases the synthesis of neuropeptide Y there, but not in brain areas outside of the arcuate nucleus [970].

It was shown that neurons in the feeding centers generate activity in correspondence with the feeding cycle [50]. The most feeding-related neurons preferentially responded to a unique phase within a cycle (hunger-satiety-hunger), but neuronal populations integrated single-unit information, which reflected the animal's motivational state across the entire cycle. Mean population firing rates can more efficiently represent phases of a feeding cycle than individual neurons. It was demonstrated that the average performance of individual neurons in the lateral hypothalamus during voluntary eating to satiety was in better correlation with the performance of the entire corresponding ensemble than the performance of neurons in the orbitofrontal cortex, insular cortex or amygdala [50]. Hence, taking into consideration that the large majority of satiety-modulated neurons preferentially responded to a unique phase of the feeding cycle, feeding behavior goes on through activity of a majority of neurons, which are organized in the system, while each one neuron executes narrow functions and does not control entire feeding cycle. This means that feeding motivation of an entire animal is based on an interaction between different neurons and is maintained by the neural network. If a whole motivation would be organized on the foundation of elemental motivations of neurons, one may expect that all phases of motivational behavior be displayed at the level of a neuron. Sometimes this is true, but this is not the rule. We had been noting a non-coincidence between behaviors of separate neurons with phases of an entire motivational behavior, describing other motivations.

Why doesn't each feeding neuron participate in the entire feeding cycle? A whole feeding behavior is probably too complex for a single cell. Even if a motivational behavior for each neuron is a closed-logic cycle and consists of detection of a metabolic flaw, completion of a goal-directed action and consumption of the reward, these phases of single neuron behavior might not coincide with the phases of an entire behavior. For instance, action of the neuron may be directed to particular side of whole behavior of the animal, while reward may be received as an opioid or GABA signal and not as a detection of true reward from the environment. Nevertheless, although whole feeding motivation (and not only feeding motivation) is organized as the activity of an entire network, elemental motivation of neurons, we will demonstrate, plays an important role in a behavior.

A large body of data indicates that neuronal damage is involved in feeding motivation. The level of motivation is connected with the status of excitation in the specific neurons. Stress and pain can also paradoxically stimulate eating [910]. Neurons in the feeding-related brain areas displayed higher activity levels during hunger and decreased activity levels during satiety [50]. Injection of low doses of excitatory amino acids into the lateral hypothalamus elicits eating [1168] at the same concentration at which they induce neuronal damage [801] and increase energy expenditure [65]. Glutamate stimulates appetite so efficiently that it is used in commercial fabrication of food, in spite of its excitotoxicity. Glutamate was ultimately removed from American baby food products, but it is still present in many other prepared foods [1045].

Initiation of feeding is dependent upon the activation of the NMDA [717] or AMPA [1168] glutamate receptors in the lateral and perifornical hypothalamus. Treatment of the nucleus accumbens core with the antagonist to glutamate NMDA receptors impairs response-reinforcement learning in the acquisition of a simple lever-press task to obtain food [615]. Once the rats learned the task, the antagonist had no effect, demonstrating the requirement of receptor-dependent plasticity in the early stages of learning. After blockage of NMDA receptors in the nucleus accumbens core, rats had normal feeding and locomotor responses and were capable of acquiring stimulus-reward associations. Hence, glutamate receptors in the nucleus accumbens core are not absolutely necessary for feeding, but their blockage weakens feeding motivation, so that animals do not make efforts to feed.

Feeding behavior cannot be explained by direct interaction of synaptic processes with an excitable membrane, as, for example, reactions to signals, and simple reflexes are explained by transmission of excitations from one neuron to others. Participation of slow metabolic processes, second and retrograde messengers is a necessary link in organization of feeding. Feeding behavior activates the cyclic AMP system, G proteins, NO, the axonal sprouting and neural-glial interactions[10]. We may suppose that feeding behavior is connected

---

[10] The origin of feeding motivation in mammalians is linked with augmentation of cyclic AMP activity in the hypothalamus [1124, 224, 1430, 1061]. However, the augmentation of cyclic AMP signaling pathway also inhibits feeding [1125] and it is not clear whether this is connected with the necessity for optimal activity of the cyclic AMP system. In the same way, Drosophila mutants with cyclic AMP signaling defects reduced taste discrimination and food intake [864]. The feeding-related substance insulin exerts its effect through cyclic AMP accumulation [761] that affects the participation of G proteins. G protein, a powerful modulator of cellular metabolism, may be essential for feeding, since after inactivation of G proteins Drosophilae were active, but stopped eating and died within 48 h [391]. Eating is also increased by the retrograde transmitter nitric oxide, whose overproduction increases damage in feeding centers, and inhibition of nitric oxide

with the metabolic pathways related to a tension of physiological functions such as those that are activated during development and damage-protection. Feeding motivation, evidently, evokes a transient cell injury in feeding centers, while starvation can be connected with irreversible neuronal damage.

3.3.6 Sexual motivation

Sexual motivation is not an obligatory attribute of individual life, as is evident from longevity after castration. Correspondingly, the origin of sexual motivation is determined by the hormonal status of the organism and by outer stimuli. Initiation of sexual motivation is not really caused by any need of the organism and does not depend on homeostasis [1344]. Its metabolic signal acts as if external to the brain: it is generated by neurosteroids, without a metabolic mismatch of vitally important demands, but creating its own mismatch and demand. Steroid hormones affect metabolism and distress homeostasis of specific neurons and the recovery of homeostasis occurs after sexual satisfaction. Nevertheless, although sexual behavior is not directly connected to any metabolic conflict, the brain sites identified as potential areas for control of sexual function have been implicated in a variety of homeostatic functions [822]. By the way, strictly speaking, with reference to an origin of motivation, sexual and defensive motivations are close enough, since defensive motivation also arises without exhausting inner resources and metabolic disturbance, but the disturbance comes later after action of an external, with respect to brain, force. May be this is the reason why both these motivations have common executive centers, the amygdala [297, 948, 279, 383, 703]. Nevertheless, although much of the scientific literature stresses the role of the amygdala in negative affects (e.g. fear, anxiety and disgust) and defensive learning, amygdala is involved also in positive and appetitive learning [552], and neurons representing positive and negative values (licking and blinking behavior) did not demonstrate clear anatomical clustering [944].

Although the basic mechanisms of sexual motivation (and any other motivation) are similar for both high and primitive animals, sometimes one considers that animals do not feel pleasure from sex and mate only for reproduction, with the exception of the high mammalian. We think this looks somewhat narcissistic. Certainly, we cannot examine what an animal experiences. May be, animals do not have pleasure from feeding, too, and feed only for filling up their supply of ATP. For instance, execution of sexual behavior can function

synthesis reduces feeding [863, 238]. Growth hormone-releasing peptide, related to damage, steadily increases food intake [692]. In snails, starvation results in the sprouting of sensory neurons in the feeding circuit [858]. This change usually occurs in adult animals as the consequence of damage. Starvation leads to early gene activation that is also related to brain damage [914, 336, 1183]. An increase in the neural-glial interaction is related to the degree of feeding motivation, too [697].

as a reward [9]. After all, one cannot even objectively check what his spouse feels, and must resort to believe her words.

Sexual behavior in mammalian is mediated by the medial preoptic nucleus, the amygdala, the piriform area, the arcuate nucleus, and the red nucleus [136, 288, 846]. The neurons of sexual centers demonstrate powerful excitation during mating [846, 1134]. The medial preoptic area is a critical regulatory site for the control of male sexual behavior [333]. Neurons in a medial preoptic area of males containing androgen receptors are active during mating and a third of these neurons contain estrogen receptors, too [802, 496], which also influence male copulatory behavior [243]. The medial preoptic area is crucially involved in consummatory aspects of sexual behavior. By contrast, ventral striatal mechanisms primarily affect appetitive sexual responses [369] and thus again demonstrate that different aspects and phases of motivations are controlled by different brain areas. Sexual motivation increases also in neuronal excitability in the hippocampus [697], which is affected also during other motivations. Male copulatory behavior in the mollusk increases excitability of neurons in the anterior lobe, enhances AP broadening and activates inwards currents that are carried by Na^+ and by Ca^{2+} ions [1154]. Powerful excitation in sexual centers may evoke excitotoxic damage.

Influences of sex hormones to neurons are in correspondence with the effects of natural sexual motivation. Sex hormones affect development of sexual motivation and influence the reception of pain and anxiety, indicating their injurious actions [431, 1053]. Steroid sex hormones modulate the activity of receptors of excitatory amino acids, GABA receptors, voltage-gated Ca^{2+} channels [118, 822], increase the number of neuronal-somatic gap junctions [441] and play a regulatory role in expression of the calcium-binding protein calbindin expression in sexual centers [1186]. The sex hormone estrogen also directly acts on damage-protection processes, such as ordinary sexual behavior. It also influences neurogenesis, the balance between neuronal survival and death, cell migration, dendritic and neurite growth, synaptogenesis, and glial differentiation and neuronal morphology, and alters membrane channel activity [1346].

Sex hormones, which are in fact multiple growth hormones, intervene in the processes of injury-protection of various brain cells and can disrupt metabolism [392, 1174, 1327]. The short-term effect of estradiol on neurons is non-protective and excitatory [118], while slower (several hours) actions of estradiol are thought to be protective in nature [1347]. Protection by estradiol follows late deleterious receptor-mediated effects [901]. Testosterone also augments neuronal damage after axotomy in young male rats, but protects neurons of female rats [1388]. Middle doses of testosterone (1 - 100 nM) also have neuroprotective actions [978].

Progestins and androgens alter sexual receptivity and anxiety through rapid membrane effects and by modulating $GABA_A$ receptors [431, 1053]. For instance, excitatory action of acute administration of estrogen (female sex hormone) is displayed as a rapid decrease ($<$ 10 min) GABA inhibition of neurons located within the insular cortex [1221], hippocampus [1013, 714] and hypothalamic ventromedial nucleus [667].

After initiation, sexual motivation advances, as for any other simple motivation, through deleterious reorganization of cell metabolism. Increase in glutamate levels in the medial preoptic area can elicit genital reflexes even in anesthetized rats (without consciousness), while glutamate receptor antagonists impair copulation [333]. Rapid (non-genomic) initial stimulation of membrane receptors by steroids alters their coupling with G protein and changes production of the second and retrograde messengers, thus changing sexual arousability [118, 200]. Nitric oxide also increases penile erection in anesthetized rats and NMDA-induced erection and yawning are mediated by increased NO synthesis in the paraventricular nucleus [1232]. Correspondingly, inhibition of nitric oxide synthesis eliminates male-like behavior and the deficit was principally in mounting, suggesting that sexual motivational systems were affected rather than consummatory mechanisms [1073]. Retrograde messenger nitric oxide [786] and excitotoxic agents [335] enhance both sexual behavior and neuronal damage in rats. Sexual behavior connected with alteration of homeostasis: in the preoptic area of males, neuronal excitability and Na^+, K^+-ATPase activity increases with the progression of mating [802]. Sexual motivation leads to astrocyte swelling [697]. In males, chronic testosterone treatment enhances the Na^+, K^+-ATPase activity in the preoptic area, whereas castration decreased it and none of these manipulations had effects in the cerebral cortex [802]. As a consequence, the complex of metabolic pathways connected with sexual arousal indicates a development of cellular damage during sexual motivation.

In some aspects, sex hormones affect differently the male and the female and this is not unexpected, in light of their different hormonal status. In the male rat brain, GABA action leads to increased cyclic AMP activity, whereas GABA action in the female brain leads to the opposite effect, while at birth, males have an increased activity of the cyclic AMP system compared to females [68]. Nevertheless, this difference needs to be explained[11]. Thus,

[11] Really, if some particular treatment (sex hormones) affects similarly the sexual function of both males and females, but differently modulate given a metabolic pathway (GABA influence to cyclic AMP activity), this pathway cannot be related to sexual motivation. In addition , the male sex hormone testosterone has a hypoalgesic effect on both the acute and tonic phases of pain after the formalin test, while female sex hormones act only on the interphase (pain decreases) and increase nociception. In particular, females have lower nociceptive thresholds than males [454]. Thus, female sex hormone increases pain just in the interphase, when pain and pain-related cell damage is ameliorated because of homeostatic compensation. Factually, male and female sex hormones evoke similar effects by different means. Male sex hormone decreases the active and tonic phases of pain, while decrease of GABA inhibits the cyclic AMP system and decreases tension. Female sex hormone leads to augmentation of pain in the interphase, during decline of pain, while GABA inhibition leads to augmentation of the cyclic AMP

in spite of fractional and scattered data related to the mechanism of sexual behavior, there are many indications to a connection of sexual motivation with neurohormone-induced cell damage.

3.3.7 Artificial motivations: drug-dependence, self-administration and self-stimulation

Besides natural motivations, there are artificial motivations, which, on the one hand, have a significant role in human life and, on the other hand, give the opportunity to investigate the mechanism of a motivation. Certainly, these phenomena are observed in animals, too. The most important of artificial motivations is drug dependence. Some substances, for instance opioids, induce euphoria, but repeating their implementation leads to the origin of dependence, the necessity of recurring their administration and to the withdrawal syndrome after refusal of further drug treatment [887]. Drugs of abuse, as a rule, produce both euphoric (reward) and despairing (negative withdrawal) reactions. Primary euphoria and subsequent abstinence syndrome are the main reasons for drug addiction. The difference between acute and chronic actions of drugs of abuse is presumably the development of overcompensation that we have considered in Chapter 2.

The syndromes of drug addiction are extraordinarily similar for any drug, although drugs of abuse are highly diverse substances. Each drug binds to its selective protein target in the brain and elicits selective physiological effects upon acute administration. The common actions on brain reward circuits include brain's endogenous opioid, GABA system, dopaminergic neurons and cannabinoid systems. Both inhibition of nucleus accumbens neurons and moderate to strong stimulation of dopaminergic transmission are distinctive features of diverse classes of abusive drugs [981]. All these metabolic pathways are interconnected. There are subregion-specific differences in the proportion of dopaminergic and GABAergic neurons [911]. Thus, the brain is non-homogenous in respect to sensitivity for drugs of abuse, as we demonstrated in reference to other motivations. However, certain drugs of abuse can induce cross-tolerance and cross-sensitization to one another with respect to their rewarding effects [887]. During a self-administration session, distinct but overlapping subpopulations of neurons in the rat medial prefrontal cortex and nucleus accumbens become active during operant responding for different rewarding substances (cocaine and heroin), which, obviously, have different

---

system and increases tension. The consequence of phases of the estradiol effect on a neuron corresponds to the standard development of excitotoxic damage: primary excitation, compensational inhibition and pathological excitation. In this way, sexual hormones come together the pain threshold in mails and females in the acute phase, interphase and tonic phase. Besides, it is not known exactly time scope of sexual motivation in male and female in different phases of mating, pain and their interaction.

targets [225]. Therefore, even if receptors for different drugs differ, some important properties of drug-dependence mechanisms are closely related.

The mechanism of drug dependence was brought to light after the discovery of endogenous opioids, endorphins and enkephalins, the central mechanism of the reward system of brain. Thus, artificially administrated morphine, an exogenous opioid, affects the same areas of a brain that are affected by endogenous opioids under natural conditions. Beside the substances that imitate action of endogenous opioids, there are stimulants of intermediate structures with the end-point link of endogenous opioids, such as cocaine, amphetamine, caffeine, alcohol, etc. For instance, the psychostimulants cocaine and amphetamine affect the dopaminergic neurons, which participate in regulation of the opioid system [282, 450].

Targets for drugs of abuse are located first of all in the nucleus accumbens, in pyramidal cells of the medial prefrontal cortex, which are primarily involved in the reward circuitry, and also in the hippocampus, in the locus coeruleus, in the ventral tegmental area and some other brain zones [1034, 911].

Just as for any other motivation, the motivation to receive the next dose of a drug may be put at the basis of an instrumental behavior. Self-administration of drugs of abuse is a well established fact. Both the euphoric effect of the drug and a withdrawal syndrome compel one to search for new doses of the drug. This corresponds to the development of any other motivation: a negative sensation leads to a search for reward.

The acute and chronic introduction of drugs acts oppositely to activity of main metabolic systems participating in processes of damage-protection, such as cyclic AMP system, Na^+, K^+-ATPase activity and GABA system[12].

---

[12] Some cyclic AMP pathways (adenylyl cyclase types I and V) were inhibited by acute opioid application and superactivated upon chronic exposure, while adenylyl cyclase type II was stimulated by acute exposure and "superinhibited" by chronic opioid exposure [889]. That is, in both cases the acute and chronic actions of the drug worked oppositely in respect to cyclic AMP activity. As well, while acute activation of inhibitory G_i-coupled receptors leads to inhibition of adenylyl cyclase, chronic activation of these receptors produces an increase in cyclic AMP accumulation [889]. In the same way, acute infusion of amphetamine in naive animals resulted in a reduction of excitability in sensitive areas, in comparison with the reaction to amphetamine in amphetamine-pretreated animals [450]. At the same time, spontaneous firing did not differ between control and amphetamine-pretreated animals and drug dependence leads to specific change in sensitivity to the drug. The difference between acute and chronic actions of drugs concerns also Na^+, K^+-ATPase. The regulation of Na^+, K^+-ATPase activity by morphine is inversely correlated with intracellular cyclic AMP accumulation and was connected with phosphorylation of the corresponding proteins. Contrary to short-term morphine treatment, long-term treatment inhibits Na^+, K^+-ATPase activity [1364]. Additionally, 1 hour following cocaine administration, the psychotropic effect decreased and brain Na^+, K^+-ATPase activity was diminished [739]. A deficit of inhibition after the acquisition of drug dependence sometimes leads to a total reorganization of inhibition. After development of opioid depen-

Chronic exposure to many drugs of abuse causes an increase in excitability of dopamine neurons and impairs the dopamine system, mediated via increases in AMPA glutamate receptor responsiveness and augmentation of the cyclic AMP system [887]. Second messenger systems participate in the development of drug addiction [850, 1296, 673, 887, 770, 851] and daily morphine administration downregulates the inhibitory G_i protein and increases cyclic AMP in motivational areas [282, 887]. Alteration of the cyclic AMP system was not uniformly distributed within the brain. Chronic decreased cyclic AMP activity in the shell of the nucleus accumbens (presumably its rewarding site) may be involved in producing positive affective states, but on the other hand, decreased cyclic AMP function in the nucleus of the amygdala (presumably the negative centers) and medial nucleus of the amygdala may be involved in producing and maintaining negative affective states (anxiety-like withdrawal symptoms) [935, 887]. Both rewarding or aversive stimuli cause short-term increases in cyclic AMP activity in the nucleus accumbens, but this is not found in other brain areas [888].

We have to remind ourselves that activation of the cyclic AMP system after damage does not mean necessarily the deleterious role of this system, but rather is evidence of tense activity, directed to protection. Injured as a result of drug dependence, neurons, probably try to survive and augment the cyclic AMP system. For example, artificial inhibition of the cyclic AMP signaling pathway increased alcohol-drinking behaviors [936].

Cyclic AMP signaling, it is supposed, plays a role in morphine withdrawal without participating in the rewarding properties induced by morphine, cocaine and natural reward (food) [1296]. Opioid dependence is heavily investigated. Acute administration of opioids, as a rule, lowers neuronal excitation, reduces cell damage, and this is pleasant for the subject, although the effect depends on the type of opioid receptor and on drug concentration [1014, 1149, 1145]. Opioids protect against ischemia and hypoxia, too. Opioid induced short-term protection and chronic damage is displayed in various brain areas, even in the visual cortex. Neurons of morphine-dependent animals are depolarized relative to the naive animal [718]. Chronic morphine withdrawal increases spontaneous activity, lowers signal-to-noise ratios and produces weaker orientation and direction selectivity in visual cortical cells that were significantly improved by morphine re-exposure [141]. Compared with the cells recorded within 3 h before morphine injection, the cells recorded within 3 h after injection showed improvement of parameter activities. Repeated administration of a narcotic drug overprotects neurons and forces the homeostasis to recover equilibrium, to counteract the protection, increase damage and depart equilibrium to injury [1002, 1149].

In response to drugs of abuse, many genes change their expression and many such genes also change their expression in response to a wide range of

---

dence, $GABA_A$ receptors in the ventral tegmental area paradoxically switch from inhibitory to excitatory signaling [1312].

other stimuli or emotional triggers such as stress, fear, attention and arousal, motor activity and aggression [1026]. Drugs of abuse are strong triggers of pathological motivation, because of loss of corresponding control and they promote pathology-related alterations in the neural tissue. The amount of striatal ribonucleic acids for voltage-dependent K^+ channels during drug dependence in addicts is reduced, the number of Ca^{2+}channel proteins in adrenal-derived cells and in the cerebral cortex is upregulated [608], the σ_1 receptors are activated [807] and NMDA receptors are supertuned. Participation of intracellular Ca^{2+} may depend on a predisposition to drug addiction. A significant increase in the high voltage activated Ca^{2+} currents during the withdrawal period was observed in the withdrawal seizure-prone mouse strain, while those of the resistant strain showed no significant enhancement [967]. All these cellular alterations increase excitability and make the neuron more sensitive to injury.

A classic explanation for addiction is that use of a drug causes compensatory physiological adaptations [1345]. Acute administration of drugs of abuse produces euphoric reactions (rewards), while their chronic administration generates despairing (negative withdrawal) reactions. Correspondingly, reactions of neural tissue to chronic and acute introductions of these drugs are also opposite, and acute action of physiologically relevant doses in naive animals is usually inhibitory, at the same time as their chronic action increases excitability. In addition, acute action of the drug in drug abusers evokes also short-term inhibition and this, evidently, is connected with the origin of a euphoric feeling.

When cellular homeostasis compensates for this overprotection, neuronal damage increases and this compels one to take new doses of opioids. This property is not an exclusive attribute of narcotic drugs. For instance, whereas a single injection of thyrotropin-releasing hormone decreases short-term food and water intake in rats, repeated daily treatments stimulate water intake but not food intake [239]. Studying of opioid addiction gives us a complete example of behavioral, physiological and biochemical patterns of drug-dependence. This convincingly complements scattered data obtained during study of other drugs.

Comparable motivational behavior may be based also on electrical stimulation of euphoria-related brain areas (self-stimulation) and these zones may be exceptionally narrow. For example, the site in the ventral tegmental area effective for self-stimulation is approximately 1 millimeter across [1137]. The brain substrates underlying stimulation-induced feeding and self-stimulation are found in close proximity, but sometimes do not coincide and the threshold for self-stimulation is usually higher [128].

Chemical substances that augment or inhibit other motivations similarly affect self-stimulation. Application of a dopamine receptor antagonist in the nucleus accumbens abolishes self-stimulation of the lateral hypothalamus [262]. Psychostimulants amplify self-stimulation in a manner suggesting enhanced reward [571]. Microinjection of neurotensin in the ventral tegmental area (which increases dopamine release in the nucleus accumbens) produces a long lasting increase in the rewarding effect of self-stimulation, whereas cholecystokinin, which decreases feeding, has the opposite action. Similarly to drug-dependence, both physical and psychological stressors, which evoke damage in neurons, facilitate the acquisition of drug self-administration [976]. Efficiency of self-stimulation, beside the location of irritation, depends only on intensity of stimulation, exactly as cell damage and LTP initiation depends on the power of excitation. Altering frequency, current, or pulse width produced almost identical changes in self-stimulation responses. These findings show that the neutral network serving self-stimulation simply integrates the amount of charge over time [571] and does not transmit or activate any information within the tissue.

Thus, metabolic pathways of drug-dependence and other simple motivations are coincidental as a whole. The number of astrocytic gap junctions increases during amphetamine withdrawal [913]. Drugs of abuse and electrical self-stimulation given for an extended period produce morphological changes in the dendritic branching and spine density in medium spiny neurons of the nucleus accumbens and in other areas involved in the reward circuitry [1034, 887]. There is evidence for a loss of neurons in many regions of the brain in chronic alcoholics [169, 373]. Even uncomplicated alcoholics who have no specific neurological or hepatic problems show signs of regional brain damage and cognitive dysfunction [518]. The enhancement of synaptic responses after ethanol withdrawal is not a result of increased transmitter release from presynaptic terminals [886], but is the consequence of change in the postsynaptic structures and display altered states of neurons. Therefore, there are a few doubts that addictive drugs in the short-term protect neurons and produce pleasure, while their chronic implementation routinely increases protection and provokes the homeostatic system to compensate this overprotection and thus injure cells and produce frustration. For that reason, this homeostatically-induced damage forces one to repeat administration of the drug in order to temporarily protect neurons and receive pleasure.

3.3.8 Motivation to sleep

The motivation to sleep arises after long wakefulness, fatigue or as a result of pharmacological treatments. When sleep begins, a consummatory stage of motivation is commenced and, if this stage is long enough, motivation to sleep ceases. Sleep comprises two main phases: deep, slow wave sleep (sleep spindles with a high amplitude of rhythmic activity in the electroencephalogram), and paradoxical sleep (rapid eye movement sleep with an active electroencephalogram).

As a result of the perceptual overload of the brain cortex, extracellular adenosine concentrations increase in parallel to cerebral oxidation, augment cell damage and trigger a sleep motivation. Sleep starts in the most loaded regions of the cortex and then eventually, after the release of the inflammatory mediator adenosine, reaches the ventrolateral preoptic area of the hypothalamus. After that, brain metabolism drops significantly with a drop in brain temperature and triggers a reaction similar to abortive or partial awakening (paradoxical sleep) [873].

The medial preoptic area is a sleep- and thermoregulatory center and lesions of this area produce a sleep decrease with an elevation of body temperature levels [575]. One of the functions of sleep is the regulation of energy consumption and the mechanism of sleep initiation is related to the mechanisms of feeding and temperature regulation: motivations that are also connected with energy exchange [575]. Sleep is dependent on feeding, because of the metabolic consequences of food ingestion. This basic signal is communicated to the vigilance-controlling centers by a cascade of peptidic and non-peptidic messengers that promote wakefulness and hunger, possibly via a hypometabolic action (as in the case of neuropeptide Y or orexins), or somnolence and satiety, possibly via a hypermetabolic action (as in the case of leptin) [891]. Different combinations of chemosensitive sites operate in wakefulness and sleep [378]. $GABA_A$ receptors, amphetamine-like stimulants, amino acids, lipids (retrograde messengers), proteins (cytokines) and the neuropeptide orexin, are known to significantly modulate sleep [897]. The consummatory phase of sleep is connected with inhibitory processes, as we have described when discussing other motivations[13].

Extended restlessness involves high general neuronal activity in the brain, sensitizes the standard pool of damage-related pathways and may promote excitotoxic damage[14].

---

[13] Sleep spindles are initiated and generated in the GABAergic neurons of the thalamic reticular nuclear. Prolonged (200 - 300 ms) hyperpolarizing potentials in spindle sequences are associated with a significant drop in the apparent input resistance, suggesting an active inhibitory process rather than disfacilitation [1179]. In order to produce spindles, inhibitory reticular neurons should be slightly depolarized [1179]. These neurons are known to be connected by gap-junctions, and neuronal coupling participates in spindle generation. An inhibitory rhythm is supported also in isolated thalamic reticular nucleus and, hence, is of local origin. Paradoxical sleep is electrophysiologically very similar to waking, but metabolic activity in both states is different [107]. Dopamine predominance and decreases glutamate in the nucleus accumbens could be the background of the well-known mental activity during dreaming [491]. However, paradoxical sleep is also supported by GABAergic mechanisms. Thyrotropin-releasing hormone, waking animals, induces membrane depolarization and increases excitability of GABAergic interneurons in the hippocampus to facilitate GABA release [310].

[14] During periods of high neuronal activity, a significant amount of oxygen is used to maintain neuronal electrical activity and other needs, which subsequently produces the free radicals and other cytotoxic reactive oxygen species [575]. A moderate increase of the cytotoxic reactive oxygen species (in the preoptic/anterior hypothalamus) during wakefulness regulates sleep and may be an initial trigger to sleep induction. Low-level brain oxidation promotes sleep via the NMDA-evoked Ca^{2+}-dependent release of sleep-inducing neuromodulators, nitric oxide

As a result, properties of all elemental motivations that we have discussed are comparable and correspond to our idea about the origin of motivation being as a result of deviation of important biological parameters from their optimum and the subsequent development of transient or sometimes permanent cell damage. If homeostatic recovery from damage by means of endogenous resources is impossible as a consequence of exhausting of energy or substances, motivational behavior is the only means for the recovery of healthy equilibrium. We consider that in this case homeostasis undertakes efforts for organization of goal-directed action, such as to establish a new set-point when maintenance of the previous set-point is unattainable. Although the majority of available data correspond to alterations in neurons during the development of motivation (and we mostly paid attention to these processes), neural-glial interaction has been shown to participate in motivational behavior, but this aspect of the problem is poorly studied.

3.4 Reward protects neurons from damage

Reward as well as motivation itself has two components, conscious and unconscious or subjective and objective. When, as a result of motivational behavior, an organism accepts reward, it receives the two components. The first is pleasure, or the conscious component of reward, which encourages further voluntary actions and metabolic improvement. The second, unconscious component of reward ameliorates homeostatic imbalance causing the current motivation. Existence of two components ensures proper functioning of a motivational system. The conscious component is accepted immediately, while the metabolic component needs time for consumption. If, for instance, an animal has received a portion of a meal, its utilization and recovering of energetic stores takes tens of minutes. However, the animal must interrupt eating earlier, otherwise it will be overfeeding, although some animals do eat until gorged and then lie down to digest over some time.

3.4.1 Place of rewards in motivational behavior

When we discussed localization of memory in Chapter 1, we compared two ideas. The first one assumed that every image or event stored in memory has a definite address in the brain. Certainly, each image is the combination of

---

and adenosine [575]. During non-paradoxical sleep, T-type Ca^{2+} channels in neurons are inactivated [107]. Sleep-related rest in Drosophila is connected with a decrease in cyclic AMP signaling [534]. In addition, the sleep-inducing lipid oleamide is gap junction blocking agent [109] and this corresponds to a change in the number of gap junctions during other motivations that we have considered. When motivation increases, the number of gap junctions increases, too. Therefore, during sleep, which is the consummatory stage of the motivation to sleep, gap junctions are blocked.

elemental features. However, not only such elemental features, but an integral image itself, as a whole, has a specific location in the brain tissue. The second idea located data in numerous neurons. Each neuron remembers the entire image, but very roughly. At the moment of identification, every neuron is involved in recognition in the degree to which its stored image corresponds to a current interrogation. We were susceptible to this second idea.

Now, when we are considering the arrangement of goal-directed behavior, this problem arises anew. If motivation and punishment are connected with excitation and neuronal injury, as we have suggested, then reward can be expected to inhibit and protect neurons. Therefore, if each neuron is the source of its own particular behavior, then motivation, action and satisfaction ought to be concentrated in each neuron. This would be perplexing if during the origin of a motivation, certain neurons are injured (in the motivational centers), while reward and, correspondingly, protection is directed to other neurons. On the other hand, there are examples of the steady functioning of collectives, where only a few persons receive rewards while the majority of other persons work hardy almost for nothing. The word "almost" here is essential. The member of a collective needs in some minimal satisfaction, even if this minimum is a continuation of its own life. Nevertheless, evidence points, at least at first sight, to the implementation of this type of brain performance. We had explained earlier that the closed logic chain of goal-directed behavior at the neuronal level may not coincide with the global motivational behavior of an entire animal. Additional explanation of this contradiction may be also found in an ambivalent conception of reward.

Conscious components of reward and motivation usually activate only partially overlapping brain areas. Thus, when a feeding reward is predictable, brain regions recruited during expectation are partly dissociable from areas responding to the actual receipt of the reward [904]. For example, μ-opioid circuits in the medial shell can stimulate food intake via either hedonistic or nonhedonic mechanisms, depending on the precise location [954]. Opioid receptors are widely distributed in the brain, close to main rewarding areas [520], whereas hedonic "liking" circuits for electrical self-stimulation may be more tightly (a single cubic millimeter) localized, for instance in the rostromedial shell of the nucleus accumbens [954] and some other areas. Many other aspects of motivational behavior also may be directed to specific neurons. Administration of the same drug in the close brain areas may elicit an opposite effect, or administration of agonist and antagonist to the same drug may evoke the same result[15].

[15] The $GABA_A$ antagonists are self-administered into the anterior ventral tegmental area, while the $GABA_A$ agonist is also self-administered into the posterior ventral tegmental area [236, 1345]. Amygdala that are considered as the center of aversive reaction, may sometimes participate in rewarding behavior. The central extended amygdala network forms a substantial part of the circuitry underlying rewarding the medial forebrain bundle self-stimulation [1320].

Nevertheless, in certain conditions, the same neurons may participate in both the motivational and rewarding behaviors. We will demonstrate later that such capability is not the inalienable belonging of some special neuron. Quite the opposite, an arbitrary neuron in special experimental conditions may demonstrate an entire motivational behavior.

Comprehension of the essence of the conscious reward has increased during the last few years [9, 115, 206, 270, 297, 299, 656, 799, 1098, 1137, 1345, 1354, 1169, 888]. Reward and reinforcement are related notions, but although reinforcements are causally concerned with rewards, it is possible to differentiate between the two phenomena. Motivation is satisfied by a reward or by avoidance of punishment, while reinforcement signals the success of a behavior. Reinforcement is a criterion for desirable increase or decrease in the rate of a beneficial behavior following modification in the environment. The mesocorticolimbic dopamine systems originating in the ventral tegmental area and their connections with the prefrontal cortex, striatum, nucleus accumbens and amygdala with their glutamatergic, GABAergic inputs and opioid receptors are responsible for reinforcement and reward. While opioids, dopamine and GABA fulfill their actions in the reward system through inhibition, glutamate accomplishes an intermittent role, as an excitatory messenger. Participation of other neurotransmitters in the organization of the reward system, for instance serotonin, is also well known.

3.4.2 Chemical mediators of a conscious reward

When one receives a reward, in some brain areas the opioids, dopamine and GABA transmitter systems are implicated. Moreover, activation of these systems evokes a euphoric sense, so that animals work in order to reach administration of these substances during self-introduction, place preference, etc. At present we know, where rewarding areas are located in the brain and which substances are rewarding. We also understand that just these brain zones and not other are rewarding; just in these areas, rewarding substances are produced and their receptors are located just in these zones, although producing areas and receptive areas are only partially coincided.

All rewarding substances affect neurons through various receptors and their outcome depends on the class of chemoreceptors. The dopamine system participate in a reward mostly through dopamine$_2$ receptors. However, this is not certain. There is a message that dopamine$_1$ receptors in the ventral tegmental area are involved in intravenous cocaine reward and food reward [1122]. Besides, there are at least fourteen serotonin receptor subtypes and many of these have been shown to interact with dopamine [22]. As well, although euphoric effects of GABA are usually achieved through GABA$_A$ receptors, some drugs of abuse act through GABA$_B$ inhibition of nucleus accumbens neurons and stimulation of dopamine transmission [981]. Similarly, μ-, δ- and κ-opioid receptors participate in the reception of reward, although μ-opioid receptors may play a more prominent role [287, 80]. Nevertheless, this knowledge does not fully supplement our comprehension of the phenomenon of reward. It would be constructive to understand how one feels pleasure when some special chemical events have happened.

Both reward and reinforcement are connected with the opioid and dopamine systems: phasic influences of endogenous opioids are more related to a hedonic impact of reward, while phasic activity of dopamine neurons is related to reinforcement, or triggers the events that instantiate the incentive salience of rewards [115]. The midbrain dopamine system is rapidly activated (50-200 ms) by rewarding events that are better than predicted and are depressed by events that are worse than predicted. Dopamine neuron activation with a slower timescale has been observed during expectation of reward for various forms of behavior: feeding, drinking, punishment, stress, sex, drug reward, and social behavior. After learning the dopamine system becomes responsive for reward predictors and unresponsive to the reward "itself" [1098]. Various classes of drugs of abuse, psychostimulants, opioids, ethanol, cannabinoids and nicotine increase dopamine transmission in limbic regions of the brain. However, only psychostimulants increase dopamine transmission clearly both necessary and sufficient to promote rewarding effects of drugs of abuse [977]. This is in a poor agreement with the decisive role of dopamine in the reception of a reward. Particularly, dopamine neurons are highly sensitive to vital aversive stimuli [1068]. There is an optimal level of dopamine in the nucleus accumbens for self-administration of cocaine and this also contradicts the role of dopamine as an endogenous reward. In drug abusers, cyclic AMP production increases and this is connected with dopamine_1 receptor enhancement and with a reduction in dopamine_2 receptors [887, 910]. This indicates a slow participation of the dopamine system in a motivational behavior and therefore also contradicts the role of dopamine as a direct endogenous reward.

It has been supposed that brain mesolimbic dopamine systems mediate 'wanting' rather than sensory pleasure [113, 206]. Both the mesocorticolimbic dopamine_1 system and the nigrostriatal dopamine_2 system involve a GABAergic mechanism [30, 656] and, according to various data, both dopamine_1 and dopamine_2 receptors are involved in the interaction with opioids during reinforcement, although dopamine_2 receptors are usually inhibitory and predominate in reinforcement [910, 22], while dopamine_1 receptor is important during acquisition of new habits [614], that is during motivational behavior. It is possible that dopamine_2 receptors participate more in the reward system, while dopamine_1 receptors in the punishment system and the abstinent syndrome. Thus, in spite of the obvious participation of the dopaminergic system in reception of rewards, dopamine is hardly likely to be the main rewarding substance that immediately protects neurons. Rather, dopamine signals the ingress of the reward into the organism.

The presumed mechanism of action of μ-opioids in the nucleus accumbens involves disinhibition of the dopamine system by inhibition of nearby GABAergic neurons that normally hold their dopaminergic neighbors under inhibitory control [1345, 80], though opioids can affect directly through specific chemoreceptors. Consequently, activation of the dopamine system is not at all the first sign of reward in the brain. By the way, just $GABA_B$ receptor stimulation is responsible for inhibition of dopamine release in the ventral tegmental region by acting on the nucleus accumbens, but just stimulation of $GABA_A$ receptors counteract anxiety. Therefore GABA, probably, can exert a rewarding action apart from dopamine pathways.

Dopaminergic system, the potent modulator of gap junctions [963] is, probably, a sensor of the GABA level at the scales of damage-protection and negative-positive feeling. Neuroleptic haloperidol, antagonist of $dopamine_2$ receptors or depletion of accumbens dopamine, prevented carrying out those instrumental tasks in rats that required large effort for successful behavior, although rats worked properly when task was simple [1067]. Considering that the dopamine system is responsive for reward predictors and unresponsive to the reward consumption, this system is a good candidate, as a participant of intentional actions.

The GABA and, to some extent, serotonin systems are tightly implicated in the development of a positive state in the brain. Pleasure is the most dependent on opioid and GABA systems [184]. While the opioid system is implicated in brain activity mostly during urgent circumstances, GABA is a continuously- acting factor. GABA is both the most widespread inhibitory transmitter and potent regulator of a background level of positive feeling. Acute positive states after receipt of reward also depend on the GABA system, its interaction with opioids and dopamine systems and counteraction with the excitatory glutamate system. So, a reduction of $GABA_A$ergic inhibition in the basolateral amygdale promotes anxiety [674], sleep [491], temperature regulation [580], an acute reaction to drugs [895, 1182, 619, 1436], sexual approval [822, 1221] and respiration [907].

One distinction of defensive motivation is an absence of physical substrate for the consummatory phase; a positive outcome in this case is the removal of danger. A real consummatory phase is absent also in related sexual motivation; only pleasure serves, as a reward. Therefore, a conscious component of reward is in given cases a single one, because a metabolic component is absent. Neuronal mediators of pleasure are endogenous opioids. Therefore, endogenous opioids are involved not only in antinociception and avoidance, but also in reward and reinforcement. Opioids may induce sensitization to the effects of other drugs [282]. The opioid system is an ultimate linkage in the reward systems for various motivations: sexual behavior, stress responses, respiration, temperature regulation, drug dependence, alcoholism, drinking, feeding, and social play behavior [270, 910, 1293, 1297]. For instance, endogenous opioid peptides are involved in the expression of physical nicotine dependence [112]; another example is that mice pups lacking specific μ-opioid receptors don't cry when separated from their mother [96]. This suggests that reward systems of various simple motivations are based on a rewarding substance in the defensive motivation, perhaps the most important motivation for the maintenance of an individual life. Besides, a rewarding substance for defensive motivation is exactly matched to avoidance from an apparent threat, whereby conscious and unconscious components of reward almost coincide. For other motivations, the same chemical means are, evidently, used as a conscious signal for future metabolic improvement.

The motivational system defends an organism from danger through adjustment of its interaction with the environment. An organism also defends itself at the expense of inner resources, without participation of consciousness. These inner protectors are homeostasis and the immune system. As we have demonstrated, homeostasis is tightly connected with motivation and it may be elevated at the conscious level. The immune system is not elevated at the conscious level, but it is also connected with motivation. Neural and immune functions tightly interact. Neurotransmitters serve as immunomodulators through their release and diffusion from nervous tissue, action to lymphocyte cell-surface receptors and the modulation of immune function, but neurotransmitters (dopamine, glutamate, opioids, etc.) can also be released from leukocytes [406]. The origin of the opioid system, probably, dates from the advent of the immune system, as a signal for avoidance of danger. Opioids reduce cellular immunity [1057]: they decrease cell proliferation and inhibit migration. Unavoidable stress affects the progression of a variety of tumors, and, in most cases, it exacerbates tumor metastasis [819], but moderate stress has a defensive function. Immunity is defense mechanism and motivations also fulfill defensive function, while opioids signal avoidance of danger. Opioids do not defend by themselves but they temporally ameliorate destructive consequences of stressful reactions, pain, fear and immune attack. When the opioid system fails in neuropathic animals [1439], they become helpless. Opioids give an opportunity to organize defense.

3.4.3 Inhibitory actions of rewards

A reward, as well as a punishment, usually evokes powerful excitation everywhere in the brain, or almost everywhere, but not in the rewarding system of brain. Satisfaction of motivations is connected with the development of inhibitory processes in the motivationally-specific areas, as for example during feeding [1164, 1375], mating [1134, 802, 605], recuperating of temperature [498], consumption of narcotic drugs [935, 887], drinking [697] and falling asleep [1179].

As to drug-induced rewards, intravenous self-administration of cocaine inhibits the firing of neurons in the nucleus accumbens [960]. Acute ethanol exposure suppressed excitability and shifted the balance between excitatory and inhibitory synaptic transmission toward inhibition [234]. The neurons of the ventral tegmental area are activated during drug-seeking behavior (heroin self-administration), but they are inhibited during drug taking behavior [638]. Passive drug injections, in contrast, cause a weak tonic increase in activity.

Natural reward-evoked reinforcement induces inhibition of activity. Consummatory feeding events [715] inhibit the firing of the nucleus accumbens neurons. A high extracellular glucose concentration, which inhibits feeding, induces hyperpolarization of orexin neurons, while a low extracellular glucose concentration excites these neurons [1375]. This inhibition, evidently, counteracts to excitotoxicity and protects cells against the damage that has give rise motivation.

A dramatic increase in the activity of a majority of the neurons in the medial preoptic nucleus (sexual center) during sexual arousal in male rats ceases abruptly after intromission (lasting for a seconds), and activity remains greatly reduced for protracted period (about 2 min) after ejaculation [1134, 802]. The neurons of the preoptic area of female rats revealed more subtle and transitory changes during sexual behavior, but half of the neurons were also inhibited after intromission [605]. Immediately after orgasm and for over 1 h following orgasm, plasma prolactin concentration increases in both men and women [679]. Prolactin secretion is controlled by inhibitory signals and the primary inhibitory input for prolactin secretion is dopamine. Mating induced activation of endogenous opioids is provided by the increase in pain thresholds following ejaculation [802]. Inhibitory influence of sexual reward may be endorsed also by a decrease in excitatory influence of glutamate. Glutamate plays a pivotal role in the consequence of events during copulatory behavior [333]. Concentrations of extracellular glutamate in mail rats, microdialysate samples from the medial preoptic area before, during, and after copulation, reveal a slight rise when the female was presented, a significant increase during periods of mounting and intromitting, and a very large increase in samples collected during ejaculation. A precipitous fall in levels occurred in the first postejaculatory sample; the magnitude of this fall was highly correlated with the length of the postejaculatory interval of quiescence. Increase in glutamate levels in the medial preoptic area before and during mating can elicit genital reflexes in anesthetized rats (without consciousness), while glutamate receptor antagonists impair copulation. Nevertheless, investigation of neuronal activity offers results that are more precise, than microdialyse. Concentration of glutamate in the medial preoptic area did not reveal short-term fall after intromission. The fall in neuronal activity after intromission was too short, a few seconds. For measurement of glutamate concentration, one need collect microdialysate samples from the medial preoptic area during couple of minutes. This consumes time and set a limit on a resolving power of the method.

Rewarding substances, as a rule, inhibit neurons. In the nucleus accumbens, $dopamine_1$ receptors mediated either inhibition or facilitation, while $dopamine_2$ receptors, which are usually responsible for rewarding effects, predominantly mediate inhibition [457]. Dopamine through the $dopamine_2$-receptor inhibits midbrain neurons on a millisecond timescale [97]. Dopamine also modulates cortical and thalamic glutamatergic signals and $dopamine_1$ receptor signaling enhances dendritic excitability and glutamatergic signaling, whereas $dopamine_2$ receptor exerts the opposite effect [1197].

Rewarding action of GABA is connected with its inhibitory action. GABAergic system is not only the same wide-spread inhibitory system that penetrates a whole brain [590]. $GABA_A$ receptors are present already in Hydra vulgaris, one of the most primitive organisms with a nervous system. Accordingly, $GABA_A$ antagonists stimulate feeding, while $GABA_A$ receptor agonists suppress ingestion [1248] and inhibit sexual behavior and drinking [939]. Inhibition can be achieved by excitation of inhibitory neurons, such as ethanol augments GABAergic transmission within the amygdala [813, 895]. Satisfaction of defensive motivation also proceeds through inhibition and participation of GABA inhibition is essential. Extinction of fear conditioning requires participation of medial prefrontal cortex and amygdala [628]. The glutamatergic efferents from the medial prefrontal cortex synapse on amygdala GABAergic neurons and contribute to the extinction of hostile experience [628, 16]. Excitation does not injure GABAergic neurons, since interneurons containing GABA are weakly vulnerable to damage [588]. Expectation of reward after learning may augment GABA inhibition in forestalling manner: the firing rate of a subpopulation of GABA neurons in the ventral tegmental area increases in anticipation of brain stimulation reward [706].

Protection against excitotoxic damage decreases pain, but $GABA_A$ receptors are not the only target for an attack against hurtful excitation of various kind. Compensatory inhibition may also be connected with taurine. Taurine is released simultaneously with or slightly later than an excess of glutamate under various neuron-damaging conditions. For instance, after passive avoidance training, both glutamate and taurine in the hyperstriatum were increased [286]. Reduction in excitability protects cell from damage and has analgetic properties. Particularly, suppression of Na^+-channels has analgesic effects [627]. Preventive treatment against the migraine, a common episodic pain disorder, include antidepressants, gap junction inhibition and antiepileptic drugs, which enhances antinociception [1142].

Opioids as well decrease negative motivations and ought to inhibit neurons. Indeed, during presentation of emotionally salient stimuli, higher baseline μ-opioid receptor binding in the brain was associated with lower regional cerebral blood flow in this region and this is consistent with an inhibitory/anxiolytic role of the endogenous opioid system [736]. Cyclic AMP system is inhibited by opioids and subsequent inhibition of voltage-dependent cation channels may be a mechanism by which opioids inhibit excitability and relieve pain [577].

Nevertheless, rewarding substances sometimes evoke excitation, dependently on concentration and localization of receptors in the brain. For example, opioid receptors can couple through stimulatory or inhibitory G_s/G_i proteins to both ion channels and adenylyl cyclase [520, 569] and, correspondingly, can exert not only inbitory, but also excitatory effects to neurons. Ultra-low concentrations (pmol) of opioids increase the AP duration and are thus excitatory. However, higher concentrations (nmol or micromol) of opioids, which evoke antinociception and are therefore physiologically relevant, decrease the AP duration and are thus inhibitory. The protective properties of opioids, sometimes attributed only to the δ-opioids protecting cells up to high concentrations (10-20 mg/kg). However, small concentrations of μ-opioids (200-400 μg/kg) [650, 859] and δ-opioids (10^{-10}-10^{-15} M) [755, 1014] also protect neurons. For instance, morphine can produce analgesia via spinal μ opioid receptors in the absence of μ-opioid receptors. [1373]. After the brief repeated exposures of μ-opioids, the excitation was enhanced, but the inhibition was desensitized and potentiating action was distinct from the inhibition in that it had a long time course (minutes to develop and last tens of minutes) [666]. This demonstrates transient inhibitory action of opioids.

Motivationally-relevant substances as well affect excitability of specific neurons. It is natural to suppose that a substance that increases motivations excites neurons, while a substance decreasing motivations has an inhibitory effect. In support of this assumption, orexins [1156, 1199], substance P [806, 957, 1001], galanin [1021], estrogen [118], oxytocin [1008], vasopressin [1323], angiotensin II [382], and estrogen [118] exert excitatory action in physiologically relevant concentrations, although androgens have less consistent effects. Conversely, substances that decrease motivations such as, alcohol [331, 334], cholecystokinin via feeding satiation-related A-receptors [294], insulin and leptin [1375] usually inhibit neurons. Receptors for satiated substances, insulin and leptin, are present in the specific hypothalamic regions that control energy homeostasis, and these hormones reduce food intake in lean rats and hyperpolarize glucose-responsive neurons in the hypothalamus of lean rats [1164]. The excitatory effect of cholecystokinin has also been reported, but it either occurs at high concentrations or was mediated mainly via B-receptors, connected with anxiety [294, 531, 1193]. Nevertheless, this assumption is not absolutely compatible with all the experimental data. Drugs may affect many different cells, many points of a cell and may act via different receptors. Drugs may also affect both excitatory and inhibitory neurons and both presynaptic and postsynaptic sites. In some cases, even GABA [295] and glycine [142] excite neurons, while excitatory amino acids sometimes induce inhibition [1101]. Besides, the dependence on the concentration of motivationally-relevant substance actions of both neuronal activity and behavior are non-linear. For instance, neuropeptide Y [621], substance P [1001], galanin [1338], and oxytocin [869], which increase motivations, inhibit neurons at higher concentrations. On the whole, the direct action of motivationally-relevant substances at physiologically relevant concentrations on neurons in motivational centers more frequently tends to be excitation, if the motivationally-relevant substance increases motivation, and tends to be inhibition if the motivationally-relevant substance decreases it.

Our consideration has demonstrated that as well as excessive excitation is a distinctive property of motivations, intense inhibition in motivational areas is connected with satisfaction of motivation and with a reward. This conclusion appears a trivial. However, excessive excitation in given case leads to neuronal damage, while intense inhibition promote homeostatic protection.

3.4.4 Specialized neurons generate motivations and accept rewards

While we considered rewarding substances, inducing pleasure, we concentrated on neuronal mechanisms of conscious reward. Much less is known about the metabolic component of reward, although this appears to be simpler.

Obviously, neurons in motivational centers are specially tuned to detect homeostatic discrepancies. They appear to have higher sensitivity to these specific factors and their resistance to damage is attenuated. Neurons that express the orexigenic peptides are exceptionally sensitive to injury [588, 1120, 1374]. They lack the calcium-binding proteins parvalbumin and calbindin [721], which normally decrease any rise in intracellular Ca^{2+} and promote cell survival following glutamate excitotoxicity [859]. Distribution of the inhibitory neurotransmitter taurine provides an example of the specific tuning of neurons in thirst motivational centers. The highest taurine concentrations are in the hypothalamic nucleus connected with water balance and the lowest - in the remainder of the hypothalamus [573]. It is possible that taurine's compensatory inhibition protects neurons during necrotic swelling and thus may prevent detection of homeostatic imbalance in hypothalamic areas not connected with water balance. High sensitivity to injury in motivationally-relevant neurons is necessary in order to detect slight disturbances in homeostasis.

The specialization of neurons in the primary motivational centers must result in a difference between the reactions to rewarding stimuli in the primary motivational centers and in the rest of the brain. An example of specific tuning of the neurons in motivational centers is the feeding motivation. A connection between motivation and hyperthermia appears to oppose the essence of the feeding motivation that relates to a fall in energy and a reduction in thermogenesis. However, in reality, primary feeding neurons detect a decrease in body energy. Neuropeptide Y and orexin neurons in the arcuate nucleus act homeostatically to restore normal energy balance; they become overactive following a critical fall in the body's energy stores [1209, 1341, 1375]. In addition, feeding neurons in the lateral hypothalamus are stimulated when the level of glucose falls, while neurons of the ventromedial hypothalamic nucleus increase their firing rate as the level of glucose rises [1341]. Correspondingly, consumption of a food reward enhances glucose and inhibits specific feeding neurons, while exciting a majority of neurons in the brain. Medulla inspiratory neurons respond to hypoxia, hypercapnia or acidic stimuli with an increase in electrical activity, while most of the other neurons in the brain are either inhibited or unaffected [613, 1224, 1319]. An increase in extracellular osmolarity depolarizes thirst neurons, but does not excite other neurons [29], while lowering of extracellular osmolarity inhibits thirst neurons but causes a reversible increase of both excitatory and inhibitory postsynaptic currents and Ca^{2+}-mediated inward currents in the hippocampus [567]. Neurons in the male medial preoptic nucleus demonstrate powerful excitation during sexual behavior, but their activities abruptly decrease after intromission and especially after ejaculation [846, 1134]. At the same time, ejaculation activates related neuronal clusters outside of the medial preoptic nucleus in both males and females and suppresses neuronal activity in the medial preoptic nucleus as a negative feedback [1299]. Behavioral positive affective reactions cause activation in corresponding brain areas, but they usually increase motivation and do not induce pleasure [113]. We therefore suggest that acceptance of a reward inhibits neurons in primary motivational centers while it excites a majority of other neurons.

Opioid-mediated conscious rewards are widely distributed in the brain, sometimes far out of primary motivational areas. For example, primary respiratory neurons in the ventral medulla, which are activated by CO_2 and by low pH [867], are opioid-insensitive, in contrast to neurons in the pre-Botzinger complex [378], which is the site for respiratory-related rhythm generation and where inspiratory bursts depend on a Ca^{2+}-activated current linked to glutamate receptors [930]. Therefore, just the neurons sensing a disturbance in oxygen homeostasis do not react to an opioid reward and control only the harmony in homeostasis itself. This is especially essential with regards to that opioids suppress respiratory rhythm [81, 960]. It is natural, that rewarding centers connected to euphoria only partially match the motivational centers. Opioid signals transiently ameliorate neuronal damage (see next subdivision), promote a positive subjective feeling and thus contribute to goal-directed behavior, but metabolic improvement cannot be replaced by opioids. We may assume that the set of neurons producing motivations and experiencing injury coincides with the neurons receiving protection, but only partially coincides with the neurons receiving conscious reward. Therefore, in producing a motivation neurons control an entire cycle of motivational behavior. Incomplete coincidence between the set of primary motivational neurons and the set of neurons that are sensitive to rewarding substances cannot prevent the proper performance of this whole unit of the neuronal system.

3.4.5 Protective actions of rewards

The first indirect indication to protective influence of rewards is their powerful inhibitory interaction with the neurons in motivational centers and what's more, this is a general rule. Protective influences exert positive feeling, decrease anxiety and produce euphoria. For example, neuroprotection evoked by the σ_1 receptor activation decreases anxiety [808], while neuronal damage, evoked by chronic ethanol is reduced by acute nicotine [959]. The substances that evoke or increase brain reward (opioids, dopamine and GABA) protect brain tissue from damage. Acute opioids not only protect cells against motivation-induced injury and against pain and anxiety, but even defend cells against ischemia and hypoxia, too [520, 141]. This concerns to activation of various opioid receptors, such as μ-opioids [630, 986], δ-opioids [151, 1107, 1189, 1425] and μ-opioids in low concentration [755, 1014]. Taking into consideration that opioids evoke dependence, protective action is observed, certainly, after their acute implementations. Chronic their action, as we already discussed lead to opposite effects: chronic consumption of opioids evokes irreversible damage of brain cells. Many other rewarding substances, such as cocaine [1153], caffeine [602], alcohol [331], and cholecystokinin [803, 1229] protect neurons from damage. However, because of the development of compensatory processes, chronic cocaine ingestion [709], chronic or high doses of alcohol [331, 334], etc. increase excitability and exacerbate damage. Serotonin also plays a pivotal role in the homeostasis of neural tissue [69], as it interacts with neurosteroids and opioids and affects the morphology of target cells [1053, 1370].

Action of motivationally-relevant substances to behavior is also can be connected with their influence to damage-protection of neurons. Usually such substances protect cells, if they decrease motivations, though this is not always the case, since their affect by a complex way depends on concentration and this is difficult task to compare effective concentrations of substances in different experiments.

Besides these relatively direct results concerning protective influences of resulting satisfaction, consumption of rewarding substances activates metabolic events in the brain, which, typically, decrease damage. Both opioids and stimulants cause a global decrease in brain metabolism [287]. Protective action of the acute μ-opioids is exhibited also in stimulation of Na^+, K^+-ATPase activity [1357] and in reduction of the Ca^{2+} current by μ-opiod agonists. Intravenous cocaine injection directly inhibits Na^+ channels [637]. The opioid antagonist naloxon unexpectedly protects cells against damage, but this protection is mediated by its inhibition of microglial activity, and its stereoisomer, inactive for opioid receptor, protects neurons with equal potency [753].

The dopamine$_2$ receptor agonists that presumable exert rewarding effect activate K^+ channels in the rat striatum and these channels open under energy-depleting conditions in the absence of a dopaminergic agonist [741]. These agonists also arrest Ca^{2+} dependent AP, hyperpolarize the membrane of growth hormone-secreting cells by means of K^+ conductance increase [1210] and affects Ca^{2+} evoked arachidonic acid release, while they produce inhibition of cyclic AMP [320, 1092]. Dopamine$_1$ agonist has the opposite effect of arachidonic acid and increases cyclic AMP activity [320, 1092]. Dopamine-containing neurons are very stable [803, 1380] and dopamine receptors are connected with the inhibitory G$_i$ protein [1422]. Dopamine$_1$ and dopamine$_2$ receptors usually have an opposite influence on intracellular events, in accordance with their opposite action to neuronal excitability and neuronal protection [887]. In the majority of cases, dopamine through dopamine$_2$-receptor protects cells against damage [789, 1107, 159]. However, both dopamine$_1$ and dopamine$_2$-receptors might contribute to the toxic action of dopamine and dopamine-induced cell death is not restricted to neuronal cells [159]. For instance, neuroleptic haloperidol, dopamine$_2$-like antagonist, induces loss of cell membrane integrity that is typical for necrosis [870].

Another important reward substance GABA besides vast inhibitory influences also protects cells against damage [593, 1300]. Protective role of

GABAergic neurons is augmented by means of gap junctions. Coupling of inhibitory GABAergic interneurons via gap junctions facilitates to integration of inhibition. The threshold for self-stimulation coincided with the threshold for electrical coupling between GABA neurons, the degree of responding for self-stimulation was also proportional to the magnitude of electrical coupling between GABA neurons, and gap junction blockers increased the threshold for self-stimulation without affecting performance [706]. The reinforcement-relevant substance cholecystokinin, too, inhibits and protects neurons. Cholecystokinin A- and B-receptors, as well as $dopamine_1$ and $dopamine_2$ receptors differently affect motivation and satisfaction and correspondingly exert a different effect states of neurons. Cholecystokinin B-receptors reduce potassium leak conductance, resulting in an excitation of the neuron and connect with stressory reactions, whereas cholecystokinin A-receptor-related inhibition was associated with a membrane hyperpolarization and a decrease in input resistance that developed 2-6 min after the arrival of the drug into the extracellular medium [161]. Inhibitory current was generated by potassium currents.

Thus, rewarding substances usually protects neurons, although this protection is insufficiently in order to substitute the natural reward itself that supplies organism by liquid, nutrients and energy, recovers metabolism and protect cell from damage. Rewarding substances ensuring a conscious component of rewards have two functions. They temporarily decrease stress after acute action of malefactors, guarantee proper interaction with an environment and signal the appearance of a metabolic component of rewards. Therefore, closed cycle of motivational behavior in neurons do not coincide with the entire behavioral course of action.

3.5 Goal-directed behavior of single cells

Motivation is a goal-directed activity of the entire brain, based on primitive goal-directed activity of an enclave of individual neurons. Specific neurons detect particular requirements, determine the goals of behavior and the means for their achievement, accomplish intentional actions, consume reward and recover metabolic equilibrium. On the other hand, broad areas of brain are involved in motivational behavior. With rare exceptions, the more subjective is an experience, the more power objective traces of neural tissue participation in this experience one may reveal. However, motivation implicates not only a large number of neurons. Diverse neural centers and different neurons perform specific functions: this is not a crowd of voting nonentities. A brain appears here as a construction having specific assemblies that are connected in the explicit architecture. Does this mean that brain, as a whole, arranges a motivation and that a motivation is the novel quality emerging in the brain, which in no way is presented at the lower level, at the level of neurons?

3.5.1 Motivationally-relevant substances distort properties of an excitable membrane

Motivationally-relevant substances act on definite receptors in specific areas of brain. The peculiarity of this interaction and other data show that on the one hand different neurons are specialized for control of restricted functions and do execute them. But on the other hand, all neurons produce a similar basic consequence of chemical reactions and each one has the capability to perform those steps of behavior that are not involved in its usual repertoire. Thus, although motivationally-relevant substances are specialized in fulfillment of narrow functions in special areas of brain (and this confirms focused participation of various brain areas in the defined motivation), the same substances can produce other motivation in other brain areas or in special circumstances and this is evidence about common mechanisms of different motivations. Exceptionally low but effective concentrations some of these substances indicate that it is possible to produce an entire motivation by means of action to very local brain regions and, may be to a single neuron. Besides, the manner of influence of various substances to different neurons is similar.

Both neuropeptides and neurosteroids reorganize properties of excitable membrane, such as this is observed during damage-protection of neurons, and they also exert a modulatory action on neurons and change their reactions to stimuli. Both neuropeptides and neurosteroids alter those properties of the neuronal membrane that determine neuronal excitability. They can reorganize excitability without a direct impact on the membrane potential, evidently influencing a critical level of AP generation. For example, motivationally-relevant substances may change firing rates without corresponding changes in membrane conductance and the membrane potential of neurons [806, 869, 1069]. They also may modulate properties of Ca^{2+} channels [331, 456, 1008] and exert an effect via modulation of K^{+} conductance [772, 1323]. These changes are usually more prominent than direct influences on a membrane, because they interact with the receptors coupled to G proteins connecting with systems of intracellular messengers. Modulatory actions have been discovered for neuropeptide Y [254, 1120], cholecystokinin [379, 1069], ethanol [895], substance P [806, 957], somatostatin [722], oxytocin [256], angiotensin II [727], insulin [1105], opioids [1293], testosterone, estrogen [118, 667, 82], orexins [1341], cocaine, and amphetamine [221]. Therefore, the cause of influence of these substances to behavior may be their ability to reorganize neuron's state. This activates homeostasis and compels it to recover equilibrium. Such an assumption is especially convincing, since motivationally-relevant substances, as we have demonstrated, really affect cell status. So a profound influence of motivationally-relevant substances on neurons appears to be somewhat strange. Neurons participating in motivation are in a critical state with respect to its individual survival. If neurons are only block of brain's construction, it would be difficult to understand, while workpieces of construction must operate at the edge of breakdowns.

3.5.2 How small may be the brain signal controlling a body?

An instrumental behavior occurs only in the presence of a motivation and therefore the contribution of a neuron to goal-directed actions represents its participation in a motivational behavior. In the majority of experiments, we study behavior of the neuron within the brain. However, sometimes one manages to devise so special methods that pinpoint the role of the only neuron in the entire behavior.

Possibly, elemental motivation is inherent in a single cell, just like chemical properties of a substance are determined by the structure of the molecules compounding this substance. However, a salient feature of any control system is the possibility to manage great material processes using a small control signal. The control system of brain is not an exception. Certainly, in different circumstances a signal controlling a body may be smaller or bigger. On the average, in a normal state only a few percentages of neurons are active and only a small share of these active neurons participates on a given behavior. In the aware state, particularly during motivational behavior, the number of active neurons increases. However, for our consideration, it is important to evaluate the low limit of the intensity of brain activity that is still capable to actuate body. Surely, such a parsimonious control is not the rule, but it outlines a power of management. Sometimes a control signal from brain may be rather fine.

Firstly, in exceptional conditions only one neuron may control the output of a macroscopic reaction. In the artificial conditions of an experiment, when environment is organized so that the result of behavior depends on the action of the single experimental neuron, any such neuron, probably, may display capability to regulate an entire behavior. For instance, it was possible to elaborate olfactory conditioning in honeybees, when the natural reward was substituted by the activation of a single identified neuron [515]. Similarly in the mollusk, stimulation of the single identified central interneurons can be used as the unconditioned stimulus for *in vitro* conditioning in both appetitive and aversive learning [616]. Behaviorally relevant neural modifications in Aplysia can be gated by a single identified cell [1358]. The cellular analog of an instrumental reflex was elaborated on isolated mollusk neurons in response to intracellular electrical stimulation [1301]. In the mammalian, stimulation of a single glucoreceptor of hypothalamus by means of microiontophoresis of glucose may serve as a functional unit of a reward [664]. Cortical neurons in tissue culture may self-administer psychostimulants and the rewarding substances ethanol, nicotine, opioids and dopamine-agonists [1173]. In the motor cortex of a monkey, neurons exceed a preset firing rate if the appearance of a spontaneously-augmented pattern is reinforced by food [385]. In addition, the same neurons may participate in triggering of goal-directed behavior and play a decisive role in the reception of reward [226].

Secondly, not only a single neuron, but a single AP in a single neuron may be responsible for action of an entire animal. For example, one AP in a single

Mauthner cell of fish can dramatically switch the behavior of the entire animal from swimming to escaping [1205]. A train of APs in a single pyramidal neuron of rat motor cortex can evoke long sequences of small whisker movements. The efficacy of individual cortical APs in evoking whisker movements is not dependent on background cortical activity and is greatly enhanced in waking rats [163]. Moreover, stimulation of single neurons in the barrel somatosensory cortex can weakly, but appreciably affect behavior of entire animals [563]. Further, properties of the selective form of excitable membrane plasticity (Chapter 1), are also evidence about the existence of adaptive interaction of individual neurons to the extracellular environment. It is possible to develop an instrumental reflex such that the discharge of a single neuron serves as an elemental instrumental action [1270]. Moreover, trials-and-errors are observed at the cellular level [1271]. It appears that motivational behavior can be determined by alterations occurring in an individual neuron, and the homeostasis of a neuron is evidently a unit of motivational behavior.

Thirdly, and this is really wonderful, sometimes brain robustly controls goal-directed behavior using so small signals that these signals are at the boundary of molecular processes, which are unstable by their nature. In small embryonic hippocampal neurons of rat (having high-resistance) spontaneous individual openings of single ion channels can trigger impulse generation in these cells [589]. As few as 200 Ca^{2+} ions and the opening of a single Ca^{2+} channel can initiate transmitter release. The action of Ca^{2+} at the release site is cooperative. More than one Ca^{2+} ion must bind to the release machinery to trigger release (from three to four ions per Ca^{2+} sensor are required). Because of 3rd or 4th power relationship, small changes in the Ca^{2+} influx will have dramatic effects on the release probability [1225].

Sensory organs possess a unique sensitivity. An individual olfactory receptor may be activated by a single odorant molecule [835], while after adaptation in darkness, toad retina cells respond to a single photon [92]. Besides vision and olfaction, such an extremely high sensitivity is characteristic also of primary neurons of hearing. 'Hair' cells of the inner ear can detect displacements of atomic dimensions [275]. Lastly, low concentrations of the motivationally-relevant substances affecting behavior, which sometimes may reach only one or a few molecules per neuron, also demonstrate sensitivity of a neuron to single molecular events. And at the same time, as we have indicated, a single neuron may, in favorable circumstances, affect behavior of an entire animal. The importance of this conclusion is determined not only by the outlining of a principal possibility of neuron behavior. It means that smoothing of activity of large neuronal populations is not the key principle of brain operation. Even if each neuron behaves independently (we know, this is not the case and neurons are undergoing powerful influences of the rest of brain), it would not exert an effect to an entire animal, since its activity would get drowned in the noise. For example, the averaging of activity of one hundred neurons will produce fluctuations in ten fold exceeding individual activity of any neuron. Therefore, averaging cannot be a basic principle of brain functioning.

3.5.3 The simplest behavior: chemotaxis

Neurons are disposed within the brain tissue and connected with other neuronal and non-neuronal cells by numerous pathways. When we study neuronal behavior, it is difficult to split the contribution of a given neuron in the behavior from completed decision of the rest of brain. We tried to solve this problem using the measurement of cell excitability, which is a local property of neuronal membrane. Comparing change in a threshold of AP in the response to given stimulus during learning with the changes of AP latency depending on both pre- and postsynaptic alterations, we could assign the share of the neuron in an entire behavior. Using this indirect means, we have demonstrated that capability to learning and remembering is an attribute of a single neuron. At least, in some artificial conditions neurons can demonstrate such capabilities. Neuron also can display a capability to primitive motivational behavior. This is demonstrated by the local instrumental reaction, which may occur within the recorded neuron, when the basic reinforcement schedule was delivered only to this neuron [1260].

Some primitive forms of learning can be demonstrated on an isolated neuron. Individual behavior of neurons could not suddenly arise in evolution because of the origin of their main distinction, Na^+ channels, since neurons use this instrument for interactions between neurons. Evolutional achievements of neural cells are directed towards the promotion of network properties. Those cell devices that endorse interactions with other cells are not likely to serve as the basis of individual cellular behavior, although they may participate in their fulfillment. Such properties are, besides excitability, secretion of neurotransmitters and the establishment of gap junctions. Interestingly, the unicellular organism Vibrio cholerae, which, it seems, does not need tight interaction with other cells, elaborates the special protein that modulates intercellular gap junctions by binding to a specific surface receptor in endothelial cells [765]. It is astonishing that even tetrodoxin, blocking neuronal Na^+ channels, was found to be synthesized by certain bacteria. Another antagonist of Na^+ channels, saxitoxin, also is produced by a primitive animal without a nervous system: flagellate protists, a kind of marine plankton. This looks enigmatic. Probably, not only neurons, but other cells are capable of executing more functions than they use in everyday life. It is impossible to suppose that some cellular mechanisms arise in evolution "in the store". Cells, perhaps, sometimes use every their option. In any case, neurons, in critical conditions, display an ample diapason of their possibilities.

One property of cells is the struggle for their own survival. An individual may achieve a many-sided being in the collective, but anyone dies personally. Therefore, neurons received from their predecessors the capability to fight for survival and, consequently, to behave. It would be interesting to evaluate the capability of single cells to behave and to modulate their behavior under influence of outer surroundings. This consideration cannot prove the individual behavior of neurons within the brain, but we may outline some possibilities of living cells to behave for survival.

To do this we may use single cell chemotaxis as an example of the simplest goal-directed behavior [486, 562, 644, 653, 782, 1010]. Chemotaxis is directional migration in response to a gradient of a chemical stimulus in ambient space. The external similarity of motivation and taxis is insufficient for concluding that the phenomena are basically similar, but there are many deeper similarities. The same intracellular mechanisms that are involved in motivation are involved in taxis, particularly intracellular Ca^{2+} augmentation [451, 1010] and participation of G proteins [460, 391, 644, 778, 940]. Repellents raise the Ca^{2+} level and induce tumbling, while attractants decrease the Ca^{2+} level and induce cell running [1231]. Second messenger cyclic AMP, besides its role as a chemoattractant, participates in organization of chemotactic movements by organization of cyclic AMP waves within the cell [304, 562]. Cyclic AMP may affect a cell through modulation of somatic hyperpolarization-activated channels [440], which represents a mechanism of homeostatic plasticity, allowing a neuron to control excitability in response to outer influences [1331].

Metabolic signals, actuating motivation, affect chemotaxis, too. Chemotaxis may be activated by salinity, pH, temperature, glutamate, cyclic AMP, cytokines, etc. Binding of glutamate to receptors results in a rapid increase in the intracellular cAMP level [562]. Motivationally-relevant substances affect motivation [350] and enhance single cell chemotaxis [344, 562, 782] at similar concentrations and with a similar time course (minutes). Motivationally-relevant substances affecting both motivation and chemotaxis are bombesin [1390], cholecystokinin [1058], insulin [719], gastrin-releasing peptide [306], neurotensin [451], somatostatin [915], neuropeptide Y [464], substance P [1151], steroid sex hormone estrogen [286], prolactin [809], and angiotensin II [1051]. GABA and taurine, inhibitory amino acids, act as chemoattractants for neurons [99]. However, motivationally-relevant substances do not stimulate chemotaxis in all cases. Testosterone decreases chemotaxis [1151], opioids [566, 814], cholecystokinin and neuropeptide Y [304, 1058] exert different effects on different cells. Neurotensin elicits non-directed migration of the microglial cells [798], while leu-enkephalin suppressed migration [440]. Not every substance display the same chemotactic effect in different cells and not every cell displays chemotaxis to a given substance, just as not every neuron displays the same response to a given motivationally-relevant substance. Mutual interaction of substances with a cell may be not linear, such as synaptic influences to a neuron may be non-linear. For instance, ligands binding to different receptors in Escherichia coli can result in a nonintegrated response and tumbling precedes the smooth-swimming response [618].

Independently of the nature of the chemoattractants and the cell type, a U-shaped response is usually observed [306, 458]. This means that intermediate concentrations of the chemoattractants are chemotactic, while very low and very high concentrations are not [778], such as we have described for many motivationally-relevant substances. This form of concentration-dependence is meaningful for chemotaxis. When the cell detects the threshold concentration of the attractant, it increases its motion. It approaches the goal little by little and attractant concentration increases. When the cell reaches the vicinity of the goal, it must reduce speed, to prevent itself from moving past the goal. However, sometimes a more complex chemotaxis dependence of concentration may be observed. Suppressing migration, leu-enkephalin has two peaks in the optimal physiological concentration range: 10^{-10}M and $10^{-13} \div 10^{-14}$ M [440].

In order to migrate directly towards an attractant, cells must be able to detect differences in signal concentrations across the cell or detect a change in concentration as a function of time [64]; that is, a cell must be capable of storing information and comparing stored information with the current input. Because cells are so small, the difference between the attractant concentrations across the cell will be insignificant compared with the spontaneous fluctuations of chemoattractant concentrations. We will demonstrate later that a goal-directed search (although not a direct migration) may be organized without participation of memory mechanisms. Nevertheless, something like memorizing exists in unicellular animals. At least something like fatigue is observed in single cells. All the more so that fatigue is a rather probable evolutionary antecedent of memory. Fatigue may concern only one fragment of a cell or one specific metabolic pathway. At this point, the cell is only one step away from memory.

Chemotaxis may be undergone to achieve adaptive influences. Taxis shows plasticity: the chemotactic system undergoes desensitization [306, 562], the axonal growth cone exhibits adaptation [847] and amoeba-like cells have the ability to find the minimum-length solution between two points in a labyrinth [876]. An Amoebae needs to distinguish itself from another species and Amoebae recognize each others as self-nonself, for example during sexual behavior and phagocytosis [728]. Many cells that carry out chemotaxis exhibit adaptation to spatially uniform step changes in the concentration of chemoattractant stimuli. An adapting cell, when stimulated by a constant dose of chemoattractant, exhibits a large transient response. In time, however, the cell's response returns to its prestimulus levels [1378]. Even a bacterium exhibits sensory adaptation in response to the continuous presence of nonsaturating stimuli (alcohol) [1131]. Bacteria are sensitive to molecular events the same as we described this for neurons. It was shown that within individual bacteria, molecular noise emerges as a tunable source of behavioural variability [658].

As a consequence, chemotaxis represents activity that, at least in outward appearance, resembles motivational behavior. Chemotaxis is activated by the same metabolic signals, is controlled by the same metabolic pathways, modulated by the same substances, in similar concentrations, with similar U-shaped dependence on concentration and depending on experience. Thus, single-cellular organisms demonstrate a capability for a goal-directed behavior that has correspondence with a motivational behavior. It would be strange, if neurons would be deprived of such capability.

3.5.4 Goal-directed behavior of single neurons

Motivational behavior at the neuronal level reorganizes membrane properties. We have demonstrated that damage distorts properties of excitable membrane. Excitability of neurons during damage lacks direct and simple dependence from membrane potential. Weak injury decreases membrane potential and excites neurons, but when damage increases, Na^+ channels inactivate and neurons fail to generate APs. Therefore, after injury, depolarization stops exciting neurons. On the contrary, a hurt neuron may be improved and activated as a result of hyperpolarization. Learning also affects excitability, which lacks direct dependence on the level of membrane potential. However, there is a principal distinction between the influences of damage and learning to excitability. An injured neuron mostly loses excitability in general, uniformly to the current signals and independently on an irritation. On the contrary, during learning, excitability is altered selectively in respect to the signal, whose significance has changed during learning. For instance, excitability in reference to a novel stimulus is higher than in respect to a habitual stimulus. Similarly, a neuron displays higher excitability within a response to a conditioned stimulus (CS^+), than in response to a discriminated stimulus (CS^-). Spontaneous activity usually does not change during habituation and classical conditioning.

If change in excitability during learning is to some extent related to changes that are observed during injury, one may compare the process of learning with a selective damage or protection in respect to specific signals. Alterations of excitable membrane as a result of learning are not stagnated. Excitability changes transiently after interaction of the signal with the neuron and is soon recovered after the passing of a signal. Homeostatic protection against damage may be specific, such as we described for long-term potentiation. According to our discussion in Chapter 2, existence of specific forms of neuronal damage is not entirely proper, but rather probable. At least, some aspects of neuronal responses to a specific signal during learning resemble interaction of the signals with the injured or protected neuron.

Let us consider how properties of excitable membrane depend on the level of membrane potential during learning. Although membrane potential usually does not change as a result of habituation or classical conditioning, it does differ between trials. This allows for comparing how neuronal reactions alter during spontaneous changes of the membrane potential. We did not implement a direct current for membrane potential modification in order to not influence the neuron by foreign signals that were not related to learning. Fig. 3.1 demonstrates how action potential amplitudes depend on spontaneous change in membrane potential during habituation of neurons of turtle primordial cortex to light flashes [1266, 1267]. These relationships were different for those APs that were generated by different causes.

The amplitude of short-latent AP in response to habitual stimulus tended to be increased, when the membrane potential of neurons was spontaneously decreased (Fig. 3.1A). Although the significance of negative correlation ($r =$

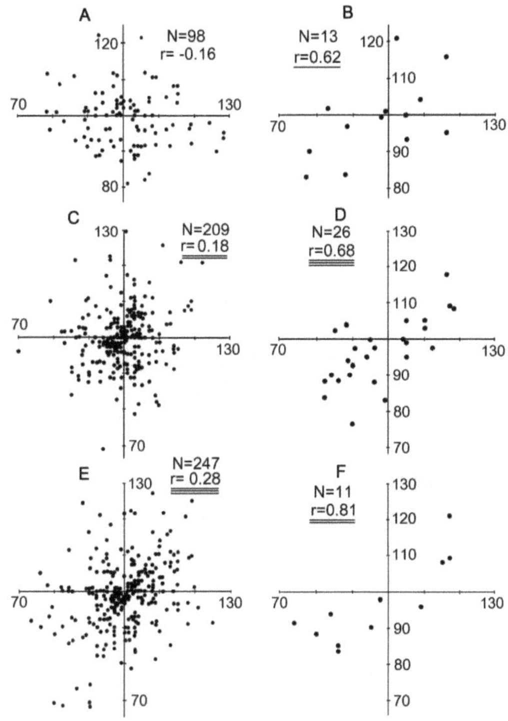

Fig. 3.1. Dependence of the spike amplitude on the resting membrane potential changes during habituation to the repeated light flashes in the primordial cortex of turtle. Ordinate AP amplitude, abscissa membrane potential, relative units. A, B - short-latent AP, evoked by a habitual stimulus (flashes, A) or by a novel stimulus (B). C, D AP arisen after the inhibitory postsynaptic potential, evoked by a habitual stimulus (C) or by a novel stimulus (D). E, F Spontaneous AP arisen during habituation (E) or after a few seconds after presentation of a novel stimulus (F). number of measurements, r coefficient of linear correlation. One, two or three lines: the significance of correlation correspondingly $p < 0.05; p < 0.01; p < 0.001$.

−0.16) between AP amplitude and membrane potential at the Fig. 3.1A was relatively low ($p < 0.1$), we ought to emphasize that this is the negative correlation, while normally this correlation is always positive. For instance, APs which have been generated after inhibitory postsynaptic potential within the response to habitual stimulus positively correlated with the membrane potential (Fig. 3.1C). This took place also for spontaneous APs (Fig. 3.1E). Corresponding APs, related to the novel stimulus (Fig. 3.1B, D, F) were connected with the membrane potential by a high positive correlation. Excitable membrane was affected only during a short period after a stimulus and the stimulus was the cause of these alterations. Correlations between excitability and membrane potential differ for APs evoked by the habitual and novel stimulus and APs generated before and after inhibitory pause. Distortion of correlation between excitability within response to the habitual stimulus and membrane potential short-term recovered (Fig. 3.1A,C) after inhibitory pause (such as we described for the examples of protections and rewards). Presentation of a novel stimulus short-term (during a few seconds) improved correlation between spontaneous spikes and membrane potential (Fig. 3.1E,F). Disturbance of properties of excitable membrane was observed the most robustly with respect to short-latent AP in the response to the habitual stimulus, whose importance changed as the result of learning.

Neuronal excitability underwent modifications also during classical conditioning, but there are distinctions from habituation. The method of elaboration of a neuronal analog of classical conditioning was described in Chapter 1 and consisted of combinations of the conditioned stimulus CS^+ and unconditioned stimulus US.

During classical conditioning [1258, 1259], a threshold o f AP within the response to CS^+ paradoxically depends on spontaneous changes of membrane potential (Fig. 3.2). On the background of spontaneous hyperpolarization, neurons generated responses at the lower level of voltage.

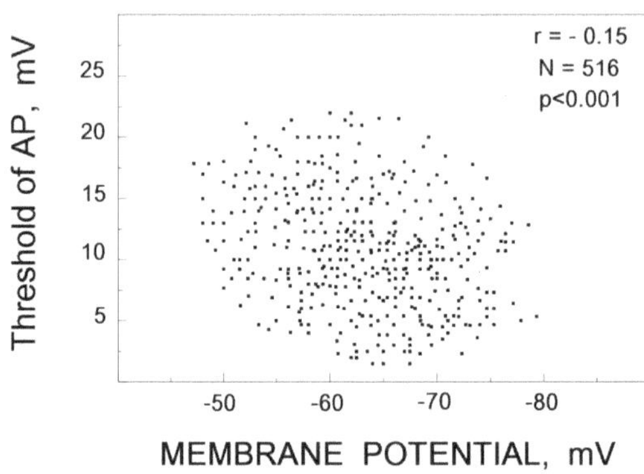

Fig. 3.2. Paradoxical dependence of the threshold within the first AP in the responses to the CS^+ during classical conditioning from the membrane potential of the command neurons of a defensive closure of pneumostome in mollusk Helix. Coefficient of linear correlation , number of measurements and significance of the correlation are indicated at the figure. We measured threshold from the level of membrane potential to the point of the maximum curvature at the front of AP, mV.

Relatively low, but highly significant negative correlation between AP threshold and spontaneous alterations of membrane potential shows that defensive classical conditioning alters properties of the excitable membrane. At the background of hyperpolarization, a neuron demonstrated enhanced excitability with reference to conditioned stimulus (CS^+), whose significance changed after pairing. Although such a change of properties of excitable membrane is evidence of damage- or protection-related alterations in cells, the change of membrane potential, by itself, cannot be the reason of modification of neuronal reactions. Really, polarization of cellular membrane ought equally to affect the responses to habitual and new stimuli or to CS^+ and CS^-.

In spite of outward similarity of alterations of excitability during habituation and classical conditioning, the signs of modifications were opposite. Firstly, selective excitability within a response to habitual stimulus decreased, whereas excitability within the response to CS^+ increased. Secondly, hyperpolarization of cellular membrane during habituation corresponded to a decrease in AP amplitude during habituation (Fig. 3.1), but during classical conditioning threshold within the response to the CS^+ decreased (Fig. 3.2) and, correspondingly, the amplitude of these APs increased (not shown in Fig. 3.2). This means that hyperpolarization lead to a decrease in excitability during habituation, but to an increase in excitability during conditioning within the responses, whose importance changed after training. However, in both cases, hyperpolarization aggravated those changes of excitability that ensued during learning. On the background of the spontaneously evoked hyperpolarization after interaction with the signal, neurons found it easier to reorganize their excitability and display learning-related modification of the responses. In the case of habituation, this is a transient decrease in excitability in the response to a habitual stimulus, while in the case of classical conditioning this is a temporary increase in excitability in the response to CS^+. The stimulus, but not the alterations of membrane potential, was the cause of short-term change in excitability. However, this change is displayed depending upon spontaneous alterations of the membrane potential. Thus, in line with our discussion, current healing of neurons promotes exhibition of learning features.

Magnitudes of neuronal responses were correlated with spontaneous alterations of membrane potential, but causal connection between these two values is not clear. Certainly, change in membrane potential may affect neuronal state and uniformly modify neuronal responses. Besides, change in membrane potential may depend on the background state of the whole brain in current time (for example, on extracellular concentration of K^+, glutamate, pH, etc.). Therefore, we studied how membrane potential changes in the neighboring neurons, whose activity was recorded simultaneously. Our experiments on Helix revealed a coordination of membrane potential alterations in different neurons during training.

Current values of membrane potential during training were measured immediately before the CS^+ and CS^-. The absolute value of the membrane potential was measured from a reference point located outside the neuron and was determined at the end of the experiment after withdrawing the electrode from the neuron. During classical conditioning, spontaneous alterations of membrane potentials in the neighboring neurons changed without outward regularities, but synchronously (Fig. 3.3). On the average, for all recorded neurons, shifts of the membrane potential during acquisition were insignificant (Fig. 3.4, bottom). However, their alterations between trials reached 20-25 mV (Fig. 3.2) and they were not chaotic.

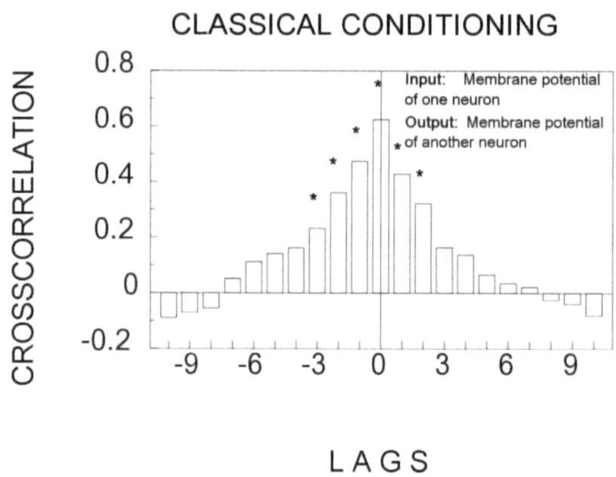

Fig. 3.3. Time delay between the membrane potentials of neighboring neurons. Ordinate: cross-correlation between membrane potentials of two simultaneously recorded neurons during acquisition of classical conditioning. Cross-correlation (bars) was calculated for each experiment and significance of the mean value (through neuronal pairs, $p < 0.05$, asterisk) for each lag was evaluated (Mann-Whitney U test).

Since alterations of membrane potentials in neighboring neurons happened synchronously, they evidently represent change of current state of the robust brain. A lot of neurons participate in classical conditioning and accept a current signal. Fig. 3.3 presents cross-correlation between membrane potentials of simultaneously recorded neurons during associative conditioning. Fig. 3.3 also demonstrates that spontaneous change of membrane potential is a slow process and potential was steady during presentation of 2-4 consecutive pairs, that is several minutes. Thus, changes in the scale of depolarization-hyperpolarization and, possibly, the scale of damage-protection proceeds in parallel in many neurons.

Interruption of pairing for 5-10 minutes after acquisition or 20 minutes after extinction increases response to the first CS^+ after the break (Fig. 3.4, top). Arrows between the last values during acquisition and the first values during extinction (or the last value during extinction and the first value during reacquisition) indicate the significance of the difference between these values. Increase in response to the CS^+ after acquisition was selective and was not observed for responses to the CS^-. This means that reorganizations of the selective excitability slowly (during minutes) continue after the end of learning and this interval corresponds to a period of reorganization of metabolic pathways during injury and the initiation of motivation.

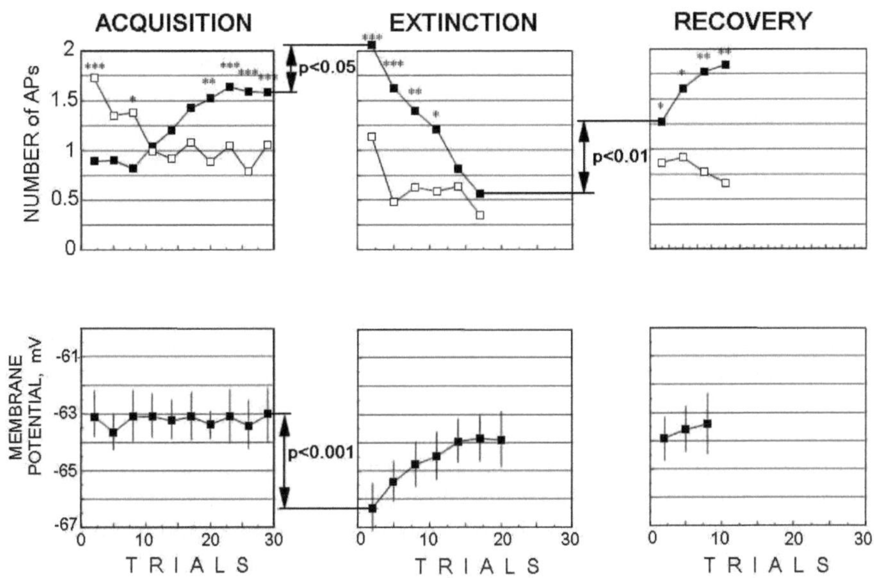

Fig. 3.4. Neuronal activity during acquisition, extinction and reacquisition of classical conditioning. At the top the number of APs in the responses to the conditioned (closed symbols) and discriminated (open symbols) stimuli v.s. trial number. Medians and significance in the difference between responses to the conditioned stimulus and discriminated stimulus are shown at the top (Mann-Whitney U test, $*p < 0.05; **p < 0.01; ***p < 0.001$). At the bottom mean membrane potential and confidence intervals ($p < 0.05$) are shown. Ordinate for membrane potential at the all figures is traditionally directed in the reverse course, since a membrane potential is always negative. Methods. The discriminated stimulus CS^- was never paired with the US and was presented every one to four combinations of CS^+ and US over the course of training. The training procedure consisted of the acquisition (25-35 combinations of the conditioned and unconditioned stimuli), an extinction series (15-20 presentations of the isolated conditioned stimulus after 5-10 min break), and a second acquisition stage consisting of repeated development of the conditioned reflex following a 20-minute break. The Fig. 3.4 was redrawn in accordance with the data [1257].

Possibly, depolarizing influences of the harmful US were homeostacally compensated. This compensation supported membrane potential during 2-4 consecutive learning trials, then changed the membrane potential for another quasi-stable level and changed again (Fig. 3.3). However, average membrane potential level during acquisition stayed stable (Fig.3.4, bottom). Regular modification in the strategy of maintenance of membrane potential during acquisition is not entirely clear. Probably, this is somehow connected with the brain strategy for search of right response: "trial-and-error", which is better displayed during instrumental conditioning (see further).

Weak average shifts of the membrane potential during habituation and acquisition of classical conditioning is in accordance with data previously reported [48, 549, 1258, 1259, 865]. However, after intensive classical defensive conditioning in the snail (160 pairs), membrane potential in identified neurons of pneumostome closuring was reported to decrease during 40 days and recovered only through 52 days following conditioning [444]. Evidently, such a protracted session of non-avoidant pain leads to profound homeostatic reorganization of ion equilibrium in the vicinity of a new set-point, as we have discussed earlier (similar events are observed in drug abusers). Moreover, prolonged depolarization may be a false result of improper measurement. In aforementioned experiments, neuronal activity from an identified neuron was recorded from the already intensively learned animals. It is not excluded that after strong defensive conditioning, defensive neurons were sensitive to damage and after impaling by the microelectrode neurons were depolarized because of physical injury of the membrane. In similar circumstances, recording from pyramidal neurons in hippocampal slices prepared from rabbits after exhaustive eyeblink conditioning revealed unspecific increase in excitability (excitability was recovered after 7 days), although no learning-related effects were observed on AP amplitude or duration, or resting membrane potential [865]. Unspecific change in excitability (without change in the membrane potential) may be observed in cerebellum after stressed learning. In Purkinje cells of the slices prepared from animals trained to classical eyeblink conditioning, increase in unspecific excitability (effect of direct current stimulation was measured) was still presented following 30 days. Although membrane potential, input resistance and AP amplitude were not different in cells from paired or control animals, a potassium channel-mediated transient hyperpolarization was smaller in cells from a trained animal [1100]. These effects were observed because, in the trained animals, part of the cells had a low threshold and might be more susceptible to impaling the cell by an electrode.

Values of responses in our experiments depended on the contingency between appearance of these stimuli (CS^+ and CS^-) and a painful US. During an extinction session, this contingency changes and responses to the signals change, too. During the classical paradigm, the conditioned response increased, but it decreased during extinction sessions and recovered rapidly during reacquisition (Fig. 3.4, top). This was not the case for responses to the discriminated stimulus.

Therefore, we investigated how during classical conditioning a completely compensated membrane potential will behave after cancellation of the unconditional painful stimulation in a Helix mollusk. For that reason, it is important to study how membrane potential changes during a whole training session, when the contingency alters.

Depolarization is an immediate result of negative reinforcement. Punishment evokes excitation and a long residual depolarization. However this depolarization was not accumulated, since membrane potential, on the average, stays the same during acquisition of the conditional reflex. Hence, during training, a homeostatic protection counteracts deleterious influences of punishment. The existence of such compensation is displayed after a break in training. Homeostasis is an inertial system and it temporarily continues its performance after a fall in the stressor. Interruption in stimulation after acquisition and after extinction leads to disturbance of the homeostatic equilibrium which had reached the steady state during training. Homeostasis, evidently, completely compensates excitations evoked by unconditioned stimuli, by means of prolonged compensational hyperpolarization. Since homeostatic compensation carries on after disruption of acquisition, this compensation influences membrane potential. Depolarization of the membrane potential through the action of unconditioned stimuli can affect ion homeostasis and disturb excitability, but is compensated via some unknown homeostatic process seen after a pause in conditioning (arrow at Fig. 3.4, bottom). After completion of the acquisition session and cancellation of the unconditioned stimulus presentation a compensational mechanism continues to hyperpolarize neurons and the value of the membrane potential increases, as can be seen from the voltage shift at the onset of extinction. The unconditioned stimuli were omitted during a session of extinction and the membrane potential then recovered, though response to the conditioned stimulus decreased. The abolition of the unconditioned stimulus evokes rapid hyperpolarization followed by slow recovering of the membrane potential. A 5-10 minute time break between acquisition and extinction approximately corresponds to a period of stability of membrane potential during acquisition (Fig. 3.4). During extinction, the compensation gradually decreases and membrane potential recovers. Although extinction, in outward appearance, resembles a habituation, the membrane potential of neurons during habituation does not change, but it does change during extinction. The main difference between extinction and habituation is abolition of the reinforcement during extinction and the absence of reinforcement during habituation.

Under motivational tension, classical conditioning, evidently, increases vulnerability to damage, although compensation recovers the normal level of membrane potential. Complete compensation of background neuronal activity is observed also after depotentiation of LTP (see Chapter 2). During habituation, classical conditioning and after completion of tetanic stimulation an appearance of outer impacts is predictable and homeostasis leads to recovery of the initial conditions. During classical conditioning, the appearance of a conditioned stimulus always brings an end to the unconditioned stimulus independently of the generation or failure of an AP in response to the conditioned stimulus. However, homeostasis cannot recover initial conditions when appearance of punishment or reward is unpredictable, since homeostatic compensation is a slow process and it needs some minutes or more for its development. Therefore, homeostasis falls to recover the normal conditions if the environment changes faster. This phenomenon is well presented in the experiments with an irregular appearance of salient irritations. For example, abrupt cancellation of narcotic doses in drug addicts leads to neuronal overexcitation and evokes withdrawal syndromes. At the same time, regular consumption of drugs is accompanied by a compensation and even overcompensation (if, of course, one does not take into considerations the remote consequences). So, long-term, during a 3 month period, ingestion of high doses of ethanol causes in mice a decrease in spontaneous firing of the Purkinje cells and impairment of motor coordination, if experiments were carried out without withdrawal. These alterations in Purkinje cell firing did not affect the ability to learn or to recall a motor coordination task [1118].

Temporary increase in neuronal response to the CS^+ at the beginning of extinction was not connected with the recovery of excitability by means of protective hyperpolarization that was observed at the same time. Such background depolarization is unspecific and cannot increase response to the CS^+, while response to the CS^- did not change. Recovering influences of the protective inhibiting during the damage of excitability arises, evidently, as an advancing homeostatic reaction to the appearance of the CS^+, which is causally connected with punishment. This is the main meaning of the selective change in excitability and, probably, selective protection. At any given moment, excitability of a neuron is its general property. However, excitability may be transiently changed in response to current stimulus. This change depends on experience and, perhaps, on having at a given moment some homeostatic protection.

Thus, we have considered how a neuron's properties change during habituation and classical conditioning in connection with motivational behavior. The level of motivation in these cases depends upon the presence of a US in the environment, that is upon a reward or punishment, which protects or injures neurons and this is displayed as a modification of the state of an excitable membrane. Besides, during these simple forms of learning, expectation of a US changes during the training session: it decreases during habituation and during extinction of classical conditioning, but it increases during a stage of acquisition. At the finish of acquisition, expectation of a US appearance after a CS^+ reaches a maximum, while after the CS^-, expectation of a US tends to decrease to a minimum. The value of expectation is important for determining the value of homeostatic compensation. Compensation that is far ahead the impending damage can be properly generated only if expectation is high enough. In this case, damage and protection may counterbalance each other and the membrane potential will acquire a stable character, as, for instance, during classical conditioning. Abolition of the US leads to uncertainty of expectation and to a loss of established equilibrium. Such evaluation of impending reward or punishment is essential also during elaboration of instrumental conditioning, but here there are some peculiarities. In classical conditioning, the properties of the environment are very simple and do not depend on animal actions. In such cases, a neuronal homeostasis can compensate for the harmful effects of an unconditioned stimuli, while instrumental conditioning is a much more complex process. Generation of voluntary actions is especially tightly connected with evaluation of coming events but evaluation of expectation is a much more difficult task, since it is additionally dependent on correspondence between the animal's own actions and future events. We will consider the problem of goal-directed behavior in the next Chapter, but now let us consider how properties of cellular membrane are altered during instrumental conditioning. During instrumental conditioning, if an experimental (trained) neuron did not respond to the conditioned stimulus, the animal received a shock, while firing of the control neuron remained unrelated to the shock. Thus, the meaning of the originally neutral stimulus was changed for the trained neuron, but not for the control neuron. The neural system had to determine which neuron was responsible for avoidance of the punishment. Fig. 3.5 portrays the results of a representative experiment. The dynamics of neuronal activity were not monotonous. Fig. 3.5 demonstrates both the correct and the incorrect responses depending on the trained neuron reaction to the CS^+. US were presented only when trained neuron RPa3 (top curve at the each frame) failed to generate an AP. US appearance did not depend on activity of the control neuron LPa3 (bottom curve at the each frame) and on responses to the CS^-. A neuronal analog of instrumental conditioning satisfied all well-known properties of instrumental behavior.

The response of the trained neuron to an originally neutral conditioned stimulus decreased at the beginning of the learning session but, at the end, an instrumental reaction was selectively generated by the trained neuron (Fig. 3.5). Only responses of the trained neuron to the conditioned stimulus did not decrease significantly after training. During elaboration of the instrumental reflex the responses of the trained neuron to a conditioned stimulus decreased during the first trials, gave rise to a harmful unconditioned stimulus (Fig. 3.5, trials N 11, 14, conditioned stimulus), but was later restored (Fig. 3.5, trials N38, conditioned stimulus).

On average, at the beginning of training, till the 25th trial, responses of the trained neuron to the CS^+ decreased, and the animal received more and more punishments. Further, the trained neuron began to generate an AP in response to the CS^+ and the number of punishments decreased, although did not cease. (Fig. 3.6).

Fig. 3.7 demonstrates average data for change in the trained and control neuronal responses during instrumental conditioning. The profiles of the learning curves had different (Fig. 3.7, top) shapes for the trained and control neuron responses to the conditioned stimulus, but similar shapes for responses to the discriminated stimulus (Figs. 1.8,1.9). Responses of a trained neuron to an originally neutral CS^+ decreased at the beginning of a session (Fig. 3.7,

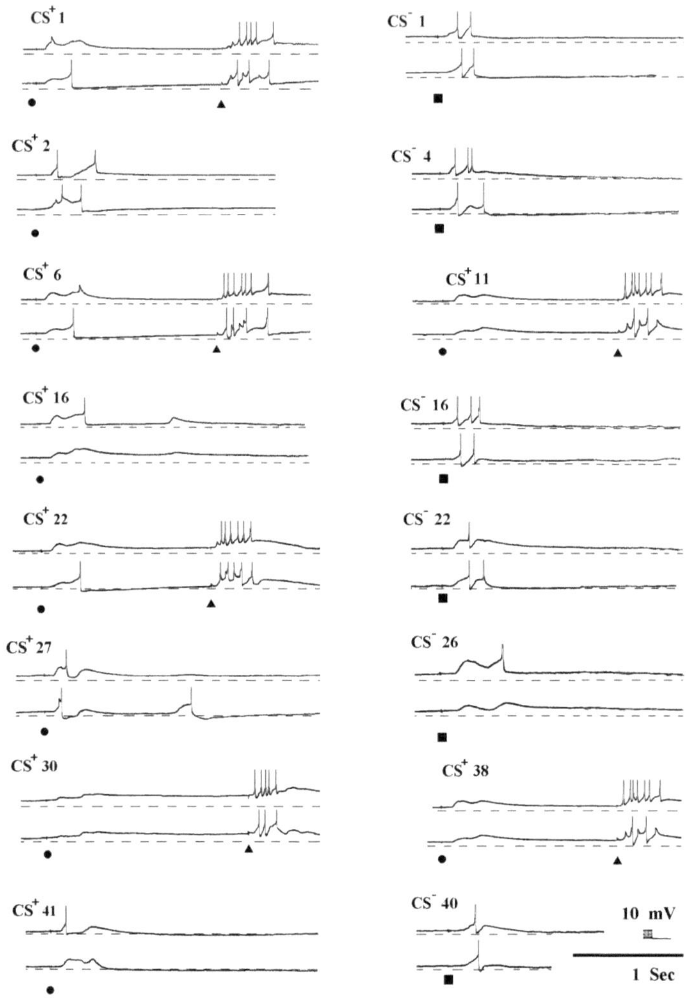

Fig. 3.5. Representative intracellular recordings of neuronal responses during elaboration of a local instrumental conditioned reflex with the trained neuron RPa3 (top in each frame) and the control neuron LPa3 (bottom) in the Helix mollusk. The number of the CS^+ is indicated for each exposure. For the responses to the CS^-, the number of the preceding CS^+ is indicated. Failure of the spike in the responses to CS^+ numbers 1, 6, 11, 22, 30, 38 (circle) in the trained neuron results in a US (triangles). Spike generation in the control neuron does not prevent delivery of a US (trials 1, 6, 22) and spike failure did not result in a US (trials 16, 41). Responses to the CS^- are shown with a shorter exposure (squares). Arrow at the trials CS^+ 16 and CS^+ 27 indicate reaction that had arisen, correspondingly, in the trained and control neurons before the responses to US (probably, reaction to time). Mean membrane potential during recording in the trained neuron was -58,8 mV and in the control neuron -60.2 mV. Dotted lines indicate level of potential -65 mV. Calibrations are shown on the plot. Tops of action potentials are not shown.

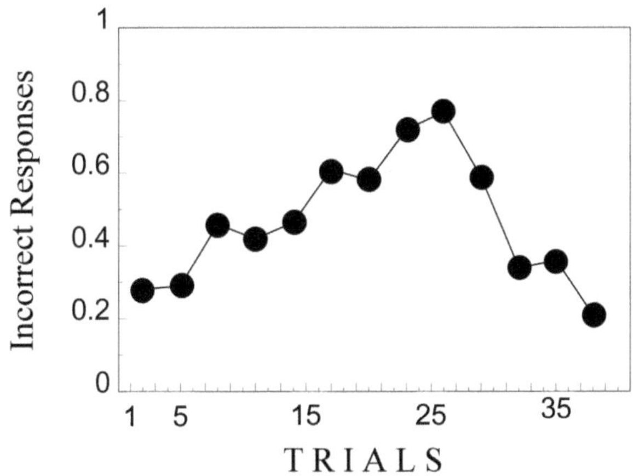

Fig. 3.6. Share of incorrect responses in response of the trained neuron to the CS^+ during instrumental conditioning. Average for all recording neurons. Standard errors are indicated. Abscissa number of trials. Ordinate probability of the failure of action potential generation in the trained neurons after presentation of CS^+.

top) and gave rise to the emergence of a harmful US. Then the neuron increased the response that served the instrumental reaction and the number of harmful stimuli decreased. The responses of the control neurons to the CS^+ decreased when the response passed a second maximum (Fig. 3.7,top, 20-30 trials). A control neuron in this period of training generates an erroneous instrumental reaction, instead of a correct instrumental reaction of the trained neuron [1271]. At the end of the learning session, an instrumental reaction was selectively generated by the trained neuron AP.

The neural system acquired information relative to the instrumental action through several stages, with varying intermediate "conclusions", relative to salient features of the ways for punishment avoidance (see Chapter 1). The response of the control neuron to the conditioned stimulus decreased in two phases on either side of a transient increase (erroneous instrumental reaction, trials 20-30; Fig. 3.7). When the control neuron generated an erroneous instrumental reaction, the trained neuron decreased its participation in the reaction. Thereafter, the trained neuron began to generate instrumental action potential in its response to the conditioned stimulus. Therefore, during instrumental conditioning, the dynamics of neuronal responses to a conditioned tactile stimulus were much more complex, consisted of several phases and point to the fact that neurons have no time in which to develop a strategy to compensate for the harmful influence of punishments. Thus, predictability of environmental alterations appears essential for the development of compensation and there appears to be a relationship between a compensational change in the membrane potential and the predictability of behavioral

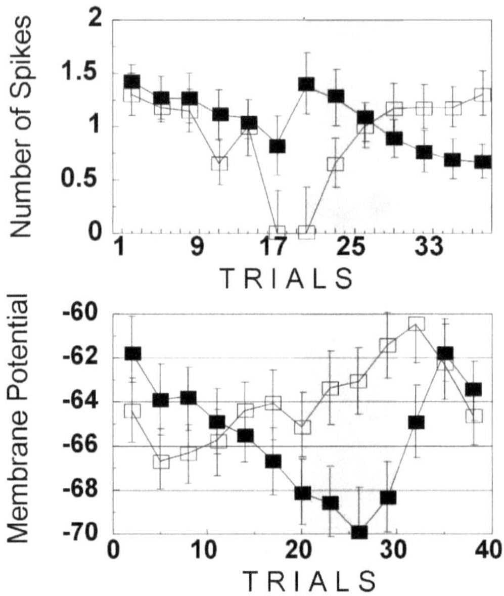

Fig. 3.7. Behavior of neuronal activity during the acquisition of instrumental conditioning. Top the number of APs in the responses to the conditioned stimulus (medians, Mann-Whitney U test). Bottom change in the average membrane potential during training, mV. Confidence intervals are shown, $p < 0.05$. Solid symbols represent the data for the control neurons and open symbols the data for the trained neurons. An interval 20-30 trial is indicated.

outcome. Homeostasis can compensate for steady damage but is unable to compensate for influences of varying unpredictably in time.

At the end of the training session, an instrumental reaction was selectively represented by the trained neuron AP, i.e. the learning process demonstrated output selectivity (Fig.3.7, top). Responses of both the trained and the control neuron to the CS^- decreased as a result of training (Fig. 3.5, cf. trials N1 $CS^-$and N38, CS^-). Therefore, the learning process also displayed input selectivity. Although the initial excitability of the control neuron in respect to the conditioned stimulus in our experiments was usually larger than the excitability of the trained neuron, the difference in excitability is not responsible for the difference in the behavior of the neuronal activity between the control and trained neuron, as previously was demonstrated for habituation and instrumental conditioning [1270, 1271].

There is little doubt that we had a satisfactory neuronal model for instrumental conditioning. The only difference between the neuronal analog of instrumental conditioning and behavioral instrumental conditioning is the utilization of an AP in an identified neuron instead of a motor action. During training, the animal learned to determine which neuronal discharge was essential for the avoidance of punishment. Instrumental conditioning was selective by both input and output. Therefore, although we observed only neuronal activity, we may consider this neuronal activation as the instrumental action of the entire mollusk. Such like-behavior is called a neuronal analog of learning, since instead of behavioral reactions there may be recorded only electrical reactions of neurons.

During instrumental conditioning, the maintenance of membrane potential is disturbed, or at least changed in a complex manner (Fig. 3.7, bottom). At the beginning of training, an animal still did not receive information about the particularity of the environment, regularities of the US appearance are not clear for the animal and neurons have no time in which to develop a strategy in order to compensate for the harmful influence of punishments. The membrane potential of both neurons at the beginning of learning slightly increased. A control neuron apparently overcompensates for harmful influences and hyperpolarizes in the region of the vicinity of a local maximum for output response (Fig. 3.7, top), but further hyperpolarization of the control neuron continued to increase and reached a maximum (-68 mV compared with the -62 mV at the beginning) approximately in the middle of the session (Fig. 3.7, bottom). Later, the membrane potential in the control neuron recovered. Membrane potential in the trained neuron decreased roughly at the same time, when the control neuron was hyperpolarized. Thus, during instrumental conditioning, depolarizing influences of a painful US were not compensated for by the protective impact of homeostasis. Just when an instrumental reaction began, the trained neuron was depolarized, membrane potential reached the minimum (in the trials 28-33), and then rapidly repolarized. Changes in membrane potential during neuronal damage (Fig. 2.1), described in Chapter 2 and during instrumental conditioning (Fig. 3.7, bottom), are similar.

Alterations in the membrane potential were not the only immediate reasons for the change in the responses. For example, Fig. 3.5 demonstrates that the control neuron generated maximal response to the US when it was hyperpolarized (response to CS^+ 22) and it generated minimal response to the US when its membrane potential decreased (response to CS^+ 38). Similarly, a trained neuron did not generate an AP in response to the CS^+ when it was depolarized (responses CS^+ 30 and CS^+ 38), while it did generate a response when its membrane potential increased (CS^+ 41). Averaged data (Fig. 3.7) also demonstrated the absence of obvious connections between membrane potential and the value of neuronal reactions.

During instrumental conditioning, the membrane potentials in the trained and control neurons also displayed coordinated changes, but cross-correlation revealed non-symmetrically significant peaks (Fig. 3.3). Trained and control neurons during instrumental conditioning fail to demonstrate highly synchronous alterations of membrane potentials (Fig. 3.8), as we have found for neurons during classical conditioning.

The origin of the correlation between the neuronal responses is impossible to explain by their similar changes during training, since cross-correlation be-

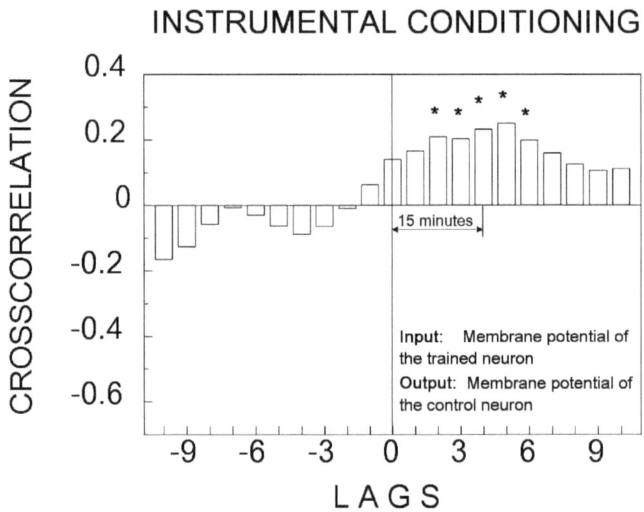

Fig. 3.8. Time delay between the factors of neuronal activity during instrumental conditioning. Ordinate: cross-correlation between membrane potentials of the trained neuron (input) and of the control neuron (output). Cross-correlation (bars) was calculated for pairs of neurons in each experiment and significance of the mean value (through neurons, $p < 0.05$, asterisk) for each lag was evaluated (Mann-Whitney U test). The interval corresponding 15 minutes is indicated.

tween the membrane potentials, preliminarily averaged through the neurons, was negative and the mean membrane potentials in the trained and control neurons changed oppositely (Fig.3.7, bottom).

Within an experiment, changes in the membrane potential of the control neuron were 3-5 trials, i.e. approximately 10 - 15 min, ahead of the corresponding changes in the trained neuron. The latencies of membrane potential changes during conditioning are too large for a spreading of the electrical signal through neural tissue, but were sufficient for cardinal reorganization of biochemical processes in the tissue and similar to both the latency of motivationally-relevant substances that act on motivation and also the latency of necrotic changes in neuronal electrical activity.

We have to remind ourselves that, all considered, here dependencies of the responses from membrane potential were received as a result of spontaneous change of the neuron state. In passing, direct artificial hyperpolarization of neurons by the current pulse can change only unspecific excitability. However, there are principal difficulties in an experimental examination of this suggestion, since neurons accept a current pulse as a signal and this procedure changes quality of the learning. And since we studied only alterations of membrane potential, which arose at the unintentional base, it was not clear what causal connection there is between changes in membrane potential and neuronal responses. We do not rule out the possibility of parallel development of these alterations or the existence of one common reason. Therefore, an increase in specific responses on the background of hyperpolarization may not be determined by the influence of the hyperpolarization itself. It is, of course, known that hyperpolarization affect differently responses during habituation and classical conditioning. Nevertheless, as we already have discussed, hyperpolarization in both cases promotes learning-evoked reorganization of responses and thus protected the organism. This is in agreement with the belief that an inhibition has a protective function, though inhibition may be directed not only to electrical events, but to metabolic processes also, for example, by means of inhibitory G_i proteins. Therefore, it is clear that the first approximation for paradoxical generation of neuronal reaction by means of protective hyperpolarization, which has been considered in Chapter 2 is deficient in the given case. Really, beneficial reorganization of neuronal plasticity during learning only in a particular case may be explained by the protective amelioration of the training-induced deterioration of membrane properties. In some cases learning decreases in excitability (habituation), whereas in other cases (classical conditioning) excitability increases and both modifications are beneficial for animal. In both cases, hyperpolarization was protective (since it promotes learning-related plasticity), but this protective influence was inhibitory for habituation, while it was excitatory for classical conditioning. Not only a hyperpolarization, but postsynaptic chemical influences also may produce ambivalent effects[16].

3.6 Paradoxical properties of instrumental reactions

We have demonstrated earlier that excitatory influences in some cases may exert a protective action. Therefore, a one-dimensional interpretation of beneficial influences as inhibitory cannot be a general rule. Similarly, when we have considered cellular damage and protection in Chapter 2, we certified that homeostasis can support a lively state of cells even if some vital parameter, for instance excitability, has deviated from the norm.

"Protective" influence of the hyperpolarization during habituation and classical conditioning was postsynaptic in nature, since it was connected with reorganization of excitable membrane. This was forestalling perfection of acquiring during learning adjustment and it was evidently directed in the right direction and supported the adequate change in responses. Inhibition retarded ineffective response to habitual stimulus and accelerated augmented response

[16] $GABA_A$ receptors sometimes exert such ambivalent actions on a neuron, when its effect is switched from inhibitory to excitatory [1312] or their agonists and antagonists produce similar behavioral consequences [236, 1345].

to the conditioned stimulus. During instrumental conditioning, inhibition and excitation of neurons lacked dependence from the level of membrane potential and reorganization of the instrumental reaction had another reason.

We compared in our discussion the behaviors of membrane potential and neuronal responses during training and have revealed inconsistency in the manners of their behaviors. Linear correlation between these characteristics during habituation and classical conditioning also differed from normal regularities. In addition, at the beginning of extinctions, hyperpolarization coincided in time with the augmentation of responses to the CS^+ (Fig. 3.4). Nevertheless, these are too indirect observations. Therefore, it would be important additionally to reveal how the efficiency to generate an AP depends on the level of membrane potential at the moment of spike generation. Taking into account that implementation of direct intracellular stimulation for evaluation of excitability may give an improper result, we compared how the level of the membrane potential in effective responses (when an AP was generated) differs from the level of membrane potential during failure of AP generation. Fig. 3.9 illustrates this comparison for the control neuron.

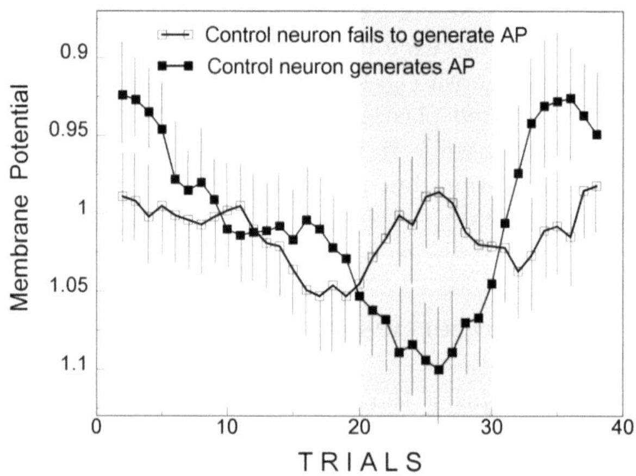

Fig. 3.9. Reorganization of properties of excitable membrane in the control neuron during a learning session (instrumental conditioning). Ordinate the value of membrane potential, on a per-unit basis. Filled symbols: membrane potential in those trials where the control neuron generated APs. Light symbols: membrane potential in the trials where the control neuron failed to generate APs. Standard errors are indicated.

At the beginning of training, the control neuron more often generated AP in the trials when it was spontaneously depolarized, in correspondence with the classical notion for dependence of excitability from membrane potential level. However, just in those trials where the control neuron generated an erroneous instrumental reaction (between 20 and 30 trials), the control neuron was hyperpolarized, whereas it was inhibited when the membrane potential decreased. Classical dependence of excitability to the level of membrane potential recovered only to the end of training (Fig. 3.9). This result is in agreement with the assumption that a protective role of inhibition endorses recovery of excitability after motivation-induced neuronal damage and this promotes the generation of output reaction. Taking this into consideration, we may evaluate the hyperpolarized state as a protection which improves neurons and thus paradoxically increases their excitability. It is most interesting to note here that between the 20th and 30th trials, the membrane potential of the control neuron, as an average, does not decrease. On the contrary, this particular neuron was hyperpolarized during this period of training. Hence, hyperpolarization can recover excitability of neurons not only as a counteraction to the excitotoxic depolarization, but also in the cases when distortion of excitability is not connected with a large fall in the membrane potential.

Such paradoxical regularities are present also for responses to the unconditioned stimulus. When the neuron tried to generate an instrumental reaction (both correct and erroneous), response to the unconditioned stimulus was larger during neuronal hyperpolarization [1257]. This phenomenon corresponds to recovery of the responses during classical conditioning, when compensatory hyperpolarization arises. We can suppose that the motivational excitation in our experiments induced non-excitotoxic damage in the control neuron. Compensational hyperpolarization decreased the damage and recovered spike generation. A similar phenomenon was discovered in the motor neuron of the Aplysia after the delivery of strong sensitizing stimuli: sensitization resulted in hyperpolarization of the resting membrane potential and a simultaneous decrease in the spike threshold [246]. Also a paradoxical reaction is described, consisting of the facilitation of spike responses of neurons to the conditioned stimulus evoked by antagonists of ionotropic glutamate transmission, which was blocked by GABA [1184]. Application of the antagonists increased baseline neuron activity, often by factors of 23 compared with the baseline, while the latency of the response to the conditioned signal decreased and its duration increased. There is constant tonic inhibitory control of the activatory spike responses of pyramidal neurons to conditioned stimuli in conscious animals. Inhibition actively involves organizing the excitatory responses of neurons in the instrumental conditioned reflex [1184]. In addition, during thirst, when neurons of the mammalian subfornical organ depolarized in response to an increase in osmolarity, the amplitude of the AP increased [35], indicating an alteration in AP-generating mechanisms. Certainly, paradoxical augmentation of responses after amelioration of damage may be connected with recuperation not only of electric characteristics of membrane, but, too, other important life factors.

This phenomenon was discovered by N.E. Vvedensky more than one hundred yeas ago [1311]. Injury of neuromuscular junction leads to development of the paradoxical state of the tissue. While responses of tissue to stimulus after injury decrease, the higher the irritation the smaller becomes the response. Thus, small damage excites, whereas large damage inhibits and when damage is large, responses increase, if strength of stimulus decreases. Nevertheless it is necessary to emphasize that paradoxical properties of neuronal damage cannot explain the selectivity of reorganization of neuronal responses during learning, if damage to a cell and especially its homeostatic recovery is an unspecific phenomenon. For instance, change in membrane potential usually spreads throughout the cell and this parameter may selectively affect only remote parts of a cell, such as different dendrite branches.

When we implemented the same method for investigation of a trained neuron, we revealed a more complex picture (Fig. 3.10).

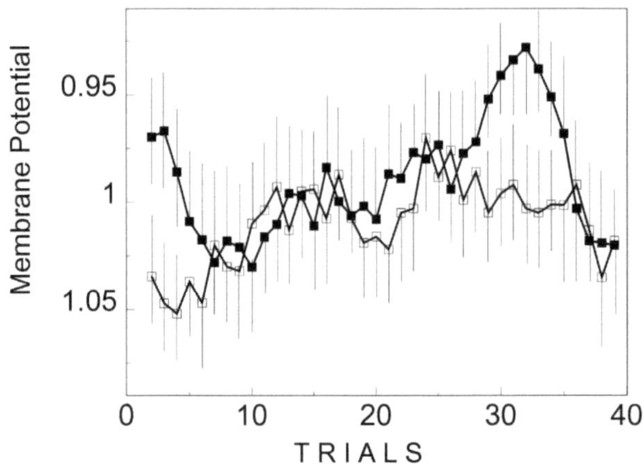

Fig. 3.10. Reorganization of properties of excitable membrane in the trained neuron during a learning session. Filled symbols: membrane potential in those trials where the trained neuron generated APs. Light symbols: membrane potential in the trials where the trained neuron failed to generate APs.

Only at the beginning of training before the influence of learning and during a short period of maximal depolarization was the trained neuron inhibited while being hyperpolarized, in correspondence with the classical rule. Spontaneous depolarizations of the membrane potential correlated with the increase in the current response to the CS^+ only in the period of trials 28-35. Here the membrane potential declines from that found in the steady state condition. Outside of these short periods, we did not reveal any dependence of excitability from the membrane potential. Although absence of correlation between membrane potential and spike generation indicates a disturbance of excitability in the trained neuron during active avoidance of punishment, we did not reveal the paradoxical phase, when inhibition protects neurons from excitotoxic damage. Moreover, although the trained neuron was depolarized just in this period of training (trials 28-35) (Fig. 3.7,bottom), its excitability was recovered through other means (Fig. 3.10) as, for instance, happens after activation of cyclic AMP system. The instrumental AP in the trained neuron arises during trials 28-35, when abundant depolarization of membrane potential began to decrease (Fig. 3.10). Just as depolarization is not the only single means for induction of neuronal damage, hyperpolarization is not the only single means for protective action of homeostatic compensation. As we have indicated, this result could be predicted, since hyperpolarization of membrane cannot control a selective augmentation of the reaction evoked by the CS^+, in opposition to the CS^-. A trained neuron, probably, generates an instrumental reaction not because of the protective influence of hyperpolarization to injury. Although the membrane potential of neurons changed during instrumental conditioning, this is not the only reason for the reorganization of the neuronal activity, if we proceed from classical neuronal properties: activation of a neuron during depolarization and its inhibition during hyperpolarization. A neuron, evidently, uses additional means (transient modification of channels, pH, second and retrograde messengers) for protection.

3.7 Trial-and-error at the cellular level during instrumental conditioning

Animal behavior looks purposeful since it is both directed to a goal and is not based on minute programming of action consequences. When an animal adjusts the current strategy of behavior it proceeds from existing conditions of environment, experience and some mysterious factor, which makes its actions non-predictable. At least, its strategy varies from trial to trial and probing actions continue even when an animal is familiar with the correct decision. However, this variability did not come about by chance.

The local nature of an instrumental action in our experiments, which we had described beforehand, affords an advantage for the investigation of instrumental conditioning, since reinforcement of a single AP in the identified neuron reduces the instrumental reaction to activation of several similar neurons and, ideally, to the only trained neuron. Therefore, it is easier to trace the development of the reaction.

The sequence of the neuron's activations resembled trial-and-error. We consider their activation as trial-and-error at the cellular level. Our data show that this assumption is plausible. During successful instrumental learning several phases can be observed, in which the trained, the control or both neurons together generate action potentials in response to the conditioned stimulus. At the beginning of training, the responses to the CS^+ of the trained neuron and, in lesser degree, those of the control neuron, decreased and were mutually correlated.

Let us consider the middle part of the training (trials 20-30), when the action of the control neuron temporarily replaces the correct action of the trained neuron. Trained and control neurons initially had a kindred function, but it was unknown a priori which neuron will play a role in the instrumental action. Due to the large level of non-stability and because of the resemblance of the two neurons, the control neuron sometimes received information that its participation in the instrumental reaction is essential. Therefore, it is natural that the control neuron AP might be considered by a neural system as an instrumental reaction. Substitution of the trained neuron AP by the control neuron AP did not correspond to the instrumental paradigm. Moreover, during substitution of the trained neuron action, the action of the control neuron depended on both the preceding unconditioned stimulus presence and its own preceding activity, while the trained neuron in this training period did not depend on either its own preceding activity (i.e. on unconditioned stimulus presence) or on the activity of the control neuron [1260]. In this short training interval (trials 20-30) the control neuron failure probably ranks erroneously as the incorrect reaction.

The classical mechanism of spike generation is evidently disturbed during origin of an action. The origin of both the correct reaction in the trained neuron and the erroneous reactions in the control neuron took place through recovery of neuron's excitability, but routes for this recovery may be different. We have demonstrated that this may be explained by the development of compensational processes in neurons. Therefore, depolarization is not the immediate cause of the AP generation, while hyperpolarization is not the immediate cause of the AP failure, when the trials-and-errors are generated. So, a logical constraint between changes in membrane potential evoked by a conditioned stimulus and output reaction is broken.

It can be concluded that in the middle part of training the control neuron action substituted the trained neuron action. The control neuron temporarily became the key neuron in the instrumental conditioning. The animal apparently uses the trial-and-error method as applied to neurons instead of instrumental movements. We come to the conclusion that alteration of membrane's properties during avoidance of punishment is the essential element of the action emission.

4

Goal-directed actions (a single neuron can behave)

The final result of brain activity is generation of actions. Action may be directed to the outer environment, inner environment of the body and to controlling interbrain operations. In all of these cases, the brain in the appropriate cases generates proper reaction. Let us consider how intentional actions arise and by what means brain directs its deed to a specific channel.

4.1 A physiological description of voluntary actions

The peculiarities of voluntary action generations are suitable for considering of the example of set experiments devoted to elaboration of local instrumental conditional reflex in Helix. In this case we precisely know which neuron is responsible for generation of an instrumental reaction.

We have demonstrated that generation of instrumental reactions may be a consequence of the recovery of neuronal excitability after damage induced by a motivational excitation, and arbitrary actions may be the result of compensational processes in neurons. Motivational excitation arises in neurons following chain of events. Discrepancy in some physiological constant leads to both processing the damage (for example, a decrease in membrane potential) and subsequent homeostatic protection (for example, an increase in membrane potential). Presentation of a conditioned stimulus (CS^+), which is connected by cause-and-effect relationship with the appearance of reward or punishment leads to intensification of compensation and thus the neuron anticipates the protection evoked by reward (satisfaction of motivation in any case, during negative or positive learning, is a reward). During training, neuron step-by-step memorizes the link between its activity and reward. We studied how an unconditioned stimulus (US) presentation affects an instrumental action in the following trial.

The US exerted a direct effect on the neurons and a latent influence, which could be detected in the next response to the CS^+. Correct and incorrect responses occurred intermittently during training and the current response depended on the presence of the unconditioned stimulus in the preceding trial (this corresponds to a failure or generation of the action potential (AP) in the trained neuron response to the preceding CS^+). As we already indicated, during learning, the responses of the control neuron to the CS^+ decreased only after overcoming a local maximum (Chapter 1). Let us consider this phenomenon from other side.

Interestingly, this temporary augmentation of the control neuron response was observed predominantly in the trial after punishment (Fig. 4.1). This means that in this period of training (approximately between 20-30 trials) the control neuron erroneously generated an instrumental reaction instead of the trained neuron. This "probing" reaction quickly disappeared. The local maximum, visible to the naked eye in Fig. 4.1, really exists. The correlation ratio and the curvilinear relationship for the next responses to the conditioned stimulus after incorrect trials are significant. The local maximum was absent, if we considered only those responses to the CS^+ of the control neuron, which were the next after correct responses of the trained neuron in the preceding trial and, hence, the animal did not receive a US on those trials.

It is possible that the control neuron (i.e. foreign neurons of the same behavior) collaborates with the trained neuron in the instrumental reaction generation during acquisition. When comparing the effect of punishment on the responses of the trained and control neuron during trials 20-30, it appears that the control neuron response substituted the actions of the trained neuron (see curves in Figs. 4.1 and 4.2). However, this scenario only coarsely describes the sequence of instrumental actions. Not only activation of different neurons, but also their combinations may be adopted as a possible instrumental reaction. The coincidence in the two events, AP absence in response to conditioned stimulus and unconditioned stimulus presence, determines the difference between the behaviors of the trained and control neuron [1271]. In accordance with our procedure, and considering that both the trained and the control neuron belonged to the same group of defensive behavior neurons, the control neuron in our experiments was representative of the kindred neurons and displayed their participation in the learning.

In our experiments, a behavior of both the trained and control neuron were inhomogeneous during training (Fig.4.1,4.2). Responses to tactile stimuli in mollusks usually increase in the next trial after detection of the harmful US. The effect of the US during the training session was ambiguous. Presence of the US led to augmentation of the control neuron response to the CS^+ just in those trials when response of the trained neuron to the same stimulus decreased.

The period of the supposed substitution of the trained neuron action by the action of the control neuron is also inhomogeneous. When the reaction of the trained neuron to the CS^+ was inhibited, this took place just after punishment: in this period, the trained neuron erroneously learned not to generate an AP in response to the CS^+. The control neuron in this period of training learned to generate a reaction and made this also after punishment, i.e. it learned by

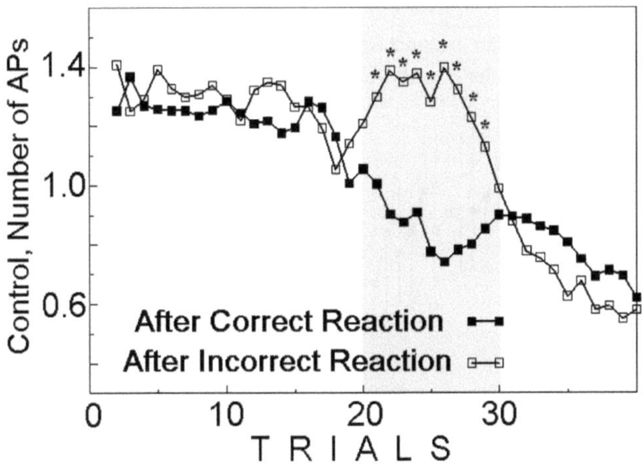

Fig. 4.1. Causal dependence of the conditioned response of the control neuron on the presence of the unconditioned stimulus in the preceding trial during elaboration of the local instrumental reflex. The data divided into two groups depending on the preliminary response of the trained neuron to the conditioned stimulus. During 20-30 trials (interval 20-30 trials is indicated.), the control neuron is disinhibited in the next trial after punishment and generates an erroneous instrumental reaction. The asterisk denotes a significant difference ($p < 0.05$). Symbols are at the Figure. Method. Generation or failure of an AP in the response of a control neuron to the conditioned stimulus is apparently similar to correct and incorrect reactions, because it would be correct or incorrect, depending on whether the control neuron is chosen as the trained neuron. Action potential failure in the trained and control neuron did not appear at the same time. However, the control neuron sometimes produced action potentials during the same trials during which the trained neuron produced an AP and received information as if its reaction was essential for the instrumental reaction. Therefore its reaction could be considered as an instrumental reaction in the earlier stage of training. An incorrect instrumental reaction in the control neuron appeared before the correct instrumental reaction in the trained neuron, since the control neuron in our experiments was initially more excitable; we chose this scheme in order to make easier acquisition of the local instrumental reflex.

its "errors" (later it ascertained that its participation is inessential). However, when the trained neuron began to generate an instrumental reaction (trials 28-35), this happened after presentation of a painful US. Trained neurons learned by their errors.

The control neuron behavior did not reveal a significant local maximum during unsuccessful learning (around a third of the experiments), when reaction of the trained neuron to the CS^+ did not recover at the finish of training and did not exceed the corresponding reaction of the trained neuron. Consequently, during unsuccessful learning, the trained neuron did not generate an instrumental reaction, while the control neuron did not generate an erroneous instrumental reaction. Both neurons participated in successful learning

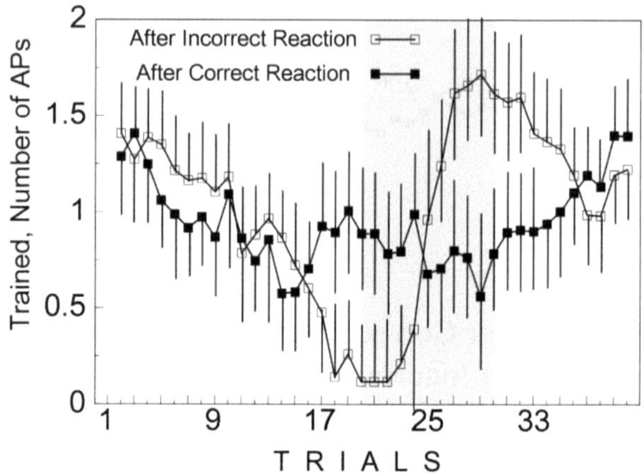

Fig. 4.2. Influence of punishment to the generation of an instrumental reaction by the trained neuron. The confidence intervals are shown ($p < 0.05$). Symbols are at the Figure.

and both neurons did not reveal relevant modifications in responses during unsuccessful learning. This means that even during elaboration of a local instrumental reflex, many neurons collaborate on learning.

4.2 An origin of agency and voluntary actions

An organism generates action to support the stability of its inner milieu. When a cat is jumping, this looks as if an animal produces energy from nothing. Certainly, an animal spends energy that it stocks up earlier. Something within the brain coerces an animal to spend its resources in the name of future success. In these cases inherent energy is converted in other types of energy, say, nervous, muscular or mechanical energy. We already considered how brain generates a goal. But what is the cellular nature of inner power, or capacity to act, or will, or agency in the philosophical sense (agency is the capacity of an object to act) that compels an animal to do something? An animal generates action if it feels needs. In this case, the animal has a mismatch in its inner characteristics, which the homeostasis cannot compensate for by metabolic recuperation and revival may be reached only through interaction with the environment. We have supposed (Chapter 2) that generation of instrumental actions for interaction with the environment also proceeds with the assistance of homeostasis. Homeostatic compensation is very appropriate for this purpose. Really, a behavioral act is directed to accomplishment of the same task that homeostasis resolves: recovery of a lost equilibrium. Necessity of behavior emerges when homeostatic efforts are at the peak of intensity. Promotion to generation of instrumental reactions is an indirect and more protracted way for maintenance of the working state of an organism. At the cellular level, homeostasis has the means in order to compel the neuron to generate an AP, since protection from damage normalizes disturbed neuronal functions. Homeostatic compensation is an energy-consuming process and it is connected with the expenditure of chemical energy and with the loss of temperature for a tissue.

Protection against damage may proceed either by means of compensation of the corrupted parameter or by storing up the distorted value of this parameter with the reorganization of the entire system using the rest of the parameters. After reorganization, homeostasis maintains a new set of optimal parameters. An animal can support any factor at a value that differs from the standard, usual value by means of the alteration of other values, but never treats a threat to its life as a reward. Such an important factor as the level of injury-reparation may serve as the mother-value of all values [321]. At the present time, we know only that the mechanism of control of a complex living system by means of a general variable does exists, but the essence of this safeguard is the urgent task for future study.

In the simplest cases, homeostasis recovers the previous optimum of the distorted variable. This improves the state of a neuron and perfects its input-output characteristics. For instance, when the process of damage is reduced to a decrease in membrane potential, the compensational increase in membrane potential protects the neuron from damage and, therefore, paradoxically activates the neuron. Goal-seeking behavior arises on the second phase of damage-induced decline in neuronal function, when compensation cannot overcome damage and threats of cell death get up. This means that homeostatic compensation of damage can generate neuronal output reactions. We may suppose that when brain stresses homeostasis in order to avoid disaster, it reaches two rather different results. On the one hand, the homeostasis recovers lost functions at the expense of internal resources, but on the other hand, protective compensation affects generation of neuronal reactions and, hence, behavior.

Hypothetically, when current, functional damage is modest and recoverable, the protection inhibits action generation that results in a passive avoidance of the danger. However, when the damage threatens to become insurmountable at the expense of domestic sources, the neurons go into a paradoxical phase and efforts of homeostasis are directed to action generation, interaction with the environment and to an active avoidance of danger. "Danger" in a given meaning may be not only a predator, pain or electric current, but also, for instance, starvation or thirst. In simple behavior, voluntary actions are, probably, initiated due to compensation of distorted excitability or other distorted parameters. The paradoxical phase is not absolutely connected with reaction augmentation after protective inhibition. Healing of other kinds also increases the capability of neurons to exhibit the appropriate behavior. In serious cases of damage, when homeostasis builds a new system of optimal values, influence of compensational processes to behavior may be more cardinal, such as, for instance, recovery of lost functions after brain damage or recuperation of forgotten habits. However, we know too little to discuss these mechanisms.

After learning, an occurrence of compensation does not come only after the instrumental reaction and receipt of the reward. The result of compensation is anticipated and leads to removal of the harmful impact. Homeostasis somehow initiates protection in the first phase of excitation, evoked by a CS^+ or even without an outer signal and at the proper time. Such a system will be serviceable only if homeostasis has access to memory stores. Some data are in agreement with such capability of homeostatic mechanism, but the problem needs serious experimental study.

However, it is not sufficient, to produce an action in order to satisfy motivation. Brain needs to produce the right actions that will lead to the appropriate goal. Certainly this problem does not exist at the level of a single neuron, having only one output. A neuron cannot choose its course of action, and the relative strength of its axonal branches is predetermined before the action. The mechanism of choice requires participation of a collective of neurons and also the involvement of memory. In passing, it should be noted that participation of some neurons in the process of choice denotes that the necessity to store diverse information needed for choice in the single neuron is absent.

Generation of intentional actions is an explainable matter. In common features, motivations emphasize protection and this recovers excitability or other essential characteristic of special neurons, which produces an AP. Turning on the compensation represents a generation of the action. When the state of an organism dictates the necessity to satisfy motivation (this is a purely material process), homeostasis exaggerates its service, promotes action generation (also a material processes) and this is perceived by outer observers as an intentional action and manifestation of will. In order to accept this description, one needs to assume existence of primitive feelings at the cellular level (injury-healthiness or worse-better). We already touched upon this problem and will return to it again in the next chapter. Voluntary actions rise as a result of intensification of compensation directed against damage. However, now it is essential to point out that we do not need to introduce a new essence, apart from primitive cellular feeling, in order to explain the appearance of will.

4.3 Common decision and the only reaction of the whole brain

A healthy brain receives information and acts as a whole in accord with a "winner-take-all" principle [252]. This assumption relates to any brain activity, complex or simple, and at any level of hierarchy. Compulsory choice of the only decision concerns, even for the highest level, apprehension, level of aware perception and intentional actions. This does not mean that at any particular moment we execute only one action. We are able, for example, at the same time to breathe, to walk, to talk and to take pleasure in a view. Besides, each of these particular performances consists of more elemental actions: coordinated work of lots of muscles, depending on signals from various senses and from past information. But what is the meaning in the statement that brain acts as a whole and in each given moment it is concentrated only on something single? We walk, or sit, stay, run, lie down, swim, crawl, clamber, etc. If we walk, then make this the determined direction and expediently react to obstacles. When some muscles contract, other, concurrent muscles, weaken. Brain can control several competing actions, but during competition for each particular action only one general decision always wins.

Originally it was proposed that the neuronal mechanism responsible for triggering the different behaviors is inhibitory interactions among the neurons responsible for different behaviors [665]. This phenomenon is traditionally explained by a lateral inhibition [781]: concurrent actions mutually inhibit each other, but a stronger action exerts a stronger inhibition to a weaker action and a weaker action only weakly inhibits the stronger action. This mechanism may be easily demonstrated at the neuronal level, but it cannot explain the generation of dominant action at the behavioral level. Particularly, the excitatory and inhibitory conductances had similar preferred orientations and tuning widths for any cell in the visual cortex [34, 997] and lateral inhibition cannot explain the origin of selectivity. A weak foreign action does not at all make weaker the dominant action and even may augment it. Unique properties of dominant behavior were described by A.A. Ukhtomsky [1284]. When, for instance, a cat is prepared to defecate, its entire organism is concentrated in this action. When an experimenter who studied neuromuscular junctions stimulated a motor nerve (he did not pay attention that the cat was prepared to defecate), the muscle did not contract, but defecation was strongly accelerated. Later it was found that the nucleation site for dominant behavior is connected with enhanced excitability of specific neurons and that excitation of foreign neurons augments the nucleation site but does not diminish competing focal activity and does not induce their own specific action. Certainly, lateral inhibition cannot explain united activity of numerous neurons for the generation of their collective dominant activity and, surely, if non-relevant neurons for a given behavior augment dominant activity, they cannot be inhibited, but ought to generate some corroborating activity.

Besides the necessity to compose a whole image from a number of neuronal messages, brain ought to bring into coincidence in time particular neuronal activities. The time when a brain receives a signal synchronizes activity of all related neurons, but activity must be synchronized also at the time of decision-making, although the speed of responses in different areas of brain may be different and the distance of signal travelling also differs. In addition, a decision may be unrelated to recent signals at all, but must be completed at the appropriate time. Therefore, in order to make a proper decision, the mechanism of decision-making needs a synchronization of many neuron activities. Though this condition is necessary, it is not sufficient by itself. Coordination of neural and glial activity is accomplished by coupling cells through an electrical synapse. Gap junctions are important for the spread of damage in the tissue, but their functioning is probably connected with the state of awareness. For instance, chemical substances affecting damage often influence motivations and gap junctions, too.

Thus, many neurons make a decision simultaneously, but they make their own elemental decisions. We discussed earlier that averaging of the current neuronal activity cannot serve as the method of choice of a biologically significant reaction. Therefore, there remains the problem, of how brain chooses a winning coalition of neurons and inhibits a losing coalition.

Lastly, if the current image is properly recognized, correctly identified with events in the memory and a common decision is elaborated, brain has to send one signal into a corresponding output channel. The problem is that each neuron has its specific output and brain must find the only right action and prevent activation of other actions, although each neuron in the interacting group makes, simultaneously, the proper decision. This problem has not arisen in the scope of representation about a brain, as a mechanism assembled from simple neurons. A construction, in contrast to a collective of individuals, can easily generate a general evaluation, generate general decision and activate a particular effector. However, we time and again noted that this hypothesis meets many other insoluble difficulties.

4.4 Gap junctions enriches a brain with a new quality

Neurons send their message through synaptic connections, constant or transient, to neurons of other types, interneurons and the next neurons in the input-output route of brain. Synaptic connections cannot unite neurons of the same levels, whether they process scattered or general information related to a key image.

On the contrary, electrical synapses do not serve for sequential communication in the brain. Rather, they increase the degree of integration at a given level and a linkage of electrical and chemical activity of almost identical neurons may, in such a way, create a new quality. This means that electrical synapses, in contrast to chemical ones, do not transmit information in a conventional way, along neuronal axes. Electrical synapses join information delivered to neurons of the same type, and they synchronize and generalize current information. As an example, in gap junction-deficient animals the solution to object recognition tasks are interrupted, since they apparently cannot distinguish between newly presented and older objects. Therefore, electrical synapses might have a role in cognitive functions [1155].

Electrical coupling may occur only between neurons of the same type, such as fast spiking/fast spiking, aspiny/aspiny, vasopressin/vasopressin neurons, parvalbumine contained/parvalbumine contained neurons, etc. Furthermore, electrical synapses in the brain often interconnect not only neurons of a similar type, but similar size, and input resistance [257] and, moreover, neocortical electrical synapses almost exclusively connect GABAergic neurons belonging to the same subtype and do not connect GABA interneurons of different types [886, 541, 1155]. However, there is evidence for electrical synapses between principal cells, too (for instance, in hippocampal pyramidal cells) [711]. A single axon linked by gap junctions to two dendrites may mediate electrical coupling even between the dendrites of the same neurons [260, 874, 1155]. This type-specific neuronal coupling may arise from expression of identical connexins in the group and different connexins between the groups of these cells. For example, matching odorant receptors project afferents to a single olfactory bulb glomerulus and spike synchrony in mitral cells is glomerulus-specific. Electrical coupling as well as correlated spiking between mitral cells projecting to the same glomerulus was entirely absent in connexin36 knockout mice [240]. Thus, cell coupling helps to gather features of images dispersed among neurons of the same type, rather then to produce neural calculations deep in the neural axe.

Regrettably, properties of gap junctions were investigated, generally, in non-neural cells, but characteristics of neuronal gap junctions perhaps do not differ principally. The highest density of gap junctions in the brain is present in glial cells that are intimately involved in neuronal function and behavior. Glial cells generate a slow negative potential shift in the cortex in the period between a conditioned stimulus and a motor response necessary to obtain a reward. Gap junctions are also present in the nervous system, and higher neural centers (cortex, hippocampus, cerebellum and hypothalamus) are especially full of gap junctions, where they connect glia and neurons [311, 697, 1326]. We have demonstrated the role of gap junctions in motivational behavior on the one hand and in processes of damage on the other hand. In addition, gap junctions and oscillations seem to participate in creation of aware states of the brain.

Properties of gap junctions with regard to the activity of neurons include the ability to unite intracellular environments of the coupled cells, permit the rapid intercellular passage of electrical signals and synchronize activity of cells in an ensemble. Intercellular channels allow also the transfer of water, ions, metabolites, messenger molecules and calcium waves between connected cells, providing a mechanism for coordinating of the chemical activities of groups of cells [311, 318, 1024, 160, 844]. A gap junction is the most reasonable device for formation of ensembles [108, 160, 844].

Electrical synapses are not fixed in their properties, and they are sometimes rapidly changeable, although new tight contacts are formed slowly. Many channels must cluster together before the first channel opens [260]. Once opened existing junctions may be modulated rapidly by means of a change in membrane potentials of coupling cells or chemical influences [180, 961]. In principle, it is possible pharmacologically to induce closing of the channels, to change preferred single channel conductance, to open channels or to keep them open [318]. Kinetics of uncoupling depends on isoforms of connexins, which form the tight junction. Heterotypic channels are sometimes composed of docked hemichannels that differ in connexin composition and hemichannels on the two sides of a gap junction plaque can be modified independently [962]. The number of connexin channels open at any given time is low (only 4-9%) and there is a huge potential for gap junction reorganizations [257].

Channels are made to open or close by means of regulatory gating mechanisms within a junction. Increasing concentrations of ions, such as Na^+, Ca^{2+}, Mg^{2+}, and H^+ can reduce acutely gap junction conductance within minutes [1066]. Gating stimuli include voltage, H^+, Ca^{2+}, NO, arachidonic acid, certain lipophilic agents, and protein phosphorylation [109, 260, 698]. Sensitivity of coupling to different ions considerably differs. The conductance of gap junction channels is reduced when intracellular Ca^{2+} increases, but Ca^{2+} must rise to pathologically high concentrations for gap junctions to close [257]. The effective concentrations of Ca^{2+} ions vary depending on cell type, type of connexin expressed and procedure employed to increase their cytosolic concentrations; however Ca^{2+} as low as 150 nM or lower have been reported to be effective in some cells [961]. Latency of Ca^{2+} elevation in astrocytes in response to neuronal activity is less short, because of Ca^{2+} release from intracellular stores, and the process is complete within two minutes [642, 964]. Spread of Ca^{2+} through gap junctions can, evidently, participate in the middle-time scale (minutes) modification of behavior, but Ca^{2+} is a much less potent regulator of gap junction conductance than intracellular H^+ and inositol trisphosphate [599, 257]. The conductance of many gap junctions is exquisitely sensitive to the pH of the cytoplasm.

Gap junction modification may proceed in a synaptic way [598] and can be modified by activation of various G protein coupled receptors [109, 874]. Coupling is dynamically regulated by the surrounding neuronal activity, excessive excitation changes the strength of the coupling and alteration may last for hours [119, 257, 701]. Activation of metabotropic glutamate receptors, via endogenous neurotransmitter or by agonist, causes long-term reduction of electrical synapse strength between the inhibitory neurons of the rat thalamic reticular nucleus [701]. Dopamine (cyclic AMP-dependent), glutamate (Ca^{2+}-dependent) and NO (cyclic GMP-dependent) can decrease or increase junctional conductance [913, 109, 963]. Nevertheless, gap junctional channels can switch between several steady conductance states within seconds or minutes or faster (see further) [317].

Rapid modulation of gap junctions allows on-line reorganization of the neuronal coalition for current needs. Gating stimuli, such as membrane potential, pH, etc. that affect coupling, are, typically, unspecific and affect a cell regardless of its particular activity. However, these stimuli change a condition of the cell as a whole in the cellular ensemble. In other words, properties of the cell's gap junctions determine the comparative wealth of the neuron's axon, or output of the cell. This is important for choice of output reaction of a neural system.

4.5 Choice of alternatives with respect to the output

There are two interrelated problems in neuroscience: how a neuron targets its reaction, since it has only one output (and if it cannot make a choice, then the brain cannot) and how a brain saves from number of competing outputs.

In our brain, many neuronal groups coexist simultaneously at any one moment and these groups actively interact with each other [428]. Sometimes it is possible to observe that separate networks operate during an unexpected change of environmental circumstances. If a specific neuronal group is beforehand predetermined by initial conditions, a choice of one group over another is not required at all. In this case, the environment chooses. Thus, during sudden switching between two reinforcers (cocaine and water) the same nucleus accumbency neurons do not respond in a phasic manner to both types of rewards. During rapid switching between cocaine and water the networks are negatively coupled and mutually inhibit each other [299]. It is easy to find the cells in an organism that specifically participate in one specific behavior and do not participate in another determined behavior. For instance, monitoring of neuronal activity affecting a muscle participating in an instrumental behavior would be the best way to collate all of the finest details of a given behavior. However, existence of separate neuronal ensembles for different behaviors does not explain the problem of free choice. Different behaviors might be controlled by different, even non-intersected groups of neurons, which are activated by separate input signals. Absence of a station for choice between two behaviors may be relevant only during unaware behavior, when the consciousness does not participates as the brain controls two behaviors simultaneously and isolated signals induce non-intersected reactions.

Usually the same neuron can participate in the different behaviors and interactions between neurons are not as simple as in the aforementioned examples. In animals trials, single neurons can intermittently participate in different computations by rapidly (from tens of milliseconds) changing their coupling without associated changes in firing rate [1291, 352]. Thus, identical sensory inputs in the leach sometimes elicit crawling and at other times swimming with roughly equal probabilities and the eventual choice may depend on the behavioral state before stimulation [166]. Some neurons discriminate earlier than others. Before the two behaviors become evident, discrimination time of a linear combination of neurons could discriminate earlier in time than any single neuron. None of the neurons with early single-cell discrimination times significantly affected the elicited behaviors. The leach neuron 208 (which is a central pattern generator neuron) can selectively bias the decision to swim or to crawl: with cell 208 hyperpolarized, the nerve shock reliably evoked swimming; with it depolarized, the nerve shock evoked crawling or delayed swimming. Neuron 208 was part of the combination of neurons that discriminate earlier in time, but was not the cell with the earliest discrimination [166]. Nevertheless, even in this wonderful example, the mechanism of decision-making that somehow modulates the membrane potential of cell 208 also slips away from our comprehension. Cell 208 receives from somewhere or other a completed decision.

One important function that cannot be completed by a single neuron, but is carried out by a brain, is a choice of particular action out of several potential possibilities. We have demonstrated that a neuron predicts the result of its reaction and makes a decision to generate or fail to generate a spike. However, a neuron only has an axonal output and does not have means for training its spike on the definite aim. In order to choose one out of, at least, two possibilities, it is necessary to foresee the results of different reactions and to possess at least two channels of actions. The only possibility to hold more than one output on the basis of neural cells is gap junctions that unite a group of neurons and glial cells, match their chemical calculations and synchronize their activity. A coupled group of cells possesses a multiple output. In order to choose a preferred action, the group must augment one output and diminish others. The means for reorganization of the output activity of the cellular group give, possibly, a high plasticity to gap junctions. When gap junctions unite the protoplasm of neurons, each one keeps its possibility to send a signal to its axon. At least, at the present time we do not know any other means for performing these two tasks: properly to recognize an entire image and to perform a suitable behavior.

At least in some cases, gap junctions allow especially rapid transmission of signals. General anesthetics reduce signal transmission at gap junctions very quickly [1030, 109, 257], to the range of a second. Single mechanosensory afferent stimulation in the lateral giant escape command neuron dendrites of crayfish results in localized increases in coupling. EPSP produced by synaptic inputs from one nerve can spread and help recruit afferents in other ipsilateral nerves and the latency of augmentation of EPSP is less than 1.0 ms. Mechanosensory stimulation spreads antidromically back through electrical junctions to unstimulated afferents [46]. Rapid dynamic modulation in baroreceptor-sympathetic coupling is also observed in the central pattern generator of the cardiac rhythm during performance of the respiratory cycle [455].

Connexin hemichannels in an on-line regime may be blocked by hyperpolarization of the plasma membrane [260] and correspondingly, weakly coupled neuronal networks can operate independently under some conditions, yet become synchronized when the neurons become depolarized close to threshold [1180, 874]. Gap junctions may be sensitive to a difference in membrane potentials or to the absolute value of membrane potentials of connecting cells [257]. Equal and simultaneous hyperpolarization of both cells caused an increase in connectivity to a maximum, whereas depolarization caused its decrease to zero. In some cases, conductivity of a gap junction depends on both the difference between membrane potentials and their absolute values together[1]. Such

[1] In the case of sensitivity to a difference in voltage between two cells, maximum conductivity of channels is observed when membrane potentials in two cells are

non-symmetric channels have different properties on two sides of the membranes. Hence, properties of gap junctions allow fine regulation of the strength of intracellular connections. For instance, the degree of coupling (measured as coherence) within the fly brain was increased when the fly was responding to salient stimuli [497]. Decreased behavioral responsiveness is associated with decreased coupling.

The axon output cannot ensure choice, whereas the possibility of even only two alternative impacts of a neuron to its neighbors via gap junctions may ensure that brain has the instrument for fine regulation of output reactions. Stimulation of individual astrocytes in hippocampal slices evokes relatively rapid (60 - 200 ms) inward currents in adjacent neurons [526] and this response occurs synchronously in multiple hippocampal neurons [380]. Yet, astrocytes themselves react more slowly to neural influences [964] and if the astrocyte uses its microdomains for choice of specific output reaction, it must make a current decision, since the route from neurons to astrocyte is too slow. The reverse route from astrocytes to neurons is rapid and, principally, may be used, say, for operation of memory removal from stores. Nevertheless, no one has observed the process of decision-making in astrocytes.

Within different branches of a cell, gap junctions may by regulated separately. Arachidonic acid, NO and their metabolites induce Ca^{2+} entry into the so-called cytosolic microdomains preferentially restricted to peripheral territories of cells, and does not necessarily invade the nuclear region in the majority of the endothelial cells analyzed [1236]. The propagating cellular signal that causes the occurrence of neuronal domains seems to be inositol triphosphate that promotes Ca^{2+} release from intracellular stores [599]. This phenomenon of spatial restriction is indicative of local Ca^{2+} signals and may depend on the geometry of the cells, the clusterization of Ca^{2+} channels and other signaling molecules and on spatial distribution of inputs. Particularly, oscillations sometimes are restricted to portions of the processes of individual cells and they do not necessarily propagate over large distances, even within one astrocyte [526]. Stimulation of one astrocyte led to a calcium response in a subset of neighboring astrocytes, but not all of them, suggesting the presence of distinct networks of astrocytes [27]. Astrocytes can integrate signals from multiple synapses and help to coordinate network events, providing a mechanism for the propagation of information [27]. Besides, glutamate re-

---

equal and, when they are unequal, conductivity decreases till 5-40%, but not to zero. At the single channel level, transitions from open to closed channel states carry on by relatively slow (10 ms) or by fast kinetics (< 1 ms). During recoupling, each channel is reopened slowly from the closed state, so that closing proceeds quicker than opening. These channels exhibit also slow kinetics (10 ms) when using chemical uncoupling agents. Uncoupling induced by arachidonic acid, high intracellular Ca^{2+}, or H^{+}, can be reversed by hyperpolarization in cells. The degree of coupling (defined as the ratio between the changes of membrane potentials of two individual neurons after injection of current into one cell) ranges from a fraction of 1% up to 50%. Most electrical synapses are bidirectional [257].

leased from astrocytes can promote synchronized activity in distinct neuronal domains [380]. Therefore, glial cells can affect local control of part of its gap junctions by means of signaling molecules and, probably can affect selectively adjacent neural cells. Such rough space selectivity on neurons and glial cells in respect to second and retrograde messengers might be sufficient for decisions regarding the problem of choice of output. Besides, neuronal ensembles may be controlled in an on-line mode (see further) and a cell may send non-symmetrical signals to its neighbors.

Synaptic potentials are more effectively spread through gap junctions than action potentials, which are much stronger but shorter. Therefore, transmission of APs through electrical synapses is significantly less efficient than low-frequency synaptic potentials or after-potentials [257, 874]. Action potentials are often followed by a relatively slow after-hyperpolarizing potential, which is much slower than the depolarizing phase of the AP and APs are attenuated to a greater degree than the slow potentials. For that reason, principally, excitation of one neuron may inhibit adjacent ceils [541]. In view of that gap junctions ordinarily connect inhibitory interneurons, excitation of principal neurons follows the excitation of interneurons, which inhibit each other by means of gap junctions. Interneurons operate with high speed and temporal precision [591]. Thus, activation of principal neurons will involve neighboring principal neurons in coordinated activity and this may ensure a non-symmetric influence to an output reaction.

Thus, gap junctions allow for the possibility that a neuron can affect non-symmetrically its adjacent cells. However and more importantly that gap junctions may ensure the plausible choice of interacting neighbors. An output reaction is produced by those neurons that gave the highest estimation of a proposed result of their actions, whereas non-relevant cells favor the relevant action. Therefore, origin of dominate states, when one specific output is augmented at the expense of relevant excitations, may be connected with gap junctions.

When coupled neurons in the ensemble generate APs, they are synchronized to a lesser degree than postsynaptic potentials in coupled cells. Coupled cells do not always generate spikes simultaneously, but the degree of synchronization increases. Correlated in time, neuronal activities effectively affect target neurons. As a result, neurons are driven more effectively by synchronous synaptic inputs, and this effect is stronger when cells are more depolarized [66]. Therefore, when one cell is more specific in respect to the current signal, this cell reorganizes its excitability and generates AP first. We may suggest that this will change cell coupling. Really, when conductances in two coupled cells are different, the coupling coefficient from a high to a low conductance cell is less than that in the opposite direction and the current is largely carried by K^+ ions [109]. At the time, as the specific cell generates a more powerful postsynaptic potential and AP, their conductance sharply increases. Therefore, the coupling coefficient from more to less specific cell becomes smaller. Taking into account that activity in coupled cells is synchronized, less specific cells will support activity of more specific cells, and this will promote the origin of a dominant state in the vicinity of high specific responses. Particularly, K^+ current directed into a specific cell will improve the state of this cell, since this increases ATP production [158] and protects it [836, 1379, 1386]. We may conclude that an ensemble promotes the establishment of dominance between neurons. In passing, the same conclusion is true when a specific cell does not generate an AP, but the subthreshold response exceeds coupling from less specific cells to the dominant cell. In this case, a less specific cell will also support activity of dominant cells and thus convert the dominant response to one of super-threshold.

4.6 Formation of neuronal ensembles during tension

A brain has a highly parallel organization and interactions between neurons may be relevant to the neural code. One assumes that the spike counts of single neurons within an arbitrary time window are the relevant features of the code [3], but others suppose that temporal structure within the spike trains is not relevant for coding [66]. In the standard, unaware or habitual situation, when the need to apply to long-term memory is absent, brain makes decisions so quickly that signals have no time to travel along the neural network in order to produce a decision. All neural calculations must be completed in milliseconds. These are, probably, chemical calculations at the interneuronal levels and a maximum of 1 - 3 consecutive neuronal switches may be involved. We regard the phenomena of application to long-term memory, motivation, awareness and ordinary behavior as connected with the temporal coding. On the other hand, recognizing current images and decision-making are not relevant for temporal coding, since these processes are too rapid compared with the spreading of neuronal excitations. Let us consider properties of neuronal ensembles.

A single neuron evidently may play a key role in a behavior of an entire animal in special circumstances, when a conditioned stimulus, reward or instrumental actions are artificially directed to this experimental neuron. However, even in such special circumstances activity of other neurons must be coordinated, in order not to prevent activity of the key neuron [1270, 1271]. In this way, in our experiments both trained and control neurons participated in acquisition of instrumental behavior, although, at the end of learning, the role of the trained neuron was dominant. A neural system usually generates one dominant action and inhibits non-relevant actions.

It is supposed that the most important function of gap junctions in the brain is to correlate the activity of coupled neurons [701]. A gap junction behaves as a conductance connecting two cells and electrical synapses operate faster than conventional chemical synapses. Certainly, this promotes development of ensembles. When current from a more positive cell depolarizes a more negative cell, the coupling excites one cell while inhibiting the other and one can characterize electrical synapses as synchronizing rather than excitatory or inhibitory [109]. Although each interneuron is coupled to 2040 neighboring interneurons, spike synchrony can be evoked between much larger cohorts of spatially extended interneurons and gap junction channels can contribute to sharpened neuronal activity by synchronizing large neuronal ensembles [1155]. Second and retrograde messengers, which control cell state, also modify cell ensembles.

In such a way also the rhythmic activity of brain can be generated. Really, the role of coupling in synchronization is well displayed, when aware brain generates oscillations[2]. Nevertheless, although gap junctions can act to synchronize network output, they, according to theoretical account, can also give rise to many other dynamic patterns including antiphase and other phase-locked states. The particular network pattern that arises depends on cellular, intrinsic properties that affect firing frequencies as well as the strength and location of the gap junctions [1084]. To a lesser degree, retrograde messengers also give the means for such synchronization, which spread slowly. These means ensure formation of neuronal ensembles and improve synchronization of activities. Strength of a coupling is a modified value and therefore the ensemble of coupled cells is changeable. A coalition of cells can be tuned in correspondence with current circumstances. As a general rule, neuronal ensembles are responsible for goal-directed actions.

Neuronal ensembles are highly changeable. One neuron is able intermittently to participate in different forms of behavior. The neurons capable of eliciting one behavior are often activated during other, sometimes conflicting, behavior [987]. The same neuron is a part of many different ensembles, suggesting that cell ensembles are overlapping sets of neurons which encode different information [1064]. From the other hand, brain networks can achieve the same

[2] Cell coupling can produce functional neuron ensembles characterized by synchronized fluctuations in the cytosolic Ca^{2+} concentration [598]. Cortical and hippocampal astrocytes, interconnecting by gap junctions, produce Ca^{2+} waves, spreading along the cortex and into other neural centers [119]. The network of astrocytes is broadly distributed in brain and astrocytes may afford merging of neurons in a system, all the more that mutual influences between neurons and astrocytes are robust enough. Glutamate and ATP also propagate in astrocytic networks in the form of regenerative waves that accompany the Ca^{2+} waves [119]. Synchronization between neurons may concern the state of channels, input resistance of cells, Na^+, K^+-ATP activity, etc. [699]. GABAergic interneurons, connecting via gap junctions, have an important role in local network oscillations [946]. Nevertheless, gap junctions are not required also for generation of oscillation, but, if present, enhance synchronization [1336]. Pharmacologically uncoupled inferior olive neurons continue to oscillate with a frequency and amplitude that were similar to those recorded in control conditions, but these oscillations were no longer synchronized across a population. Gap junctions are not necessary for generating subthreshold oscillations, rather, they are required for clustering of coherent oscillatory activity [730]. Correspondingly, gap junction blockers depress synchronization without blocking electrical activity [874].

behavioral goals through different patterns [208]. Population activity has been implemented to encode both the sensation and the movement abilities that emerge from concerted activity of individual neurons and many neurons are active during various movements and during various signals [1110]. It has been demonstrated that co-activation of neural cells is not permanent and neurons may rapidly, within a fraction of a second, switch to a new ensemble [1292] and are spontaneously changeable [1391, 1114], changeable during reception of a visual signal [699], during a motor response [893, 1234] and while performing behavioral tasks [1292, 1133, 143, 497].

A rapid dynamics of ensemble reorganizations gives evidence that the reason for the origin of the coordinated behavior of neurons is difficult to explain by the formation of stable synaptic connections between the cells. We have demonstrated in Chapter 1 that the same synapse can participate in different input patterns that exert even an opposite effect on neuron excitability. Therefore it is not surprising that the same neuron participates in different actions and, particularly, the correspondence between current behavior and current activity of neurons in many cases is very tight. However, it is scarcely likely that even high-speed modulation of synaptic efficacy may explain plasticity of neuronal ensembles.

Activation of a specific synaptic pattern may compel neurons to participate in dedicated behavior. And if overall complexity of the brain would multiply its concentration in each neuron, formation of a transient functional system for current behavior [43] could be explained as formation of the ensemble with the strongest, quickest and the most relevant reaction to the key stimulus. At least the "binding problem", decision-making and choice of precise output would be solved. However, even in this ideal, but unreal case, it would be unclear why the relevant neuronal coalition for a particular behavior turned out to be stronger than non-relevant coalitions for all other behaviors.

On the one hand, the rise of cellular ensembles, supposedly, facilitates the gathering of dispersed traits of current signal in the common image. Really, the members of an ensemble represent related cells. They usually have shared receptive fields and are disposed in a close proximity. This ensures continuity of the image. Synchronization of cellular activity allows evaluating of all traits of the image simultaneously. Joint activation of cells by the current image increases conductivity, augments gap junctions and thus creates regenerative process, enhances synchronization and, most importantly, unites participating cells into a whole. Moreover, because coordination of neuronal calcium changes by gap junctions is independent of rapid electrical signals, the function of gap junctions between neurons may be synchronization of biochemical rather than electrical activity [599] and this should further increase synchronization.

The dynamics of neuronal populations usually better determines choice than individual neurons as this was demonstrated in the aforementioned experiment of the leach with the choice between crawling and swimming [166]. Similarly, although the degree of correlation between the firing of single neurons and movement parameters of monkeys was non-stationary, stable predictions of arm movements could be obtained from the activity of neuronal ensembles [208]. These facts are trivial and they do not clarify the mechanism of decision. In particular, this does not confirm averaging, as a possible mechanism of decision-making. Ensemble coding by a population of neurons does not imply that many neurons with equivalent properties construct a population and summate their firings simultaneously just to overcome the unreliability of single neuronal activities [1064]. The activity of neurons sometimes partially and sometimes completely corresponds to behavior. As an example, firing rates of some single cells of somatosensory cortex of monkey fluctuated along with the monkeys' decisions during discrimination of correct and incorrect reactions in trials with identical sensory stimuli [1038]. Studies in vision have also revealed a tight association between sensory and motor activity during decision-making [1038]. Cell coupling in ensembles may support a merging of complementary features of images.

Two possible reasons may explain the good correspondence usually observed between neuronal and behavioral reactions. Different neurons within the same brain have similar experience, produce similar decisions and, further, thanks to a large number of neurons they could overcome their chaotic behavior. And, as an alternative, neurons in an ensemble may merge not only "equivalent properties", but rather "complementary properties" of images. The general image in a neuronal ensemble may emerge as a result of a combination of complementary but scattered features. Integration of cells into ensembles through gap junctions could be better at producing a general evaluation and making decisions and actions, than synaptic connections. Therefore, properties of gap junctions may allow one to come to a conclusion regarding the "binding problem". In addition, gap junction may ensure a choice of the right action.

Nevertheless, there are some doubts concerning the plausibility of some attributes of ensembles to participate in current behavior. It is unknown if calcium waves may be considered as the instrument suitable for physical integration of neurons in vast brain areas, although the temporary merger of neurons in functional units does exist. Coupling between pairs of cells falls off quickly with distance and is generally negligible for separations greater than about 200 μm [257], but Ca^{2+} can propagate through larger distances in astrocytes [27, 526], although rarely expand beyond 300400 μm from their point of origin [885]. In cell cultures, Ca^{2+} waves often travel for millimeters, possibly underlying the spatial transfer of information and coordination of distant circuits of the neuronglia network [119]. The velocity of Ca^{2+} signal propagation is in the order of tens of micrometers per second, i.e. two to three orders of magnitude slower than AP propagation [160] and therefore calcium waves cannot serve for current control of aware behavior. At the same time, brain electrical activity during seizures seems to be effectively propagated via gap junction channels [1066].

When one describes neuronal ensembles, some different phenomena are implied[3]. However, for all particular types of ensembles, the probability of participation of neurons in the ensemble depends on the distance between them and this distance does not exceed 1 mm. So, correlation of membrane potentials in striatal spiny neurons in anaesthetized rats [1180] and in the cat visual cortex [699], is displayed better in the short distance between neurons. During the active waiting stage, interneuronal functional connections are observed between the adjacent cortical neurons and variable connections between tremote neurons [452, 143]. At the local level, within neurons, proximity of synaptic contacts also determines synchronization. Such synaptic integration in cortical neurons depends on the distance between the inputs and this interplay of space and time is modulated by voltage-gated and leak conductances [70]. Correlation between activities of two cells is related to rather slow processes, while APs do not correlate. Synchronization for γ rhythms, which is the same rapid rhythm in the electroencephalogram, is too slow compared with an AP duration. However, for instance, nearby retinal ganglion cells often fire action potentials in near synchrony and this depended on electrical coupling [167].

Neurons servicing communal function are usually coherent. In the cat visual cortex, fluctuations in membrane potentia of neurons were asynchronous, but they revealed significant correlations within the pair of cells with overlapping receptive fields, while the highest correlation occurred among cells with similar receptive fields and correlations increased when both cells responded to the stimulus and decreased when neither cell responded [699]. At the moment of visual stimulation, synchronization was displayed even in unrelated neurons. The neurons involved in the detection of odors show temporally correlated activity and pairs of principal cells in the olfactory bulb that project to the same glomerulus display highly synchronized spike activity, whereas pairs projecting to different glomeruli do not [1443]. This synchronous firing depended on gap junctions.

Dynamic reorganization of functional groups with similar response patterns takes place also during acquisition of a conditioned reflex [950]. According to our data [1257], during pairing, the processes in the two recorded neurons occurred concurrently. In spite of instability, the responses of two recorded neurons during classical conditioning change synchronously, whereas

[3] One such phenomenon is synchronous oscillation in the γ diapason (20-70 Hz) [725, 699, 167, 1180, 964, 380, 1019, 946]. On the other hand, neuronal members of an ensemble may display similar responses to the signal [1110, 699, 208, 950, 1443]. Such phenomena are frequently related to each other, but correlations in membrane potential between the cells may be observed without cell coupling [699] and synchronization within the ensemble with similar response patterns may be absent [950]. An ensemble may also be determined as the presence of functional connections between neurons [452, 143, 3] and this implies participation of synaptic switches. Functional connections are displayed as cause-and-effect relations between activities of neurons in the ensemble.

during instrumental learning, synchronization between responses of trained and control neurons was observed during some periods of training, as if the neurons were members of the same neuronal ensemble. Although the dynamics of the average change in the neuronal responses for two recorded neurons during instrumental conditioning were different in respect to the conditioned stimulus and were similar to the discriminated stimulus, their activities displayed a spatial synchronization and were correlated for both the conditioned and discriminated stimulus. Therefore, the origin of the correlation between the neuronal responses is impossible to explain by a similar change in activity of two neurons during training. Periods of concerted actions of the neurons were replaced by periods of asynchronous changes. Participation of the trained and the control neuron in the same neuronal ensemble is usually kept during a few minutes, but may depend on their activity in the preceding trial.

Merging of neurons in ensembles proceeds during tension in the neural system, connected with an active behavior (with the exception of pathological synchronization during epileptic attacks). As a rule, neuronal ensembles support the activity of specific neurons and such ensembles are responsible for goal-directed actions. When a neural system accepts significant information, synchronization between effective neurons increases. Thus, synchronous fluctuations within whisker and auditory cortical columns exceed fluctuation between different, even neighbor, cortical columns [1019]. The number of neuronal and glial gap junctions during motivation increases and metabolic coupling between the glia and neurons is also enhanced. For instance, functional connections between foreign neurons, too, increase during feeding motivation [143], whereas unconsciousness may depend on inhibition of the formation of ensembles [393]. Oscillation within a neuronal group is coherent, but may be non-coherent between the groups. Neuronal communication between two neuronal groups depends on coherence between them and the absence of neuronal coherence prevents communication. Both communication through coherence and non-communication through non-coherence are essential [428], since this will augment the dominant behavior and inhibit non-relevant reactions. We may conclude that formation of the ensembles of various natures is connected with intensification of brain function. Ensembles, probably, unite efforts and informational resources.

4.7 Physiology of free will

4.7.1 Instability of neuronal reactions

An animal acts if it knows how to obtain a useful result. If it does not, it uses the trial-and-error method [582]. Both behavioral and neuronal reactions alter from trial to trial. Variability of neuronal and macroscopic reactions is similar [1261, 1361] and this means that neuronal activities are timed in coordination. Particularly, dispersion of neuronal responses almost does not exceed dispersion of evoked potentials and motor reactions. This important observation illustrates that averaging of neuronal activities is not the principle of brain function, similar to the absence of averaging of synaptic reactions within a neuron. Averaging ought to result in a decrease of instability of evoked potentials and, moreover, to disappearance of behavioral instability, but this is not the case. We observe in the brain the well-controlled chaos. Instability is an inherent property of the active brain [1086]. For example, synaptic transmissions between identified neurons in the central nervous system of the freshwater mollusk, Lymnaea stagnalis were variable in sign, when compared from animal to animal, when environmental conditions changed and spontaneously within individual preparations over a matter of minutes [779]. In higher animals and in higher neural centers, instability is also higher. Cortical neurons in brain exhibit especially highly variable responses during repetition of the same stimulus, but the latency of unconditioned movements is much more stable than the latency of conditioned neuronal reaction of cerebral cortex neurons [188]. The very different neural processes of perception and action are supposedly limited by the same sources of noise [920].

Healthy systems are self-regulated to reduce an accidental change of the state and maintain physiological constancy. Contrary to the predictions of homeostasis, however, the output of various systems fluctuates in a complex manner, even under resting conditions, although under harassed conditions fluctuations are larger [482]. Variability is higher for aware brain, comparing with unconscious brain and in accidental reactions compared with beforehand intentional reaction [349]. In aware brain, instability is rather connected with processing of proper information, than with inexact calculations in the brain. Training is not only a gradual memorizing of the correct instrumental reaction, but is the series of consistent attempts to achieve a healthy result.

Instability is observed at the same primitive levels of life and it has beneficial meaning. There are specific molecular events that cause temporal behavioral variability in an individual bacterium of Escherichia coli. The temporal variability depended on the relative concentration of a key chemotaxis protein and when the concentration slightly increased, the cell exhibited a steadier behavior [658]. This might promote chemotaxis, since instability is decreased when the subject approaches the goal and this prevents loss of the goal.

The dynamics of the formation of conditioned reactions are extremely unstable and protracted and the neuronal firing pattern is especially irregular during instrumental learning [1260]. Nevertheless, there are only sporadic data on the instability of neuronal reactions during motivational excitation. It is known that synaptic noise increases during activation of Na^+ sensitive neurons in the thirst center, which is responsible for primary reception of thirst [1307]. Response variability does not disturb synchronization in neuronal ensembles and high temporal precision of neuronal responses [71]. Another example is the activity of microglia. Variability of microglial activity is not a reflection of stimulus strength or persistence; rather, it is determined largely by the nature and context of the stimuli and by the intracellular signal transduction pathways that they activate [1109].

Therefore, a noise is not always and without fail a negative phenomenon. Instability is a characteristic of the aware brain. Neural activity in lightly or moderately anesthetized animals is relatively quiet, regular, and even stately. In the waking animal the activity becomes livelier, especially in the presence of sensory stimuli and motivating antecedents such as food deprivation [412]. Correlation of noise between neuron activities may have potentially beneficial effect.

For understanding how unified neuronal reaction arises, it is valuable to consider the nature of instability. Really, neuronal reactions are variable, but in the normal state they alter synchronously in many, especially neighbor, neurons. Therefore instability for some reason or other does not prevent the generation of unified reactions. Visually-evoked neuronal firing rates and spike latencies in cortical neurons are only weakly correlated with spontaneous or visually-evoked fluctuations in the mean membrane potential, but were found to be linearly correlated with the magnitude of membrane potential fluctuations in the γ (20-70 Hz) frequency band, which generated with active participation of GABAergic inhibitory interneurons, tightly connected via gap junctions [71]. Response variability is attributable largely to coherent fluctuations in cortical activity preceding the onset of a stimulus and to variations in AP threshold. Hyperpolarization resulted in a general enhancement of membrane potential correlation immediately before and after the onset of evoked response [71], and, hence, according to our preceding discussions of the protective role of inhibition, synchronous variability is an attribute of working brain.

The degree of variability is, probably, connected with alteration in the state of damage-protection of cells. Correspondingly, when in multicellular animals a goal is approached and the state of the organism becomes more positive, the GABA-inhibitory system is augmented and instability decreases. So, the response evoked at regenerating nerve sprouts usually shows erratic behavior, i.e. the discharges are irregular and not well reproducible with repeated stimulation [1439]. Correspondingly, GABA synapses formed *in vitro* by local axon collateral of nucleus accumbent neurons decreases noise activity of these neurons [1129]. At the same time, low pH and agonists to metabotropic glutamate receptors increase noise [1028]. Instability also consistently increases during activation of AMPA/ kainate receptors, which increases damage [686]. In the limits of low or middle injury, which is easily recovered by passive inhibitory protection, factors of damage, perhaps, increase, while direct protection decreases noisy instability. At the same time, dopamine, which usually affect cell metabolism and protects neurons, is a "disperser" in the network of cortical cells. Dopamine increases the instability of neuronal activity and both $dopamine_1$ and $dopamine_2$ receptors are involved in the dispersing effect of dopamine [372]. This is, possibly, connected with the antagonistic correspondence between GABA and dopamine systems in the nucleus accumbent. Likewise, in olfactory receptor cells of rat in the absence of smells, large random fluctuations in membrane potential were induced by elevating the intracellular cyclic AMP concentration, which usually promotes stressed protection [764]. Noisy variability is, probably, an attribute of damage, whereas synchronous variability is connected with the process of tense protection through reorganization of optimal values of variables.

4.7.2 Instability and trial-and-error

Correspondence between damage-protection and cell instability may affect the origin of trial-and-error during motivational behavior, which, as we have shown in Chapter 3, arises as the result of homeostatic disbalance in a brain. Let us evaluate how instability of neuronal reactions and instability of membrane potential during instrumental conditioning depends on presentation of a US in the previous trial.

The general outlines of the neuronal analogs of instrumental and classical conditioning, which are demonstrated in Figs.4.1 and 4.2, do not display the fine dynamics of elaboration, in particular the high level of trial-to-trial instability of responses. In order to evaluate the trial-to-trial variability of response, we calculated the differencing between two neighboring responses from the time series data during instrumental conditioning (Fig. 4.3). Although the mean differencing during training did not exceed significantly the zero level, the mean absolute differencing of the responses, which manifests instability of the responses, changed significantly during training. This value represents attempts to obtain a correct result. Therefore, Fig. 4.3 shows the current trial-to-trial variability of the neuronal activity. Uncertainty of choice correlates with an intensification of the instability of neuronal reactions, which decreases only at the end of training and after avoidance of danger.

The initial variability of the trained neuron responses slightly exceeded the variability of the control neuron responses, since initial excitability of the trained neuron was chosen so as to be lower. At the beginning of training session, in naive animals, both values of responses and their instability did not depend upon the appearance of a painful US. Soon after the beginning of training, presentation of a US did not affect the following values of neuronal reactions to the CS^+ (Figs. 4.1, 4.2), but affected their instability (Figs. 4.4, 4.5). Instability of larger and stable reactions of the control neuron increased, while instability of weak and more variable reactions of the trained neuron decreased.

Presentation of a US when a trained neuron generated an incorrect reaction affects instability of both the trained and control neurons but in a non-uniform manner. Fig. 4.3 demonstrates values of instability separately for the responses following correct and incorrect reactions. Instability of the trained neuron responses was observed when it generated a probing reaction, before the decrease in the responses in the middle of training (trials 5-15) and during recovering of its responses (trials 25-30) (Fig. 4.4). During trials 15-25, instability of control neuron responses reached a maximum, whereas instability of trained neuron responses failed to reach its minimum (Fig. 4.5).

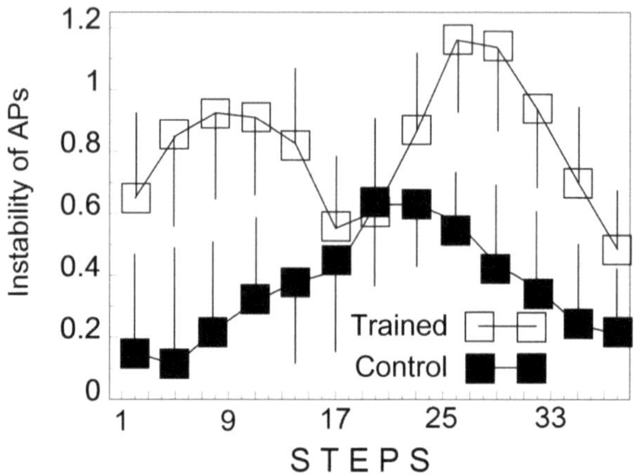

Fig. 4.3. Mean absolute value of differencing (further - differencing) between two successive responses to a CS^+ (a measure of variability) in the trained neuron and control neurons. Designations and confidence intervals ($p < 0.05$) are shown in the figure.

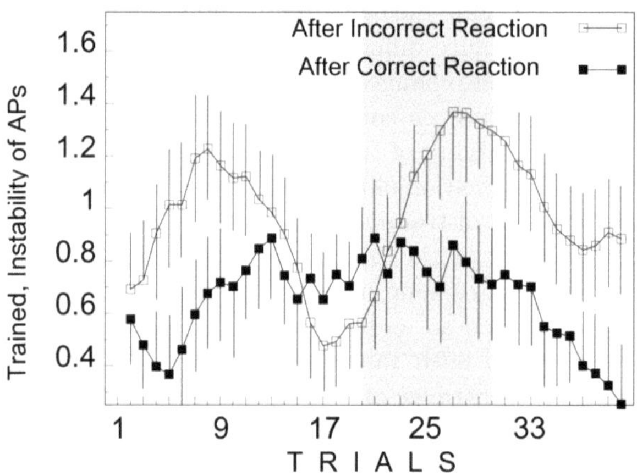

Fig. 4.4. Differencing between two successive responses in the trained neuron, dependently on its previous reaction. The data received after correct and incorrect reactions are shown separately (designations in the figure). Abscissa - trial number. Confidence intervals are indicated.

Instability of the control neuron responses appeared when it began to generate an erroneous instrumental reaction, just, when instability of the trained neuron responses temporarily decreased. This happened approximately in the same periods of training, when we observed an inverse dependence of neuronal reactions from the membrane potential. Non-stability of the trained neuron behavior seemed as if it turned into non-stability of the control neuron behavior. Evidently, a non-stability of the neuronal reaction increases when the neural system tries to use this reaction as an instrumental reaction, since both correct and erroneous instrumental reactions arose during augmented instability of the responses.

Presentation of a US increased instability of following responses of the trained neuron to the CS^+ everywhere, besides a short period of origination of substitution of correct instrumental reaction by incorrect instrumental reaction of the control neuron (Fig. 4.1). In this period of substitution, the US inhibited the instability of responses, but this effect was much more significant only for the control neuron, which play the main role during this short period of training (Fig. 4.5).

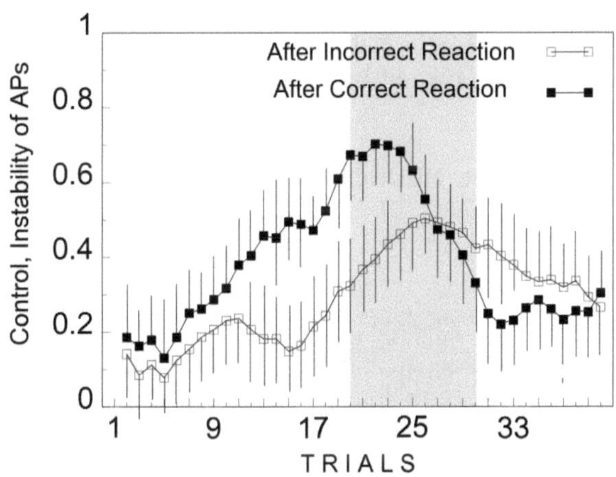

Fig. 4.5. Instability of responses in the control neuron. Designations as at the Fig. 4.4

Thus, dispersive influence of the US to the following response concerned specifically only the trained neuron, when it reorganized the instrumental response. Instability of an incorrect instrumental reaction of the control neuron differently depended on the US presentation.

It is interesting to compare the influence of a US on the generation of instrumental reactions (Figs. 4.1, 4.2) and its influence on instability of these reactions (Figs. 4.4, 4.5). Both an incorrect instrumental reaction of the control neuron and a correct instrumental reaction of the trained neuron increased after presentation of the US and this looked like a correction of error undertaken by the neural system after punishment. When the control neuron began to generate an erroneous reaction in the next trial after a US presentation, instability decreased and augmented reaction generated in 2-3 trials in series. Thus, the aftereffect of a US (2-3 trials) takes 5-10 minutes and this is a characteristic time, attributable for development of simple protection and motivation. A trained neuron, in contrast, during the same period of substitution of the instrumental reaction decreased its reaction after a US presentation and also during 2-3 trials in series.

In contrast, when a trained neuron began to generate an instrumental reaction, though this happened in the next trial after a US appearance (as was observed for the control neuron, when it generated an erroneous probing reaction), instability of the trained neuron responses increased. Learning was almost completed and the trained neuron quickly generated the correct reaction after punishment. At the end of the training session, sporadic appearance of a US did not influence the value of the following responses in the control neurons, but continued to increase instability of trained neuron responses and thus its response recovered.

Instability of membrane potential decreased after presentation of the US and, but this was observed only at the time of reorganization of response in the corresponding neuron (Fig. 4.6).

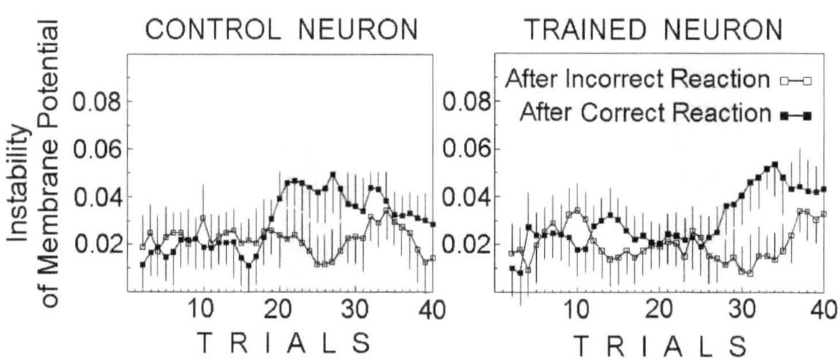

Fig. 4.6. Instability of membrane potentials in the control (left) and trained (right) neurons during instrumental conditioning, dependent on the presence of the unconditioned stimulus in a preceding trial. Confidence intervals ($p < 0.05$) are shown.

In the control neuron, instability of the membrane potential decreased during generation of an erroneous instrumental reaction (after appearance of a US). In the trained neuron, similar regularities were observed when the neuron began to generate an instrumental reaction (also after appearance of a US). However, stabilization of membrane potential in the control neuron was observed when it was depolarized, whereas stabilization of membrane potential in the trained neuron was observed when it was hyperpolarized in the next trial after US presentation.

In view of data concerning instrumental conditioning from Chapter 3, many pieces of the overall puzzle are becoming increasingly better understood. During generation of correct and incorrect instrumental reactions, correspondingly, by the trained and control neurons, responses rise following presentation of the painful unconditioned stimulus and instability of responses in both neurons increases. However, generation of a correct instrumental reaction in the trained neuron rose when this neuron was protractedly depolarized and this stagnated depolarization was weakened after US presentation. In contrast, the control neuron generated an erroneous instrumental reaction in the period of time when it was hyperpolarized, but a US decreased its membrane potential. Following the logic of our discussion, we may conclude that the origin of instrumental reaction generation is connected with disturbance of excitable membrane properties and was actuated by a protective increase in membrane potential of a trained neuron. The control neuron generated an erroneous instrumental reaction when its overprotection was reduced by the excitatory influence of a US. At the same time, although in both cases instability of responses increased during generation of probing reactions, this unsteadiness was not determined by the variations of membrane potential level and, evidently, was related to modulation of excitability. The level of membrane potential cannot control specific processes, while a change in excitability may be selective in respect to a CS^+ and a CS^-.

Such simplifying considerations may explain some outline features of instrumental behavior, although on the basis of plasticity of electrical activity of neurons lies regulation of many chemical processes within cells. Specifically, in our description, we silently implied that a painful US promotes generation of those reactions in the next trial that may prevent the next appearance of a US. When a control neuron was inhibited, a US decreased membrane potential and promoted generation of probing reaction (although it turned out to be erroneous reaction) by the control neuron. However, when the trained neuron was hyper-depolarized, the US increased membrane potential, recovered excitability and thus also promoted generation of a true probing reaction by the trained neuron. Certainly these are experimental facts. We may consider that neurons counteracted to detrimental effects by means of homeostatic protection. But we here run into the same problem encountered previously: a cell somehow works against the harmful quality of a factor instead of counteraction with particular characteristics.

Instabilities of AP generation and, especially, of membrane potential are slow processes with characteristic times in the range of minutes. Besides, there is a rapid change of neuronal state in the range of milliseconds. Short-term instability arises on the short-term scale during spike generation. Fig. 4.7 demonstrates extracellular spikes have been evoked by the whisker stimulations in the barrel cortex of a rabbit [445].

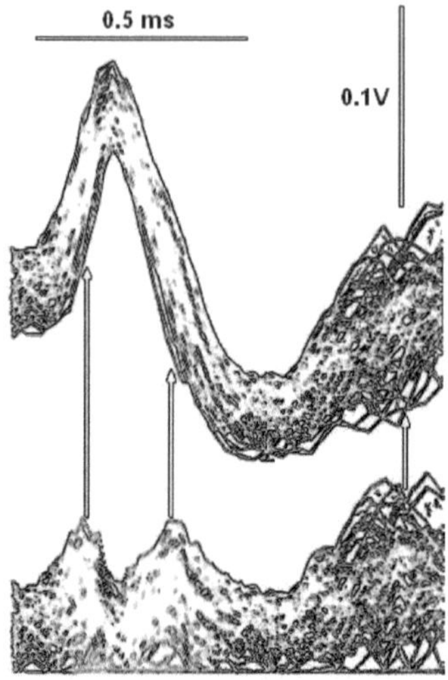

Fig. 4.7. Instability of the local extracellular potential, connected with single spike generation. At the top - extracellular somatosensory cortical single unit activity in response to whisker stimulation. Waveforms of around 100 extracellular APs, which were generated under identical conditions, are shown. Spikes were matched according to waveforms, since their latency varied. Calibrations at the Figure. The Fig. 4.7 (top) was redrawn in accordance with the data [445]. At the bottom variability of the potential during spike generation. Distance across combination of the beams reveals three maxima of instability (indicated by the arrows).

Waveforms of APs altered spontaneously, and maximum instability was observed in the critical times of AP generation, indicated by the arrows in Fig. 4.7. The extracellular AP roughly matches the first derivative of an intracellular AP. Therefore, the first maximum corresponds to the point of maximum curvature at the front of an intracellular spike, or to the AP threshold, second maximum beginning of intracellular spike decrease (zero of first derivative corresponds to maximum of intracellular AP) and third maximum coincides with the period of absolute refractoriness. The distance between these critical points is only a fraction of a millisecond. This means that intraneuronal processes may affect instability without the participation of the rest of the network.

We may conclude that instability of neuronal reactions is tightly connected with the working brain. Instability increases through instrumental behavior, during augmentation of motivational states, in the aware brain, after harmful influences and other hostile circumstances.

4.7.3 Organization of choice

Living systems initiate behavior spontaneously, or at least it looks spontaneous to the outside observer and for the person himself. The enigma of free will consists of the necessity to combine the single initial condition with an existence of different decisions. Relevant research in physiology of the free will does not yet exist. However, free will is regarded to be compatible with causal determinism [875]. If we do not take into consideration quantal effects (which are absent or, at least are not analyzable for macrosystems), only mathematical randomness possesses such properties. Random events are non-predetermined. Moreover, chance may result from identical initial conditions to different outcomes, with the matter of a goal being absent. In the physical world, occasional events are observed in non-linear chaotic systems. Chaotic behavior, strictly speaking, is predetermined, but this predetermination so abruptly depends on initial conditions that any small fluctuation in these conditions may rearrange the result to be unrecognizable. Therefore, the practical result of chaotic behavior is similar to a random event. Free will cannot be reduced to chaos, since animal actions are goal-directed. Neither instability nor chaos can ensure free will. This is a freedom of the chip in a rocky stream, although by the outward impression, advance of the chip looks worth while. We will show how neuronal activity diminishes a motivation by affecting the process of goal attainment. Activity need not be coordinated or directed toward the satisfaction of a motivation by some external control system. Let the goal be an optimal value of a given metabolic factor and let neurons at the center of the motivation detect a deviation from this optimal value as an error signal. Neuronal reaction to this deviation is a corrective action, so let the sign of a given output response be determined by synaptic noise, while the amplitude of a response is determined by neuronal excitability and depends on the deviation of the metabolic quantity from its optimal value. In accordance with properties and the order of the origin of instability in a neural system, we may assume that at each step, the value of the output signal is proportional to the error signal and has a random direction [1254, 1256, 1257]. This algorithm was proposed as the basis of explanation of a neural network tuning during learning [74] and allows one to solve the system of algebraic equation [1256]. This method of search demonstrates the plausibility of free will existence. Two simple examples are shown in Fig. 4.8. When a primitive animal - woodlouse - searches for a moist place, it continues to move in a random direction until it finds the moist place under a cobble-stone (Fig. 4.8A). Here the woodlouse calms down. Another example is the condensation of dissolved substances in the cold region of a solution (Fig. 4.8B). The last example brightly demonstrates successfulness of the search even when memory is completely absent, such as in molecular particles. Fig. 4.9 explains the mechanism of search of the resolute value of a function. This firm magnitude is characterized by the error signal between this value of function and its current level. Such a system acts by chance at each instant but finally chooses the reaction that satisfies its need. We shall

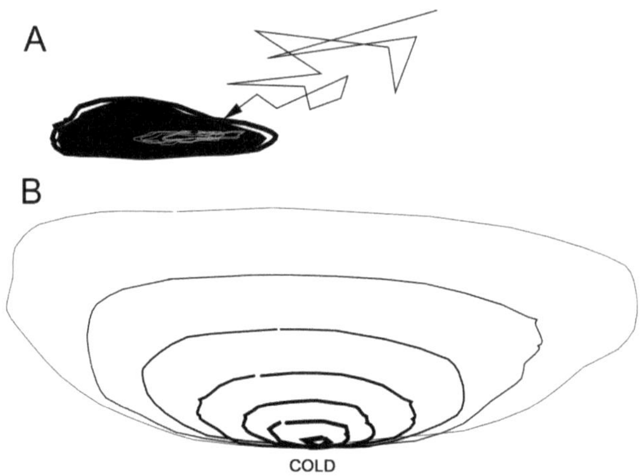

Fig. 4.8. Examples of the simplest goal-directed search. A - woodlouse is finding a moist place. The woodlouse moves in on accidental direction, but intensity of its movement decreases when humidity increases. B - diffusion is directed in the pole of cold.

now show that goal-directed behavior can arise from non-directed actions of individual neurons. Indeed, let there be an efficiency function $Q(x)$, which describes the dependence of an inner metabolic factor of the system Q on the external variable x (Fig. 4.9).

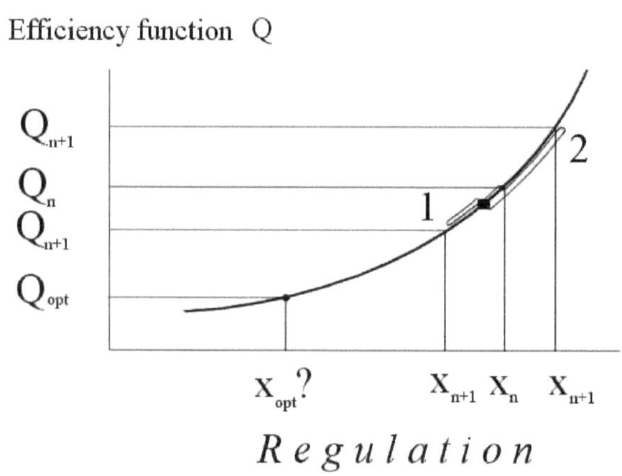

Fig. 4.9. The mechanism of choice. Two trajectories are shown (1 and 2) after two steps of the search for the optimal value of the efficacy function. Moving along the axis changes the value. See text for explanations. Redrawn from [1256].

This firm value is characterized by a value of error signal between this value of function and its current value. Although a physiological response is usually proportional to an error signal, dependence of the error signal on external variables may be very complex. We suggest that the outer factor x immediately affects an inner metabolic factor Q as, for instance, a change in glucose concentration influences damage to neurons. This is clearly a simplification, because duration of the influence also affects the inner variable $Q(x)$. But this suggestion is acceptable, if the function $Q(x)$ is continuous and differentiable.

Two versions of behavior are thus shown. Let a random signal at the n-th step (trajectory 2) result in a further deviation from the metabolic optimum Q_{opt}. For example, one shift from x_n to x'_n intensifies hunger (increases the efficiency function Q_n to the magnitude of Q'_n) and enhances the search. At the next step, the magnitude of the signal will increase and its direction will be random. If this signal is in the opposite direction, the resulting value will approach the optimum, i.e. will draw near the goal. The same result is obtained after version 1 of the behavior. By these actions, the system only uses information about its actual state and uses neither short- or long-term memory.

The numerical experiments for one-dimensional non-linear cases demonstrated that the proposed mechanism ensures overcoming the obstacle [1256]. The search continues in the vicinity of the local minimum and there is some probability, higher than zero, for overcoming the local maximum and for attainment of the optimal value. The mechanisms postulated here enable the search for a required value Q_{opt} without exploitation of any memory, short- or long-term. Therefore, such a system is very simple. The system searches optimum and not local minimum, in correspondence with the peculiarity of setpoint search by homeostasis.

The numerical experiments for one-dimensional nonlinear cases demonstrated that the proposed mechanism ensures overcoming the obstacle. The search continues in the vicinity of the local minimum, and there is some probability for overcoming the local maximum and for attainment of the optimum. Increasing the error evokes enhancement of the correction, whereas in the vicinity of the optimum the correction decreases. During a search, the current value of the function at the beginning sinks to the local minimum, raises to the right branch of the functions and here exceeded the level of local maximum, after that sinks again at the local minimum, overcomes the local maximum and after that quickly finds the optimum. Certainly the efficacy of a search is sharply increased if the system possesses memory with a nesting level of only one step.

4.7.4 Free will without mysticism

The sense of the random search, which becomes goal-directed, may by delineated in the simple rule: if you feel unwell, do it, but if you feel tolerably well, calm down. Our data show that by such method it is possible to tune large systems, such as neural networks. Our model furnishes an explanation for the generation of goal-directed behavior without its reduction to the stimulus-reaction method. Absence of a steady cause-and-effect connection between an input signal and an output reaction favors such reactions to be similarly emitted rather than elicited. Creation of a model of motivation will in the future make it possible to design completely autonomous machines that don't require programming and are controlled by the user through establishing connections between the user's aim and the machine's aim. Hence, the existence of non-predetermined actions does not contradict logic and we have proved this assumption rigorously for linear one-dimensional cases. Certainly, we do not know exactly whether a neural system uses this method to search for optimal states. We have only demonstrated that the existence of free will does not contradict rigorous logic, since the search may be simultaneously both non-predetermined and goal-directed. Some indirect observations make this mechanism more or less plausible. Really, erratic behavior rises during various forms of injury, and within separate cells it may have a molecular cause. Motivation must somehow overcome mismatch of an inner variable. During such a tense state, directed to overcoming a transient injury, this instability becomes coordinated in neighboring neurons or in neurons with related function. This synchronous instability especially increases when the brain controls tense behavior and further increases, when a neural system generates probing reactions and after punishment, that is after an increase in motivation. At the same time, when motivation decreases or after avoidance of punishment, instability fails. Therefore, a simple homeostatic search for the optimal state and goal directed behavior during satisfaction of elemental motivation has a basic feature that we have postulated in our mechanism of search: instability increases during augmentation of motivation and this simple rule is sufficient for pursuit a goal. The numerical experiments for one-dimensional non-linear cases demonstrated that the proposed mechanism ensures overcoming the obstacle to an optimal state. Augmentation of the disturbance evokes enhancement of corrective actions, resembling the phobic reaction to missing the goal of a single cell during taxis [782]. These choices will not be predetermined and each one will produce its own goal-directed action. However, such a quest is too slow for tuning of multi-variable functions and may be used by living creatures only in exceptional cases, when memory is not developed. Even so, in principal, goal-directed behavior can arise from non-directed activity of individual neurons. Nevertheless, the algorithm of search that we have considered may be, without modification, implemented in a neural system, but only for a one- or two-dimensional search, since in these cases the rapidity of search may be increased. To additionally increase the dimension sharply slows down the procedure. Homeostasis during recovery of mother values exerts regulation by sensor proteins through a two-dimensional search (Chapter 2). An organism and even separate cells simultaneously regulate a variety of variable. There are a few ways to make such a regulation more effective. Adjustment of complex function may proceed via several steps and the control would be satisfactory using a dichotomy, since a two-dimensional search is within reach. Besides, random movement generation takes more time than choice of one action out of several possibilities [707]. Gap junctions in the brain synchronize the activity of neurons and ameliorate the formation of ensembles. This, inevitably, decreases a dimension of the system, permitting it to increase the value of step and accelerate the search. Besides, fundamental perfection of a search may allow the participation of memory. We have considered an extremely simple system, which does not employ past information, neither long- nor short-term memory. Our search was slow, but the only knowledge that the system uses is the value of deviation from the optimum in the time of evaluation. Availability of memory will harshly complicate the system, but robustly accelerate a search. Certainly a neural system generally uses this complication. There are many examples of how exploitation, even of the most primitive memory (when the system remembers only its experience at the last step), vigorously accelerates the search. The most basic improvement of regulation is the possibility to regulate only one general variable instead of a lot of particular variables. Cells do have the ability to control one very general value: the level of damage-protection. However, we do not know how a cell performs this. We have assumed that the emergence of primitive cellular feeling "worse-better" as the basis of evaluation of "damage-protection" was dictated by the evolutional necessity for homeostatic regulation of one general value instead of a variety of variables. Correspondingly, a brain has the ability to control conclusions of all cells in the ensemble that are related to dominant behavior. Maybe the main cause for the origin of consciousness is the necessity to decrease the dimension for decision-making. We have only preliminary notions about how brain achieves this. Knowledge of this problem will allow us to understand how we feel, will, decide and embark on efforts. Only after that may one try to install an artificial mind. Although decision-making is a local process, this is not a sporadic occurrence. Decision-making takes time and is a dynamic process: the leech nervous system can start to make a decision and subsequently change to an alternative choice. The dynamics of neuronal populations can determine choices, but individual neurons in these populations can profoundly influence decision-making [166]. A similar phenomenon was observed also in the monkey [872]. In the mammalian, the presupplementary motor area was found to be tightly associated with the free selection of responses [707]. Nevertheless, rapidity of decision-making and the large number of steps needed for a search is inconsistent with the assumption of random exhaustion in a neural network even for a one-dimensional case. However, decision-making may proceed through molecular exhaustion, as was supposed by E. Liberman [735]. Awareness converts the result of behavior to be unpredicrable and exerts an influence on motor memory inconsistencies [292], but our considerations have demonstrated that even a relatively simple molecular system generating an error signal in a random direction will display unpredictable behavior while it searches those conditions that remove mismatch.

4.8 The emergence of higher-level organizations from the interactions of lower-level units

A neuron, like as any other cell, for instance glia, is, for sure, a self-contained system. However, brain behavior is immeasurably more plentiful than neuronal behavior. Mutual activities of many neurons and glial cells give rise to new qualities that are not habitual for a single cell. These new qualities are originated by means of the raise of collective behavior of cells. Nevertheless, neurons are not fully free in the choice of partners for interactions since this is restricted by the construction of the brain. This restriction on the one hand imposes difficulties on the rules of interactions between cells, but on the other hand, collective behavior does not exclude participation of neurons in the behavior in the capacity of a constructive element of the brain.

Each neuron is both a self-contained system and an element for the formation of the steady brain construction. Certainly, a neuron can not be considered as a conductor, summator, coding apparatus or any other kind of brain piece [1141]. Nevertheless, different regions of a brain are non-equivalent in respect to the functions (sensory, associative, motor, motivations, etc.). Besides, different neurons, which support the same behavior, care for specific sides of this behavior. The meaning of our statement that elements of collective neuronal behavior play a decisive role, compared with spatial organization of brain consists in the capability of brain to organize new effective construction when previous construction is injured. In addition, during such reorganization, consciousness is not disturbed or disturbed only for a short period. Neuronal homeostasis can properly function in distorted conditions. At the same time, during disturbance of collective behavior of cells, the brain stops to function as the center of awareness, such as during a blockage of cell coupling during narcosis or slow-wave sleep. Awareness is blocked also during a pathological change in the collective behavior of cells, for instance during the period of an epileptic attack. Learning, memory, motivation and intentional action generation are not, probably, an emergent property arising from a whole brain, but are the attribute of a neural cell, as a unit. We may ascribe to neurons many amazing features, but brain is not the sum of its neurons.

Brain possesses qualities that are absent in single cells; the world of a is multi-dimensional. Brain monitors outer variables and inner constants, and it can distinguish the first from the second. Besides spatial and time variables, brain evaluates conditions in many other scales: degree of danger, hunger, thirst, tiredness and sometimes it even uses a moral scale. As well, in each given moment, brain for any of its scales evaluates a whole, although limited, interval, instead of one point on the scale. For example, when one observes traffic at the highway, he or she may follow several cars simultaneously. This capability is impossible to explain by a rapid switch of attention from one car to others, since neurons cannot function so swiftly. Similarly, when brain evaluates a time interval at the current moment or in the past, it considers a whole restricted period. Just this property allows it to make a prediction.

Brain also can execute several (usually unconscious) actions simultaneously, chooses the best variant of behavior and exerts multi-dimensional influences on the environment. Brain can bring together current information dispersed in its neurons and moreover can extract an essential ingredient of past information.

All these peculiarities of brain arise from the possibility to unite the resources of many neurons. Neurons altogether accept more information than any one neuron, remember more previous information, have a variety of micro goals and control any output of the brain. When brain combines neurons, it, as a minimum, keeps all partial complexities of its neurons, as if each neuron stores the possibility to become dominant in order to fulfill its partial function. However, possibilities of the brain may considerably exceed the common capability of its neurons, since a collective decision may create new information, tasks and options unknown to the neurons. The problem for us is to consider and understand is how neurons unite their information and produce a collective decision.

When neurons are organized into the neuronal ensemble in order to carry out a definite task, they began communications and their interaction somehow lead to collective decision-making. There is a famous example of collective decision: the "invisible hand" of Adam Smith, which maintains spontaneous social order without a central controlling station. The models of collective behavior of aware neurons would, probably, be in the future, an important direction of investigations. The needs for such an approach are perceptible.

It has been thought that a neuron ensures its demands at the expense of substances arriving from other neurons, particularly from neurotransmitters [21]. This rather reminds one of the Adam Smith ideas. Nevertheless, a neuron cannot make several different decisions and a collective of neurons cannot function, according to free market laws. Also, a presynaptic neuron must know the demands of postsynaptic neurons and whether or not these occurs is debatable. Besides, it is not clear, how a collective of neurons deals with the needs of individual neurons for, say GABA, opioids, or substance P. In addition, a majority of neurons "need" glutamate that injures them. In passing, the unconscious, unplanned forms of cooperation and intelligence among intentional agents are not necessarily good and beneficial, although they can be self-organizing and stable [216].

There is an extreme position suggesting that our self each time is concentrated in some single neuron [353]. In this case, the collective behavior of neurons is reduced to a confrontation between them and in spite of the fact that similar phenomena sometimes may be observed in exceptional conditions (such as a local instrumental reflex), the neurons that control a whole behavior in normal conditions would be too overcomplicated. It has been supposed, also that neurons of the brain may principally produce a coordinated decision without confrontation, since within the same brain they experience the same impressions [1255].

The next example illustrates this principle. At the daybreak of *perestroika*, in 1986, two famous TV journalists Vladimir Pozner from the USSR and Phil

Donahue from the USA organized a TV show, and they co-hosted 'A Citizen's Summit', a bilateral, televised discussion (or "*spacebridge*") between audiences in the Soviet Union and the US, carried via satellite [264]. This was the first example of free and public personal contact between citizens of two countries. The Soviet audience affirmed that they completely agree with the politics of the Communist Party, they voluntarily support their government, have all their rights to object against government but everyone has the same opinion as the communists have. They declared that nobody intimidated them and nobody coerced them to lie. The American audience could not believe them, since they heard about restriction of liberty in the Soviet Union. Phil Donahue was also struck. However, he inclined to believe. Phil explained to American citizens, that the Soviet audience was chosen by chance. He himself cached people in the street and invited them to participate in the *spacebridge*. The *KGB* did not participate in the selection and therefore it was impossible to assume that these people were slipped under his attention by the officials. It was necessary to accept that the Soviet people intentionally supported their government. Nevertheless, following the developing of eminent events has proved that agreement in Soviet society was not so firm. But why would randomly-picked audience demonstrate so strong a consolidation? The reason was completely clear for the local on-looker. It was not necessary to choose a special audience and it was not required to intimidate the participants. In the USSR, everybody had the same information about existing threats and each one knew what is permitted and what is not. Hence, everybody (or almost everybody) behaved similarly in public conditions.

Within a brain the information is spread among many neurons. Various neurons in the same brain during a lifetime receive similar information concerning the biological significance of the received signals. The more essential the event, the wider the information is spread among neurons. They might make a similar decision and thus ensure one general reaction of the brain. This idea reduces complexity of the brain to complexity of the sum of neurons, whereas collective behavior leads to creation of a new quality. The most known example of collective behavior of neurons is their transient participation in ensembles under tense conditions. The immune system also operates in a distributed manner and displays the ability to learn, the ability to reason and the ability to deal with situations both known and unknown [488].

Interacting groups of people and some animals also create emergent organizations at a higher level than the individual. At present, models of collective behavior of neurons are poorly developed. The use of computational models of collective behavior has grown impressively in automation theory [1249], sociology, psychology and behavioral sciences, but not in neuro-modelling, although there are some good attempts [1020]. The meaning of the problem of collective behavior is prominently represented in the title of the well-known book devoted to collective behavior, "Orchestra plays without a conductor" [1322]. Communal decisions of interacted artificial individuals, it was demonstrated, can rise spontaneously without leaders ordering the organization [487]. The model individual was relatively simple (several traits and few features) and therefore this model would be possible to implement for neuronal interactions, but the process of decision-making is slow. From tens to thousands of consecutive interactions, in which neighboring agents interacted, is required for this. Let us consider how an interaction between subjects leads to the appearance of new properties at the level of the collective. We only briefly take into account this problem in order to reveal what principles may be implemented to ensure collective behavior of neurons. For example, there is a parallel between the individual intention and social function in sociology and the interaction of neurons in brain. Each neuron fires at its own "will" and apparently for its own "gain", but, together, a network of neurons accomplishes complex functions unknown to individual neurons [1194].

Collective behavior of neurons and glial cells is carried out within the morphological structure of the brain and may also connect some cells into a temporary structure. Similarly, when unicellular animals experience deficiency of nutrients, they unite into a fruiting body that has some features of primitive animals, has a more complex behavior and thus cells protect themselves in critical circumstances. Thus, appearance of a repellent or nutrient depletion both produces cellular morphogenesis. A repellent compels cells to convert from rod-shaped vegetative cells to spherical spores, environmentally resistant against starvation. Negative chemotaxis plays an important role in the development of aggregation [1130]. A cellular colony is the result of self-organization of individuals without external planning. The individual dissolves into the myriads cells of a colony. On the one hand, cells are the fundamental building blocks of all organisms, but on another hand, cells are individuals. Cytokineses, nitric oxide and arachidonic acid regulate colony formation. These substances are retrograde messengers; they regulate gap junctions, participate in the processes of damage-protection and take part in development of motivational behavior.

Sometimes, collective behavior of cells is observed without the anchor of a firm location of cells but only via chemical connection through the extracellular environment. The chemotaxis network that governs the motion of Escherichia coli exhibits some essential characteristics of biological complexity, such as adaptation and response to environmental signals [658]. Movement of cell is realized by a molecular motor, which switch rotation and thus change the direction of movement. There are specific molecular events that cause temporal behavioral variability in an individual cell. The switching events of individual flagellar motors from non-stimulated cells were monitored in a medium. The chemotaxis network is controlled by the concentration of the phosphorylated form of a signalling molecule, which is produced by the cells, reaches other cells and binds preferentially to the cytoplasmic base of the motors. When the concentration of these molecules increases, motors spend more time spinning clockwise. If the relative concentration of a key chemotaxis protein had been slightly higher the activity of cells would have been synchronized and each cell would have exhibited a steadier behavior than when external stimuli are absent [658]. At the level of populations, the chemotaxis network produces a steady output in the absence of external stimuli, while the output from individual cells is noisy. The appearance of an external stimulus changes the production of signaling molecules and affects collective behavior. Social insects also use chemical signals (pheromones) for organization of collective behavior [309].

There are two kinds of decision in a group. Animals may decide individually, but in a manner that is dependent on the behavior of other group members. The combined results of these individual decisions usually affect the group as a whole. The members of a group may also choose between two or more mutually exclusive actions (for instance, movement direction) with the aim of reaching a consensus [258]. Certainly, only the first kind of decision may be similar to a neuron's behavior, since a neuron can only generate or fail to generate an output reaction. Another example is a group choice of a swarm.

The collective decision of a swarm of bees seems to arise through a self-organized process driven by the dynamics of interacting individuals following simple rules based on local information [1306, 1113]. If the differences in direction between the preferred goals are not too large, the whole group moves in the most preferred direction. Examples of such a collective decision include homing in migrating birds and fishes and nest site choice in social insects, which involves relatively little conflict of interest because the goal (to find the best nest site) is similar for all individuals. Nest-site choice is a well studied example of a decision-making process [258]. A small portion of the dispersing bees ('scouts') explores the surrounding region. Each scout contributes equally and independently to the decision, but the remaining bees do not contribute to the decision. Scouts that have found a good site dance longer and with a higher intensity. As a result, more scouts are recruited to these better sites (positive feedback). At first, dancing scouts communicate the location of many sites, but within a few days all dances focus on the same high-quality site. Each scout follows a relatively simple rule: individuals communicate only locally, do not compare the nest sites directly and no one has access to all information. These simple rules ensure a communal decision, which does not require complex mechanisms or advanced cognitive abilities, even in large groups of individuals. In mammals, this path of the integration process can occur on the basis of purely internal signals, such as vestibular or proprioceptive signals [367]. We have considered recently a similar mechanism of involvement of neighboring cells in the dominant activity of neurons, when the strongest activity of some neurons involves the reaction of neighboring cells. In this way, cells may merge into an ensemble and generate a specific reaction. For instance, after learning, specific elevations of the firing rate of neuronal activity in the primary motor cortex [951] or in the middle temporal visual area [252] were only observed in a subpopulation of cells with directional properties that was near the training one sub-population. In this case, the cells that are united in the ensemble also have very little "conflict of interest".

Division of labor, based on the decision of an individual worker to perform a task, is one of the most basic and widely studied aspects of colony behavior in social insects [116]. Variation of internal thresholds for responding to task-specific stimuli among workers in a colony generates division of labor. All workers have some threshold for a task and therefore higher stimulus results in the recruitment of additional workers into a task group. Interestingly, division of labor can be generated by the effects of experience, since the successful performance of a task increases the probability of performing that task again. Therefore, individual thresholds may be altered by experience [116]. These results seem unrelated to the collective behavior of neurons, since a neuron always participates in the same local task, though this local task may be included in various behaviors. Nonetheless, neurons acquire different thresholds to different signals during learning. For that reason, the division of labor and choice of a preferred task corresponds, with respect to neurons, to a division of input signals into significant and insignificant signals for each neuron and participation of each neuron in the response to its significant signal. In addition, when a neuron acquires specific excitability in respect to a given signal, this may promote incorporation of neurons into an ensemble through gap junctions and thus a neuron might introduce its assistance into a behavior that under other circumstances does not coincide with its standard function.

Collective behavior may also be based on morphological distinctiveness of individuals. For instance, in an ant-hill the ant-soldiers have huge jaws and they are the best adapted to battle. Morphological distinctiveness sometimes plays a role also in the collective behavior of cells. The sperm of most murid rodents are characterized by an apical hook of the sperm head which is used to form sperm groups, or "trains" that exhibited increased swimming velocity and thrusting force compared to individual sperm. Species with more strong sperm competition have an elongated apical sperm hook [576]. Previously determined features, undoubtedly, may influence the collective behavior of neurons; for instance the type of connexines determines the establishment of gap junctions, but such characteristics do not possess a plasticity potential.

We know a little concerning properties of collective behavior of neurons and their operation in neuronal ensembles. However, this knowledge may be tremendously important for understanding the high neural functions. Self-erection of neuronal ensembles, evidently, depends on complementarity of information that is stored in the neurons and this may ensure constancy of our Self, while their plasticity may depend on the interaction of Self with current information and with the contents of long-term memory.

5

Death as an awareness-rising factor (a single neuron can suffer and delight)

Consciousness is like the wind, you don't see it, what you see are the effects of its. (Joe Bogen)

The essence of a mind symbolizes a subjective, that is non-objective, experience and this is directly opposite to the focus of investigations in natural history. Certainly, mind is, in a common sense, a non-physical phenomenon, but the primary mission of neurobiology is discovery of brain-psyche relations. Our reasoning proceeds from the fact that till now all known phenomena in the world have received natural explanations or they are on route to elucidation. Therefore, it is logical to try to find the material roots of psyche. Any study that has been undertaken in the field is dedicated to elucidation brain's capability to perceive the present, to recall relevant features of the past and to produce an agency in order to impinge on the proposed future, even if authors do not declare these tasks directly. For a long time it was unclear whether this task is within reach of analysis by objective experiment. Thanks to the authority of Sir Francis Crick [276], the problem of brain-psyche relations has been enjoying full rights in the scientific literature of the last years. In passing, it was revealed that the consciousness problem is knotted much more firmly, than the double helix of DNA.

The definition of consciousness is different in different fields of knowledge, in different scientific schools and with different authors. Moreover, sometimes one considers consciousness as an epiphenomenon, which does not exist at all. To be conscious one stands for being awake and receptive to one's environment. Consciousness is connected with various qualities such as subjectivity, self-awareness, the ability to perceive and even with sapience. Certainly, we cannot define consciousness and consciousness-related phenomena on the basis of material concepts until we understand these phenomena in depth. Existence of consciousness in animals is continuing to be under discussion [535]. The primary property of consciousness is a state with qualia. The philosophic concept of qualia, or qualities of the personal attitude (warmth, redness, happiness, suffering) is more simple than consciousness, but also debatable. Beside qualia, we have also the capability to act consciously or to will and therefore there are two aspects of aware states: to feel and to will. The notion of consciousness or awareness we will consider in its simple form, and will not address the philosophical question of its nature. If consciousness, and, in the first turn, qualia and will are fundamental essences and are irreducible to other substances, for example the qualia to energy, or charge and the will to force or momentum, this means that consciousness has emerged abruptly or it existed for all time, independently of life. Humanity never met such an irreducible phenomenon earlier and we might accept this explanation at the last turn, if other explanations will be exhausted.

5.1 Physiological access to consciousness

A number of data evidence that all mental events, even the most complex, arise from material events in the brain [597]. The first and the most uncomplicated question is to find parallels between material events in the brain and its aware activity. What proceeds in the brain, when we are reflecting? Myriads of neurons enter into ensembles with other neurons and with glia cells. However, even when we are doing nothing a conscious (and unconscious too) brain continues to operate. So, during silence, the functionally defined speech-sensitive auditory cortex is characterized by intermittent episodes of significantly increased activity in a large proportion (in some cases $> 30\%$) of its volume [572]. It is tempting to suppose that such idle activity of cells is a substrate for the development of hallucinations or, say, for mental reasoning. However, we do not know which material events correspond to mental processes. The minimal set of brain activities required for consciousness is commonly called the "neural correlates of consciousness" [223]. It would be convenient to know what activity is more essential for consciousness, neural or glial, synaptic or spike activity, chemical or ionic processes and so on.

Both perception and agency may be aware and unconscious. Moreover, the same event or action may be sometimes conscious and in other cases unaware. Although conscious and unconscious behaviors look as though they belong to separate categories, there are also quantitative differences between these two phenomena and, hence, transitions between aware and unaware states give us opportunity to reveal which brain processes are responsible for origin of the conscious status of a brain. Actually, some signs of the aware brain have been found. For instance, intentional instrumental actions after a number of repetitions become automated, and, on the other hand, everybody can convert their autonomous vegetative reactions into aware state, for instance, breathing. All mental events begin unconsciously and unconscious cerebral processes precede subjective sensory experience [737], so that conscious actions may be predicted on the basis of cerebral activity. Cortical activation in higher motor areas clearly preceded the initiation of movement and is therefore not confined by motor activity associated with movement execution. This activity (which is known as the readiness potential) begins to increase up to 2 s prior to voluntary self-initiated movement [283]. A progressive rise in neuronal activity in the motor area precedes the subjective intention by more than a second, and is larger when subjects judge the time of their desire to move rather than the movement itself [349]. Readiness potential corresponds to intention and not to the action itself. This means that some material processes in the brain control our aware decisions. Particularly, it is possible to decode and to predict expected intended actions of animals on the base of their neuron activities [872, 166]. Preliminary prepared intentional action proceeds in a faster rate than spontaneous voluntary movement. Therefore when the police inspector sits in your car in order to examine whether you had an adequate amount of time for extreme inhibition before a sudden obstacle, he always will accuse you of exceeding the speed limits: he has prepared beforehand for breaking and wins the fraction of a second.

Neuronal mechanisms of consciousness were attributed to the most advanced cortical area, the prefrontal cortex [277], although responses to the stimuli associated with its conscious perception are simultaneously distributed to a large subset of brain regions, including occipital, temporal, and frontal areas [1237]. If neocortex is really necessary for consciousness, primitive animals, such as mollusks, amphibians and reptilians cannot be conscious in any degree at all. However, even if intact cortex is important for the aware state, it is not necessary. Congenitally decorticated children necessarily remain indefinitely in a developmental vegetative state. However, in a condition of total or near-total absence of cerebral cortex they, nevertheless, possess discriminative awareness: for example, distinguishing familiar from unfamiliar people and environments, social interaction, functional vision, orienting, musical preferences, appropriate affective responses, and associative learning [1127].

One may observe correlation between brain and mind as on the one hand change in the brain activity in the state of awareness and on the other hand, as a possibility to elicit or change aware reactions after influence to brain. There is a large quantity of data that the aware behavior of any subject can be modified through physical impact to its brain. Changes of behavior are sometimes intense and sometimes moderate, they may be transient and may be persistent, but, importantly, we can produce modification of aware behavior via manipulations with the brain structures. So, the surgery that severs the corpus callosum, separating the two hemispheres, produces two consciousness within the same skull, as if the soul could be cleaved in two with a knife [980]. This harsh surgery is implemented for treatment of epilepsy. Many evidences of brain/mind relations were received after traumatic brain damage or after stroke. After right-hemisphere stroke, some people see things in their world as having no left side. In the acute stage immediately following the stroke, such patients may fail to orient themselves to things disposing from their left, cannot recognize their left side of their body as their own and they eat from the left side of their plate [2]. Perception may be mistaken even in healthy persons. Truthfulness of this conviction is easy demonstrated in experiments with optical illusions.

Neural activity in different brain structures can be correlated with specific perceptual and cognitive functions, whereas direct electrical microstimulation, which allows the experimenter to manipulate the activity of small groups of neurons with spatial and temporal precision, can be used to demonstrate causal links between neural activity and specific cognitive functions [252]. Cortical microstimulation alone may produce feelings of the same kind as this is usually evoked by natural stimuli. Neuronal activity elicited by electrical microstimulation in the secondary somatosensory cortex is localized and can trigger perceptual experiences [1038]. Moreover, microstimulation of cortex can affect forthcoming decisions of the animal, for instance, the choice of the face in the face selective sites in the inferior temporal cortex of primates [8]. Direct stimulation of sensory cortex may be indistinguishable from natural stimuli. For example, the primary somatosensory cortex contains a continuous map-representation of the whole body surface. When animals were presented either centrally or peripherally with the same CS^+, conditioned animals were not able to discriminate between peripheral (vibrissae) stimuli and stimuli presented to the corresponding cortex (vibrissae) area, but they were able to discriminate between conditioned stimuli presented to different areas of cortex [710]. It is interesting to note that after stimulation with a central CS^+, the animals acquired their conditioned reflex faster.

On the other hand, an effect of conscious events may lead to a change in the material substrate of brain, such as when neuronal damage and death arise after emotional stresses. When we declare that consciousness can control our body, there are some material processes controlling body, and this unidentified process is somehow transformed into a conscious impression. For instance, the greater use of particular muscles (piano exercise) causes a greater estate in the motor cortex, while mental practice in the brains of volunteers who imagined playing the music resulted in a similar reorganization in the brain [98]. It was established also that motor pathways are activated not only during the obvious task of producing voluntary movement but also during motor imagery and during action observation [625]. Single unit recordings from a cortical premotor area in the macaque monkey have disclosed neurons that produce a similar firing pattern when the monkey performs a goal-directed hand action and when it observes another monkey or a human experimenter perform a similar action ("mirror" neurons). These findings suggest that during observation, an excitability of motor pathways is modulated subliminally reproducing with high temporal fidelity the motor commands needed to execute the observed movement. Excitability of such neurons, measured as changes in amplitude of either the H-reflex or motor-evoked potentials (evoked by transcranial magnetic stimulation), is modulated in correspondence with the same time course as a real movement or an observed movement [152].

Another example is the implementation of the placebo effect. The placebo effect means the beneficial response of a subject to a substance or to any procedure known to be without any therapeutic consequence for the specific condition being treated. Nocebo is a phenomenon opposite to placebo. The nocebo effect occurs when a subject is given a substance known to be non-hyperalgesic and is told that the substance will induce a pain increase. Nocebo, in extreme cases, leads to severe pathological impairments. Oppositely, the placebo effect represents the expression of a positive mindbody link that evokes protective response. The cognitive stimulus in such cases elicits unaware protective response. Belief in regard to a therapy may stimulate physiological processes, enhancing naturally occurring compensatory processes by augmenting their level of performance. For instance, placebo analgesia is mediated by endogenous opioids: enhanced levels of enkephalins that have been associated with a sense of well-being. Placebo also may initiate NO release and, as a result, prevents the expression of potentially deleterious cytokines [106, 1171]. This is the clear demonstration that consciousness can actuate body.

Active neuronal firing is necessary (but probably not sufficient) for consciousness. Brain activity is normally coupled with a state of alertness and is primarily active during vigilance and primarily inactive during slow wave sleep [575]. Nevertheless, in pathologic states both high electrical activity (epileptic attack) and a failure of electric activity (deep narcosis) may be linked with the unconscious state. The relationship between global levels of brain function and the presence or absence of awareness is not absolute: some awake healthy volunteers have global brain metabolism values comparable to those observed in some patients in a vegetative state. Inversely, some well-documented vegetative patients have shown close to normal global cortical metabolism [708]. Consequently, the aware state cannot be directly correlated with the intensity of brain functioning.

Yet, usually the simple sign of aware activities is their more powerful exhibition in the brain when compared with unaware activity. Neural firing may need to be sufficiently strong and sustained through time to support a conscious experience. For instance, an attended stimulus elicits a much larger sensory response than did an identical unattended stimulus [20]. When an animal performs the task and presumably pays attention to the stimuli, the evoked firing rates are stronger: responses to visual, auditory, and somatic stimuli are increased when animals use them to initiate arm movements [1038]. The neuromagnetic responses (connecting with the level of brain metabolism) evoked by a stimulus are also stronger when the subjects are conscious of the stimulus than when they were not conscious of it [1237]. Cortical microstimulation studies in sensory cortical areas in humans show that for conscious perception, stimulation of sufficient duration and intensity is usually required [252]. The technique of neuromagnetic imaging demonstrates that activity of unperceived stimuli in the cortex is usually weaker than activity of conscious sensitivity [933]. And, correspondingly, neural activity produced by stimuli will be either enhanced or suppressed depending on whether the stimulus is attended or not. Emotionally arousing pictures (enjoyable and unlikable) elicit larger late positive potentials (300-1000 ms) in neocortex than neutral pictures during passive perception [248]. Thus, not every stimulus can enter into consciousness. The input signal interacts with neuronal ensembles, which depend on the brain's knowledge about the past. Strong and sustained firing evoked by the stimulus makes it more likely that the corresponding representation will be part of the winning neuronal coalition and, therefore, there is an association between consciousness and strength and stability of neuronal firing [781].

Nevertheless, strong brain activities are not an absolutely necessary prerequisite for beginning of awareness. Stimulation of single neurons in the barrel somatosensory cortex affects behavioural responses during detection of signals. Although stimulation effects varied greatly between cells, and on average only in 5% of trials was a response induced, the effect was significant [563]. Reaction times for single-cell stimulation were long and variable, such as is usually observed when one provokes an aware reaction.

The occurrence of states of consciousness depends on the formation of active neuronal ensembles, whereas unaware actions are connected with a local synchronization. Conscious access is accompanied not only by increased activity in perceptual areas, but by the specific increasing of functional connectivity [1116]. Consciousness depends on the rate at which large active ensembles are generated. The various causes of unconsciousness (e.g., anesthetics or brain stem lesions) have a common denominator: they directly or indirectly inhibit the formation of ensembles [393, 394, 361]. Furthermore, recovery from a vegetative state is associated with a reconnection of functional connectivity [28]. Certain mental disorders, such as schizophrenia, epilepsy, autism, Alzheimer's disease, and Parkinson's are associated with abnormal neural synchronization, [1281].

Nevertheless, consciousness does not require a compact synchronization. Neurons are selectively involved en ensembles. For example, only about 9% of all secondary somatosensory cortex neuron pairs fired more synchronously when the monkeys paid attention the tactile stimuli [1038]. At the same time, emotional awareness is developed as synchronization across neural systems [729]. Therefore various levels of awareness probably exist.

Another feature that usually correlates with the state of consciousness is generation of γ-oscillation. The recurrent thalamo-cortical connections organize oscillations in the γ-band frequency and consciousness is often considered as a corticothalamic resonance in 40 Hz activity [202, 375]. Nevertheless, strong activity, synchronization of large neuron groups and oscillatory activity are not always connected with awareness. For instance, strong synchronous and oscillatory activity during epileptic seizures is associated with loss of consciousness [139].

Conscious behavior is rather time-consuming and considerably variable. Latency of aware reactions significantly exceeds the latency of unaware, automatic, actions [163]. There is a delay (up to 0.5 s) in the generation of mental operations [20, 737]. The first neural correlates of conscious perception for somatosensory stimuli are already found during the earliest stages of cortical processing from 30 to 150 ms after stimulus onset [933]. Somatosensory stimuli usually need rapid responses, but perception of visual stimuli is more sluggish. When identical visual stimuli are fully perceived or not detected at all, the early visual potential (the latency < 180 ms), evoked by seen and unseen visual stimulus did not differ, either in amplitude or in topography; a larger divergence occurred around 270 ms [1116]. On the other hand, when after learning animals differentiate input signals to the CS^+ and CS^-, responses to these signals began to differ already through 7-10 ms (see Chapter 1) after stimulus. Rapid discrimination does not, evidently, require fast access to long-term memory. The first stage of perceptual analysis allows for the rapid efficient classification of items and events. The second "attentional" stage is necessary for the scrupulous identification of the conscious report of events [795]. Slow development of aware reactions indirectly indicates possible participation of glial processes in conscious behavior, which is propagated in similar time scales. On the other hand, retrograde messengers may serve as organizers of slow forms of collective cell behaviors, for instance, while it is necessary to choose the current motivation or to reorganize the cell ensemble with regards to information in long-term memory (Chapter 2).

We may conclude that material impacts to brain affect consciousness and this influence satisfies the standard conditions of scientific experiments: effects are measurable, replicable and objective. On the other hand, consciousness by some mysterious manner influences the body and this influence is also measurable, replicable and objective. One may ask: if the mind can actuate the body, why it is impossible to admit that telepathy or telekinesis exists as well? They are possible if you consider that your immaterial mind actuates your body. However, this is improbable if your mind emerges as the traces of survival in your individual brain. In this case your mind can actuate only your body.

Aware states are characterized by the augmented activity in some neural centers, synchronization of neuronal activity and their incorporation into ensembles, generation of high frequency oscillations, slow development of aware reactions and their variability. Nevertheless, these features of aware states are not absolute. Powerful brain activity may be observed in an unconscious state. Ensembles and oscillations are extremely variable even in the same aware state and aware action may become very stable. Latency of aware action cannot be short, but this is too feeble a feature for insight into the problem.

5.2 Merging of odd information in aware perception

Perception is definitely not a local process. We perceive a scene in the environment using many sense organs. Every image consists of numerous components, such us shape, sound, smell, pain, etc. Information from the sense organs enters the neural system through different channels, with each informational flow being continuous. Besides, the image is constantly changed during reminiscence and acquiring new memories. We are not only retrieving the initial memory. We also remember the last time we thought about it and image in the memory can be slightly distorted. When a synchronously-generated pattern spreads at a large distance, it must dissipate in time. All these components are somehow bound in our awareness into a whole. Simultaneous access to information located in different sites is possible only in the quanta world, but this approach is inapplicable for large objects, such as the brain, the neurons, proteins and even simple chemical substances. Some forces, such as gravitational, magnetic, electrical and forces in the solid body can interact at long distance, but binding of simultaneously-activated features by means of such forces seems improbable because of specific conditions in neural tissue, which is inhomogeneous, heterogeneous, discontinuous, damp, opaque and non-stationary. Binding of all informational flows is easier to explain by a close-distance interaction. Within a single neuron, in the area of tight interactions, reunion of substantial influences and information is potentially realistic by means of chemical interactions, diffusion, ionic and covalent forces operating at a close distance.

The binding problem may be defined as the absence of an explanation for the integration of many sensory elements into one output reaction [1237]. This difficulty is known also as the "perceptual unity". We can sense complex images and can recognize them in millisecond times. It would be attractive to understand how we do it.

J. Edwards [353] argues that "there is no more reason why information should be shared between two cells a hundred microns apart in a single brain than between two cells in two brains a meter apart". Binding in the whole image must be the property of an individual cell, not of a group of cells, and each neuron has a version of our consciousness. Therefore, the single subjective 'soul' may be a confabulation. Nobody can suggest a mechanism for access to information held in several neurons other than through signals converging on a single neuron. J Edwards pointed out that "we are colonies of sentient cells, each with a hermetic unshared consciousness". He even believes that in any moment, certain of our neurons each think that he is our Self. This extreme conclusion looks paradoxical, but let us not forget that another extreme conclusion is the merging of information within a homunculus. Which is the more rational theory?

Which is better, a conscious neuron or a homunculus? Both are terrible. And both look like something supernatural. However, our existence means that the problem is real and necessitates an answer. If we do not assume that integration between cells is accomplished using an unknown physical force, the only means to connect different neurons in a physical and informational whole is coupling through gap junctions: electrical synapses that are faster than chemical synapses.

Union of cells via gap junctions makes easier a decision regarding binding problem, but does not close this question definitively. Coupling between pairs of cells falls off rapidly with distance and is generally negligible for separations greater than about 200 μm. A union of neurons in an ensemble is too slow and relatively weak; an electrical current generated in one cell usually declines by a factor in the coupled cell. Ensembles may explain generalization of those events that already were united in neurons.

A related option for explanation of the binding problem is collective decision-making of neurons. We will consider this possibility further. Nonetheless, binding of heterogeneous information even within a single neuron by chemical means is difficult to explain because of the slow rate of diffusion even for relatively small molecules, such as the amino acids and nucleotides that often play a role as second messengers. Only diffusion between neighboring synapses is relatively rapid. Integration of such neighboring synapses is especially essential, as near pathways in the neural system contain usually related information and their integration ensures permanence of perception. Intracellular signals may be also rapidly mechanically transmitted through channels of a cytoskeleton [696, 230].

5.3 Changeability of consciousness

We may keep in the short-term memory a few objects at one time, but aware brain is usually concentrated and perceives only one input image, which may be complex and variable in space and time. Current intentional behavior is also single and it dominates other possible actions, although dominated behavior also may be rather complex. Correspondingly, the image either enters the consciousness or not, and the action is performed or not, in correspondence with the all-or-none principle. Nevertheless, degree of awareness may vary gradually, or, at least, several levels of awareness exist. There is no agreement with this notion since, by the way, not everybody even considers awareness as an object for investigations. Nonetheless, the problem of awareness levels exists and, for instance, this problem acquires practical and even a tragic sense, when one must decide whether the patient is dead if its body is alive.

Attention notion is close to awareness and the level of attention is, certainly, changeable, for instance, during habituation. Particularly, variability of neuronal responses in an aware state increase, whereas the evoked firing rates become less variable across trials when the animals performed the task and, presumably, paid attention to the stimuli [1038]. Arousal appears to be a graded state depending on the degree of coupling between brain regions [497]. During anesthesia variability additionally decreases, because in individual units the discharge rate in awake, quietly resting rats was found,to be almost three-fold more variable than during anesthesia [1343]. However, attention and consciousness are different phenomena that may occur or not occur together. Subjects can become conscious of an isolated object despite the near absence of top-down attention (iconic memory, gist of a scene) and, conversely, subjects can attend to perceptually invisible objects (priming, adaptation)

[648]. The mechanism of attention can be easily explained by a selective decrease in excitability in response to the repeated signal compared with the excitability in response to the novel stimulus (Chapter 1).

There are states of the living organism, which for all general purpose are considered as unconscious: coma, narcosis and slow-wave sleep. The electroencephalogram during slow wave sleep exhibits higher coherence, but this can be explained by the spindle activity and its frequency is lower then the γ-oscillation, whereas surface-evoked potentials in the sensory cortex during slow-wave sleep is less spatially coherent than in the waking state [1019]. Differently, paradoxical sleep (sleep with rapid eye movement) is very similar to waking. In both states, neuronal activity has the similar features of depolarizing membrane potential and increasing membrane resistance, thereby making neurons more active and responsive. Acetylcholine levels are high during waking and paradoxical sleep, while during slow-wave sleep acetylcholine levels drop to a minimum [988]. At the same time, waking and paradoxical sleep are different states. The slight differences in neuronal activity, metabolic activity, and acetylcholine release between waking and paradoxical sleep could be significant. During paradoxical sleep there is a dramatic decrease in release of norepinephrine and serotonin, in comparison with waking [107]. Late temporal responses in the sensory cortex, which are usually connected with aware states, were enhanced in waking and paradoxical sleep, compared with quiet sleep, while the size and shape of the short latency postsynaptic potentials (¡ 50 ms) were larger, but less complex during quiet sleep [1019]. The state of consciousness during paradoxical sleep is not complete and differs from the aware state, but during slow-wave sleep awareness is absent.

The state of narcosis may remind us of the slow-wave sleep, but during surgical operations unconsciousness is more profound. The goal for the anesthetic implementation is immobility, anesthesia, calm and loss of memory. Anesthetized animals do not produce alert reactions to painful stimuli. At the same time, such stimuli evoke unconscious reactions. Not every neuron is switched off when consciousness is absent. Activity during anesthetation involves not only to the neurons regulating life-important functions and the neurons ensuring inhibition and protection (inhibitory interneurons), but also involves neurons servicing ordinary signals. Noxious stimuli in anesthetized animals causes enlargement of receptive fields of spinal neurons, but nonpainful stimuli were without effect although they cause enlargement of receptive fields in conscious animals [539]. Augmentation of narcosis-inducing drugs has led to coma and death, although acute effects of these drugs are protective. For instance, overdosing of benzodiazepines has led to self-induced poisoning and coma. Flumazenil, a specific and competitive antagonist at the central benzodiazepine receptor, can be given to prevent patients from relapsing into coma. The use of antidote flumazenil results in complete awakening [1329]. Recovery from anesthesia after introduction of antidotes is a rapid process and it takes only 1-3 minutes [163]. Thus even deep states of narcosis are reversible. Nevertheless, overprotection may be also dangerous.

States of awareness of sick brain may be deeper than during slow-wave sleep and during narcosis. During coma, neuronal activity is totally absent. Patients in the vegetative state are almost completely unconscious. However, the vegetative state of an organism is not the same as minimally conscious. During the vegetative state, patients 'awaken' from their coma but show no 'voluntary' interaction with their environment. They move in a meaningless "automatic", non-purposeful manner and in addition, large-scale network activation is, normally, not observed in vegetative patients [708]. At the same time the minimally conscious state, evidently, exists. Recently, a condition of severely altered consciousness has been described, which characterizes the borderzone between the vegetative state and so-called "normal" consciousness. This condition, referred to as the minimally conscious state, is distinguished from the vegetative state by the presence of minimal but clearly discernible behavioral evidence of Self or environmental awareness [466].

5.4 Recurring change of consciousness during bipolar disorder

Bipolar disorder is known as a manic-depressive illness. Patients are characterized by recurrent fluctuations of depression and mania (from days to years). Mania is not the opposite of depression. Both mania and depression are described by irritability, difficulty in concentration and restlessness. However, during episodes of depression, patients experience loss of pleasure, sadness and thoughts of death and low energy. During episodes of mania, patients demonstrate a sense of euphoria, abnormally increased activity and aggressiveness. Both mania and depression can be modelled in animals based on observation of dominant and submissive behavior and drugs used to treat mania inhibit the dominant behavior [784]. In bipolar disorder, GABAergic neurons expressing protective calcium binding proteins are diminished in number. Molecular and cellular targets of current mood stabilizers include inositol pathways, sodium channels, adenosine receptors, cyclic AMP, arachidonic acid [493]. Mood stabilizers improve both mania and depression. Anticonvulsant mood-stabilizing drugs are effective in preventing hyperactivity [51]. In depressed patients or patients with bipolar affective disorder, postmortem studies demonstrate decreased levels of cyclic AMP system activity as well as decreased level of the stimulatory G_s-protein involved in cyclic AMP synthesis [931, 358, 271].

Mania, probably, is related to the capacity to act, or agency, and is related also to an augmented level of generalized (non-specific) motivation. For example, chronic administration of the antimanic agents reduces AMPA glutamate receptors in hippocampal neurons [338] and thus decrease damage-related factors. A reduced adenosinergic activity, mostly via A1 receptors, is related to manic behavior [773, 51]. At the same time, extracellular concentration of adenosine increases upon stress and depression. The accumulation of adenosine, it is known, induces damage and triggers sleep, as a result of the perceptual overload of the brain cortex [873]. Adenosine through A1 receptors coupled to inhibitory G_i proteins protects and inhibits neurons, and a decrease in the transmitter release (glutamate, dopamine, etc.) inhibits cyclic AMP and increases inositol-trisphosphate formation, that in turn stimulates the release of intracellular Ca^{2+} from internal stores [242, 432, 17, 1102]. Adenosine is generated at sites that are subjected to 'stressful' conditions [522]. Thus, during manic episodes and dominant behavior, neurons are imposes a strain on system connected with both the protection and the damage: cyclic AMP pathways strengthen, Ca^{2+} release decreases, but excitability and transmitter release increase. This could promote action generation and may increase will, but the problem of bipolar disorders still waits new developments.

5.5 Properties of the alive, but unconscious brain

States of brain during consciousness and unconsciousness are different. A condition of deep unconsciousness is produced by diverse chemical substances and, particularly, by means of acute narcotics and general anesthetics. We already have considered inhibitory, protective and euphoric effects of acute narcotics. Augmentation of inhibition through specific receptors evokes loss of consciousness. The mechanism of action of general and especially volatile anesthetics is less clear.

Physically and chemically dissimilar anesthetics produce similar effects, which usually are the profound inhibition of neuronal activity and acute cell protection. In spite of more than a century of anesthetic treatment, mechanisms of volatile anesthetic action are insufficiently clear. Nitrogen narcosis may be caused by altered permeability of neural cell lipid membranes and earlier this mechanism was considered as typical. However, there is a long list of molecular targets of volatile anesthetics, they may be different for different agents and in some cases molecular targets are unknown. So, the global effects of anesthetics on protein dynamics, particularly neurotransmitter-gated ion channels, rather than on lipids may be in a base of their action [408, 1217]. Lipid membranes, most voltage-gated ion channels (Na^+ and K^+ channels), many second messenger systems and axonal conduction are less sensitive neuronal targets to general anesthetics. Voltage-gated Ca^{2+} channels are probably the most anesthetic-sensitive of all voltage-gated channels. The inhibitory $GABA_A$ channel and glycine receptors are strongly potentiated by the majority of general anesthetics, while excitatory glutamate receptors are insensitive to many anesthetics, especially volatile agents [394, 408, 52, 1050]. Nonvolatile anesthetics, for instance barbiturates, also promote GABA-inhibition [491]. General anesthetics may also induce inhibition by escaping synaptic inhibition. Thus, several simple anesthetics may induce general anesthesia by means of membrane hyperpolarizations through special two-pore-domain K^+ channels and the opening of these channels protects neurons against both ischaemic and epileptic insults [407]. The most important target for general anesthetics is, probably, blockage of cell coupling. Gap junctions are strongly reduced by general anesthetics [1030, 257, 109, 434, 1332]. The effect of some (but not all) inert gases, such as argon and xenon, is especially amazing, since they cannot participate in chemical reactions. Binding kinetics of anesthetics is rapid and all volatile anesthetics are rapid in onset, but xenon gas is extremely prompt in anesthesia onset. Blockage of cell coupling by volatile anesthetics indicates that they may fulfill the role of a cork in gap junctions, but this is a controversial subject. The thalamus, the center of γ-oscillation generation is, supposedly, a key target of anesthetic effects on consciousness [28].

The complexity of the anesthesia mechanism is a consequence of its connection with the most mysterious aspect of neuroscience, with the consciousness. Moreover, general anesthetics act exclusively on the consciousness, do not suppress local oscillatory activity, do not affect other tissues, do not suppress global cerebral metabolism and are, frequently, not toxic. Cell uncoupling disturbs their accorded activity, and this may be a may reason for loss of consciousness.

5.6 Inhibition in the brain and consciousness

As a rule, brain receives outer information from the environment through excitation. Principal neurons accept input signals via excitatory synaptic connections and send their resulting excitation along the main neural axis. Inhibition has many important functions in the brain, but now we would like to emphasize that, as a rule, inhibition is an inner reaction of brain. A brain in such a way responds to the impact of the environment, changing a local state of neural tissue and thus preventing a signal transmission along the neural axe. In accord with aforesaid discussion, roughly speaking, excitation is the signal of a possible danger and a forthcoming damage, while inhibition signals safety and relaxation. Inhibition is intimately connected with cell protection and it is necessary for homeostatic healing.

Among neural cells, only the inhibitory neurons are, mainly, connected via gap junctions and this indicate an exclusive role of inhibition in a union of cells into a temporal organ, ensemble for maintenance of the current function. Inhibition tends to combine neurons rather than emphasize the contrast between their activities through 'lateral inhibition'. Coupling ought to prevent lateral inhibition. As we know, coupling increases during active brain states, just when collaborative activity of many neurons is usually observed. Neuronal ensembles may unite scattered information into a whole image, create a stable core of excitation or inhibition during a dominant state, and thus promote generation of a coordinated reaction of the entire brain. Information in the brain is merged correctly by means of an association of inhibitory neurons and, may be, by means of glial cells, which are also tightly connected into a chemical-electrical conglomerate.

GABA, the most prevalent inhibitory transmitter in the brain, exerts protective effects although protection-related inhibition is not restricted to the GABA system; other abundant inhibitory transmitters including the opioids, glycine, taurine and adenosine also produce protection. Nevertheless, GABA-inhibition plays a special role. Particularly, $GABA_A$ receptors are present in membranes of Hydra, one of the most primitive organisms with a nervous system.

Only GABAergic interneurons are intensively coupled through gap junctions and this assists an integration of inhibition and synchronizes neurons. Therefore, the GABAergic system, probably, promote to cell organization in ensembles and γ-oscillation generation. GABAergic neurons protect cell against excessive excitation and their coupling indicates the participation of protective compensatory processes in an image perception and in a reaction generation. Particularly, the GABA system is diminished during bipolar disorders: a disease of intentional actions. Unexpectedly, the GABA system modulates even microcirculation of blood flow in the brain. An evoked discharge of single cortical GABAergic interneurons was sufficient to either dilate or constrict neighboring microvessels [193].

This cardinal role of GABA-inhibitions in the neural system may be determined by their effects on aware behavior. Ionotropic $GABA_A$ receptors modulate both slow-wave sleep and paradoxical sleep occurrence [491]. Evidently, excessive inhibition through the synapse, gap junction blockage or hyperpolarization through ionotropic and potential-dependent channels decreases awareness. Protective action of inhibitory neurons is correlated with the development of a positive state in the brain, achieved usually through $GABA_A$ receptors. Stimulation of $GABA_A$ receptors counteracts anxiety while reduction of $GABA_A$ergic inhibition promotes anxiety [509]. Activation of $GABA_A$ receptors exerts euphoric effects and, probably, can serve as a reward [706]. Therefore, the GABAergic system is one of the powerful regulators of awareness.

Interneurons operate with high speed and temporal precision [591] especially ionotropic $GABA_A$ receptors. The ability to generate precise spike times, with millisecond resolution, depends upon the generation of higher-frequency oscillations in the membrane potential [235]. Fast-spiking inhibitory GABA neurons more easily maintain synaptic transmission at higher frequencies than regular spiking pyramidal cells and endorse synchronization of neuron firing. Therefore, inhibitory potentials may exhibit a higher degree of synchrony than excitatory potentials and this is expected to occur in awake, behaving animals [521]. Here why γ-frequency oscillations are so important for aware state: this is mutually connected system of inhibitory GABA-interneurons. We may suppose that fast modulation of inhibition in large neuronal populations may quickly adjust subjective feeling at the scale of positive-negative senses and in such a way modulate selective attitude to one's actual level of being.

Inhibitory postsynaptic potentials exhibit more power than did the recurrent excitatory network and are more synchronized in neighboring (within 1 mm) pyramidal cells. Local stimulation of single inhibitory neurons in the barrel cortex affects aware behavior more effectively than stimulation of excitatory neurons. Whereas stimulation of excitatory neurons led to weak biases towards responding, stimulation of inhibitory neurons led to more variable and stronger sensory effects [563]. During synchronization, GABA-inhibition simultaneously appears in a large number of neurons and, probably, concurrently induces in these neurons an elemental euphoric feeling. A single interneuron influences thousands of postsynaptic principal cells and this agree with their effect on cell conditions rather than on current transmission of information. Synchronization is, evidently, important for parallel but not for sequential structures. The sense of this inhibition-related cellular euphoria is clear: it coincides with convalescence of the cell. At the same time, synchronization of elemental feelings may produce general sensitivity in the brain.

GABA-inhibition sometimes participates in the psychoactive action of many substances. Anabolic androgenic steroids elicit a decrease in anxiety and an enhancement the sense of well-being via allosteric enhancement of forebrain $GABA_A$ receptors [1053, 102, 533]. Correspondingly the dopamine reward system is also connected with GABA inhibition and dopamine application, which supports self-stimulation and activating coupling between GABA neurons. The threshold for responding to selfstimulation was the threshold for electrical coupling between GABA neurons, the degree of responding to self-stimulation was proportional to the magnitude of coupling between these cells, and gap junction blockers increased the threshold for selfstimulation without affecting performance [706]. However this is not a general rule: ethanol inhibits electrical transmission via inhibition of NMDA receptor-mediated excitation, not via enhancement of GABA receptor-mediated inhibition [895, 1182].

Psychotropic substances exert their effects not only through postsynaptic inhibition. Inhibition-related protection may be achieved in several ways. For instance, intravenous cocaine injection directly inhibits Na^+ channels [637], activates the σ_1 receptor within neurons [807] and thus protects cells. Many psychoactive substances are directed towards stimulation or inhibition of Na^+, K^+-ATPase and sometimes low concentrations of the substance activate, while high concentrations inhibit this enzyme, as happens with opioids [1362], cyclic AMP pathways [105] and nitric oxide [677] operates. Particularly, this may be one of the causes why low concentrations of many psychoactive substances evoke euphoria, middle concentration produce narcosis and large amount of the same substance leads to death. For that reason, inhibition is connected with a pleasures sense, but extensive inhibition produces an unconscious state.

Homeostatic efforts wasting to protection lead to a positive feeling (if somebody wants it is easy to replace 'positive feeling' with 'breaking through injury') and emergence of a feeling is an equivalent to the appearance of homeostatic protection. The functional sense of a relationship between protection and inhibition is a cessation of efforts after a recovery of equilibrium. A satisfied neuron does not impel other neurons to activity, and, moreover, an inhibited neuron itself stops homeostatic recovery, since deepening protection leads to narcosis and to the disappearance of euphoria. Blockage of consciousness may be determined by a decrease in any sensitivity in the inhibited neurons. Such neurons do not perceive damage, do not feel danger and cannot enjoy avoidance of danger. This guarantees a shutdown of the homeostatic machine after reaching the protection.

Thus, homeostasis somehow (we do not know how) directs its efforts to recovery of one main neuron's parameters: intactness. Nothing will change in this management, if we entitle this main parameter "sense of being". Correspondingly, neurons evaluate their current state as the positive or negative dependence on their condition of being. Inhibitory neurons promote integration of these elemental protective efforts into goal-directed action.

5.7 Chemical modulation of consciousness

Chemical influence may affect aware states. We have already demonstrated that perception is modified after administration of anesthetics, stimulants, antidepressants, narcotics, neurotransmitters and some other substances. We cannot be sure that gap junctions control awareness, but we suspect this is so. There are substances that decrease [104, 918], increase [989, 645] or modify awareness, though the signs of the effects depends on concentration, time of administration and extent of substance introduction. These substances may affect the electrogenesis of cells, systems of second and retrograde messengers and gap junctions.

There is a variety of psychotropic correlation of material events. Psychotropic substances act via damage-protection related mechanisms. Mental feelings are infinitely diverse, but let us consider the simplest ones, suffering and pleasure. Treatments that protect neurons usually inhibit their activity and exert psychotropic actions related to relief. For instance, opiates, dopamine and $GABA_A$ agonists have anxiolitic and antiepileptic effects [297, 247, 832, 1349], neuropeptide Y, vasopressin and oxytocin have antidepressant properties [1303, 972], whereas $GABA_A$ receptor antagonists are convulsants [590], and antagonists to dopamine receptors and to opiates evoke withdrawal [910]. Positive correlations were found between cytokine secretion and anxiety [1023]. σ_1 receptor agonists protect against oxidative stress [182] and have an antidepressant effect [1288]. Many other substances inducing cell damage also increase anxiety-like behavior, aversion and epileptic status, [948, 87], whereas protective substances produce anxiolytic, analgesic, anticonvulsant, sedative, hypnotic and anesthetic states [989, 631, 1016, 95, 102].

Aware behavior directly depends on exposure to metabolic pathways of cells, connected with damage-protection. The precursor of inositol-trisphosphate, inositol, controlling intracellular Ca^{2+} influx, is anxiolitic [651]. Excitatory amino acids provoke aversive reactions, while antagonists to glutamate and to substance P receptors act as antidepressants [906, 832, 1016, 932, 517]. Positive influences proceed also through inhibition of the voltage-dependent Na^+ currents [1349] and increase the production of new neurons [218], while negative influences act through the cyclic AMP system [87], through G protein-coupled metabotropic (mGlu) receptors [743], etc.

Consciousness changes when active neuronal ensembles are restructured. Formation of ensembles depends not only on gap junctions, but also from properties of constituting neurons and glial cells. We demonstrated many examples such as during altered awareness and when excitability and chemosensitivity of neurons are modified and responsiveness of neurons to signals changes. These examples indicate transient reorganization of the homeostatic machines of neurons during aware behavior.

Even so delicate a sensation as a love has a material basis and can undergo chemical modification. When a female prairie vole is mated with a male, she forms a distinct preference for this partner; however, when a dopamine agonist is infused into the nucleus accumbens, she begins to prefer a male present at the time of infusion, even if she has not mated with this male [472, 757]. Psychotropic substances usually affect cells on the scale of damage-protection and they alter various aspects of awareness: perception, will, sense of happiness, anxiety, etc. We discussed these facts earlier. Let us demonstrate the chemical means for modification of awareness at the examples of place preference and state-dependent learning, phenomena that allow the repetition of experiments and the comparison of their results.

Conditioned place preference and place aversion compel animals to prefer or to avoid some circumstances in the environment, when in the past these animals received pleasant or unpleasant chemical impact during acquaintance with the same circumstances. Chemical impact may be directed to various points of a body. For example, if a rat received oral administration of a bitter solution in a determined corner of the experimental camera, this rat afterwards will stop visiting this corner (place aversion). It is interesting to note that similar results may be obtained, if some chemical substances are introduced in certain brain areas, although brain tissue does not have receptors responding to outer signals. If brain administration of the substance produces place preference or aversion, this means that this substance, through interaction with the brain tissue, exerts a pleasant or unpleasant influence that interacts with qualia. Brain sites that have been implicated in the mediation of drug-induced place conditioning include the 'traditional' brain reward sites and some additional areas [1280]. Conditioned place preference may be induced by psychostimulants, dopaminergic, cholinergic, GABAergic, adrenergic, adenosinergic, serotonergic and glutamatergic drugs, opiates, substance P, cholecystokinin, hormones, and calcium channel blockers [1298, 1324, 513, 1230, 545, 1420, 325]. All these substances exert an effect on cell damage or cell protection.

At the same time, corticosterone (steroid hormone released by the adrenal cortex in response to acute stress) is neither rewarding nor aversive in either behavioral phenotype and fails to produce conditioned place preference or conditioned place aversion in rats [325]. Exposure to high concentrations of corticosteroids, rather than killing neurons directly, typically induces a physiological state that renders neurons more susceptible to subsequent neurological insults [186]. Therefore only chronic corticosterone administration may lead to a significant atrophy of neurons in the hippocampus [400]. Corticosterones probably do not interact with qualia, since they do not affect damage directly, but reduce resistance to damage and, perhaps, decrease a will. Particularly high corticosterone levels are observed in subordinate rats and monkeys that experience decreased dominance [784].

Another related phenomenon is state-dependent learning(or dissociated memory): if an animal acquires habit at a time when the concentration of certain active substance in the brain is modified, this animal will not recollect the habit when the concentration of the substance is normalized. However, repeated administration the same active substances immediately recovers the habit [877, 53]. The retrieval of responses learned before chemical treatment is temporally blocked by the substance, but the ability of the animals to acquire new learning tasks remains intact. Typically, the same substances may participate in elaboration of place preference or aversion and in state-dependent learning. The impossibility to reproduce a memory trace after a change in the chemical background of brain confirms that the current conscious state may or may not include particular information from the past. In other words, the Self is not always the same.

5.8 Materialization of the SELF

A common topic in all theories of subjective awareness is incorporation in the theory a self-related "witness" (a Homunculus), or "conscious" neuron without which awareness can not emerge. Both premises look equally non-relevant and both have some experimental and theoretical roots. Meanwhile, these two approaches are not mutually contradictory. The core of consciousness is the existence of the object, or the Self, or center of awareness. One understands a "Self" as a generation of subjectivity with its unique experience. The Self is experienced as distinct from any other person and may be described as the feeling that one is the same person across time who is the author of our thoughts and actions and who is distinct from the environment [634]. This definition implies the necessity to perceive others, before one might percept himself. Surely, one is seeing others directly. In accordance with Friedrich Nietzsche, "Older is the pleasure in the herd than the pleasure in the ego. The THOU is older than the I".

An inner model of body is, evidently, also close to the Self, although it does not coincide with it. Protected territory also is close to the Self. For a human, property is part of the Self. Interestingly, Lev Trotsky described a talk with the Swiss ambassador in 1917, when Trotsky was foreign minister.

The ambassador protested against the confiscation of cars from Swiss citizens in Moscow. Lev Trotsky related that the ambassador did not understand that the social revolution in Russia expropriated property, and that change of the forms of property is the meaning of revolution, and that property is not something alive, but is only the result of interaction between people. "However", sarcastically wrote Trotsky, "the ambassador perceived the confiscation of cars as an amputation of apart of the human body" [1247].

Naturally, a Self exists only in an aware state. Consciousness is most damageable in respect to chemical or physical impacts. Patients with amnesia can show selective impairment in conscious memory in comparison with preserved skill learning [178]. Quite the reverse, the self is most permanent when consciousness is recovered. Even after severe amnesia, the Self is usually kept intact.

Why is "Self" better than "Homunculus" in our description? Surely, the term is not essential. What is important, is whether Self (or Homunculus) is living as an independent being within the brain or is an on-line product of the brain processes.

It is interesting that in spite of the sense that we are the same person across our lifetime, the Self changes during experience. The Self alters under influence of narcotic and psychostimulants. There is psychotic-like mental activity of dreaming [491] and the Self does not coincide during waking and paradoxical sleep, probably, because of background difference in concentration of biogenic amines affecting state-dependence of brain [107]. The Self reversibly changes during mental illnesses and periods of remission. Patients with bipolar disorder experience a cyclic alterations of their Self in the periods of mania and depression [271, 784]. There are also patients who feel that their Self is located outside of the physical body, or people even experience seeing a double of themselves in their extrapersonal space. During a first illness, patients generally describe their impressions as highly realistic experiences, whereas during a second illness the subject immediately realizes the hallucinatory nature of the experience. Visual details of the "autoscopic" body differ from the patient's actual appearance (such as clothes, age, hair cut, size, and coloring of the body). The main diagnosis for illness connected with autoscopic hallucinations is epileptic seizure and may be accompanied by damages to various brain areas [133]. Results of investigation of neuroimaging experiments show a segregation of activity during perception and during introspection. Moreover, regions that showed enhanced activity during introspection underwent a robust inhibition during the perceptual task [1291, 352, 484]. Thus, during the same episode of awareness the active coalition of neurons may be robustly reorganized. Weak reorganization of neuronal ensembles during a single episode of awareness is easily observed in any experiment.

A special interaction between the consciousness and the Self is observed in the state of hypnosis. Persons in this state are receptive to such a complex signal as speech, but participation of their Self in their behavior is restricted although it does not disappear completely. The material traces of a hypnotic state are possible to reveal during an objective experiment. The hypnotic state was found to lead to significant reorganization of spatial synchronization of brain electric activity. Their dynamics were opposite in the hypnotic and conscious states, which appears to result from the temporary exclusion in the hypnotic state of the functions of the frontal areas of the cortex responsible for conscious self-control and regulation of ongoing activity [1211].

During a lifetime, the model of outer world and inner environment updates. Moreover, restricted stability of the Self allows supposing that elements forming the Self are assembled anew on every occasion during waking hours, so that it is not identical each time it emerges with the emergence of awareness. Self-assembling each time produces similar core of the Self by means of combination of the most essential personal elements of the personal experience. Such a process is observed in many biological systems: proteins out of amino acids, viruses out from proteins and nucleic acids, organism from cells, ensembles from neurons and society from individuals. This complex process needs time and this is apparently why recovery of the Self after an unconscious state takes seconds, a period of time that is several orders of magnitude greater, than the characteristic timespan of neuronal events such as spike generation and interaction with the next neuron. The time, may be, is necessary, for instance, in order to assemble the relevant coalition of neurons. Peculiarity of the Self may depend on current circumstances and on related information in long-term memory. A large latency in time of aware reactions may be connected with the necessity for reorganization of the Self with participation of permanent memory.

We do not know whether gap junction conductivity may depend on long-term memory. Traces of connectivity are permanently working and are quickly recovered by neural control and serve for association of cells into an ensemble. In fact, gap junctions are organized slowly, but rapidly modulated. This is precisely the property that is necessary for reproduction of images. Coupling arises between the cells that are frequently activated together and serve for uniting cells in the same ensemble during reproduction of memory. Gap junctions are controlled by a normal physiological activity and by means of messengers of damage, protection, motivation and awareness. Blockage of gap junctions prevents normal reproduction of memory, while gap junction recovery, for instance after narcosis, recovers memory. Self-erection of the most reliable connections between neurons evidently recovers personality after loss of consciousness. Chemical memory, possibly, is not the material substance that may exist independently of brain structure. Information that is scattered within neurons must be assembled and amalgamated within the Self. It is impossible to exclude that the same chemical code has a different sense in different neurons. This topic needs elucidation. It is likely that the Self ensures easy access to long-term memory. Interaction between a Self and long-term memory is the subject of instrumental examination, although the answer to many questions lie a long way ahead of us.

Assembling the Self by means of forming cellular ensembles entails two components, objective and subjective. Firstly, the dominant ensemble involves the neurons which, in correspondence with their inner resources of memory, are evaluating their current input signal as important and predict forthcoming events; they have tuned their excitability to spike generation or spike failure. This is connected with mobilization of the homeostatic machinery: second and retrograde messengers, Na^+, K^+-ATPase, the system of G proteins, modification of potential-dependent channels, chemoreceptors and gap junctions. Thus, the ensemble concentrates the most relevant information at any given moment with every recovery of awareness. This information represents objective past experience and objective current data. Secondly, the dominant ensemble incorporates a plethora of individual estimations of performance criteria for the conditions of being of each involved cell. This evaluation is one-dimensional for every member of the ensemble: overload or rest, injury or protection or even pleasure or frustration. The term is unessential. In any case, these criteria will have a subjective coloring for individual cells, although they are established on the basis of objective evaluation of error signal in performance criterion of homeostasis. As a result, the dominant ensemble will incorporate a broad appraisal of personal qualities of the cells. How individual cellular quality is dissolved within the Self is a problem that we put off until the future because of a shortage of data even for a mental picture.

The development of objective and subjective components of Self during a lifetime is different. Until brain is healthy, subjective components of Self change slightly and is perceived as an unaltered essence. Even at a venerable age, a human being perceives itself as the same person that in the past was the child. At the same time, the brain alters, weakens and grows old, even being healthy in the limits of age. The Self may become sick, but does it ever become decrepit? The brain, together with the traces of individual experience, is the objective component of Self. It is a measurable substance and it change gradually. According to our description, the subjective component of Self may change only qualitatively. It is either alive or dead, either is or is not. A performance criterion, an objective appraisal is quality of life.

Thus the duality of the brain/mind relation we do not transfer from the brain into the neuron. Instead, we are formulating an impending task: how homeostasis generalizes imbalances in a variety of variables into one joint performance criterion. This general criterion allows for a common interpretation at the levels of brain and mind. At least after implementation of this criterion, the material consequences will not depend on its interpretation, as, say, cell damage or cell dissatisfaction. An organism is built of unreliable elements and self-regulation is directed to protecting them against damage. Does this mean that an artificial mind that is capable of feeling could be created on the base of unreliable elements? To our thinking, the response is no. The main point is rather a possibility of elements of self-recovery instead of their option to be damaged.

5.9 Discrete time steps in perception

A discontinuous threshold for conscious access is said to exist. A subject may focus on only one thing at a time [1117, 301]. Every few seconds there is an exchange in perceptual dominance without special voluntary intention. For instance, in one technique, called binocular rivalry, vertical stripes are presented to the left eye and horizontal stripes to the right. The eyes compete for consciousness, and the person sees vertical stripes for a few seconds, then horizontal stripes, and so on: although both stimuli are presented, person will report alternately seeing only one or the other. While the visual system receives signals from both stimuli, only one at a time is accessible to conscious experience [1237, 980]. Thus, in spite of the subjective impression of our infinite multivaluedness, the aware brain operates as a whole. At any given moment we pay attention to only one signal. In addition, identical stimuli are sometimes perceived and sometimes not. When two visual targets are presented successively, conscious identification of the first target hinders the detection of the second target, whereas unconscious detection of the first stimulus does not mask the second stimulus [795, 1116]. Pair presentation of auditory stimuli also leads to suppression of a response to the second stimulus. Response to a second tone in the amygdala was inhibited during seconds and the greater inhibition occurred after the animal had experienced a stressful stimulus [279]. The greater the impression, the more is refractoriness in the consciousness.

The time interval for conscious perception of sequential images is in the tens or hundreds of milliseconds. Conscious perception of sequential stimuli without identification of each stimulus is much quicker than that but not unbelievably fast. The temporal quantum of consciousness for auditory perception is approximately 1215 ms: when two 1 ms auditory clicks were played in succession, subjects heard them both as long as the clicks were separated by at least 13 ms, but heard only one click (of normal duration) when the inter-click interval was less than 12 ms [375]. Cortical dynamics is not continuous but operates in steps that resemble frames in a cinema [412]. During more protracted time (seconds or minutes) we can hold in short-term memory several stimuli, but the number of such stimuli is rather small [794].

Not only perception, but executive function of the brain occupies all resources of consciousness. We have discussed possible mechanisms of dominant behavior (Chapter 4). Just as we can barely attend to more than one object at a time, we also cannot perform two conscious tasks at once, and when a test subject reports conscious seeing of an image, this may represent a competition for two or more conscious actions.

Information that does not enter consciousness tends to decay quickly [300], probably because it does not enter into a winning coalition of neurons. Aware memory compared with unconscious memory is more vulnerable also in respect to brain damage and in relation to resistance to exogenous chemical influences. Consciousness, probably, allots direct access to long-term memory.

5.10 Common currency for choice: between displeasure and pleasure

There are a number of data that a biological system regulates some general parameter (Chapter 2). This is related to various levels of complexity: a collective of individuals, a whole organism and subsystems of an organism or a cell and system of cells. For instance, a neuron maintains the preferable level of firing instead of an exhaustive control of a variety of particular types of channels. A single neuron segregates input signals into biologically significant or not, in order to decide to generate or fail to generate a spike. Also, in order to choose the most effective output reaction of the brain, it is necessary to compare attractiveness of supposed actions. Similarly, for choice of dominant motivation, motivational states must to be compared to each other by means of a common currency. It has been argued [197, 198, 1137] that this common currency is pleasure/displeasure. Particularly, the unidimensional nature of a common currency is displayed as an interaction of different motivations and their similar mechanisms. However, reception of metabolic mismatches and location of the primary motivational centers in the brain is differed (Chapter 3). The choice of the general performance criterion is important for effective control not only because of difficulty of multivariate management, though this consideration is also essential. A hierarchy of general criteria indicates the scale of their worth of choice of the most imperative goal and at the top of the hierarchy is positioned the being itself. What norm could guide adjustment in such a case, if not survival [321]? If a mother-value of all values exists (and it does exist) then it could play a role in the sense of life. An organism supports those values of its inner constants that do not damage its neurons.

Certainly, there is one essential flaw in all these methods of self-control. How can brain find its performance criteria as 'preferable', 'attractive', 'biologically significant' or 'pleasurable', if we do not apply to the sovereign opinion of Homunculus? Surely, it is not necessary to evaluate each of these subjective criteria, if the universal criterion, or mother-value of all values, had been determined. Rather, it is possible that those physical characteristics that are implemented for evaluation of the quality of being take the role of parameter at the subjective scale of pleasure-displeasure. We have demonstrated existence of such a parameter already at the level of a cell. Confluence of elemental cellular feelings into consciousness is not a trivial process, but it does not require introducing a new category of items: consciousness can be explained by means of assembling cells (physical substrates of feeling) into ensemble.

When cells are uncoupled and cellular merger is disturbed, awareness is absent, but some features of sensitivity are present as they are during vegetative states. Although neuronal coupling, brain connectivity, synchronization of neuronal activities and oscillation in the γ-diapason are interrupted during vegetative states, neurons stay alive. Global brain metabolism and electrical activity of an individual neuron may be normal in such patients. Also, broken brain connectivity prevents development of synchronous oscillation, but within individual neurons oscillations are easily recorded. Vegetative patients do not display attributes of awareness, such as an intentional communication with their surroundings, but may be aroused from their coma and move in an automatic manner without the appearance of synchronization in large neuronal ensembles. A healthy level of individual cells during the vegetative state is evident. Particularly, sometimes even though exceptionally seldom, consciousness recovers after a long period of vegetative existence and after recovery functional connectivity is reconnected.

The unique capability of neurons and some other cells to survive in microenvironment makes their system of self-regulation attractive for investigations. Neurons can live for up to tens of years, but the ion channels, ferments and receptors that underlie electrical signaling, even under normal conditions, turn over in the membrane rather rapidly, from minutes to weeks. Neurons express a number of different voltage-, chemical- and even mechanical time-dependent processes. Awareness protects a non-linear multi-dimensional system from chaotic behavior and without a general criterion or elemental feeling a neuron may become uncontrollable. Then, what is the ultimate figure of merit connected with the quality of being? It is unknown which parameter is really used as a common currency, but some parameters are more probable while others are less related. A true parameter cannot be particular and has to be tightly connected with cell being (which cell, neuron, glia, or their union is unknown; in this book we have paid little attention to glial cells: the reason is a deficiency of data, not an insignificance of the glia role in behavior). Regulation of such a parameter cannot be replaced by regulation of other parameters.

Intactness of a cell, as a whole, depends on several variables: membrane potential, input resistance, excitability, cell volume, tension of cellular membrane, pH of inner environment, concentrations of intracellular Ca^{2+}, cyclic AMP or ATP, free radical concentration, metabolism power, some important proteins, such as Na^{+}, K^{+}-ATPase, etc. Naturally, all pretenders affect the cell in the scale of damage-protection. What is the demand that the neuron aspires to satisfy? On the one hand, although a neuron needs to support many important factors, it has only one output, it can make only one decision and only one integral factor has to be essential. On the other hand, homeostasis also needs a general criterion for proper functioning. A neuron regulates its major demands, such as the avoidance of injury and the aspiration to life. Excitability is intimately connected with cellular damage-protection and is robustly regulated by the homeostatic machine. On the other hand, excitability depends on both the current signal and on the past. Homeostatic protection operates sluggishly on the minute time scale and cannot affect mind at the on-line mode, but excitability may interconnect homeostasis with memory apparatus. The interruption of excitotoxic death by means of compensational protection improves cell state, is converted to a generation of action potentials without a specific-purpose signal, and reiteration of these actions inevitably leads to a search for favorable conditions (Chapter 4).

Excitation may be evaluated as the sign of a danger in the environment and it predicts future damage. Elemental subjective coloring may be as unique to a cell as is its shift towards death or towards life and we suggest, as the first approximation, that excitation and inhibition are perceived by a neuron as negative and positive sensations. Modulation of neuronal excitability satisfactorily describes principal regularities of neuronal behavior and might correspond to an elemental feeling, if, of course, elemental feeling is the privilege of a neuron and other unexcitable cells cannot feel at all. We cannot be sure, whether this is true. At least, so heterogeneous phenomena as chemotaxis, colony formation, development, damage, protection, homeostasis, stress, motivation and mental disorders involve similar critical intracellular processes. The common point of these phenomena is their connection with life and death. There are general criteria associated with other, more universal properties of a cell.

The most important function of homeostasis is, probably, running cell repairs and its being itself. In principle, the quality of homeostatic defense or tension of homeostatic compensation may be put at the basis of an ultimate criterion of being. For instance, compensational activity of homeostasis may be presented as a consumption of ATP molecules in a unit of time that is an intensity of energy consumption. Obviously, this parameter needs the measurement of time. Cells possess means for evaluation of time intervals by protein and ATP only, in the absence of transcription, and the molecular components of the clock are conserved in all living forms from plants to vertebrates [654]. Energy consumption, surely, cannot be interchangeable with any other parameter or with their combination. Besides, running homeostasis combines two vital properties of awareness: evaluation of the quality of being and a will to operate, which is increased together with an increase in metabolic mismatch. A possible example of such homeostatic machinery is Na^+, K^+-ATPase. For instance, the content or production of ATP in hypothalamic neurons during control of feeding behavior, temperature and osmolarity may be a signal for energy-sensing error, which in turn may trigger the appropriate neural, endocrine and appetitive responses [775]. Goal-directed behavior is connected with temperature regulation [635]. Brain, as any physical body, is cooled when it produces mechanical energy.

5.11 What is bad and good for a neuron?

What are distinctive features of an injured neuron and what compels it to suffer? Suicidal brains tell us about suffering. Investigations of the postmortem brains of suicide victims and those suffering from a mood disorder allow an evaluation of physical traces of negative emotional states.

There is much evidence for an association of lower serotonergic function and suicidal behavior [788, 49, 916, 129]. Particularly, antidepressants augment the level of serotonin in the brain and this was considered as evidence of a direct connection between serotonin and mood. However, antidepressants exert their mood-elevating effects only after prolonged administration, indicating that enhanced 5-HT neurotransmission per se is not responsible for the clinical actions of these drugs [1305, 186], but that these drugs elevate production of new neurons. During stress-evoked depression neurogenesis decreases [492, 342], that is the brain stops to recover. Hence, in this case suicidal behavior is also connected with inhibition of cell protection.

The brains of suicide victims reveal neuronal atrophy, cell loss and a disturbance of intracellular biochemical machinery, expressed as imbalance in G protein function and altered regulation of intracellular signal transduction pathways [271]. The concentration of inositol decreases in the brains of the suicide victims [651]. Brains of subjects who committed suicide showed a pronounced increase in excitatory amino acid binding [410], stimulatory G_s protein concentration [269] and in the level of cannabinoids [1305]. Particularly, cannabinoids decrease the level of cyclic AMP in the brain [402]. There is also imbalance between diverse neurotransmitter systems [501]. In the brains of other suicide victims, affinity of glycine sites on the glutamate receptor complex was reduced [972], as well as density of $GABA_A$ receptors [1162]. Naturally, high concentrations of corticotropin-releasing factor, secreted during stress, were observed in postmortem samples of cerebrospinal fluid obtained from severely depressed suicide victims [213]. However, although corticotrophin-releasing factor usually stimulates the second messenger cyclic AMP production [839], cyclic AMP levels decreased in suicide victims [269, 347] and this indicates to a decrease in protective homeostatic processes. These data demonstrate that excessive excitation of neurons and their irreparable injury is apparently a negative signal for a brain and may induce suicidal behavior. However, in order to commit suicide the suffering is, probably, not enough. These indirect evidences indicate that in the suffering brain of suicide victims a resistance to neuronal damage is attenuated. An organism stops resisting. Reduction of the cyclic AMP system and the $GABA_A$ protective pathway and increased corticosteroid levels evidence a decrease in the will to resistance.

The defensive motivation and negative feeling is, evidently, the most important sense for the protection of an individual life [1206] and reward systems of various motivations are based on a rewarding substance of defensive motivation, opioids (Chapter 3). A motivation arises as the result of a shift of inner constants from their optimal values and is related to transient injury or to the threat of injury. Any misbalance in an organism is unpleasant. When we feel any motivation, this goes up against negative feeling, while deliverance from motivation ascends a relief. Moreover, treatment that induces damage intensifies motivations and increases negative feeling, while treatment that protects neurons usually induces positive feeling and satisfies them. In contrast, treatment that usually increases negative feeling also induces damage and that which decreases negative feeling, protects neurons. Cellular damage intensifies instability, while damage-induced motivation generates trial-and-error search and raises awareness. Satisfaction of motivations recovers distorted equilibrium, protects neurons, decreases instability, and brings relief. This leads to a decline in the level of awareness. Therefore, instability is an exhibition of a tensely running brain. Awareness usually indicates trouble about the future.

Negative information is processed more thoroughly than positive, and the Self is more motivated to avoid a bad self-sensation than to pursue a good one [90]. In response to unknown, not only harmful, impact of the outer environment the brain, for any case, generates a defensive reaction. The biological system is readily activated, responding to even a slight chance of danger and it is oriented toward readiness for action [1206]. One-fold adverse experiences (for instance, handling) in early life can induce neurochemical and morphological changes that may underlie modifications in emotionality and cognition in the adult life: change the density of calcium binding proteins and inhibit acquisition of complex behavior [1252, 465]. Early one-fold favorable experience does not lead to controllable change of brain and mind. Only a protracted enriched environment leaves a positive trace [1040] (broadening the possibilities of the brain), but this is not the trace of a positive sensation.

Brain generates actions in order to receive positive sensation: pleasure or avoidance of damage, while in an indifferent state brain does not undertake any efforts. Attainment of equilibrium gives satisfaction. Hence, it is not inconceivable that action generation itself passes into positive feeling when it has an effect on the environment. When a resultant action becomes pleasure, this may promote an emergence of dominant behavior. Positive feeling appears as the overcoming of impediments, while overprotection leads to loss of consciousness and then promotes damage. Therefore, some brain traces of a nasty experience are possible to describe, while a pleasant experience is correlated, mainly, with inhibition, that is counteraction to excitation.

The ability to feel is built on the fact of mortality and we are able to feel pleasure only because we are "deadly". The biblical sages had expressed this idea more than 3000 years ago and Zarathustra had spoken later: "The good grows through evil, as the purpose grows out of chance". Really, our consideration has demonstrated that the defense one could make out of harm, attainment is created out of chance, will could be created out of homeostatic efforts, pleasure out of fear and prepare memory out of fatigue. The categories of 'good' and 'evil' are not within the boundaries of this scrutiny. This is demonstrated by the ordinary fact that the United Nations cannot agree on a definition the notion of 'terrorism'. We can explain mortality but not morality.

Part II

Mathematics of feeling

6

Introduction to fuzzy logic

6.1 Phenomenology of a neural cell's behavior and fuzzy logic

A neural cell is a very complicated system, where information processing is closely related to life-sustaining processes. It is obvious now that there are a number of the molecular links between intracellular activity that keeps up the cell's being and the process of a spike's generation. Unfortunately, nowadays knowledge about these processes is very fragmentary and it seems unlikely that an adequate description of neuron behavior on molecular level will be formulated in the near future. Some details of neuron activity, like spike's propagation or the mechanism of the ion channels functioning, are well understood, while others, like molecular mechanisms of a neuron's decision-making, neuron's memory-formation and many others are still ambiguous. In such a situation any attempt to develop "microscopic" theory of a neuron's functioning, which will be based on concrete bio-chemical and physical processes in the neural cell, will inevitably contain too many simplifications and hardly verified assumptions. In the best case scenario, some of the assumptions can be justified *a posteriori*, but in most situations they remain unjustified. Moreover, in such a complexly structured system as a neural cell, different (and sometimes opposite) assumptions often lead to the same conclusions, so even *a posteriori*-proof assumptions remain doubtful.

This circumstance, however, open the way to another theoretical approach, which had been proposed by L.Landau [700] and was named *Phenomenological Theory*. More then half century ago L.Landau had been considering phenomenon of phase transitions in crystal and liquid matter. Instead to trying to replace lacking knowledge in microscopic interactions with arbitrary assumptions, he proposed introducing *phenomenological macro-variables*, which could describe macro-behavior of the system regardless of its microscopical meaning and after that he studied dependence of general characteristic of the system on these phenomenological variables by using well-defined experimental data and plausible reasons for analytical and symmetry properties of the variables and the system's characteristics. This approach resulted understanding the behavior of important features of the phenomenon, without precise knowledge about microscopical mechanisms of the phase transitions. Afterward, the phenomenological approach was successfully applied to many complicated problems in physics and in some other fields as well.

In Chapter 8 we will use the phenomenological paradigm for the study of a neural cell's behavior. To do this, we will introduce phenomenological variables such as neural cell's *excitability*, the cell's *damage* or *protection*, the neuron's *expectation of the environmental reaction* and the damage and protection *compensation*, which have definite biological meaning and could explicitly describe a neuron's behavior. Some of these variables, like *excitability* are experimentally measurable. Other ones, like *damage* and *protection* are not measured but correlated with measurable features, like membrane-potential, while *damage and protection compensation* can not be obtained from the experimental data, but is very useful in understanding the implicit trends of a neural cell's behavior.

It should be noted, however, that we cannot directly use the mathematical apparatus, which has been developed in theoretical physics, because description of living systems as well as experimental data in the neuro-physiological experiments considerably differ from corresponding ones in the physical system. In experiments with a living cell, only a few parameters of the systems are observable and controllable, while many others remain "hidden" or out-of-control. This leads to considerable instability in experimental results from trial to trial, so real knowledge, which can be extracted from the experiments, relates to tendencies in the cell's behavior rather than to numerical values of the experimental data. This means that a fair description of the behavior of a complex system like a neural cell is expressed in a natural-language form with "perceptive" notations like *high excitability, weak damage, low expectation of punishment* and so on, which cannot be directly numerically assigned. On the other hand, the variables attributed to perceptions like *high, low, weak, ets.* still have clear linguistic meaning (so they were called *linguistic variables*[1398]). It should be emphasized that this form of description of the cell's behavior, cleverly corresponds to the biological meaning of the neuron's functioning, because neural cell is an elementary unit in processing of perceptions.

So, if we want to develop an adequate apparatus for the phenomenological theory of a neural cell's functioning, in a some sense, we are compelled to search for a mathematical tool, which could directly operate with "perceptions" as with mathematical objects. Fortunately, such mathematical apparatus was proposed by L.Zadeh [1392] and has been further developed during the last decades. It is called *Fuzzy Logic* and *Fuzzy Set Theory.*

In the next two Chapters we will develop phenomenological theory of a neural cell's behavior based on the apparatus of theoretical physics, which will be generalized in order to be compatible with the Fuzzy Logic approach. Note, that this Chapter doesn't contain comprehensive introduction to the

Fuzzy Logic and Fuzzy Sets Theory and only the features necessary to understand the next Chapters will be explained. It should be emphasized that way of explanation points the way to intuitive clarity and useability, rather than to mathematical rigourousness. Readers who want to find more rigorous mathematics and more information about Fuzzy Sets and Fuzzy Logic can refer to some good books published during the last decades in particular to [568, 1372, 900, 156, 830]. The original papers of L.Zadeh [1392]-[1416] are highly recommended as well.

6.2 Perceptions as a Mathematical Object

6.2.1 Possibility and Fuzzy Set

Commonly, a mathematical description of a system's behavior requires representation of the system's state as a "point" in the abstract space of the system's variables. In our case however, such representation is ineffective, because imprecision of our knowledge makes closely spaced points hardly distinguishable, or even completely undistinguishable. Actually, an expression like *low excitability and low protection* corresponds to some domain in the system's state space rather than to precise points (Fig. 6.1). Moreover, such

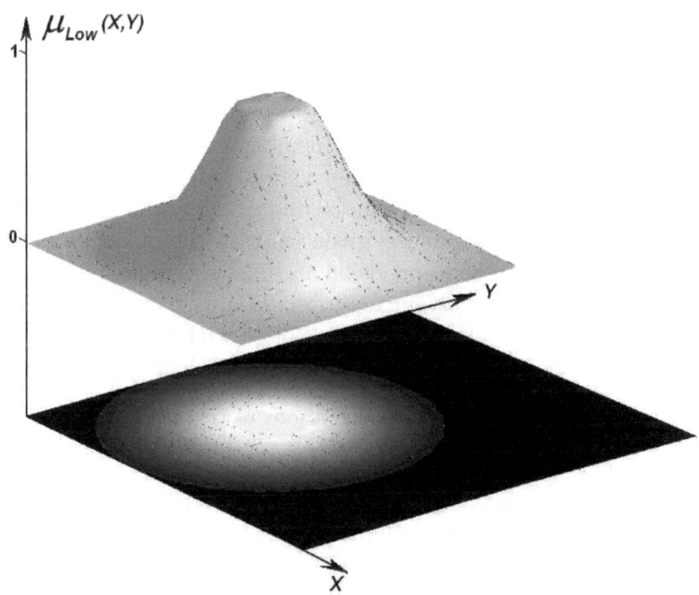

Fig. 6.1. Down plane - "Fuzzy" domains in the system's state space, which represent the perceptively described variables like "low X" and "low Y". Upper part - relative membership of the each point to belong to the domain "low X and low Y".

a domain should be "cloud-like" and cannot have a crisp boundary, because in this case close points inside and outside of the border will be well distinguishable, which isn't so our case. In order to attach a mathematical meaning to such "cloud-like domains", L.Zadeh has proposed to generalize the mathematical notation of the crisp *sets* and has introduced a notation of *fuzzy sets* [1392].

The proposed generalization can be explained as follows. Let us assume that our system is described by the some parameters $\{x_1, x_2, ..., x_n\}$ such that each state of the system is represented by a point $\boldsymbol{x} = \{x_1, ..., x_n\}$ in the abstract space $\mathcal{X}$ (we have used vector notation (bold $\boldsymbol{x}$) for simplification of the subsequent formulas). An ordinary, "crisp" domain or set[1] $\mathcal{D}$ of the space $\mathcal{X}$ can be defined by its *Identity function* $Id_D(\boldsymbol{x})$:

$$Id_D(\boldsymbol{x}) = \begin{cases} 1 & \text{if } \boldsymbol{x} \in \mathcal{D} \\ 0 & \text{otherwise} \end{cases} \tag{6.1}$$

Assume, now, that our knowledge about the system's states is imprecise, so for a point near the border of $\mathcal{D}$ we cannot say with certainty whether it belongs or does not belong to the domain. In this case instead of *Identity function* $Id_D(\boldsymbol{x})$ we could introduce some *Membership function* - $\mu_D(\boldsymbol{x})$ $(0 \leq \mu_D(\boldsymbol{x}) \leq 1)^2$, which indicates our estimation of *possibility* that the given point $\boldsymbol{x}$ belongs to the set or domain $\mathcal{D}$ (see Fig. 6.1). If $\mu_D(\boldsymbol{x})$ is a continuous function, which tends to zero out of some region, the nearest points of the space are still being hardly distinguishable, while the distant ones are well distinguishable, which fits our intuitive imagine about "cloud-like" sets and domains.

On the other hand, since the *fuzzy set* can be considered as a mathematical representation of our perception of a system's state, we could, in turn, represent the linguistic variables by the membership functions, which reflect uncertainty of our knowledge about the system's characteristics. For example, the linguistic variable "*Low X*" can be described by the function $\mu_{Low}(x)$, which estimates *possibility*[3] that the given value of the quantity x is low.

This idea allows to handle imprecision and incompleteness by "equipping" a system's parameter with certain functions of this parameter and operating with the pairs $\{x, \mu_A(x)\}$ rather than solely with the parameter x. It is not a completely new situation. Considering the stochastic processes, we also

---

[1] In fact, there is a difference between definitions of the *domain* and of the *set* in the abstract space, but for our purposes this difference is irrelevant, so we will use these words as synonyms.

[2] Generally speaking, membership function is not necessary restricted by the interval $[0, 1]$ and even may be not a real number [854, 543], but in our book we will use the above-mentioned restriction only.

[3] It should be noted that there several semantical interpretations of the membership function: it can be interpreted as a *grade of belonging* of X to the fuzzy set *Low*, or as *truth value* of the expression "*X is low*", or as the possibility that "*X is low*" and, generally speaking, there are delicate differences between these semantics (see a comprehensive analysis of this problem in [346]).

use probability distributions of the parameters' values instead their values by themselves. In our situation however, imprecision doesn't have stochastic nature, so the membership functions should not possess properties of the probability distribution, and we must define operations with the membership functions that correspond to the logical and mathematical operations to the linguistic variables.

For example, we should define how to calculate the membership functions for a *composite linguistic variable* like "X is Low OR X is Middle" - $\mu_{L \vee M}(x)$ if we know $\mu_L(x)$ and $\mu_M(x)$. So we must define mathematical representations for the logical connectives *OR, AND, NOT* and so on.

Another question, which arises, is: let us x and y are fuzzy numerical quantities. That means that we know only possibilities that "X is Large" and that "Y is Small" - $\mu_L(x)$ and $\mu_S(y)$. What can we say about $x+y$, $x \cdot y$, $\sqrt{x}$ and so on?

It was assumed above that $\mu_A(x)$ is a function from a point-wise argument. There are, however, more sophisticated situations. For example, how do we find the possibility that "*MOST* of X from the interval I_x are *Large*"? Obviously, that in this case, the whole interval $x \in I_x$ should be considered as an argument of the function $\mu_{MostLarge}[I_x]$ and we must define how to calculate the interval membership functions. Below, we will discuss these problems.

6.2.2 Logical Connectives and Triangular Norms

Defining representations of the logical connectives in the fuzzy case, we want them to be compatible with common human logic. So, they should satisfy certain conditions. In particular, the possibility that "X is Low *OR* Y is Moderate" and the possibility that "Y is Moderate *OR* X is Low" should be equal (symmetry of the *OR*-representation). On the other hand, if Y is certainly Moderate ($\mu_M(y) = 1$) then the possibility that "X is Low *OR* Y is Moderate" should be equal *1*, while if Y is certainly not Moderate ($\mu_M(y) = 0$) then possibility that "X is Low *OR* Y is Moderate" should be equal to the possibility that "X is Low", e.i. to $\mu_L(x)$. Let us represent the possibility that "X is L *OR* Y is M" as $S\{\mu_L(x), \mu_M(y)\}$. Then in accordance with the above-mentioned, it should satisfy to:

$$S\{\mu_1, \mu_2\} = S\{\mu_2, \mu_1\}, \tag{6.2}$$

$$S\{\mu, 0\} = \mu, ; \quad S\{\mu, 1\} = 1, \tag{6.3}$$

$$\text{If } \mu_2 \leq \mu_3 \text{ then } S\{\mu_1, \mu_2\} \leq S\{\mu_1, \mu_3\}, \tag{6.4}$$

$$S\{S\{\mu_1, \mu_2\}, \mu_3\} = S\{\mu_1, S\{\mu_2, \mu_3\}\} = S\{\mu_1, \mu_2, \mu_3\}. \tag{6.5}$$

Conditions (6.4) and (6.5) reflect monotonicity and associativity of the *OR*-representation.

Correspondingly, if we denote representation of the possibility that"X is L *AND* Y is M" as $T[\mu_L(x); \mu_M(y)]$ we should have:

$$T[\mu;0] = T[0;\mu] = 0,; \quad T[1;\mu] = T[\mu;1] = \mu, \tag{6.6}$$
$$\text{If } \mu_2 \le \mu_3 \text{ then } T[\mu_1;\mu_2] \le T[\mu_1;\mu_3], \tag{6.7}$$
$$T[T[\mu_1;\mu_2];\mu_3] = T[\mu_1;T[\mu_2;\mu_3]] = T[\mu_1;\mu_2;\mu_3]. \tag{6.8}$$

Note that unlike $S\{\mu_1,\mu_2\}$, we do not require that $T[\mu_1;\mu_2]$ will be necessarily symmetrical for all μ_1,μ_2, because sometimes the possibility of the event "A *AND* B" could be different from the possibility of "B *AND* A". For example, "Rotation on Angle α *AND* Rotation on Angle β" leads to a different state than "Rotation on Angle β *AND* Rotation on Angle α", if β and α are rotations around the different axis.

Mathematical operations, which are satisfied to (6.2)-(6.5), are well known and were intensively studied during the last decades (see [191, 830] and references there). In mathematics literature they were called "*triangular conorm*" (t-conorm). Correspondingly, the operations are satisfied to (6.6)-(6.8) with symmetrical $T[\mu_1;\mu_2] = T[\mu_2;\mu_1]$, were called "*triangular norm*" (t-norm). In the Fuzzy Sets Theory t-norms and conorms define "intersection" and "union" of the fuzzy sets, which is quite natural, because condition "x belongs to S_1 *AND* x belongs to S_2" ($x \in S_1 \bigwedge x \in S_2$ for short) defines the intersection - $S_1 \bigcap S_2$, while "x belongs to S_1 *OR* x belongs to S_2 ($x \in S_1 \bigvee x \in S_2$ for short)" defines the union - $S_1 \bigcup S_2$.

Many examples and applications of the t-norms and conorms can be found in [855] and [830]. For our consideration two cases are most important:

Max-Min:

$$T_{min}[\mu_1;\mu_2] = \min(\mu_1;\mu_2) \tag{6.9}$$
$$S_{max}\{\mu_1,\mu_2\} = \max(\mu_1,\mu_2), \tag{6.10}$$

and *Pseudo-arithmetic Sum* [829, 937], which covers a wide class of the t-norm:

$$T[\mu_1;\mu_2] = g^{-1}[g(\mu_1) + g(\mu_2)], \tag{6.11}$$
$$g(0) = \infty, \quad g(1) = 0,$$

where for $0 < x \le 1$, $g(x)$ is a continuous monotonic function.

Note, that $\min(\mu_1;\mu_2)$ is the strongest t-norm, so for any t-norm $T[\mu_1;\mu_2] \le \min(\mu_1;\mu_2)$. Indeed,

$$T[\mu_1;\mu_2] \le T[1;\mu_2] = \mu_2,$$
$$T[\mu_1;\mu_2] \le T[\mu_1;1] = \mu_1,$$

therefor

$$T[\mu_1;\mu_2] \le \min(\mu_1;\mu_2). \tag{6.12}$$

So, an arbitrarily t-norm is looking as in the Fig. 6.2. Another useful relation, which follows from monotonicity of the t-norm, is:

$$T[\mu_1;\max\{\mu_2,\mu_3\}] = \max\{T[\mu_1;\mu_2],T[\mu_1;\mu_3]\}. \tag{6.13}$$

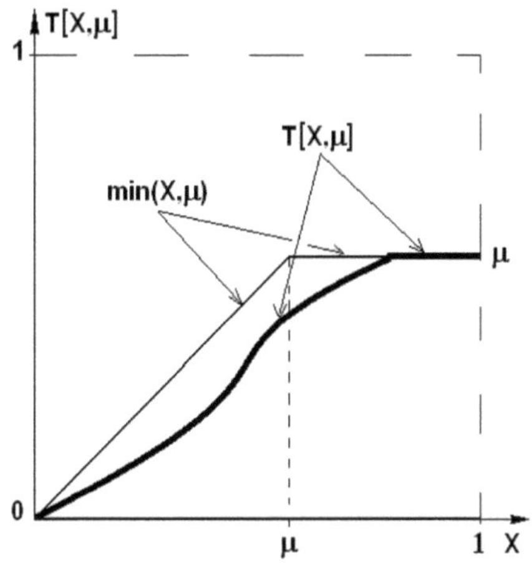

Fig. 6.2. T-norm as the function of the one of its arguments.

Of course, there are infinite number of the different t-norms and conorms, but some additional conditions on them could lead to unique representation of $S\{\mu_1, \mu_2\}$ or $T[\mu_1; \mu_2]$. For example, let us assume that $S\{\mu, \mu\} = \mu$. We can write for $0 < \mu_2 < \mu_1 < 1$:

$$\mu_1 = S\{\mu_1, 0\} \leq S\{\mu_1, \mu_2\} \leq S\{\mu_1, \mu_1\} = \mu_1,$$

so,

$$S\{\mu_1, \mu_2\} = \mu_1 \text{ for } \mu_1 > \mu_2,$$

In the opposite case - $0 < \mu_1 < \mu_2 < 1$, the same arguments lead to:

$$S\{\mu_1, \mu_2\} = \mu_2 \text{ for } \mu_2 > \mu_1.$$

Therefore, we conclude, that only

$$S\{\mu_1, \mu_2\} = \max\{\mu_1\ \mu_2\}. \tag{6.14}$$

is the suitable expression for $S\{\mu_1, \mu_2\}$ in the case $S\{\mu, \mu\} = \mu$ (this result will be used in the next Chapter).

"Negation" - "*NOT* A" ($\bar{A}$) can be represented by any monotonic decreasing function $N(\mu)$ with $N(0) = 1$, $N(1) = 0$, i.e.:

$$\mu_{\bar{A}} = N(\mu_A). \tag{6.15}$$

Negation, which satisfies $N(N(x)) \equiv x$ is called an *involution*. An obvious and mostly useable example of the negation is

$$N(\mu) = 1 - N(\mu), \tag{6.16}$$

Negation (6.16) is obviously involution, but there are many others as well. For example, involuting negation, suggested by Sugeno, is:

$$N(\mu) = \frac{1-\mu}{1+a\mu}, \quad (a > -1).$$

6.2.3 *Consistent t-norms

There is an important class of t-norm: "t-norm qualitatively consistent with Φ-transformation" (*Q-consistent* for short) , where $\Phi(x)$ is a monotonic function, such that:

$$\Phi(0) = 0, \quad \Phi(1) = 1. \tag{6.17}$$

The norm $T[x, y]$ is called *Q-consistent*, if

$$\Phi(T[x, y]) = T[\Phi(x), \Phi(y)]. \tag{6.18}$$

The practical importance of Q-consistence follows from its logical meaning. Let us consider two experts assigned possibilities of the events A,B and A *AND* B as:

First expert: Possibility of A is μ_A, Possibility of B is μ_B and Possibility of A *AND* B is $\mu_{A\wedge B} = T[\mu_A, \mu_B]$;
Second expert: Possibility of A is μ'_A, Possibility of B is μ'_B and Possibility of A *AND* B is $\mu'_{A\wedge B} = T'[\mu'_A, \mu'_B]$.

Two experts will be qualitatively consistent if their assignments will be differed by a monotonic increasing function, i.e. by a Φ-transformation:

$$\begin{aligned} \mu'_A &= \Phi(\mu_A) \\ \mu'_B &= \Phi(\mu_B) \\ T'[\mu'_A, \mu'_B] &= \Phi(T[\mu_A, \mu_B]). \end{aligned}$$

If we want to use the same representation of the logical connectives for both experts, we should have:

$$T'[\mu'_A, \mu'_B] = T[\mu'_A, \mu'_B],$$

which imply that

$$\Phi(T[\mu_A, \mu_B]) = T[\Phi(\mu_A), \Phi(\mu_B)].$$

So $T[\mu_A, \mu_B]$ should be *Q-consistent.* The similar condition, of cause, must be held for the conorm, which represent the connective *OR*:

$$\Phi(S\{\mu_A, \mu_B\}) = S\{\Phi(\mu_A), \Phi(\mu_B)\}. \tag{6.19}$$

If (6.18) or (6.19) is held for any monotonic function Φ with $\Phi(0) = 0$, $\Phi(1) = 1$, the norm is called *self-consistent* . An obvious example of the self-consistent t-norm and conorm are *Min* and *Max* norms:

$$\Phi(\min(x,y)) = \min(\Phi(x),\Phi(y))$$
$$\Phi(\max(x,y)) = \max(\Phi(x),\Phi(y)).$$

Note, that if the t-norm is explicitly defined, Eq. (6.18) can be considered as a functional equation for the function Φ. As an example, consider the *pseudo-arithmetic sum* (6.11). In according with (6.18) one has:

$$\Phi(g^{-1}[g(\mu_1)+g(\mu_2)]) = g^{-1}[g(\Phi(\mu_1))+g(\Phi(\mu_2))]. \tag{6.20}$$

Altering both sides of Eq. (6.20) by the function g and denoting:

$$\mathcal{P}(z) = g(\Phi(g^{-1}(z))), \tag{6.21}$$

and, therefor,

$$g(\Phi(\mu)) = \mathcal{P}(g(\mu)),$$

one obtains:

$$\mathcal{P}(g(\mu_1)+g(\mu_2)) = \mathcal{P}(g(\mu_1)) + \mathcal{P}(g(\mu_2)), \tag{6.22}$$

which means that $P(z)$ is a linear function:

$$\mathcal{P}(z) = mz + q, \tag{6.23}$$

By using (6.21), restrictions (6.17) and condition $g(1) = 0$, we obtain $P(0) = 0$ that leads to $q = 0$. Therefor, the pseudo-arithmetic sum (6.11) is *Q-consistent* with:

$$\Phi(\mu) = g^{-1}(mg(\mu)). \tag{6.24}$$

In particular, for $g(\mu) = \ln(1/\mu)$ we have

$$T[\mu_1;\mu_2] = \mu_1 \cdot \mu_2,$$

and

$$\Phi(\mu) = \mu^m.$$

Practically, however, an opposite, more important question arises: Let us assume that we know class of the transformations $\Phi(\mu)$. What are the t-norms that will be *Q-consistent* with this class? It can be shown, that for Φ from (6.24) t-norm (6.11) is *Q-consistent*. For the others concrete Φ, the question remains open.

6.3 Mathematical Operations with Fuzzy Quantities and Zadeh's Extensional Principle

If the linguistic variables belong to numerically valued quantities, we should define mathematical operations with these variables. In accordance with the fuzzy logic paradigm, this means that we should define mathematical operations with the membership functions, which correspond to the mathematical operations with the imprecise numerical data.

6.3.1 Extensional Principle of L.Zadeh

We will begin with a simple case and will then make a generalization to the more complex situations. Assume that we deal with two numerical parameters x and y, which are described by two linguistic variables "X is approximately L" and "Y is approximately M", i.e. the membership functions $\mu_L(x)$ and $\mu_M(y)$ are known. What is the possibility that $x-y$ is z? It is obvious that $x-y=z$ for a fixed $z=z_0$ if:

$$\begin{array}{l} X = x' \;\; AND \;\; Y = z_0 - x' \\ OR \\ X = x'' \;\; AND \;\; Y = z_0 - x'' \\ OR \\ \text{... and so on, for the all possible values of } X \end{array}$$

Let us, now, assume that connective OR is represented by the Max-conorm, while the connective AND is an arbitrary t-norm (this representation will be widely used in the next chapters). Then the possibility for $x-y$ to be z_0 can be calculated as:

$$\mu_{L-M}(z_0) = \max\{T[\mu_L(x'); \mu_M(z_0 - x')], T[\mu_L(x''); \mu_M(z_0 - x'')], \\ \text{... and so on for all possible vales of } x\}. \tag{6.25}$$

This expression can be rewritten in the more compact form:

$$\mu_{L-M}(z) = \sup_{y-x=z} T[\mu_L(x); \mu_M(y)], \tag{6.26}$$

(sup designates *supremum*) which means that we should take maximal value of the quantities

$$T[\mu_L(x'); \mu_M(y')], T[\mu_L(x''); \mu_M(y'')], \text{ ... and so on} \\ \text{for all possible vales of } x \text{ and } y \text{ such that } y\text{-}x\text{=}z.$$

A definition such as (6.26) has been suggested by L.Zadeh and is named as *Zadeh's Extensional Principle*. Expression (6.26) is easily generalized to other operations with two variables. For example, for $x+y$ or $x \cdot y$ we should, simply, replace $y - x = z$ to $x + y = z$ or $x \cdot y = z$. In fact, for any mathematical operation with x and y: $z = x \circ y$, we can write:

$$\mu_{L \circ M}(z) = \sup_{x \circ y = z} T[\mu_L(x); \mu_M(y)], \tag{6.27}$$

(6.26) can be generalized to the case of more than two variables as well:

$$\mu_{A \circ B \circ \ldots \circ C}(z) = \sup_{x_1 \circ x_2 \ldots \circ x_n = z} T[\mu_A(x_1); \mu_B(x_2); \ldots; \mu_C(x_n)], \tag{6.28}$$

Further consideration of the operations with the fuzzy data can be found in [84].

6.3.2 Fuzzy functions

In the next chapters we will deal with functional dependence between linguistic variables. In traditional mathematics the functional dependence is a mapping of the space X to the space Y - $\{F : X \to Y\}$, such that to each point x from X corresponds to some point y form Y. Practically, y is obtained by some mathematical operation, like $y = x^2$, $y = \ln(x)$, $y = \sum_k^\infty A_k x^k$ and so on. However, if our knowledge about a problem is incomplete, we cannot exactly define the functional dependence $y = F(x)$, but often we may reckon that this dependence is close to some concrete function, say, "$F(x)$ is approximately αx^2" and one estimates the possibility that our guess is true. For example, it could be (see Fig. 6.3):

$$\mu_F(y|x) = \exp - \left(\frac{y - \alpha x^2)}{\sigma}\right)^2, \tag{6.29}$$

which indicates the possibility that for certain x, $F(x)$ is equal to the certain y.

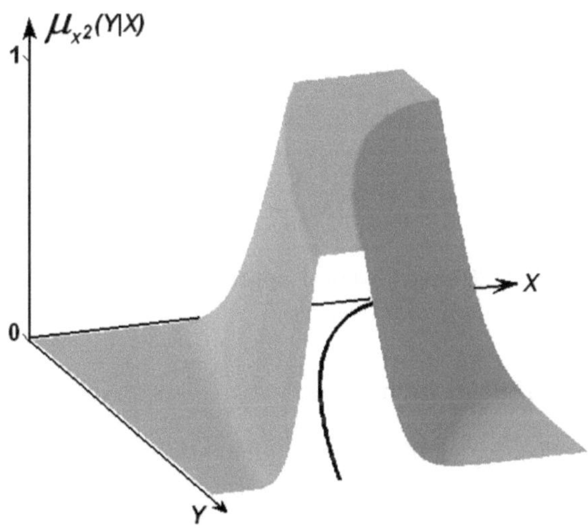

Fig. 6.3. Example of the fuzzy function "$F(x)$ is approximately αx^2".

In (6.29) it is assumed that x is an ordinary numerical variable. It may be however, that x is the linguistic variable, say "X is Large" - $\mu_L(x)$. In this case (6.29) should be extended in accordance with the Extensional Principle:

$$\mu_F(y|L) = \sup_x T[\mu_F(y|x); \mu_L(x)]. \tag{6.30}$$

Fuzzy functions are closely related with somewhat more general membership functions of the linguistic variable, for example as "Most X from the interval $I = [x_l, x_u]$ are Large"[4]. Let us assume that $\mu_I(x)$ is a membership function, which indicates the possibility that x belongs to the interval I and $\mu_L(x)$ is the possibility that "X is Large". Then, the possibility that "X belongs to I *AND* X is Large" will be:

$$T[\mu_I(x); \mu_L(x)]. \tag{6.31}$$

The possibility that "Most X from the interval I are Large" can be defined as average value of (6.31):

$$\mu_{Most-Large}(I) = \frac{\int_{x_l}^{x_u} T[\mu_I(x); \mu_L(x)]dx}{\int_{x_l}^{x_u} \mu_I(x)dx}, \tag{6.32}$$

There is, however, a more elegant definition of the perception "*Most*", which is close to that suggested by L.Zadeh in [1413]. In accordance with Sec. 6.2.1 we can consider the linguistic variables "*Large*" and "*Interval*" as the fuzzy sets, so $T[\mu_I(x); \mu_L(x)]$ corresponds to the intersection between these sets. It is obvious, that if we randomly take, the several points $\{x_1, x_2, ..., x_N\}$ from the interval and if this intersection covers most of the interval I, most of these points will belong to the intersection, so we can well approximate $\mu_{Most-Large}(I)$ as:

$$\mu_{Most-Large}(I) \simeq \frac{1}{N} \sum_{i=1}^{N} T[\mu_I(x_i); \mu_L(x_i)]. \tag{6.33}$$

Note, that $N = 10 \div 20$ gives a good approximation of $\mu_{Most-Large}(I)$, so Eq. (6.33) is convenient for practical use.

6.3.3 Fuzzy differential inclusions

In the dynamics theory, solution of the differential equations plays a central role:

$$\frac{d\boldsymbol{x}}{dt} = \boldsymbol{V}(\boldsymbol{x}, t), \tag{6.34}$$

where t is the time and $\boldsymbol{V}(\boldsymbol{x}, t)$ has meaning of velocity of changing of the variables $\boldsymbol{x}(t)$. In many real situations, however, exact form of the function $\boldsymbol{V}(\boldsymbol{x}, t)$ is unknown and we can say only that the system's velocity belongs to some fuzzy set, i.e. we know only fuzzy function $P(\boldsymbol{V}|\boldsymbol{x}, t))$, which defines the possibility that at the time t and near the point $\boldsymbol{x}$ the system has velocity $\boldsymbol{V}$ (see Fig.6.4). In this situation the function $\boldsymbol{V}(\boldsymbol{x}, t)$ in (6.34) is not certainty

[4] $[x_l, x_u]$ can be an ordinary *crisp* interval, i.e. $x_l \leq X \leq x_u$, or a fuzzy interval, so we know only the possibility that the concrete x belongs to $[x_l, x_u]$

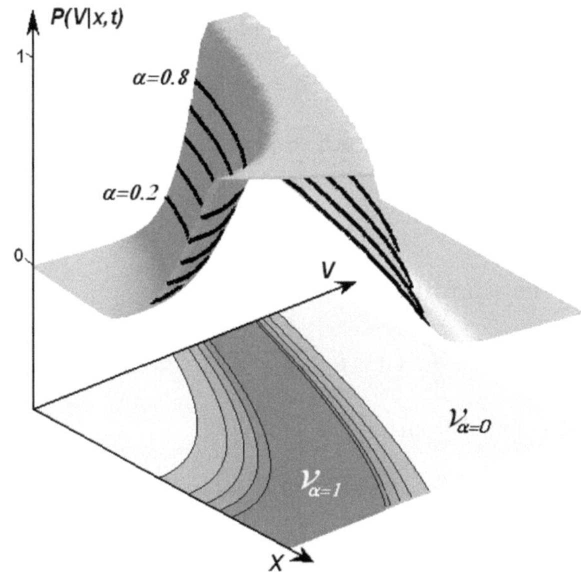

Fig. 6.4. Membership function of the fuzzy velocity $P(\boldsymbol{V}|\boldsymbol{x},t))$ and its α-cuts.

defined, but belongs to the fuzzy set $P(\boldsymbol{V}|\boldsymbol{x},t)$ in the velocity space. Let us fix some value of P, say $P = \alpha$, and one finds $\mathcal{V}_\alpha(\boldsymbol{x},t)$ as a solution of an inequality:

$$P(\mathcal{V}_\alpha|\boldsymbol{x},t) \geq \alpha. \tag{6.35}$$

Solution (6.35) gives us some crisp domain of the velocity space[5]. Then we can say that for all $\boldsymbol{V}_\alpha \in \mathcal{V}_\alpha(\boldsymbol{x},t)$, the trajectories $\boldsymbol{x}(t\alpha,)$, which are satisfied to:

$$\begin{aligned} \frac{d\boldsymbol{x}}{dt} &= \boldsymbol{V}_\alpha, \\ \boldsymbol{V}_\alpha &\in \mathcal{V}_\alpha(\boldsymbol{x},t), \end{aligned} \tag{6.36}$$

are possible with the possibility more or equal to α. Since $\boldsymbol{V}_\alpha$ is any point of the set $\mathcal{V}_\alpha(\boldsymbol{x},t)$, Eq. (6.36) is called fuzzy *differential inclusion*[6], instead differential equation, like (6.34). Note, that in the literature (6.36) is often written in the form:

$$\frac{d\boldsymbol{x}}{dt} \in \mathcal{V}_\alpha(\boldsymbol{x},t). \tag{6.37}$$

Solution of the fuzzy differential inclusion can be interpreted in terms of time-dependent α-cuts of an evolving fuzzy set. In order to describe it in an explicit

[5] In the literature $\mathcal{V}_\alpha$ is called α-cut of P.

[6] See [67], [570], [322]-[323] for details and additional information.

form, denote as $\mu(\boldsymbol{x},t)$ the possibility that a system is in the point $\boldsymbol{x}$ at the time t and one considers the points correspond to the constant value of $\mu(\boldsymbol{x},t)$

$$\mu(\boldsymbol{X}(t),t) = \alpha = constant. \tag{6.38}$$

Giving total derivative of (6.38) with respect to t one obtains:

$$\frac{d\mu(\boldsymbol{X}(t),t)}{dt} = \frac{\partial\mu}{\partial t} + \left(\frac{d\boldsymbol{X}}{dt}\cdot\nabla\mu\right) = \frac{d\alpha}{dt} = 0. \tag{6.39}$$

By using (6.36) one can write

$$\frac{\partial\mu}{\partial t} + (\mathcal{V}(\mu,\boldsymbol{x},t)\cdot\nabla\mu) = 0, \tag{6.40}$$

where $\mathcal{V}(\mu,\boldsymbol{x},t)$ is found from

$$P(\mathcal{V}|\boldsymbol{x},t) = \mu. \tag{6.41}$$

In order for the Eq. (6.40) to have an unique solution, we should add an initial value $\mu(\boldsymbol{x},0) = \mu_0(\boldsymbol{x})$. Partial fuzzy differential inclusion (6.40)-(6.41) describes evolution of a fuzzy set under a fuzzy flow, whose velocity is defined by $P(\boldsymbol{V}|\boldsymbol{x},t)$. Note, that Eqs. (6.40)-(6.41) are *self-consistent* and invariant under any Φ-transformation (see Sec. 6.2.3). Fuzzy differential inclusions in the form 6.40)-(6.41) will be widely used in the next chapters.

6.3.4 *Fuzzy integral

Integral[7] of an ordinary function $f(x)$ can be defined as:

$$\int_a^b f(x)dx = \lim_{n\to\infty}\sum_{k=1}^{n} f(a+k\varepsilon)\varepsilon, \tag{6.42}$$
$$\varepsilon = \frac{b-a}{n}.$$

We define *fuzzy integral* - $\mu_f(I|a,b)$ as the possibility that integral $\int_a^b f dx$ has a certain value I, if the fuzzy function is defined by the $\chi(f|x)$. In order to calculate $\mu_f(I|a,b)$ we can use Eq. (6.28) and one writes:

$$\mu_f(I|a,b) = \lim_{n\to\infty}\sup_{\Sigma_k^n f_k\varepsilon = I} T[\chi(f_1|a);\chi(f_2|a+\varepsilon);\ldots$$
$$\ldots;\chi(f_{n-1}|b-\varepsilon);\chi(f_n|b)], \tag{6.43}$$

which means that the possibility that integral $\int_a^b f dx$ is equal to I is the possibility that:

[7] So-called Riemann integral

$f(a) = f_1$ *AND* $f(a+\varepsilon) = f_2$ *AND* $f(a+2\varepsilon) = f_3$ *AND*...
AND $f(b-\varepsilon) = f_{n-1}$ *AND* $f(b) = f_n$ - such that $\sum f_i \varepsilon = I$
OR
$f(a) = f'_1$ *AND* $f(a+\varepsilon) = f'_2$ *AND* $f(a+2\varepsilon) = f'_3$ *AND*...
AND $f(b-\varepsilon) = f'_{n-1}$ *AND* $f(b) = f'_n$ - such that $\sum f'_i \varepsilon = I$
OR
... and so on, for the all possible values of f

Since for any f we have $\int_a^a f dx \equiv 0$, we obtain that:

$$\mu_f(I|a, a) = \delta\{I, 0\}, \tag{6.44}$$

where

$$\delta\{i, j\} = \begin{cases} 1 \text{ if } i = j \\ 0 \text{ otherwise} \end{cases}$$

is the Kroneker symbol[8]. Note, that if t-norm in (6.43) is *Q-consistent* the fuzzy integral is *Q-consistent* as well.

The fuzzy integral (6.43) is the fuzzy analog of the well-known Weiner path integral in the stochastic theory [1330] and it has many similar properties. In particular, the fuzzy integral can be calculated by solving the associated differential equation.

To show this, let us rewrite (6.43) by using (6.8) and (6.13):

$$\mu_f(I|a, x) \approx \sup_{f_n} \sup_{\Sigma_i^{n-1} f_i \varepsilon = I - \varepsilon f_n} T[T[\chi(f_1|a); \chi(f_2|a+\varepsilon); ...$$

$$...; \chi(f_{n-1}|x-\varepsilon)]; \chi(f_n|x)] = \sup_f T[\sup_{\Sigma f_i \varepsilon = I - \varepsilon f} T[\chi(f_1|a); \chi(f_2|a+\varepsilon); ...$$

$$...; \chi(f_{n-1}|x-\varepsilon)]; \chi(f|x)] \approx \sup_f T[\mu_f(I - \varepsilon f|a, x-\varepsilon); \chi(f|x)]. \tag{6.45}$$

It is obvious from Fig. 6.2 (see proof in [418]) that the supremum of the right side of Eq. (6.45) is:

$$\sup_f T[\mu_f(I - \varepsilon f|a, x-\varepsilon); \chi(f|x)] = \mu_f(I - \varepsilon f^s|a, x-\varepsilon), \tag{6.46}$$

where the point $f^s(I, x)$ satisfies:

$$T[\mu_f(I - \varepsilon f^s|a, x-\varepsilon); \chi(f^s|x)] = \mu_f(I - \varepsilon f^s|a, x-\varepsilon). \tag{6.47}$$

Therefore Eq. (6.45) can be rewritten as:

$$\mu_f(I|a, x) \approx \mu_f(I - \varepsilon f^s|a, x-\varepsilon). \tag{6.48}$$

Taking the limit $\varepsilon \to 0$ and keeping in Eqs. (6.45)-(6.47) the lowest non-trivial terms, we can see that the fuzzy integral (6.45) can be obtained by solution of the partial differential equation[9]::

---

[8] In the literature, the Kroneker symbol often is designated as δ_i^j

[9] or partial differential inclusion, if solution of (6.50) is a set-valued function

$$\frac{\partial \mu_f}{\partial x} + f^s(I,x)\frac{\partial \mu_f}{\partial I} = 0, \tag{6.49}$$

where $f^s(I,x)$ should be found from:

$$T[\mu_f(I|a,x);\chi(f^s|x)] = \mu_f(I|a,x). \tag{6.50}$$

If Eq. (6.50) has several solutions, we must take the solution, where:

$$-f^s\frac{\partial \mu_f}{\partial I} \Rightarrow \max, \tag{6.51}$$

that corresponds to the supremum of the (6.46) (see page 341). An initial condition to (6.45) - $\mu_f(I|a,a) = \delta_0^I$ is taken from (6.44).

Consider an example, where $\chi(f|x)$ has the form:

$$\chi(f|x) = \Psi\left(\frac{f-\xi(x)}{\rho(x)}\right), \tag{6.52}$$

where Ψ has a single maximum, $\Psi(0) = 1$ monotonically decreases at both sides of the maximum and $\Psi(\pm\infty) = 0$. In this case the equations (6.49)-(6.50) can be solved analytically.

It is convenient to seek a solution in the form:

$$\mu_f(I|a,x) = \Psi(\Phi(I,x)). \tag{6.53}$$

Consider, first, the *Min* t-norm: $T[\mu;\chi] = \min(\mu;\chi)$. Then (6.50) gives:

$$f^s = \rho\Phi + \xi, \tag{6.54}$$

while (6.49) leads to

$$\frac{\partial \Phi}{\partial x} + [\rho(x)\Phi + \xi(x)]\frac{\partial \Phi}{\partial I} = 0, \tag{6.55}$$

with the initial condition:

$$\Phi(I,a) = \begin{cases} 0 & \text{if } I = 0 \\ \infty & \text{otherwise} \end{cases}, \tag{6.56}$$

which can be represented in the form

$$\Phi(I,a) = \lim_{\sigma\to 0}\Phi_\sigma(I,x=a) = \lim_{\sigma\to 0}\frac{I}{\sigma}, \tag{6.57}$$

where limit $I \to 0$ should be taken before limit $\sigma \to 0$.

We will seek the solution of (6.55) in the form:

$$\Phi_\sigma(I,x) = u(x)I + v(x) + \phi(I,x) \tag{6.58}$$

where $u(x), v(x)$ and $\phi(I,x)$ are unknown functions. Substituting (6.58) for (6.55) one obtains the system of ordinary differential equations:

$$\frac{du}{dx} + \rho u^2 = 0, \tag{6.59}$$

$$\frac{dv}{dx} + \rho uv + \xi u = 0, \tag{6.60}$$

and the equation:

$$\frac{\partial \phi}{\partial x} + (\rho u I + \rho v + \rho \phi + \xi) \frac{\partial \phi}{\partial I} + \rho u \phi = 0, \tag{6.61}$$

with the initial conditions:

$$u(a) = \frac{1}{\sigma}, \quad v(a) = 0, \quad \phi(I, a) = 0.$$

The system (6.59)-(6.61) has the solution:

$$\begin{aligned} \phi(I, x) &\equiv 0, \\ u(x) &= \left(\sigma + \int_a^x \rho(x) dx \right)^{-1}, \\ v(x) &= -u(x) \int_a^x \xi(x) dx. \end{aligned}$$

By using (6.58) and (6.53), we obtain finally:

$$\mu_f(I|a, x) = \lim_{\sigma \to 0} \Psi \left(\frac{I - \int_a^x \xi dx}{\sigma + \int_a^x \rho dx} \right) = \Psi \left(\frac{I - \int_a^x \xi dx}{\int_a^x \rho dx} \right). \tag{6.62}$$

This result looks quite reasonable: most possible value of the fuzzy integral is equal to integral of the most possible value of the fuzzy function:

$$\mu_f(I|a, x) \Big|_{I = \int_a^x \xi dx} = 1, \tag{6.63}$$

while uncertainty of the fuzzy integrating - $\int_a^x \rho dx$ increases with increase of the interval $[a, x]$.

Let us however, take:

$$T[\mu; \chi] = g^{-1}[g(\mu) + g(\chi)] \tag{6.64}$$

(see (6.11)). In this case Eq. (6.50) leads to:

$$\begin{aligned} g^{-1}[g(\mu) + g(\chi(f^s|x))] &= \mu, \\ g(\chi(f^s|x))] &= 0, \\ \chi(f^s|x)) &= 1, \end{aligned}$$

therefore, in accordance with (6.52) we have:

$$f^s(I,x) = \xi(x). \tag{6.65}$$

Solution of (6.49) can be obtained immediately by observing that (6.65) corresponds to (6.54) for $\rho \equiv 0$. So, by putting in (6.62) $\rho \equiv 0$, one obtains:

$$\mu_f(I|a,x) = \lim_{\sigma \to 0} \Psi\left(\frac{I - \int_a^x \xi dx}{\sigma}\right) = \delta\left\{I, \int_a^x \xi dx\right\}. \tag{6.66}$$

The result (6.66) seems somewhat strange, because it means that integration of the imprecise fuzzy function gives a precise number. Explanation of this fact however is quite simple.

For t-norm from (6.64) we can rewrite (6.43) by using (6.8) and (6.13) in the form:

$$\mu_f(I|a,b) = \lim_{n\to\infty} \sup_{\Sigma_k^n f_k \varepsilon = I} T[\chi(f_1|a); \chi(f_2|a+\varepsilon); ...; \chi(f_{n-1}|b-\varepsilon); \chi(f_n|b)] =$$

$$= g^{-1}[\sup_{\lim \Sigma_k^n f_k \varepsilon = I} \lim_{n\to\infty} \sum_{k=1}^{n} g(\chi(f_k|a+k\varepsilon))]. \tag{6.67}$$

It is obvious that if for almost all x from $[a,b]$ we have[10] $g(\chi(f|x)) \neq 0$, the limit in (6.67) will be infinite and, therefore, $\mu_f(I|a,b) = 0$, because $g^{-1}(\infty) = 0$. On the other hand, if $g(\chi(f|x)) \equiv 0$, i.e. $f = \xi(x)$, the sum in (6.67) will be zero and, therefore, $\mu_f(I|a,b) = 1$, since $g^{-1}(0) = 1$. This result exactly corresponds to (6.66)[11]. In fact, for any t-norm, which leads to the infinite sum or product, the result (6.66) will be held. Apparently, only *Min* t-norm is natural for the definition (6.43).

Note, that there are other definitions of the fuzzy integral as well (see [494]). An example of another definition of a fuzzy integral is considered in the Appendix A.2, where t-norm (6.11) is used for the case of non-atomic fuzzy measures.

[10] That is $f \neq \xi(x)$, which implies that $\chi(f|x) \neq 1$.

[11] The case of similar collapse of the path integral, when probability of all trajectories is neglected in comparison with the most probable one, is well known in Stochastic theory and in Quantum Mechanics, where the path integrals are being used.

7

Evolution of Perceptions

7.1 Fuzzy Dynamics: Evolution of a System with Vague Parameters and Uncertainty in the Dynamics law

Learning and adaptation of a neuron are evolutionary processes, so in order to explore fuzzy logic for description of a neuron's behavior we should extend the fuzzy approach to the dynamical process. There are different ways to do it. The first attempt to describe the dynamics of fuzzy systems was done in 1973 by Nazaroff [884] who investigated fuzzy topological poly-systems. In the next two decades, several approaches to describing fuzzy dynamics were proposed. For instance, Levary [724] has applied Fuzzy Logic to the description of system dynamics developed by Forrester [399], Kloeden [643] defined and considered abstract fuzzy dynamical systems; fuzzy differential equations for description of fuzzy evolution were studied by Kaleva [594]. Application of fuzzy differential inclusions to the dynamical problems was considered in [322],[323], [570].

In this book, however, we will use a different approach [417]-[427], [1077]-[1268], where the apparatus of theoretical physics was generalized for systems with vague parameters and uncertainty in dynamics laws.

In order to better understand our approach, we begin from a classical approach to dynamical problems. Let us consider a system described by a set of time-depended parameters $x_1(t), ..., x_n(t)$. Like in the previous Chapter, we will use vector notation $\boldsymbol{x}(t) = \{x_1(t), ..., x_n(t)\}$ and will call $\boldsymbol{x}(t)$ - dynamical variables. For a short-time interval dt these variables are changed as:

$$d\boldsymbol{x} = \boldsymbol{V}dt + o(dt^2), \tag{7.1}$$

where $o(dt^2)$ designates high-order-smallness terms. If dt is very small these terms can be neglected and we can write:

$$\frac{d\boldsymbol{x}}{dt} = \boldsymbol{V}(\boldsymbol{x}, t). \tag{7.2}$$

The quantities $\boldsymbol{V}$ have meaning velocities of changing of the system's evolution parameters and they are determined by the forces influence. For a system without long-term memory, the velocities $\boldsymbol{V}(\boldsymbol{x}, t)$ are functions of the dynamics variable and time.

The system's evolution is described by functions $x_1(t), ..., x_n(t)$, which constitute a solution to first order ordinary differential equations (7.2). As an example, we consider a very simplified model of a neuron adaptation under instrumental conditioning, in which only neuron excitability (possibility of spike generation) - x_1, neuron damage - x_2 and the "conditional environment's reaction" - $r(x_1)$ are taken into account. In this case, the rules of adaptation (see page 361) are reduced to:

1. If damage is large, neuron's excitability increases
2. If damage is small, neuron's excitability decreases
3. If neuron is punished, damage increases
4. If neuron isn't punished damage decreases
5. If neuron's activity is low, there is an environmental punishment

This means that:

i) Velocity of change of the neuron excitability - $V_{ex}(x_2)$ is increasing function of the damage
ii) Velocity of change of the neuron damage - $V_{dam}(r)$ is increasing function of the punishment- r
iii) The punishment is decreasing function of the neuron activity

Assuming that a neuron's activity is proportional to its excitability, we can try to approximate velocities of change in the neuron activity and damage as:

$$\begin{aligned} V_1 &\simeq a x_2 + a_0, \\ V_2 &\simeq b r + b_0, \\ r &\simeq -c x_1 + c_0, \end{aligned} \tag{7.3}$$

where a, b, c are some constants. So, Eq. (7.2) takes the form of:

$$\begin{aligned} \frac{dx_1}{dt} &= a x_2 + a_0, \\ \frac{dx_2}{dt} &= -b c x_1 + b c_0 + b_0, \end{aligned} \tag{7.4}$$

In order to solve these linear differential equations , we should add an initial conditions. For example, we can choice $x_1(0) = A_0, x_2(0) = 0$. The solution of Eqs. (7.3) is well known (see [595]):

$$x_1 = X_0 + (A_0 - X_0)\cos(\omega t), \tag{7.5}$$

where

$$\omega = \sqrt{abc}, \quad X_0 = \frac{c_0 + b_0/b}{c},$$

which means that neuron activity oscillates around X_0 with frequency ω and magnitude $A_0 - X_0$.

7.1.1 Fuzzy logic setup of the evolution problems

In order to obtain a more realistic solution, we could add non-linear terms to V_1, V_2 and r or to consider more complicated models (see Appendix A.1). For example, in order to taking into account natural instability, methods of Statistical Mechanics could be used of a neuron reaction and appearance of the "strange attractors" and other phenomena of chaotic dynamics can be expected [1274, 1275]. Indeed, many interesting results were obtained in this way, but such an approach requires a great information deal of the physics and biochemistry of neuron functioning. Such information is insufficient currently and it is immoderately need of details, in order to be suitable for the phenomenological approach. So, in order to avoid an attempt "to explain the unintelligible through use of an unknown", we, instead, will utilize the fuzzy logic approach to the dynamical problems.

In order to do this, we need to consider a few general features of system evolution from the fuzzy logics point of view. Let us assume that the system is moving in some "space", whose "points" are described by coordinates $\boldsymbol{x} = \{x_1, ..., x_n\}$. We assume that it is unable to obtain an exact system's location and exact value of its velocity, so we can say only that there is some possibility that at time t, the system is close to the point $\boldsymbol{x}$ and its velocity is close to $\boldsymbol{V}$. In such a case the system's movement can be described as follows. If we denote by $\boldsymbol{V}', \boldsymbol{V}'', \boldsymbol{V}''', ...$ the possible values of the velocity $\boldsymbol{V}$, we can say that:

- If at the time $t+dt$, the system is located in the vicinity of the point $\boldsymbol{x}$, then at the previous time t, the system could be near the point $\boldsymbol{x}' \approx \boldsymbol{x} - \boldsymbol{V}'dt$, or near the point $\boldsymbol{x}'' \approx \boldsymbol{x} - \boldsymbol{V}''dt$, or near the point $\boldsymbol{x}''' \approx \boldsymbol{x} - \boldsymbol{V}'''dt$, or ..., and so on, for all possible values of the velocity $\boldsymbol{V}$.

Let us denote the possibility that the system is in a small domain $\Delta_{\boldsymbol{x}}$ around the point $\boldsymbol{x}$ at the time t as $m(\Delta_{\boldsymbol{x}}, t)$, and the possibility that the system's velocity is near the point $\boldsymbol{x}$ at the time t is approximately $\boldsymbol{V}$ as $P(\Delta \boldsymbol{V} | \Delta_{\boldsymbol{x}}, t)$. Then the above expression can be symbolically written as

$$\begin{aligned} m(\Delta_{\boldsymbol{x}}, t + dt) = \Big\{ & \Big[P(\Delta \boldsymbol{V}' | \Delta_{\boldsymbol{x}'}, t) \bigwedge m(\Delta_{\boldsymbol{x}'}, t)\Big] \bigvee ... \\ & ... \bigvee \Big[P(\Delta \boldsymbol{V}'' | \Delta_{\boldsymbol{x}''}, t) \bigwedge m(\Delta_{\boldsymbol{x}''}, t)\Big] \bigvee ... \\ & ... \bigvee \Big[P(\Delta \boldsymbol{V}''' | \Delta_{\boldsymbol{x}'''}, t) \bigwedge m(\Delta_{\boldsymbol{x}'''}, t)\Big] \bigvee ... \bigvee \\ & \text{and so on}\Big\}, \end{aligned} \tag{7.6}$$

where $\{A \bigvee B \bigvee C \bigvee ...\}$ designates logical connective A *OR* B *OR* C *OR* ... and $[A \bigwedge B]$ designates logical connective A *AND* B. Expression (7.6) is nothing more than the previous natural-language expression, which is written in the symbolic form. In order to translate it into an equation of a system's dynamics, we should define mathematical representation of the expressions $m(\Delta_{\boldsymbol{x}}, t)$, $P(\boldsymbol{V} | \Delta_{\boldsymbol{x}}, t)$ and the logical connectives $\bigvee$ and $\bigwedge$.

It is understood, that $m(\Delta_{\boldsymbol{x}})$ correspond to some measure of the domain $\Delta_{\boldsymbol{x}}$. In accordance with Fuzzy Logic paradigm, $m(\Delta_{\boldsymbol{x}})$ can be considered as a *truth value* of the fact that the system is in the domain $\Delta_{\boldsymbol{x}}$. Since this domain is assumed to be small, it is reasonable to consider a limit, where the domain is collapsed to point:

$$\lim_{\Delta_{\boldsymbol{x}} \to \boldsymbol{x}} m(\Delta_{\boldsymbol{x}}, t) = \mu(\boldsymbol{x}, t). \tag{7.7}$$

Theoretically there are two cases[1]:

$$\mu \neq 0,$$

and

$$\mu \equiv 0.$$

The first case is the main case of our study, while the second one is equivalent, actually, to the stochastic or quantum approaches to the dynamical problems and will be briefly considered in Appendix A.2. For our purposes it is enough to consider $\mu(\boldsymbol{x}, t)$ and $P(\boldsymbol{V}; \boldsymbol{x}, t)$ as continuous, bounded functions: $0 \leq \mu, P \leq 1$, so value 0 corresponds to the minimal possibility, while value 1 to corresponds the maximal one.

As we have shown in the Sec. 6.2.3 the connectives $\{\mu_1 \bigvee \mu_2\}$, $[\mu \bigwedge P]$ can be represented by the various t-norms and conorms. It is remarkable, however, that the natural properties of the state space local topology drastically restrict the available choice of the representations of the connective $\bigvee$. To demonstrate this, let us consider two nearest-neighbor domains Δ_1, Δ_2 of the system's state space (see Fig. 7.1). It is obvious that the possibility that the system is in the joint domain $\Delta_1 \bigcup \Delta_2$ is equal to the possibility that it is in the domain Δ_1 *OR* it is in the domain Δ_2, so we can write:

$$m\left(\Delta_1 \bigcup \Delta_2\right) = \left\{m(\Delta_1) \bigvee m(\Delta_2)\right\} \tag{7.8}$$

Now, if both domains are collapsed to the same point: $\Delta_1, \Delta_2 \to \boldsymbol{x}$, we have

$$m(\Delta_1) \to m(\Delta_2) \to m\left(\Delta_1 \bigcup \Delta_2\right) \to \mu(\boldsymbol{x}),$$

which imply that:

$$\mu(\boldsymbol{x}) = \left\{\mu(\boldsymbol{x}) \bigvee \mu(\boldsymbol{x})\right\}. \tag{7.9}$$

As it was shown in Sec. 6.2.2 (see page 323), in this case only

$$\left\{\mu_1 \bigvee \mu_2\right\} = \max\{\mu_1\ \mu_2\}. \tag{7.10}$$

[1] In the first case $m(\Delta_{\boldsymbol{x}})$ can be considered as a common additive measure, while in the second one $\mu(\boldsymbol{x})$ corresponds to the so called *atomic measure* of the domain $\Delta_{\boldsymbol{x}}$.

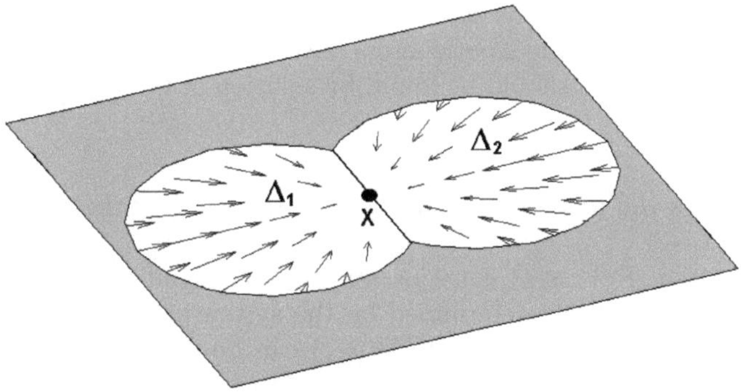

Fig. 7.1. Local topology of space of the system's parameters (the system's "state space")

have to be used as representation of $\bigvee$. This result is crucial for our study and it is important that this representation for $\bigvee$ is dictated by local topology of the system's state space rather than is a result by our mathematical taste, convenience, *etc.* Similar arguments, however, cannot be employed to the connective $\bigwedge$. The point is, that this connective can include membership functions, which depend on the system's variables that belong to different state spaces. For example, in the expression (7.6), the possibility $m(\Delta_{\boldsymbol{x}})$ depends on domain of the system's state space, while the possibility $P(\Delta \boldsymbol{V})$ depends on domain of its tangential space. In this case collapsing both of the domains to the same point is impossible and, therefore, the above-mentioned argumentation becomes invalid. So, mathematical representation of the connective AND is still arbitrary[2].

7.2 Master-Equation of fuzzy dynamics

By using (7.7) for the above-mentioned mathematical representations of the system location and (7.10) for representation of the connective $\bigvee$ and some t-norm[3] $T[...;...]$ for representation of the connective $\bigwedge$, we can rewrite (7.6) as an equation:

$$\mu(\boldsymbol{x}, t + dt) = \sup_{\boldsymbol{V}} T[P(\boldsymbol{V}; \boldsymbol{x}, t); \mu(\boldsymbol{x} - \boldsymbol{V}dt, t)], \tag{7.11}$$

which can be considered as "Master-Equation" of the fuzzy dynamics.

[2] In the case: $\lim_{\Delta_{\boldsymbol{x}} \to \boldsymbol{x}} m(\Delta_{\boldsymbol{x}}, t) = 0$, however, mathematical representation of the AND connective can be found by using topological properties of the possible system's trajectories (see Appendix A.2).

[3] See Sec. 6.2.2 for the definitions and details

The system's evolution is described by the function $\mu(\boldsymbol{x},t)$, which reflects possibility that the system's parameters have the values $x_1, ..., x_n$ at the time t. The function $\mu(\boldsymbol{x},t)$ should be found by solution of the Eq. (7.11) with the initial condition:

$$\mu(\boldsymbol{x},0) = \mu_0(\boldsymbol{x}), \tag{7.12}$$

where $\mu_0(\boldsymbol{x})$ is the possibility that the system was near the point $\boldsymbol{x}$ at the initial time $t = 0$.

The function $P(\boldsymbol{V};\boldsymbol{x},t)$ contains information about specificity of a system's dynamics and it is determined by the system's dynamics laws. In the next Chapter we will show how $P(\boldsymbol{V};\boldsymbol{x},t)$ can be derived from the natural language expressed dynamics laws, but now we want to consider some general properties of the "Master-equation" (7.11).

First note, that as it is seen from the Fig. 7.2:

$$\sup_{\boldsymbol{V}} T[P(\boldsymbol{V};\boldsymbol{x},t);\mu(\boldsymbol{x},t-\boldsymbol{V}dt)] = T[P(\boldsymbol{V}_m;\boldsymbol{x},t);\mu(\boldsymbol{x},t-\boldsymbol{V}_m dt)] =$$
$$= \mu(\boldsymbol{x},t-\boldsymbol{V}_m dt), \tag{7.13}$$

where $\boldsymbol{V}_m$ is the velocity corresponds to the maximal value of the right side of (7.13). So we can write:

$$\mu(\boldsymbol{x},t+dt) = \mu(\boldsymbol{x},t-\boldsymbol{V}_m dt). \tag{7.14}$$

If we are interested in the time intervals much more than dt: $t \gg dt$, it is reasonable to take the limit $dt \to 0$. For the small dt we can expand

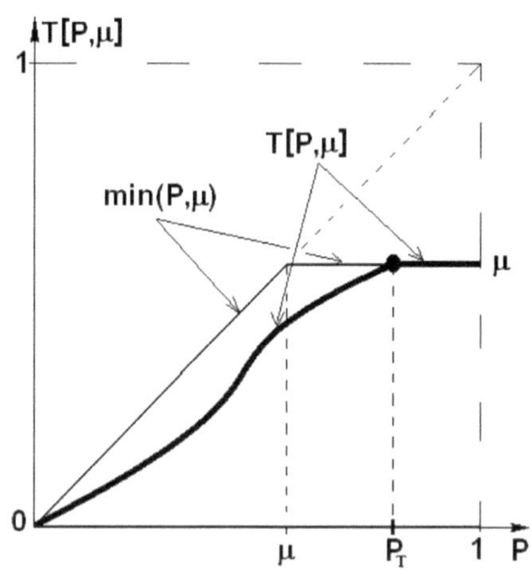

Fig. 7.2. Since function $T[P,\mu]$ is monotonically increasing, $T[P,\mu] \leq \min(P,\mu)$ and $T[1,\mu] = \mu$ (see Sec. 6.2.2), $T[P,\mu]$ should looked like in the diagram. It is seen that for given μ maximum of $T[P,\mu]$ is equal to μ.

$\mu(\boldsymbol{x} - \boldsymbol{V}dt, t)$ with respect to dt. In the first approximation one obtains:

$$\mu(\boldsymbol{x},t) + \frac{\partial \mu}{\partial t}dt = \mu(\boldsymbol{x},t) - (\boldsymbol{V} \cdot \nabla\mu)dt + o(dt), \tag{7.15}$$

where $\nabla\mu$ designates *gradient* μ and ($\cdot$) means *scalar product*:

$$(\boldsymbol{V} \cdot \nabla\mu) \Leftrightarrow \left(\boldsymbol{V} \cdot \frac{\partial \mu}{\partial \boldsymbol{x}}\right) = V_1\frac{\partial \mu}{\partial x_1} + ... + V_n\frac{\partial \mu}{\partial x_n}.$$

$\partial/\partial x_i$ designates partial derivative with respect to x_i. So, for $dt \to 0$ Eq. (7.15) leads to:

$$\frac{\partial \mu}{\partial t} + (\boldsymbol{V}_m \cdot \nabla\mu) = 0, \tag{7.16}$$

where $\boldsymbol{V}_m$ has to be found as the function of $\boldsymbol{x}, t$ and $\mu, \nabla\mu$.

To do this, note, that in order for $\mu(\boldsymbol{x},t) - (\boldsymbol{V}_m \cdot \nabla\mu)dt$ to be maximal, $(\boldsymbol{V}_m \cdot \nabla\mu)$ should be minimal. On the other hand for $dt \equiv 0$ we have:

$$T[P(\boldsymbol{V}_m;\boldsymbol{x},t);\mu(\boldsymbol{x},t)] = \mu(\boldsymbol{x},t). \tag{7.17}$$

Therefore, $\boldsymbol{V}_m(\mu,\nabla\mu;\boldsymbol{x},t)$ can be found by minimization of

$$(\boldsymbol{V}_m \cdot \nabla\mu) \to \min \tag{7.18}$$

under the restriction :

$$P(\boldsymbol{V}_m;\boldsymbol{x},t) \geq P_T, \tag{7.19}$$

where P_T is solution of the equation $T[P_T;\mu] = \mu$ (see Fig. 7.2). Solution of the system (7.18),(7.19) is a well known problem and it is solved by the method of the *Lagrange-multiplier* :

$$\lambda\frac{\partial P}{\partial V_{mi}} = \frac{\partial \mu}{\partial x_i}, \tag{7.20}$$

$$P(\boldsymbol{V}_m;\boldsymbol{x},t) = P_T(\mu), \tag{7.21}$$

where $\lambda > 0$ is a Lagrange-multiplier.

7.3 Fuzzy trajectories

Equations:

$$\frac{\partial \mu}{\partial t} = -(\boldsymbol{V} \cdot \nabla\mu), \tag{7.22}$$

$$\lambda\frac{\partial P}{\partial \boldsymbol{V}} = \nabla\mu, \tag{7.23}$$

$$P(\boldsymbol{V};\boldsymbol{x},t) = P_T(\mu), \tag{7.24}$$

with the initial condition (7.12) are equivalent to (7.13) for long-time evolution $t \gg dt$ (for simplicity of notations we have omitted index m).

The first equation (7.22) looks as a partial differential equation of the first order. In physics of dynamical systems such a type of equations is called a *Liouville-equation*. In our case, however, the equation (7.22) is more general than a common partial differential equation, because solution of the Eqs. (7.23),(7.24) can be a *set value function* that means that all values of $\boldsymbol{V}$, belonging to some set - $\boldsymbol{V} \in \mathcal{V}(\mu, \nabla\mu; \boldsymbol{x}, t)$, can be solutions of the Eqs. (7.23)-(7.24). Such types of equations are named *differential inclusions* and they have been considered in Sec. 6.3.3.

Let us consider the trajectories $\boldsymbol{X}(t)$, which satisfy to:

$$\frac{d\boldsymbol{X}}{dt} = \boldsymbol{V}(\mu, \nabla\mu; \boldsymbol{X}, t). \tag{7.25}$$

Besides $\boldsymbol{X}$ and t, the right side of the Eq. (7.25) depends on unknown functions $\mu(\boldsymbol{X}, t)$ and $\nabla\mu(\boldsymbol{X}, t)$. Dependence on $\mu(\boldsymbol{X}, t)$ doesn't cause problems, because $\mu(\boldsymbol{X}(t), t)$ is constant along the trajectories (7.25). Indeed, let us find the total derivative of $\mu\left(\boldsymbol{X}(t), t\right)$ with respect to t:

$$\frac{d\mu}{dt} = \frac{\partial\mu}{\partial t} + \left(\frac{d\boldsymbol{X}}{dt} \cdot \frac{\partial\mu}{\partial\boldsymbol{X}}\right) = \frac{\partial\mu}{\partial t} + (\boldsymbol{V} \cdot \nabla\mu) = 0.$$

Therefore

$$\mu(\boldsymbol{X}(t), t) = \mu_0(\boldsymbol{x}), \tag{7.26}$$

where $\boldsymbol{x} = \boldsymbol{X}(0)$ is the initial value of $\boldsymbol{X}(t)$.

Unlike this, $\nabla\mu(\boldsymbol{X}(t), t)$ is not preserved along the trajectory and, at first appearance, it should be found from the solution of the Eq. (7.22). There is, however, another elegant way, where fuzzy trajectories can be found without explicit solution of the Liouville-equation.

To do this, let us consider $\nabla\mu(\boldsymbol{X}(t), t)$ as new unknown functions of t:

$$q_i(t) = \frac{\partial}{\partial X_i}\mu(\boldsymbol{X}(t), t) = \nabla_i\mu, \tag{7.27}$$

and let us take total derivative of $\boldsymbol{q}(t)$ with respect to t. We have[4]:

$$\frac{dq_i}{dt} = \frac{\partial}{\partial t}\nabla_i\mu + \left(\frac{d\boldsymbol{X}}{dt} \cdot \nabla\mu\right) = \frac{\partial}{\partial t}\nabla_i\mu + (V_k\nabla_k\mu)\,. \tag{7.28}$$

On the other hand, we can differentiate Eq. (7.22) with respect to X_i:

$$\frac{\partial}{\partial X_i}\left(\frac{\partial\mu}{\partial t} + (V_k\nabla_k\mu)\right) = \frac{\partial}{\partial t}\nabla_i\mu + \left(V_k\frac{\partial^2\mu}{\partial X_i\partial X_k}\right) + \\ + \left(\frac{\partial V_k}{\partial\mu}\nabla_k\mu\right)\nabla_i\mu + \left(\frac{\partial V_k}{\partial X_i}\nabla_k\mu\right) + \left(\frac{\partial V_k}{\partial\nabla_m\mu}\frac{\partial\nabla_m\mu}{\partial X_i}\nabla_k\mu\right) =$$

4 For the sake simplicity of notations, here and later, we will omit the sign of sum $\sum$ for the repeated indexes, i.e. we will write $C = A_iB_i$ instead $C = \sum_i A_iB_i$ (so called "*Einstein's rule of summation*")

$$= \frac{\partial}{\partial t}\nabla_i\mu + (V_k\nabla_k\mu) + \left(\frac{\partial V_k}{\partial \mu}\nabla_k\mu\right)\nabla_i\mu +$$
$$+ \left(\frac{\partial V_k}{\partial X_i}\nabla_k\mu\right) + \left(\frac{\partial V_k}{\partial \nabla_m\mu}\frac{\partial \nabla_m\mu}{\partial X_i}\nabla_k\mu\right) = \tag{7.29}$$
$$= \frac{\partial \nabla_i\mu}{\partial t} + (V_k\nabla_k\mu) + \left(\frac{\partial V_k}{\partial X_i}q_k\right) + \left(\frac{\partial V_k}{\partial \mu}q_k\right)q_i + \left(\frac{\partial V_k}{\partial q_j}q_k\right)\frac{\partial \nabla_j\mu}{\partial X_i} = 0.$$

It is easily seen[5] that the last term in (7.29) is equal to zero, so if one compares (7.28) and (7.29), one obtains:

$$\frac{dq_i}{dt} = -\left(q_k\frac{\partial V_k}{\partial X_i}\right) - \left(q_k\frac{\partial V_k}{\partial \mu}\right)q_i. \tag{7.30}$$

So, for trajectories of the fuzzy evolution ("*fuzzy trajectories*") we have:

$$\frac{d\boldsymbol{X}}{dt} = \boldsymbol{V}\left(\mu_0, \boldsymbol{q}; \boldsymbol{X}, t\right), \tag{7.31}$$
$$\frac{d\boldsymbol{q}}{dt} = -\frac{\partial(\boldsymbol{V}\cdot\boldsymbol{q})}{\partial \boldsymbol{X}} - \frac{\partial(\boldsymbol{V}\cdot\boldsymbol{q})}{\partial \mu_0}\boldsymbol{q}, \tag{7.32}$$

where the function $\boldsymbol{V}(\mu_0, \boldsymbol{q}; \boldsymbol{X}, t)$ should be obtained from the equations (7.23) and (7.24) with replacement $\nabla\mu$ to $\boldsymbol{q}$. Note that $\boldsymbol{V}(\mu_0, \boldsymbol{q}; \boldsymbol{X}, t) \equiv \boldsymbol{V}(\mu_0, \boldsymbol{k}; \boldsymbol{X}, t)$ depends only on direction[6] $\boldsymbol{k} = \boldsymbol{q}/|q|$ of $\boldsymbol{q}$, but not on its magnitude $q = |\boldsymbol{q}|$.

Since the equation (7.31) for fuzzy trajectories is determined only by direction of $\boldsymbol{q}$, it is useful to obtain an explicit equation for $\boldsymbol{k}(t)$. This is easy:

$$\frac{d\boldsymbol{k}}{dt} = \frac{d}{dt}\frac{\boldsymbol{q}}{q} = \frac{1}{q}\frac{d\boldsymbol{q}}{dt} - \frac{\boldsymbol{q}}{q^3}\left(\boldsymbol{q}\cdot\frac{d\boldsymbol{q}}{dt}\right) =$$
$$= \frac{1}{q}\left[-\frac{\partial(\boldsymbol{V}\cdot\boldsymbol{q})}{\partial \boldsymbol{X}} - \boldsymbol{q}\frac{\partial(\boldsymbol{V}\cdot\boldsymbol{q})}{\partial \mu_0}\right] - \frac{\boldsymbol{q}}{q^3}\left[-\left(\boldsymbol{q}\cdot\frac{\partial(\boldsymbol{V}\cdot\boldsymbol{q})}{\partial \boldsymbol{X}}\right) - q^2\frac{\partial(\boldsymbol{V}\cdot\boldsymbol{q})}{\partial \mu_0}\right] =$$
$$= -\frac{\partial(\boldsymbol{V}\cdot\boldsymbol{k})}{\partial \boldsymbol{X}} + \boldsymbol{k}\left(\boldsymbol{k}\cdot\frac{\partial(\boldsymbol{V}\cdot\boldsymbol{k})}{\partial \boldsymbol{X}}\right). \tag{7.33}$$

This means that the fuzzy trajectories $\boldsymbol{X}(t)$ can be found from the system of the ordinary differential equations (or inclusions, if $\boldsymbol{V}$ is a set-valued function):

$$\frac{d\boldsymbol{X}}{dt} = \boldsymbol{V}\left(P_T, \boldsymbol{k}; \boldsymbol{X}, t\right), \tag{7.34}$$
$$\frac{d\boldsymbol{k}}{dt} = -\frac{\partial(\boldsymbol{V}\cdot\boldsymbol{k})}{\partial \boldsymbol{X}} + \left(\boldsymbol{k}\cdot\frac{\partial(\boldsymbol{V}\cdot\boldsymbol{k})}{\partial \boldsymbol{X}}\right)\boldsymbol{k}, \tag{7.35}$$

---

[5] Differentiating (7.24) with respect to q_i and using (7.23) one obtains:

$$\frac{\partial P}{\partial q_i} = \frac{\partial P}{\partial V_k}\frac{\partial V_k}{\partial q_i} = \frac{1}{\lambda}\left(q_k\frac{\partial V_k}{\partial q_i}\right) = \frac{\partial P_T(\mu)}{\partial q_i} = 0$$

[6] For $\nabla\mu \neq 0$ it can be seen by dividing both sides of (7.23) on $|\nabla\mu|$ and assigning $\lambda/|\nabla\mu|$ as some new Lagrange-multiplier.

with the initial conditions:

$$\boldsymbol{X}(0)=\boldsymbol{x}, \quad \boldsymbol{k}(0)=\frac{\nabla \mu_0(\boldsymbol{x})}{|\nabla \mu_0(\boldsymbol{x})|}, \tag{7.36}$$

where $\boldsymbol{V}\left(P_T, \boldsymbol{k}; \boldsymbol{X}, t\right)$ is found from the system of algebraic equations:

$$\lambda \frac{\partial P}{\partial \boldsymbol{V}}=\boldsymbol{k}, \tag{7.37}$$

$$P(\boldsymbol{V}; \boldsymbol{X}, t)=P_T(\mu_0). \tag{7.38}$$

It should be noted, that Eqs.(7.35) and (7.37) are valid for $\nabla \mu_0(\boldsymbol{x}) \neq 0$. In the case $\nabla \mu_0(\boldsymbol{x})=0$ it follows from (7.32) that $\boldsymbol{q}(t) \equiv 0$, so $\boldsymbol{V}$ doesn't depend on $\nabla \mu$ and $\boldsymbol{X}(t)$ should be found from the equations (7.34) and (7.38) only.

Since $\mu(\boldsymbol{X}, t)$ is a constant along each fuzzy trajectory $\boldsymbol{X}(t)$ (see Eq. (7.26)), results of the fuzzy dynamics approach can be represented in intuitively clear and "visible" form: all possible ways of a system's evolution are presented as a set of trajectories, where each trajectory is indicated by a certain value of its possibility. It is important that, since the fuzzy dynamics preserve the notion of trajectory, many methods of the classical dynamics can be directly applied to the fuzzy dynamics as well.

Eqs. (7.34)-(7.38) are the main equations of our theory and in the next Chapter we will employ them to describe of neural cell behavior, while in this Chapter we will continue to discuss their mathematical properties.

7.3.1 The most possible and impossible trajectories of the fuzzy evolution

The system of equations (7.22)-(7.24) possess several important features. Firstly, the region of the most possible ($\mu(\boldsymbol{X}(t), t)=1$) and the most impossible ($\mu(\boldsymbol{X}(t), t)=0$) trajectories of the fuzzy evolution don't depend on concrete representation of the connective $\bigwedge$, because $P_T(1)=1$ and $P_T(0)=0$ for any $T[P; \mu]$ (see Fig. 7.2). In addition, the most possible trajectories do not depend on subjectiveness in assignment of the intermediate values of possibility for the initial system's states. Therefore, these trajectories contain most reliable information about the system's behavior and are the most interesting in the fuzzy dynamics approach.

Moreover, let us assume that $T[P, \mu]$ is *Q-consistent* representation, i.e. it satisfies:

$$\Phi\left(T[P, \mu]\right)=T[\Phi(P), \Phi(\mu)]. \tag{7.39}$$

where Φ is a monotonically increasing function with $\Phi(0)=0$, $\Phi(1)=1$ (see Sec. 6.2.3 for details). In this case transformation:

$$\begin{aligned} \mu &\rightarrow \Phi\left(\mu\right) \\ P &\rightarrow \Phi\left(P\right), \end{aligned} \tag{7.40}$$

does't change the equations (7.34)-(7.38), their initial values (7.36) and the value of P_T. Therefore, the picture of the fuzzy trajectories doesn't change either. Besides, for initial condition - $\mu'_0(\boldsymbol{x}) = \Phi(\mu_0(\boldsymbol{x}))$, the solution of the equations (7.22)-(7.24) will be $\mu'(\boldsymbol{x},t) = \Phi(\mu(\boldsymbol{x},t))$.

It should be emphasized that freedom under *Φ-transformation* - (7.40) reflects the nature of the fuzzy approach. Numerical values of the events are strongly "expert's-dependent" and are based, mainly, on the expert's intuitive perceptions and experience, rather than on exact measured values. As a rule, several experts will be consistent about the statements: "*At an initial time the system's state* $\boldsymbol{x}_a$ *is preferable to the state* $\boldsymbol{x}_b$" or "*If the system is in state* $\boldsymbol{x}_a$ *then after a short time the state* $\boldsymbol{x}_b$ *will be preferable to the state* $\boldsymbol{x}_c$", but if we will ask them to assign possibilities for each state or transition, they will come up with different numbers and the possibilities of one expert may not add up to those another. This means that in the considered situations, only correlation of preferability of events contains objective information in the expert's description of a problem.

It is obvious, that robust dynamics theory should be invariant under subjectivity in the system description. Mathematically it means that $\mu_0(\boldsymbol{x}) \to \Phi(\mu_0(\boldsymbol{x}))$ should lead to $\mu(\boldsymbol{x},t) \to \Phi(\mu(\boldsymbol{x},t))$, while the picture of the fuzzy trajectories should be invariant under *Φ-transformation*, that is really held in our approach.

7.3.2 *Fuzzy dynamics of "oscillator"

Let us assume that σ characterizes uncertainty of our knowledge about total velocity $V = \sqrt{V_1^2 + V_2^2}$. Then, the possibility of a system's velocities could be chosen in the form:

$$P(\boldsymbol{V};\boldsymbol{x},t) = f\left(\frac{|\boldsymbol{V} - \boldsymbol{U}(\boldsymbol{x})|^2}{\sigma^2}\right), \tag{7.41}$$

where $f(u)$ is a continuous, monotonically non-increasing function with $\max f = 1, \min f = 0$.

Consider a simple fuzzy version of the example, which has been considered in (7.4). In order to avoid cumbersome expressions it is convenient to renormalize the system's variables and velocities as:

$$\begin{aligned}
&x_1 \to \sqrt{bc}x_1 + \frac{bc_0 + b_0}{\sqrt{bc}}, \quad x_2 \to \sqrt{a}x_2 - \frac{a_0}{\sqrt{a}} \\
&V_1 \to \sqrt{a}V_1, \quad V_2 \to \sqrt{bc}V_2,
\end{aligned}$$

so, in accordance with Eqs. (7.3) we will have:

$$\begin{aligned}
U_1(\boldsymbol{x}) &= -\omega x_2, \\
U_2(\boldsymbol{x}) &= \omega x_1,
\end{aligned}$$

where $\omega = \sqrt{abc}$.

In the case: $\nabla\mu_0(\boldsymbol{x}) \neq 0$ (the case $\nabla\mu_0(\boldsymbol{x}) = 0$ should be considered separately (see later)), velocity $\boldsymbol{V}(P_T, \boldsymbol{k}; \boldsymbol{X}, t)$ is an ordinary point-valued function and equations (7.34)-(7.38) are the ordinary differential equations.

From the Eqs. (7.37),(7.38) one finds:

$$\lambda f' \frac{V_1 + \omega x_2}{\sigma^2} = k_1,$$
$$\lambda f' \frac{V_2 - \omega x_1}{\sigma^2} = k_2, \tag{7.42}$$
$$f\left(\left(\frac{V_1 + \omega x_2}{\sigma}\right)^2 + \left(\frac{V_2 - \omega x_1}{\sigma}\right)^2\right) = P_T(\mu_0),$$

where f' designates the derivative of f. These equations are simply solved and we obtain:

$$\lambda = \frac{\sigma}{f'\sqrt{f^{-1}(P_T)}},$$

and

$$V_1 = -\omega x_2 + \alpha\sigma k_1 \tag{7.43}$$
$$V_2 = \omega x_1 + \alpha\sigma k_2, \tag{7.44}$$

where

$$\alpha = \sqrt{f^{-1}(P_T)}, \tag{7.45}$$

and f^{-1} is the reverse function: $f^{-1}(f(u)) = u$. By using (7.34),(7.35) together with (7.43),(7.44) one obtains:

$$\frac{dX_1}{dt} = -\omega X_2 + \alpha\sigma k_1, \tag{7.46}$$
$$\frac{dX_2}{dt} = \omega X_1 + \alpha\sigma k_2, \tag{7.47}$$
$$\frac{dk_1}{dt} = -\omega k_2, \tag{7.48}$$
$$\frac{dk_2}{dt} = \omega k_1, \tag{7.49}$$

The initial conditions for these equations are:

$$X_i(0) = x_i,$$
$$k_i(0) = \frac{\nabla_i\mu_0(\boldsymbol{x})}{|\nabla\mu_0(\boldsymbol{x})|}. \tag{7.50}$$

Since the equations for $\boldsymbol{k}(t)$ do not depend on $\boldsymbol{X}$, they can be solved separately and by using (7.48)-(7.49) one finds:

$$k_1(t) = \cos(\omega t + \phi),$$
$$k_2(t) = \sin(\omega t + \phi), \tag{7.51}$$

where ϕ is found from $\cos\phi = k_1(0),\ \sin\phi = k_2(0)$.

With known $k_1(t), k_2(t)$, Eqs. (7.46)-(7.47) are easily solved (see [595]):

$$X_1(t) = X_{10}(t) + \alpha\sigma\, t\cos(\omega t + \phi),$$
$$X_2(t) = X_{20}(t) + \alpha\sigma\, t\sin(\omega t + \phi). \tag{7.52}$$

where

$$X_{10}(t) = x_1\cos(\omega t) - x_2\sin(\omega t),$$
$$X_{20}(t) = x_2\cos(\omega t) + x_1\sin(\omega t). \tag{7.53}$$

are trajectories of the classical oscillator (where dynamics law is exactly known).

Deviation of the classical trajectory from the fuzzy trajectory increases during the time in "resonance manner". This "drift" of the trajectories results from the vagueness of dynamic law, but the "cost" that we pay for this vagueness is reasonable, because the deviation does not grow so quickly.

If the system's velocity depends on $\boldsymbol{x}$ in the non-linear fashion, the resonance (7.52) could lead to chaotic behavior [1418], which had been observed experimentally for the fuzzy systems in [1219]. Chaos phenomenon in the fuzzy evolution is a very interesting subject, but it not within the framework of this book.

7.3.3 *Splitting of the fuzzy trajectory into a bundle

Solution (7.51),(7.52) is valid for $\nabla\mu_0(\boldsymbol{x}) \neq 0$. In the case $\nabla\mu_0(\boldsymbol{x}) = 0$ it follows from (7.32) that $\nabla\mu(\boldsymbol{x}, t) \equiv 0$. So, two first equations in (7.42) are satisfied identically by choosing $\lambda = 0$ and can be omitted, but solution of the last equation in (7.42) becomes a set valued function:

$$\boldsymbol{V}(\boldsymbol{x}) \in \left\{ \mathcal{V}_V : f\left(\left(\frac{V_1 + \omega x_2}{\sigma}\right)^2 + \left(\frac{V_2 - \omega x_1}{\sigma}\right)^2 \right) = P_T \right\}.$$

It is easily seen that $\boldsymbol{V}(\boldsymbol{x})$ can be represented in the form:

$$V_1 = -\omega x_2 + \alpha\sigma\cos\varphi \tag{7.54}$$
$$V_2 = \omega x_1 + \alpha\sigma\sin\varphi, \tag{7.55}$$

where φ is an arbitrary function of time. Substituting (7.54)-(7.55) to (7.34) leads to:

$$\frac{d\widetilde{X}_1}{dt} = -\omega\widetilde{X}_2 + \alpha\sigma\cos\varphi, \tag{7.56}$$
$$\frac{d\widetilde{X}_2}{dt} = \omega\widetilde{X}_1 + \alpha\sigma\sin\varphi, \tag{7.57}$$

which is very similar to (7.46)-(7.47), but instead the certain function $\boldsymbol{k}(t)$, Eqs. (7.56)-(7.57) have within an arbitrary function $\varphi(t)$. Actually, Eqs. (7.56)-(7.57) are differential inclusions, but in our case, their solution can be obtained as solutions of the deferential equations:

$$\begin{aligned}\widetilde{X}_1(t) &= X_{10}(t) + \frac{\alpha\sigma}{\omega}\int_0^{\omega t}\left[\cos\varphi\cos\tau - \sin\varphi\sin\tau\right]d\tau,\\ \widetilde{X}_2(t) &= X_{20}(t) + \frac{\alpha\sigma}{\omega}\int_0^{\omega t}\left[\sin\varphi\cos\tau + \cos\varphi\sin\tau\right]d\tau.\end{aligned} \tag{7.58}$$

It is seen that instead of the trajectory (7.52) we obtain a bundle of the fuzzy trajectories $\widetilde{\boldsymbol{X}}(t)$, which is determined by an arbitrary function φ.

What is the boundary of the bundle? In general, this is a sophisticated question, but for (7.58) we can find the bundle's boundary in explicit form. To do this, it is convenient to introduce the "polar coordinates":

$$\begin{aligned}\widetilde{X}_1(t) - X_{10}(t) &= \rho(t)\cos\theta(t),\\ \widetilde{X}_2(t) - X_{20}(t) &= \rho(t)\sin\theta(t),\end{aligned}$$

where

$$\rho = \sqrt{\left(\widetilde{X}_1 - X_{10}\right)^2 + \left(\widetilde{X}_2 - X_{20}\right)^2}, \tag{7.59}$$

$$\tan\theta = \frac{\widetilde{X}_2 - X_{20}}{\widetilde{X}_1 - X_{10}}, \tag{7.60}$$

Trajectories $\widetilde{\boldsymbol{X}}_B(t)$ belonging to the bundle's boundary can be found by maximization of $\rho(t)$ with respect to φ for fixed θ and t. It is a simple problem of the variational calculus and it is done by variation of $\rho(t)$ with respect to φ:

$$\frac{\delta\rho}{\delta\varphi} = 0, \tag{7.61}$$

where $\delta\rho/\delta\varphi$ designates the variational derivative . It is easily seen that (7.58),(7.59) and (7.61) lead to

$$\tan\varphi = \frac{\sin\tau - \tan\theta\cos\tau}{\cos\tau + \tan\theta\sin\tau}, \tag{7.62}$$

therefore,

$$\varphi = \theta - \tau. \tag{7.63}$$

Substituting (7.63) to (7.58) and using (7.59) one finds:

$$\rho_{\max}(t) = \alpha\sigma\, t, \tag{7.64}$$

which leads to:

$$\widetilde{X}_{1B}(t) = X_{10}(t) + \alpha\sigma\, t \cos\theta(t),$$
$$\widetilde{X}_{2B}(t) = X_{20}(t) + \alpha\sigma\, t \sin\theta(t),$$

with an arbitrary function $0 \leq \theta(t) \leq 2\pi$. This means that the bundle contains all the trajectories, which satisfy to:

$$\left(\widetilde{X}_1(t) - X_{10}(t)\right)^2 + \left(\widetilde{X}_2(t) - X_{20}(t)\right)^2 \leq \alpha^2\sigma^2 t^2. \tag{7.65}$$

Since the bundle expands during the time, whole domain of the state space will correspond to the same possibility of the system's location, so uncertainty in prediction of the system's state increases during the evolution (Fig. 7.3).

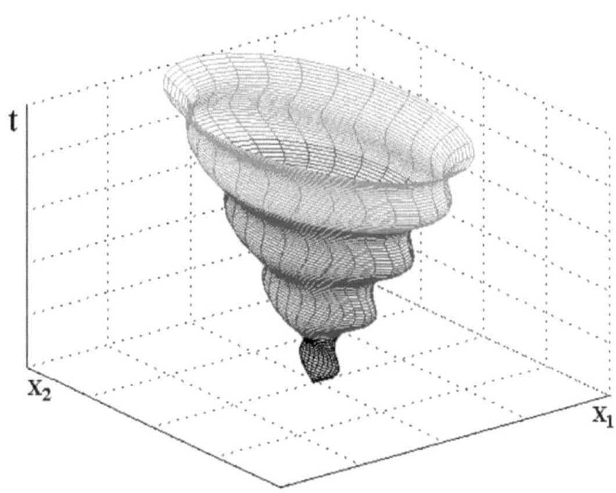

Fig. 7.3. Bundle of the fuzzy trajectories for fuzzy oscillator.

It should be emphasized, that because arbitrariness of the function $\theta(t)$, the bundle contains also non-periodic trajectories, like the one shown in Fig. 7.4.

7.3.4 *Some fundamental solutions of the fuzzy dynamics equations

Fuzzy dynamics of a linear system

The result (7.52) can be generalized to the case of general linear dependence of the system's velocity on its dynamical variables $\boldsymbol{X}$:

$$\boldsymbol{V} = \hat{\Gamma}\boldsymbol{X} + \boldsymbol{\Psi}(\boldsymbol{k}), \tag{7.66}$$

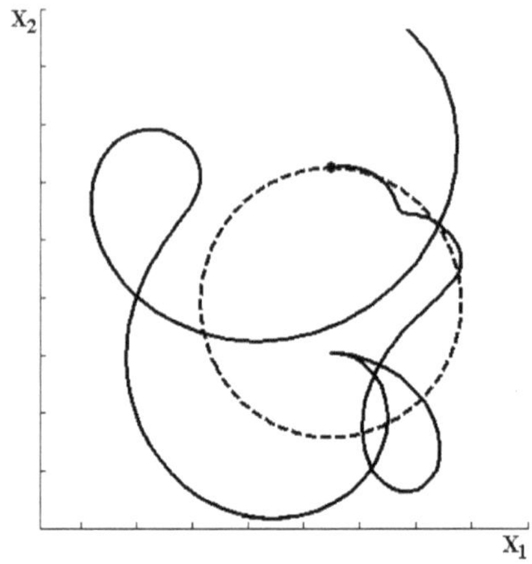

Fig. 7.4. Possible trajectory of fuzzy oscillator in the bundle. Dashed line - periodic trajectory of the classical oscillator, solid line - non-periodic trajectory of the fuzzy oscillator in the bundle.

where $\hat{\Gamma}$ is a constant matrix, $\boldsymbol{\Psi}$ is a vector function, which doesn't depend on $\boldsymbol{x}$ and we've used the matrix notation: $\hat{\Gamma}\boldsymbol{x} \leftrightarrow \Gamma_{ij}x_j$ (dimension of the system's state space is assumed an arbitrary, i.e. $\boldsymbol{X} = \{X_1, ..., X_N\}$). In this case equations (7.34)-(7.35) take the form of:

$$\frac{d\boldsymbol{X}}{dt} = \hat{\Gamma}\boldsymbol{X} + \boldsymbol{\Psi}(\boldsymbol{k}), \tag{7.67}$$

$$\frac{d\boldsymbol{k}}{dt} = -\hat{\Gamma}^\top \boldsymbol{k} + (\boldsymbol{k}^\top \cdot \hat{\Gamma}^\top \boldsymbol{k})\boldsymbol{k}, \tag{7.68}$$

where $\hat{\Gamma}^\top, \boldsymbol{k}^\top$ designate Hermitian-transposition. It can be directly checked, that solution of (7.67)-(7.68) is:

$$\boldsymbol{X}(t) = e^{t\Gamma}\boldsymbol{X}(0) + \int_0^t e^{(t-\tau)\Gamma}\boldsymbol{\Psi}(\tau))d\tau \tag{7.69}$$

$$\boldsymbol{k}(t) = \frac{e^{-t\Gamma^\top}\boldsymbol{k}(0)}{\sqrt{(\boldsymbol{k}^\top(0) \cdot e^{-t(\Gamma^\top + \Gamma)}\boldsymbol{k}(0)))}}, \tag{7.70}$$

where $e^{t\Gamma}$ is the matrix exponent . It is expedient to expand $\boldsymbol{X}(t)$ and $\boldsymbol{k}(t)$ with respect to eigenvectors of the matrix $\hat{\Gamma}$: $\hat{\Gamma}\boldsymbol{n}_\alpha = \gamma_\alpha \boldsymbol{n}_\alpha$

$$\begin{aligned} \boldsymbol{X}(t) &= \chi_\alpha(t)\boldsymbol{n}_\alpha \\ \boldsymbol{k}(t) &= \kappa_\alpha(t)\boldsymbol{n}_\alpha, \end{aligned} \tag{7.71}$$

Assuming that $\{\boldsymbol{n}_\alpha, \boldsymbol{n}_\alpha^\top\}$ are orthogonal-normalized (i.e. $(\boldsymbol{n}_\alpha^\top \cdot \boldsymbol{n}_\beta) = \delta_\beta^\alpha$) and substituting (7.71) to (7.70) we obtain:

$$\kappa_\alpha(t) = \frac{e^{-\gamma^*_\alpha t}\kappa_{\alpha 0}}{\sqrt{\sum_\beta e^{-(\gamma_\beta+\gamma^*_\beta)t}\kappa_{\beta 0}\kappa^*_{\beta 0}}}, \tag{7.72}$$

where $\kappa_{\alpha 0} = (\boldsymbol{n}^\top_\alpha \cdot \boldsymbol{k}(0))$ and there isn't summations on the repeated indexes in Eq. (7.72). Correspondingly, from (7.69) one follows:

$$\boldsymbol{X}(t) - \boldsymbol{X}_0(t) = \sum_\alpha \int_0^t \Big[e^{\gamma_\alpha(t-\tau)}(\boldsymbol{n}^\top_\alpha \cdot \boldsymbol{\Psi}(\tau))\boldsymbol{n}_\alpha + \\ + e^{\gamma^*_\alpha(t-\tau)}(\boldsymbol{\Psi}^\top(\tau)\cdot \boldsymbol{n}_\alpha)\boldsymbol{n}^*_\alpha \Big]\, d\tau. \tag{7.73}$$

where

$$\boldsymbol{X}_0(t) = \sum_\alpha \left[e^{\gamma_\alpha t}\chi_{\alpha 0}\boldsymbol{n}_\alpha + e^{\gamma^*_\alpha t}\chi^*_{\alpha 0}\boldsymbol{n}^*_\alpha \right],$$

and $\chi_{\alpha 0} = (\boldsymbol{n}^\top_\alpha \cdot \boldsymbol{X}(0))$.

Consider a long-time asymptotic of (7.72) and (7.73). Let us assume that ν_α and ω_α are the real and imaginary parts of the eigenvalue γ_α, so $\gamma_\alpha = \nu_\alpha + i\omega_\alpha$. Since $\hat{\Gamma}$ is a real valued matrix there are, also, conjugated eigenvalues $\gamma^*_\alpha = \nu_\alpha - i\omega_\alpha$. It is convenient to arrange the eigenvalues in decreasing order with respect to magnitudes of their real parts: $\nu_1 \geq \nu_2 \geq ...\nu_N$. In order to avoid unnecessary complication, we assume that $\nu_1 > \nu_2$. It is seen from (7.72) that for a long time $t \gg 1/(\nu_1 - \nu_2)$ the main contribution to (7.69) gives the terms corresponding to the eigenvalues with maximal real part: γ_1, γ^*_1.

$$\boldsymbol{k}_m(t) \sim \Re\left[e^{-i(\omega_1 t+\phi_1)}\kappa_{10}\boldsymbol{n}_1 \right] + o\left(e^{-(\nu_1-\nu_2)t} \right). \tag{7.74}$$

where $\Re$ designates the real part and $\cos\phi_1 = (\kappa_{10} + \kappa^*_{10})/2\sqrt{\kappa_{10}\kappa^*_{10}}$. For $t \gg 1/(\nu_1 - \nu_2)$ the last term in (7.74) can be neglected and we can expand $(\boldsymbol{n}^\top_\alpha \cdot \boldsymbol{\Psi})$ in the Fourier-series with respect to $\exp i(\omega_1\tau + \phi_1)$:

$$(\boldsymbol{n}^\top_\alpha \cdot \boldsymbol{\Psi}(\tau)) \sim \sum_p \psi^\alpha_p e^{ip(\omega_1\tau+\phi_1)}. \tag{7.75}$$

Substituting (7.75) to (7.73) and integrating over τ, one obtains asymptotic of (7.73):

$$\boldsymbol{X}(t) - \boldsymbol{X}_0(t) \sim \sum_{\alpha,p} \psi^\alpha_p \Bigg[\frac{e^{-ip(\omega_1 t+\phi_1)} - e^{\gamma_\alpha t+ip\phi_1}}{ip\omega_1 - \gamma_\alpha}\boldsymbol{n}_\alpha - \\ - \frac{e^{ip(\omega_1 t+\phi_1)} - e^{\gamma^*_\alpha t-ip\phi_1}}{ip\omega_1 + \gamma^*_\alpha}\boldsymbol{n}^*_\alpha \Bigg]. \tag{7.76}$$

Since Ψ is the real-valued, we have $\psi^\alpha_{-p} = \psi^\alpha_p$, so (7.76) can be rewritten as:

$$\boldsymbol{X}(t) - \boldsymbol{X}_0(t) \sim \sum_{\alpha=1}^{N}\sum_{p=1}^{\infty} \frac{\psi_p^\alpha}{\nu_\alpha^2 + (p\omega_1 - \omega_\alpha)^2} \Big\{ \{\nu_\alpha \left[\cos p(\omega_1 t + \phi_1) - \right.$$
$$- e^{\nu_\alpha t} \cos(\omega_\alpha t + p\phi_1)\big] - (p\omega_1 - \omega_\alpha)\left[\sin p(\omega_1 t + \phi_1) + \right.$$
$$+ \left. e^{\nu_\alpha t} \sin(\omega_\alpha t + p\phi_1)\right]\big\} \boldsymbol{e}_{1\alpha} + \{\nu_\alpha \left[\sin p(\omega_1 t + \phi_1) + \right.$$
$$+ e^{\nu_\alpha t} \sin(\omega_\alpha t + p\phi_1)\big] + (p\omega_1 - \omega_\alpha)\left[\cos p(\omega_1 t + \phi_1) + \right.$$
$$+ \left. e^{\nu_\alpha t} \sin(\omega_\alpha t + p\phi_1)\right]\big\} \boldsymbol{e}_{2\alpha} \Big\}, \tag{7.77}$$

where $\boldsymbol{e}_{1\alpha}$ and $\boldsymbol{e}_{2\alpha}$ are unit vectors in the directions $\boldsymbol{n}_\alpha + \boldsymbol{n}_\alpha^*$ and $i(\boldsymbol{n}_\alpha - \boldsymbol{n}_\alpha^*)$. The expression (7.77) is somewhat cumbersome, but it can be simplified by noting that resonance terms with $\omega_{\alpha_p} = p\,\omega_1$ contribute mainly to (7.77). This leads to:

$$\boldsymbol{X}(t) - \boldsymbol{X}_0(t) \sim \sum_{p,\,\alpha_p} \psi_{\alpha_p} \left[\boldsymbol{e}_{1\alpha_p} \cos p(\omega_1 t + \phi_1) + \right.$$
$$\left. + \,\boldsymbol{e}_{2\alpha_p} \sin p(\omega_1 t + \phi_1)\right] \left(\frac{e^{\nu_{\alpha_p} t} - 1}{\nu_{\alpha_p}}\right). \tag{7.78}$$

*Fuzzy dynamics of the non-linear system

Consider now the non-linear dynamical system, where the possibility of a system's velocities has the form of (7.41) with[7]:

$$\boldsymbol{U}(\boldsymbol{x}) = \nu(|\boldsymbol{x}|)\boldsymbol{x} + \hat{\Upsilon}\boldsymbol{x}, \tag{7.79}$$

where $\hat{\Upsilon}$ is an antisymmetric matrix: $\Upsilon_{ij} = -\Upsilon_{ji}$, so $(\boldsymbol{x} \cdot \hat{\Upsilon}\boldsymbol{y}) = (\boldsymbol{y} \cdot \hat{\Upsilon}^\top \boldsymbol{x})$ and $(\boldsymbol{x} \cdot \hat{\Upsilon}\boldsymbol{x}) \equiv 0$. As above, the dimension of the system's state space is assumed as arbitrary. In the same way as in Sec. 7.3.2 one has:

$$\lambda f' \frac{\boldsymbol{V} - \boldsymbol{U}(\boldsymbol{x})}{\sigma^2} = \boldsymbol{k}, \tag{7.80}$$

and

$$f\left(\frac{|\boldsymbol{V} - \boldsymbol{U}(\boldsymbol{x})|^2}{\sigma^2}\right) = P_T(\mu_0),$$

The last equation leads to:

$$\lambda = \frac{\sigma}{f'\sqrt{f^{-1}(P_T)}}.$$

Therefore,

$$\boldsymbol{V} = \nu(|\boldsymbol{x}|)\boldsymbol{x} + \hat{\Upsilon}\boldsymbol{x} + \alpha\sigma\boldsymbol{k} \tag{7.81}$$

[7] In physics such a system is called "oscillator with non-linear friction"

where α has been defined in (7.45). By using (7.81) and (7.35)-(7.35) one obtains:

$$\frac{d\boldsymbol{X}}{dt} = \nu(|\boldsymbol{X}|)\boldsymbol{X} + \hat{\Upsilon}\boldsymbol{X} + \alpha\sigma\boldsymbol{k}, \tag{7.82}$$

$$\frac{d\boldsymbol{k}}{dt} = -\hat{\Upsilon}^\top\boldsymbol{k} - \nu'(|\boldsymbol{X}|)(\boldsymbol{X}\cdot\boldsymbol{k})\frac{\boldsymbol{X}-(\boldsymbol{X}\cdot\boldsymbol{k})\boldsymbol{k}}{|\boldsymbol{X}|}, \tag{7.83}$$

with the initial conditions being the same as in (7.50).

Since $\nu(|\boldsymbol{X}|)$ depends on magnitude of $\boldsymbol{X}$ and not on its direction, it is convenient to put: $\boldsymbol{X} = \varrho(t)\boldsymbol{n}(t)$ and $(\boldsymbol{n}\cdot\boldsymbol{k}) = \cos\Omega(t)$, where $\boldsymbol{n}(t)$ is a unit vector. By substituting these expressions to Eqs. (7.82)-(7.83) and performing somewhat clumsy, but simple transformations, one obtains:

$$\frac{d\varrho}{dt} = \nu(\varrho)\varrho + \alpha\sigma\cos\Omega \tag{7.84}$$

$$\frac{d\Omega}{dt} = \sin\Omega\left(\nu'(\varrho)\varrho\cos\Omega - \frac{\alpha\sigma}{\varrho}\right), \tag{7.85}$$

and

$$\frac{d\boldsymbol{n}}{dt} = \hat{\Upsilon}\boldsymbol{n} + \frac{\alpha\sigma}{\varrho}(\boldsymbol{k} - \boldsymbol{n}\cos\Omega), \tag{7.86}$$

$$\frac{d\boldsymbol{k}}{dt} = -\hat{\Upsilon}^\top\boldsymbol{k} - \nu'(\varrho)\varrho\cos\Omega\,(\boldsymbol{n} - \boldsymbol{k}\cos\Omega), \tag{7.87}$$

Eqs. (7.84)-(7.85) can be solved independently on the next ones, afterward Eqs. (7.84)-(7.85) become the linear differential equations.

Consider long-time asymptotic behavior of the system (7.84)-(7.85) in this particular case:

$$\nu(\varrho) = \nu_0 - \nu_1\varrho, \tag{7.88}$$

where ν_0 and ν_1 are some positive constants. By expanding right sides of the equations (7.84)-(7.85) with respect to $\delta\varrho = \varrho - \varrho_*$ and $\delta\Omega = \Omega - \Omega_*$ one has:

$$\frac{d\delta\varrho}{dt} = [\nu_0\varrho_* - \nu_1\varrho_*^2 + \alpha\sigma\cos\Omega_*] + (\nu_0 - 2\nu_1\varrho_*)\delta\varrho - \alpha\sigma\sin\Omega_*\delta\Omega + \\ + o(\delta\varrho^2, \delta\Omega^2), \tag{7.89}$$

$$\frac{d\delta\Omega}{dt} = \left[\sin\Omega_*\left(\nu_1\varrho_*\cos\Omega_* - \frac{\alpha\sigma}{\varrho_*}\right)\right] + \frac{\sin\Omega_*}{\varrho_*^2}(\nu_1\varrho_*^2\cos\Omega_* + \alpha\sigma)\delta\varrho - \\ - \frac{1}{\varrho_*}(\nu_1\varrho_*^2\cos 2\Omega_* - \alpha\sigma\cos\Omega_*)\delta\Omega + o(\delta\varrho^2, \delta\Omega^2). \tag{7.90}$$

In order to $\delta\varrho$ and $\delta\Omega$ tend to zero as $t\to\infty$, expressions in the square brackets should be equal zero. This leads to:

$$\varrho_* = \frac{\nu_0 + \sqrt{\nu_0^2 + 4\nu_1 \alpha \sigma}}{2\nu_1}, \tag{7.91}$$

$$\Omega_* = 0. \tag{7.92}$$

Indeed in this case we have from (7.89)-(7.90):

$$\delta\varrho \sim e^{-\sqrt{\nu_0^2+4\nu_1\alpha\sigma}\, t},$$
$$\delta\Omega \sim e^{-\nu_0 t},$$

so $\delta\varrho$ and $\delta\Omega$ asymptotically disappear and we have:

$$\varrho \to \varrho_*, \tag{7.93}$$

$$\Omega \to 0. \tag{7.94}$$

Since $(\boldsymbol{n} \cdot \boldsymbol{k}) = \cos \Omega \to 1$, we have asymptotically $\boldsymbol{k} \to \boldsymbol{n}$ and Eq. (7.86) is reduced to:

$$\frac{d\boldsymbol{n}}{dt} \sim \hat{\Upsilon}\boldsymbol{n}. \tag{7.95}$$

As $\hat{\Upsilon}$ is anti-symmetric, this means that $\boldsymbol{n}$ is rotated with frequencies equal to eigenvalues of the matrix $\hat{\Upsilon}$. For example, in the two dimensional case, where:

$$\hat{\Upsilon} = \left\| \begin{matrix} 0 & -\omega \\ \omega & 0 \end{matrix} \right\|,$$

we obtain:

$$X_1 \to \varrho_* \cos(\omega t + \phi),$$
$$X_2 \to \varrho_* \sin(\omega t + \phi),$$

where ϕ is a constant. Therefore, in a non-linear system like (7.79), (7.88), the fuzzy trajectories tend asymptotically to a limited cycle .

7.3.5 *Behavior of fuzzy system near critical points

The existence of an analytical solution in the case of linear dependence of the system's velocity on its dynamical variables is important, because it allows to analyze the system's evolution near the points, where its velocity is close to zero, so the system "spends" a lot of time in their vicinity. In dynamics theory such the points are called *critical points*. In this case expression (7.66) can be considered as a first approximation for the velocity with respect to deviation from the critical point.

Correspondence of critical behavior of a fuzzy system and its classical analog is shown in Fig. 7.5. Correspondence of critical behavior of a fuzzy system and its classical analog is shown on the Fig. 7.5 We see that near stable and unstable centers, unstable focus and the limited cycles of the classical system,

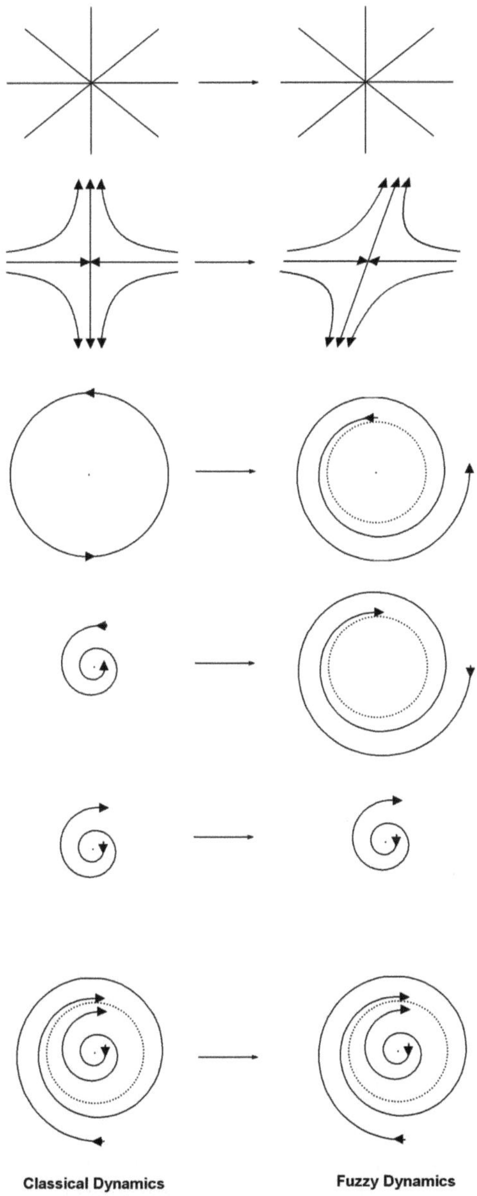

Fig. 7.5. Behavior of fuzzy and classical systems near the critical points. 1st row - stable and unstable centers; 2nd row - saddle point, 3rd row - cycle in classical dynamics is transformed to unstable limited cycle in fuzzy dynamics; 4th row - stable focus is transformed to stable limited cycle; 5th row - unstable focus; six row - stable limited cycle.

fuzzy dynamics predicts the similar behavior. Unlike, near the cycle of the classical system, the fuzzy system behaves like an unstable limited cycle, while near the stable focus, the fuzzy dynamical system shows the stable limited cycle behavior. These relations between classical and fuzzy dynamical systems are quite useful practically, because they allows to predict qualitative behavior of the fuzzy system if we know the critical behavior of the correspondent classical system.

7.4 Evolution of uncertainty

In order to understand qualitative behavior of uncertainty in the system's evolution, consider two trajectories, which begin at nearby points $\boldsymbol{x}$ and $\boldsymbol{x}+\Delta\boldsymbol{x}$, where $\Delta\boldsymbol{x}$ is small. For clearness, we consider a two-dimensional case, so in accordance with (7.78) we have:

$$\boldsymbol{X}(t,\boldsymbol{x}) = e^{\nu t}\Big\{ (x_1\cos\omega t + x_2\sin\omega t)\boldsymbol{e}_1 + (x_2\cos\omega t + x_1\sin\omega t)\boldsymbol{e}_2 \\ + \overline{\psi}(\mu_0)\left(\frac{1-e^{-\nu t}}{\nu}\right)[\cos(\omega t+\phi)\boldsymbol{e}_1 + \sin(\omega t+\phi)\boldsymbol{e}_2]\Big\}. \quad (7.96)$$

Let us assume that $d(t)$ is the distance between the above-mentioned trajectories at the time t:

$$d^2(t) = (\boldsymbol{X}(\boldsymbol{x}+\Delta\boldsymbol{x},t) - \boldsymbol{X}(\boldsymbol{x},t))^2. \quad (7.97)$$

Since $\Delta\boldsymbol{x}$ is small, we can approximate

$$\overline{\psi}(\mu_0(\boldsymbol{x}+\Delta\boldsymbol{x})) - \overline{\psi}(\mu_0(\boldsymbol{x})) \simeq \overline{\psi}_{,\mu}(\nabla\mu_0\cdot\Delta\boldsymbol{x})$$

where $\overline{\psi}_{,\mu} = \partial\overline{\psi}/\partial\mu_0$. By using (7.96) and (7.97) one obtains:

$$d^2(t) \simeq e^{2\nu t}\left[(\Delta x_1)^2 + (\Delta x_2)^2 + \overline{\psi}^2_{,\mu}(\nabla\mu_0\cdot\Delta\boldsymbol{x})^2\left(\frac{1-e^{-\nu t}}{\nu}\right)^2 + \\ + 2\overline{\psi}_{,\mu}(\nabla\mu_0\cdot\Delta\boldsymbol{x})[\Delta x_1\cos\phi + \Delta x_2\sin\phi]\left(\frac{1-e^{-\nu t}}{\nu}\right)\right]. \quad (7.98)$$

This expression can be written in a compact form by introducing the initial distance - $d_0 = \sqrt{(\Delta x_1)^2 + (\Delta x_2)^2}$ and the unit vector - $\boldsymbol{u} = \Delta\boldsymbol{x}/|\Delta\boldsymbol{x}|$:

$$\frac{d(t)}{d_0} \simeq e^{\nu t}\sqrt{1 - (\boldsymbol{k}_0\cdot\boldsymbol{u})^2 + (\boldsymbol{k}_0\cdot\boldsymbol{u})^2\left(1 + \overline{\psi}_{,\mu}|\nabla\mu_0|\left(\frac{1-e^{-\nu t}}{\nu}\right)\right)^2}, \quad (7.99)$$

where $\boldsymbol{k}_0 = \nabla\mu_0/|\nabla\mu_0|$ (see, also, (7.50)-(7.51)).

Qualitative behavior of $d(t)$ and the corresponding distance in classical dynamics - $d_c(t) = d_0\exp(\nu t)$, is shown in the Fig. 7.6. We see that asymptotical

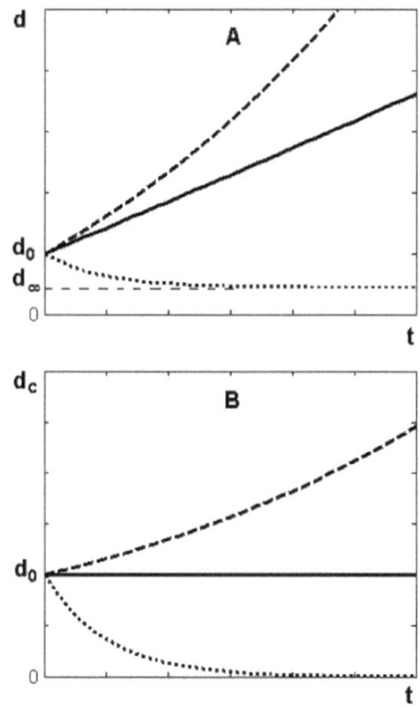

Fig. 7.6. Distance between nearby trajectories. A - fuzzy dynamics, B - classical dynamics. Dashed line - unstable center or focus ($\nu > 0$), dotted line - stable center or focus ($\nu < 0$), solid line - cycle ($\nu = 0$).

behavior of these distances in fuzzy and classical dynamics is qualitatively similar, but there are differences. If the classical trajectories are converged and asymptotically $\lim_{t\to\infty} d_c(t) = 0$, distance between the fuzzy trajectories decreases as well, but $\lim_{t\to\infty} d(t) = d_\infty > 0$. If the classical trajectories are diverged, the fuzzy trajectories are diverged, but $d(t)$ increases more quickly than $d_c(t)$. However, in the marginal case, where $d_c(t) \equiv d_0 = const.$, the fuzzy trajectories are diverged as:

$$d(t \to \infty) \propto d_0 \overline{\psi}_{,\mu} |(\nabla \mu_0 \cdot \boldsymbol{u})|\, t.$$

Evolution over time of the distance between initially nearby trajectories, qualitatively reflects dynamic of uncertainty of the system's evolution. In the classical dynamics, this uncertainty is determined by imprecision of initial location of the system solely, while in fuzzy dynamics vagueness in the dynamics laws influence the uncertainty's evolution as well. Qualitatively, it can be seen on the Fig. 7.7 by comparing (7.99) and (7.52). We see that coefficient $\overline{\psi}$ is proportional to vagueness of the dynamics law (σ in (7.52)). It is quite reasonable that uncertainty's evolution depends on this vagueness and on $|\nabla \mu_0|$, which reflects the rate of change of an expert's evaluation of an initial system's location. More unexpectedly, that evolution of uncertainty is strongly anisotropic and depends on mutual orientations of $\nabla \mu_0$ and $\Delta \boldsymbol{x}$. We see that uncertainty is maximal if $\nabla \mu_0$ and $\Delta \boldsymbol{x}$ are parallel (or anti-parallel)

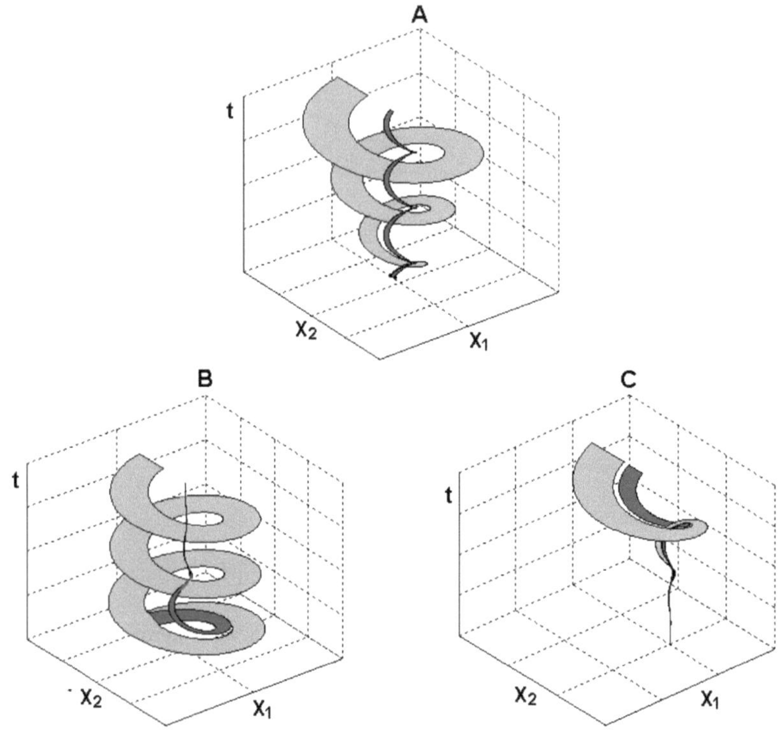

Fig. 7.7. Evolution of the uncertainty. Dark ribbon - uncertainty in the classical case, where dynamics law is exactly known. Light ribbon - uncertainty in the case of vagueness in the dynamics laws.(B - corresponds to a classically stable system, C - corresponds to a classically unstable system and A - corresponds to the marginal case.)

and it is minimal if they are orthogonal[8]. The last circumstance is important for applications of fuzzy dynamics, since it allows to find an optimal way for prediction of behavior of the fuzzy systems.

7.5 Evolution of perceptions

Assume that we know fuzzy trajectories of a system's evolution. What we can say about evolution of our perception of the system's properties? For example, what has happened with the truth value of the statement "system's parameter X is *large*" during a long time evolution? Our approach to fuzzy evolution allows us to answer to this question.

[8] In fact, in this case evolution of uncertainty in the fuzzy approach is the same as in the classical dynamics: $(\boldsymbol{k}_0 \cdot \boldsymbol{u}) = 0 \rightarrow d_{fyzzy}(t) = d_{classic}(t)$.

For example, consider the statement "A is large" with membership functions $M_L(A)$ (see Fig. 7.8). Truth value of the statement "X is large at the

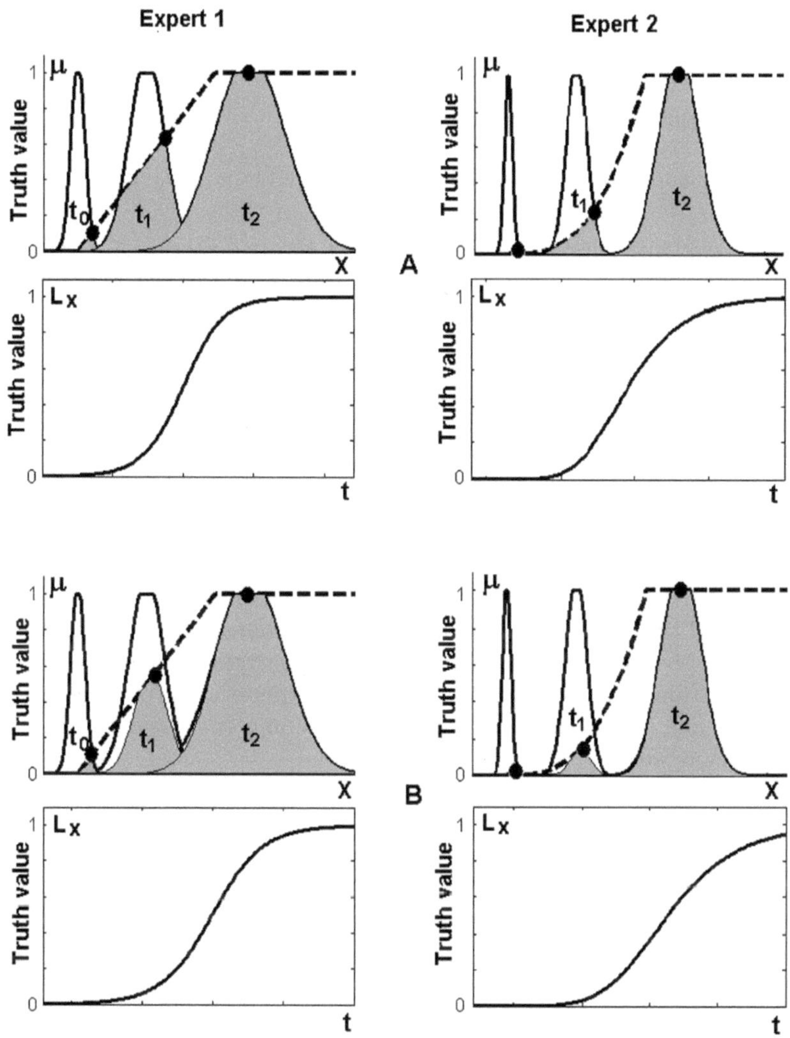

Fig. 7.8. Evolution of the perception "large" for two Q-consistent experts with $\Phi(\mu) = (\mu)^3$. A - *min*-representation of the connective $\bigwedge$, B - *product* representation of the connective $\bigwedge$. Dashed line - membership function of the statement "A is large" $M_L(A)$. Solid lines - $\mu(X,t)$ for the different times $t_0 < t_1 < t_2$. Gray areas show $T[M_L, \mu]$, solid points show $\sup T[M_L, \mu]$. First rows in A and B show evolution of $\mu(X,t)$, while the second rows show evolution of the perception "X is large".

time t" - $L_X(t)$ is equal to truth value that:

$X = x'$ *and* x' is possible at the time t
or
$X = x''$ *and* x'' is possible at the time t
or
... and so on, for all values of x.

So, the value of $L_X(t)$ is calculated as:

$$L_X(t) = \sup_{\boldsymbol{x}} T\left[M_L(x), \mu(\boldsymbol{x}, t)\right] = \sup_{\boldsymbol{x}} T\left[M_L\left(X(t, \boldsymbol{x})\right), \mu_0(\boldsymbol{x})\right], \qquad (7.100)$$

where we explicitly designate dependence of the trajectories $X(t, x)$ on its initial values: $X(0) = x$. It should emphasized, that the last expression in (7.100) allows to find evolution of perceptions without explicit solution of the Liouville equation[9].

An example of calculation of $L_X(t)$ for two consistent experts:

Expert 1: $\mu_1 = \mu(\boldsymbol{x}, t)$; $M_{L1}(X) = M_L(X)$
Expert 2: $\mu_2 = \Phi(\mu(\boldsymbol{x}, t))$; $M_{L2}(X) = \Phi(M_L(X)$,

with $\Phi(\mu) = (\mu)^3$ is shown in the Fig. 7.8. The calculation is shown for *min* and *product* representations of the connective $\bigwedge$: $\bigwedge \to \min[M_L, \mu]$ and $\bigwedge \to [M_L \cdot \mu]$. Note, that both representations are *Q-consistent.*

In spite of the numerical values of $L_X(t)$ depend on an expert's estimation of $\mu_0(\boldsymbol{x})$ and on concrete representation of the connective $\bigwedge$, this ambiguous, however, becomes irrelevant, if we return to natural-language-description of the evolution process (of cause, if all the experts and $\bigwedge$ representation are *consistent*). This conclusion is quite general and it is valid for the other perceptions as well.

[9] Note that expression (7.100) is the fuzzy analog of the average value of "*observable*" in the Statistical and Quantum Mechanics

8

Fuzzy dynamics of a neuronal behavior

8.1 Linguistic variables and linguistic rules of a neuron's behavior

For convenience in later discission, we repeat here the laws of a neuron's behavior that have been discussed early[1]. :

Rule 1 Activity of a neuron increases with neuronal damage and decreases with neuronal protection (Chapter 2).
Rule 2 Activity increase, if expectation of punishment after absence of AP is high, and decrease if expectation of the punishment after AP is high (Chapter 1).
Rule 3 The damage increases with a punishment or with a high activity of the surrounding neurons and decreases with a reward (Chapter 3).
Rule 4 Reward is high, if expectation of the punishment is high, but the punishment is absent. Otherwise, reward is small(Chapter 3).
Rule 5 Expectation of punishment increases if corresponding correlation between AP generation and punishment is high, and decreases otherwise (Fig. 1.10).
Rule 6 Compensation is activated either after AP generation or after AP absence depending upon in which of the cases (**a** or **b**, see below) the expectation of punishment is higher. If these expectations are close, the compensation is activated for both types of the neuronal response (Chapters 1,4)).
Rule 7 Compensation of damage decreases the damage, and compensation of protection decreases the protection (Chapter 2).

Phenomenological theory of learning of a single neuron represents the neuron behavior as "trajectory" in a 5-dimensional state space: $\{\boldsymbol{x}, \boldsymbol{\Theta}\}$:

---

[1] For simplicity, we will not consider very high neuronal damage, which shutting down activity of a neuron.

$x_1(t)$ – **Excitability**: measure of probability of *AP* generation
$\pm\, x_2(t)$ – relative value of the **Protection/Damage**
$\pm\, x_3(t)$ – relative value of the protection/damage **Compensation**
$\Theta_\nu(t)$ – relative value of **Expectation** of the environmental reactions $(\nu = a, b)$.

In order to reduce number of the variables, we have jointed *Damage* and *Protection* to one variable x_2, where negative x_2 corresponds to the relative value *Damage*, while positive x_2 corresponds to the relative value of *Protection* and near-zero x_2 corresponds to the normal state of a cell. The same has been done for the *Compensation*, where negative x_3 corresponds to the compensation of *Damage*, while positive x_3 corresponds to compensation of *Protection*.

The variables Θ_ν describe neuron's expectation of correspondence between the neuron's response on the coming up signal and the environmental reaction. Later, we will consider two situations:

a. expectation of (*AP leads to punishment*) is high/low,
b. expectation of (*absence of AP leads to punishment*) is high/low;

An environment is described by the variable r, which is determined by the environmental reactions: punishment, reward or absence of the reaction. In the considered case, r is a function of *AP* generation and expectation of a punishment:

$$r = r(A(x_1), \Theta_\nu), \tag{8.1}$$

where $A = 1$ for the generated spike and $A = 0$ otherwise, depending on the level of the neuron's excitability. Negative r corresponds to the punishment and the positive one to the reward. Dependence of spike-generation on the excitability $A(x_1)$ can be considered in two approximations: 1) deterministic and 2) quasi-stochastic:

1) $A = 1$ if $x_1 > \theta$ and $A = 0$ otherwise;
2) probability that $A = 1$ is high if x_1 is larger then θ and it is low if x_1 is smaller then θ,

where w is some threshold. The second approximation appears to be more realistic, so $A(x_1)$ is calculated by the Metropolis Algorithm :

$$A = \begin{cases} 1 & \text{if } \ p(x_1) > m_{rand} \\ 0 & \text{if } \ otherwise \end{cases}, \tag{8.2}$$

where m_{rand} is a random number between 0 and 1. If x_1 is a point-wise variable, the function $p(x_1)$ is chosen, usually, as:

$$p(x_1) = (1 + \exp \beta(\theta - x_1))^{-1}.$$

In fuzzy dynamics, however, we deal often with set-valued variables $[x_1^{min}, x_1^{max}]$ (see the previous Chapter), so whole interval $x_1^{min} < x_1 < x_1^{max}$ of the values of x_1 should be considered as an argument of the function $p(x_1)$. In this case the condition $p(x_1) > Q_{rand}$ in (8.2) should be somewhat generalized. It appears to be natural to think that *the spike will be generated if* $p([x_1^{min}, x_1^{max}]) > m_{rand}$ *for MOST of* x_1 *from the interval* $[x_1^{min}, x_1^{max}]$. In "fuzzy sense", the expression "*MOST of* x_1" can be simulated by repeated random choice of x_1 from the interval $x_1^{min} < x_1 < x_1^{max}$ and taking of the median value of $p_{med}(x)$, i.e. if we repeatedly calculate $p(x_1^i)$ with random x_1^i for $i = 1...2k+1$ and will put the results in the order $p_1 \leq P_2 \leq ... \leq p_{2k+1}$ than $p_{med}(x) = p_k$. Indeed, if we randomly choose an element from a set $A = \bigcup_i B_i$, the most probable element will belong to the maximal subset of A (see, also, [1413] .

Quantity $\beta \geq 0$ reflects level of randomness of a neuron's response: the large β corresponds to the almost deterministic response, while the small β corresponds to the almost random response.

It should be emphasized that the variables $\boldsymbol{x}, \boldsymbol{\Theta}$ are differently related to the experimental data. Variable x_1 can be directly estimated from the experimental data. Variable x_2 cannot be directly measured, but its behavior correlates with behavior of the membrane potential, which is directly measured. Unlike this, the variables x_3 is phenomenological and can not be observed experimentally. Its analysis, however, is very helpful in understanding of the implicit motives of a neural cell's behavior.

As we have shown in the Chapter 7, behavior of the adapting neuron is determined by the possibility function $\mathcal{P}(\boldsymbol{V}, \boldsymbol{x}, t)$ for fuzzy velocity $\boldsymbol{V}$ and by the Eqs. (7.20)-(7.21). An advantage of this approach is that the function $\mathcal{P}(\boldsymbol{V}, \boldsymbol{x}, t)$ can be obtained almost "algorithmically" from the dynamics law expressed in a qualitative form.

First of all, we should define the *linguistic variables* (see Chapter 6), which describe qualitative values of the dynamics variables. This is a subtle point, because we must keep balance between desired accuracy of the theoretical description, real precision of the experimental data and unwanted sophistication of the dynamical equations.

The high instability of trial-to-trial neuron's response makes it excessive to use too fine a classification of the variables' values. So, it is reasonable to restrict ourself to certain distinguishable terms in the experimental data qualitative categories, like *Positive*, *Negative*, *Positive Large*, *Negative large* and *Small.* Let us symbolize them as:

PoL - *Positive Large,*
Po - *Positive,*
Sm - *Small,*
Ne - *Negative,*
NeL - *Negative Large.*

Then, the rules **1-7** can be symbolically represented as in Table 8.1, where the quantities v_i and v_α designate velocities of change of the variables x_i and Θ_α. In case of instrumental conditioning environmental reaction - r is symbolically represented as:

1) for the trained neuron

$$\begin{aligned} r \text{ is Ne} &\leftarrow A(x_1) = 0 \\ r \text{ is Po} &\leftarrow A(x_1) = 1 \wedge \Theta_a \text{ is PoL} \\ r \text{ is Sm} &\leftarrow A(x_1) = 1 \wedge \Theta_a \text{ is Sm} \end{aligned}$$

2) for the surrounding (control) neurons:

$$\begin{aligned} r \text{ is Ne} &\leftarrow A(x_1) = 0 \\ r \text{ is Po} &\leftarrow (A(x_1) = 1 \wedge A(y_1) = 1 \wedge \Theta'_a \text{ is PoL}) \vee (A(y_1) = 0 \wedge \Theta'_b \text{ is PoL}) \\ r \text{ is Sm} &\leftarrow A(x_1) = 1 \wedge \Theta'_a \text{ is Sm} \wedge \Theta'_b \text{ is Sm}, \end{aligned}$$

where the arrow represents fuzzy relation: $A \to B =$ [*The Case "X is A" Corresponds to the Case "Y is B"*].

Rule No	*Dynamics Rules*
1.1	V_1 is Po $\leftarrow x_2$ is Ne $\vee$ $(A(x_1) = 0 \wedge \Theta_b$ is PoL$)$
1.2	V_1 is Ne $\leftarrow x_2$ is Po $\vee$ $(A(x_1) = 1 \wedge \Theta_a$ is PoL$)$
2.1	V_2 is Po $\leftarrow ((r$ is Po $\vee$ r is Sm$) \wedge y_1$ is Sm$) \vee$ $(x_2$ is Ne $\wedge$ x_3 is Ne$)$
2.2	V_2 is Ne $\leftarrow r$ is Ne $\vee$ y_1 is PoL $\vee$ $(x_2$ is Po $\wedge$ x_3 is Po$)$
3.1	V_3 is Po $\leftarrow x_2$ is Po $\vee$ $(A(x_1) = 0 \wedge \Theta_a - \Theta_b$ is Po$)$ $\vee$ $(A(x_1) = 1 \wedge \Theta_a - \Theta_b$ is Ne$)$
3.2	V_3 is Ne $\leftarrow x_2$ is Ne $\vee$ $(A(x_1) = 0 \wedge \Theta_a - \Theta_b$ is Ne$)$ $\vee$ $(A(x_1) = 1 \wedge \Theta_a - \Theta_b$ is Po$)$
a.1	V_a is Po $\leftarrow A(x_1) = 1 \wedge r$ is Ne
a.2	V_a is Ne $\leftarrow A(x_1) = 1 \wedge (r$ is Sm $\vee$ r is Po$)$
b.1	V_b is Po $\leftarrow A(x_1) = 0 \wedge r$ is Ne
b.2	V_b is Ne $\leftarrow A(x_1) = 0 \wedge (r$ is Sm $\vee$ r is Po$)$

Table 8.1. Symbolic representations of the dynamics law for the trained neuron. For the surrounding neurons (control neuron in our case) $\{\boldsymbol{x}, \boldsymbol{\Theta}; \boldsymbol{V}\}$ should be replaced to $\{\boldsymbol{y}, \boldsymbol{\Theta}'; \boldsymbol{V}'\}$.

Because sub-rules **n**.1,**n**.2 in the Table 8.1 are related by the connective $\bigvee(OR)$, while the rules **n**.1-2 for different **n** are related by the connective $\bigwedge(AND)$, total truth value of the all rules in the Table 8.1 is:

$$\begin{aligned} \mathcal{T}_\mathcal{V} = &\left(\mathbf{1}.1 \bigvee \mathbf{1}.2\right) \bigwedge \left(\mathbf{2}.1 \bigvee \mathbf{2}.2\right) \bigwedge \left(\mathbf{3}.1 \bigvee \mathbf{3}.2\right) \\ &\bigwedge \left(\boldsymbol{a}.1 \bigvee \boldsymbol{a}.2\right) \bigwedge \left(\boldsymbol{b}.1 \bigvee \boldsymbol{b}.2\right) \bigwedge \left(\boldsymbol{c}.1 \bigvee \boldsymbol{c}.2\right) \\ &\bigwedge \left(\boldsymbol{d}.1 \bigvee \boldsymbol{d}.2\right). \end{aligned} \tag{8.3}$$

Now we should define mathematical representation of the logical connectives and the linguistic variables. To do this we represent the linguistic variables *Positive, Negative, Small, ets.* as membership functions $P(z), N(z), S(z)$, *etc.*, which correspond to the *truth values* of the expressions: "z is positive", "z is negative", *etc.* Remember, that these functions must be restricted as:

$$0 \leq P(z), N(z), S(z), ets. \leq 1,$$

Further simplification of the theory can be achieved by representation the membership functions in term of the one function, say, $P(z)$ (see Fig. 8.1):

$$N(z) = P(-z),\ PL(z) = P(z-a),\ NL(z) = PL(-z),\ S(z) = P(a - |z|), \tag{8.4}$$

where a is value of z where $P(z = a) = 1$. Since all considered quantities are limited, is is assumed that $P(z > z_{max}) = 0$ and $P(z < z_{min}) = 0$.

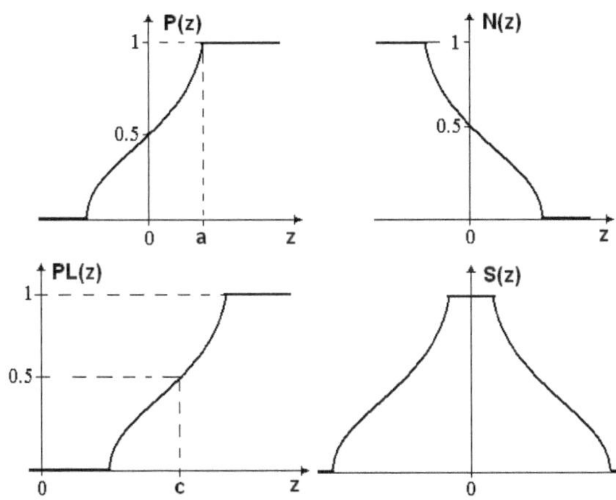

Fig. 8.1. Membership functions for the linguistic variables. $PL(z), N(z)$ and $S(z)$ have been obtained by shifts of the function $P(z)$.

Now we can find *truth value* of each rule in Table 8.1 by using (8.3) and representation of the logical connectives. As we have figured out (see Chapter 7), the connective $\boldsymbol{a} \bigvee \boldsymbol{b}$ must be represented as $\max\{\boldsymbol{a}, \boldsymbol{b}\}$, while representation of the connective $\bigwedge$ and the function $P(z)$ remains for a free choice and should be adjusted to the considered problem.

The truth value $\mathcal{T}_{\boldsymbol{n}}$ of the rule $\boldsymbol{n}.1, 2$ is:

$$\mathcal{T}_{\boldsymbol{n}} = \max\{T[P(V_n); I_{\boldsymbol{n}.1}(\boldsymbol{x}, \boldsymbol{\Theta})], T[N(V_n); I_{\boldsymbol{n}.2}(\boldsymbol{x}, \boldsymbol{\Theta})]\}, \tag{8.5}$$

where $I_{\boldsymbol{n}.k}(\boldsymbol{x}, \boldsymbol{\Theta})$ for the trained neuron are:

$$
\begin{aligned}
I_{\mathbf{1}.1} &= \max\{N(x_2), [1 - A(x_1)]PL(\Theta_b)\}, \\
I_{\mathbf{1}.2} &= \max\{P(x_2), A(x_1)PL(\Theta_a)\}, \\
I_{\mathbf{2}.1} &= \max\{T[\max\{Pr, Sr\}; S(y_1)], T[N(x_2); N(x_3)]\}, \\
I_{\mathbf{2}.2} &= \max\{Nr, PL(y_1), T[P(x_2); P(x_3)]\}, \\
I_{\mathbf{3}.1} &= \max\left\{[1 - A(x_1)]P(\Theta_a - \Theta_b), A(x_1)S(\Theta_a - \Theta_b), P(x_2)\right\}, \\
I_{\mathbf{3}.2} &= \max\left\{[1 - A(x_1)]N(\Theta_a - \Theta_b), A(x_1)P(\Theta_a - \Theta_b), N(x_2)\right\}, \\
I_{\mathbf{a}.1} &= A(x_1)Nr, \\
I_{\mathbf{a}.2} &= A(x_1)\max\{Sr, Pr\}, \\
I_{\mathbf{b}.1} &= [1 - A(x_1)]Nr, \\
I_{\mathbf{b}.2} &= [1 - A(x_1)]\max\{Sr, Pr\},
\end{aligned}
$$

while for the surrounding (control) neurons we should replace $\boldsymbol{x}$ to $\boldsymbol{y}$ and $\boldsymbol{\Theta}$ to $\boldsymbol{\Theta}'$.

Note, that in (8.5) fuzzy relation $\{\leftarrow\}$ is represented as t-norm. The reason is that the truth value of the rule, like "*From X is something Follows that V is something specific*" is close to the truth value of the expression "*X is something AND V is something specific*".

Definition of the quantities Nr (punishment), Pr (reward) and Sr (absence of environmental reaction) depend on conditions of the neuron adaptation.

For the conditioning signal these quantities are defined as:

1) for instrumental conditioning of the trained neuron

$$
\begin{aligned}
Nr_T &= 1 - A(x_1), \\
Sr_T &= A(x_1)S(\Theta_a), \\
Pr_T &= A(x_1)PL(\Theta_a),
\end{aligned}
\tag{8.6}
$$

2) for instrumental conditioning of the surrounding neurons:

$$
\begin{aligned}
Nr_C &= 1 - A(x_1), \\
Sr_C &= A(x_1)T[S(\Theta_a'); S(\Theta_b')], \\
Pr_C &= A(x_1)\{A(y_1)PL(\Theta_a') + [1 - A(y_1)]PL(\Theta_b')\}.
\end{aligned}
\tag{8.7}
$$

3) For classical conditioning for all neurons Nr, Sr, Pr are defined as:

$$
\begin{aligned}
Nr &= 1, \\
Sr &= 0, \\
Pr &= 0,
\end{aligned}
\tag{8.8}
$$

4) For habitation the environmental reactions are:

$$
\begin{aligned}
Nr &= 0, \\
Sr &= T[S(\Theta_a); S(\Theta_b)], \\
Pr &= A(x_1)PL(\Theta_a) + [1 - A(x_1)]PL(\Theta_b).
\end{aligned}
\tag{8.9}
$$

for all neurons as well.

For the differentiating signal the environmental reaction is the same as in (8.16). Note, that in (8.5)-(8.16) it is taking into account that A is 1 or 0 solely.

Finally, function $\mathcal{T}_{\mathcal{V}}$ is found as:

$$\mathcal{T}_{\mathcal{V}}(\boldsymbol{V},\boldsymbol{x},\boldsymbol{\Theta}) = T[\mathcal{T}_1;\mathcal{T}_2;\mathcal{T}_3;\mathcal{T}_a;\mathcal{T}_b;\mathcal{T}_c;\mathcal{T}_d]. \qquad (8.10)$$

$\mathcal{T}_{\mathcal{V}}(\boldsymbol{V},\boldsymbol{x},\boldsymbol{\Theta})$ itself, however, cannot be used as a possibility measure for $\boldsymbol{V}$, because the possibility measure - $\mathcal{P}(\boldsymbol{V},\boldsymbol{x},\boldsymbol{\Theta})$ must be normalized: $\sup_{\boldsymbol{V}}\mathcal{P}(\boldsymbol{V},\boldsymbol{x},\boldsymbol{\Theta})$ $=1$, while for some $\boldsymbol{x}$ and $\boldsymbol{\Theta}$ we have:

$$\sup_{\boldsymbol{V}} \mathcal{T}_{\mathcal{V}}(\boldsymbol{V},\boldsymbol{x},\boldsymbol{\Theta}) < 1.$$

So, in order to obtain $\mathcal{P}(\boldsymbol{V},\boldsymbol{x},\boldsymbol{\Theta})$ we should normalize $\mathcal{T}_{\mathcal{V}}(\boldsymbol{V},\boldsymbol{x},\boldsymbol{\Theta})$. The simplest way to do this is:

$$\mathcal{P}(\boldsymbol{V},\boldsymbol{x},\boldsymbol{\Theta}) = \frac{\mathcal{T}_{\mathcal{V}}(\boldsymbol{V},\boldsymbol{x},\boldsymbol{\Theta})}{\sup_{\boldsymbol{V}} \mathcal{T}_{\mathcal{V}}(\boldsymbol{V},\boldsymbol{x},\boldsymbol{\Theta})}. \qquad (8.11)$$

Concretization of an expression for $T[U,W]$ and defining of the function $P(z)$ complete the set-up of the fuzzy dynamics model of a neuron's adaptation/learning process.

Representation of $T[U,W]$ as $\min[U,W]$ gives an "optimistic" description of the neuronal dynamics; it is *self-consistent* and somewhat simplifies the calculation, but the other representations (for example, *pseudo-arithmetic sum*[2]) may be used as well, because in the framework of the phenomenological approach choice between these options can not be justified *a priori*. Such an arbitrariness, however, is admissible, because it does't change drastically qualitative picture of a system's evolution (see the next Section).

8.2 Fuzzy dynamics of a neural cell's learning

Construction of any fuzzy logic model begins from definition of the membership function for the linguistic variables. In our case it means that we have to define the function $P(z)$. *Q-consistent* allows to simplify definition of $P(z)$ by representation of this function as $P(z) = \Phi(L(z))$, where $L(z)$ is a piecewise linear function (see Fig. 8.2) and Φ is a monotonic function with $\Phi(0) = 0$ and $\Phi(1) = 1$.

$$L(z) = \begin{cases} 1 & \text{if } a \le z - c \le z_{max} - c \\ (1 + a^{-1}(z-c))/2 & \text{if } \max(-a, z_{min}) < z - c < a \\ 0 & \text{otherwise} \end{cases} \qquad (8.12)$$

[2] see Sec. 6.2.2

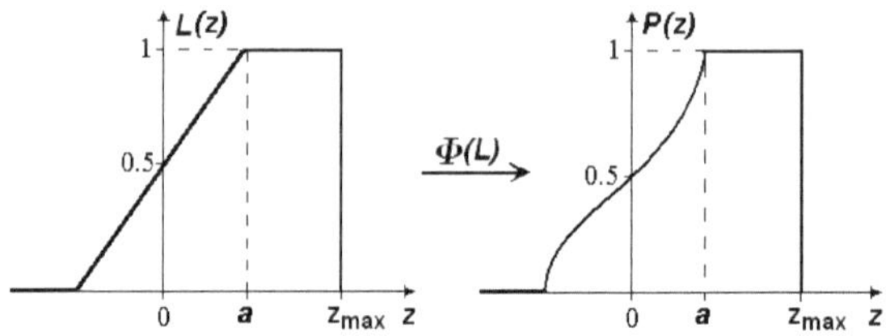

Fig. 8.2. Definition of the function $P(z)$

Parameters a, c, z_{max}, z_{min} are phenomenological parameters of our theory and they can be different for the different variables. In being considered case qualitatively reasonable results are obtained by using the functions shown in Fig. 8.3. It will be seen later how changing of these functions influence the obtained results. We have noted above (see Sec. 7.3.1) that the most possible trajectories (trajectories with $\mu(\boldsymbol{X}(t), t)) = 1)$ contain the most reliable information about the system's behavior and are the most important in the fuzzy dynamics approach.

Very often, however, for the most possible trajectories the equations (7.20)-(7.21) become *differential inclusions* (see page 342), so whole interval of the values of the system's variables: $x_i^{min}(t) \leq x_i(t) \leq x_i^{max}(t)$ (in short $x_i(t) \in [x_i^{min}(t), x_i^{max}(t)]$) will be their solution. In this case arguments of the membership functions (8.4) become set-valued variables and we should define how these functions have to be calculated. Two definitions seem to be reasonable: theoretically one can think that the neuron chooses value $x^{(1)} \in [x_i^{min}, x_i^{max}]$ or value $x^{(2)} \in [x_i^{min}, x_i^{max}]$ or ... and so on for all $x^{(k)}$ from the interval $[x_i^{min}, x_i^{max}]$. In this case:

$$P([x^{min}, x^{max}]) = \max_{x \in [x^{min}, x^{max}]} P(x). \tag{8.13}$$

On the other hand, random choice of x from the interval $[x^{min}, x^{max}]$, could be a reasonable approximation as well, because there are "smaller order" factors that influence the considered quantities, but they are not being taken into account in our theory. In this case $P([x^{min}, x^{max}])$ can be calculated as:

$$P([x^{min}, x^{max}]) = P(x^{min} + \zeta_{rand}(x^{max} - x^{min})), \tag{8.14}$$

where ζ_{rand} is a random number between 1 and 0. Interestingly that in fuzzy logic sense the last approximation is close to the statement that "*quantity x is Positive (Negative, Small, ets.), if MOST of x from the interval $[x^{min}, x^{max}]$ are Positive (Negative, Small, ets.)*" (see page 8.1). For short, we will call the first - (8.13) and the second - (8.14) definitions as "*Max*[,]" and "*Most*[,]" approximations. Both of these approximations will be used later and we will see a corresponding difference in a neuron's behavior.

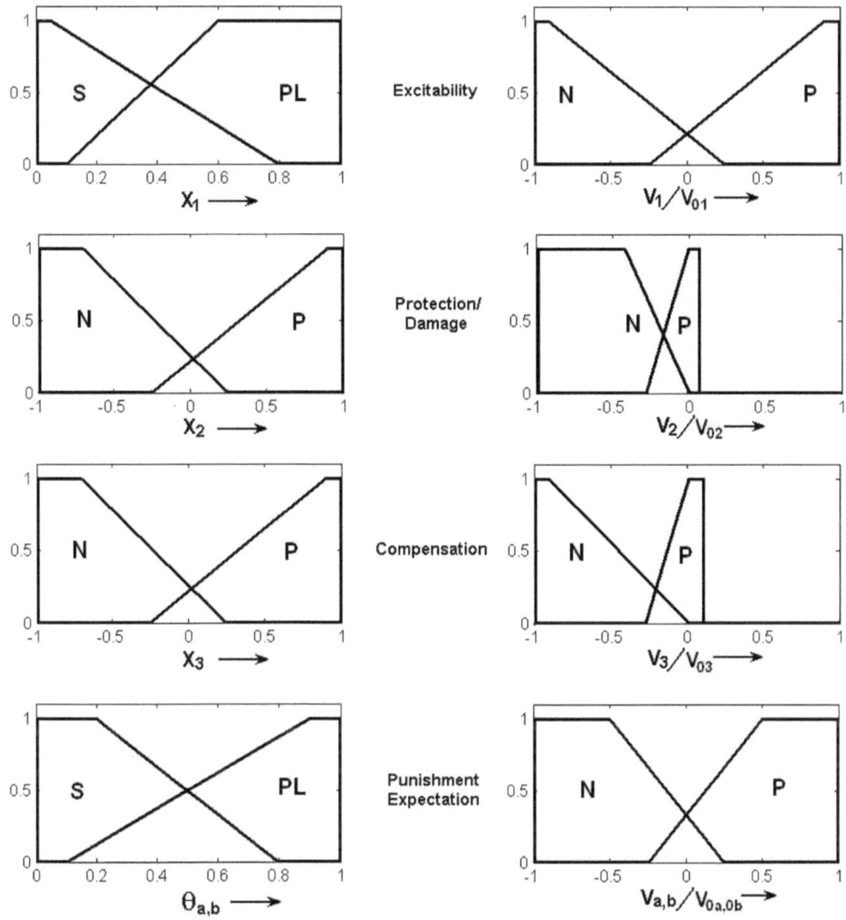

Fig. 8.3. Functions $L(x)$ for the linguistic variables "x_1 is S", "V_a is P", *etc.* We would like to call attention to the fact that unlike to $P(V_1), N(V_1)$ and $P(V_{a,b}), N(V_{a,b})$, relations between $P(V_{2,3})$ and $N(V_{2,3})$ are strictly asymmetric and $V_{02} \sim 2V_{03}$.

8.2.1 A simplified model of the neuron's learning

In order to better understand the influence of each rule on neuron behavior, we begin from the simple model, which is described by rules 1 and 3 only (see page 361). Symbolically, these rules are written in Table 8.2. This model doesn't take into account differences between training and surrounding (control) neurons, as well as differences between the conditioning and the differentiating signals, but it qualitatively describes "basic engine" of the adaptation process and allows to understand some general properties of the solution of the fuzzy dynamics model.

1.1	V_1 is Po $\leftarrow$ x_2 is Ne
1.2	V_1 is Ne $\leftarrow$ x_2 is Po
2.1	V_2 is Po $\leftarrow$ r is Sr
2.2	V_2 is Ne $\leftarrow$ r is Nr

Table 8.2. Dynamics law for the simplified model, where the environmental reaction is: $Nr = 1 - A(x_1)$, $Sr = A(x_1)$.

Truth value $\mathcal{T}_V(\boldsymbol{V}, \boldsymbol{x})$, which is described in Table 8.2, is:

$$\mathcal{T}_V(\boldsymbol{V}, \boldsymbol{x}) = T\left[\max\{T[P(V_1); N(x_2)], T[N(V_1); P(x_2)]\}\right.; \\ \left.\max\{A(x_1)P(V_2), (1 - A(x_1))N(V_2)\}\right]; \tag{8.15}$$

So, in accordance with (8.11) and (8.4) possibility-function for the velocity $\boldsymbol{V} = \{V_1, V_2\}$ is:

$$\mathcal{P}(\boldsymbol{V}, \boldsymbol{x}) = \max\left\{\frac{T[P(V_1); P(-x_2); A(x_1)P(V_2) + (1 - A(x_1))P(-V_2)]}{\max\{P(x_2), P(-x_2)\}},\right. \\ \left.\frac{T[P(-V_1); P(x_2); A(x_1)P(V_2) + (1 - A(x_1))P(-V_2)]}{\max\{P(x_2), P(-x_2)\}}\right\}, \tag{8.16}$$

Corresponding differential inclusions for the most possible trajectories (see Sec. 7.3) are:

$$\begin{aligned} \frac{dX_1}{dt} &= \tilde{V}_1, \\ \frac{dX_2}{dt} &= \tilde{V}_2, \\ X_1(0) = x_1 \ &; \ X_2(0) = x_2, \\ \mu_0(\boldsymbol{x}) &= 1, \end{aligned} \tag{8.17}$$

where $\tilde{V}_1, \tilde{V}_2$ are the set valued functions, which is found from the condition $T[\mathcal{P}(\boldsymbol{V}, \boldsymbol{x}); \mu_0(\boldsymbol{x})] = \mu_0(\boldsymbol{x})$. Since we consider only the most possible trajectories we have $\mathcal{P}(\tilde{\boldsymbol{V}}, \boldsymbol{X}) = 1$, which leads to :

$$\max\{T[P(\tilde{V}_1); P(-x_2); A(x_1)P(\tilde{V}_2) + (1 - A(x_1))P(-V_2)], \\ T[P(-\tilde{V}_1); P(x_2); A(x_1)P(\tilde{V}_2) + (1 - A(x_1))P(-\tilde{V}_2)]\} = \\ = \max\{P(x_2), P(-x_2)\}. \tag{8.18}$$

The last equation is easily solved in the implicit form:

$$\begin{array}{ll} P(\tilde{V_1}) = 1 & \text{if } P(-x_2) > P(x_2), \\ P(-\tilde{V_1}) = 1 & \text{if } P(x_2) > P(-x_2), \\ P(\tilde{V_2}) = 1 & \text{if } A(x_1) = 1, \\ P(-\tilde{V_2}) = 1 & \text{if } A(x_1) = 0, \end{array} \tag{8.19}$$

where $P(V_1), P(V_2)$ are shown in Figs. 8.2 and 8.3. We see that whole intervals of the velocities, corresponding to the top plateaus of the functions P, satisfy the solution (8.19). Differential equations (in fact, inclusions) can be solved analytically, but the final analytical expression is clumsy and hardly interpretable, so the better way is to present the solution in a graphical form (see Fig. 8.4).

We see that the simple model demonstrates "oscillating" behavior, which is not amazing, because dynamics law in Table 8.2 "linguistically" equivalent to the fuzzy oscillator has been considered in Sec. 7.3.2. Now, however, oscillations are strongly non-linear which corresponds to the non-linear character of system of the equations (8.17)-(8.18). Qualitatively, behavior of the model for *Max*[,] and *Most*[,] approximations are quite similar, but *Max*[,] approximation leads to a more "sharp" picture of a bundle of the most preferable trajectories, and the neuron's response (AP) becomes a little bit more stable. As we will see in the next section this is a quite general situation, so both of the approximations are qualitatively appropriated.

Adding for the effect of compensation to the simple model, almost doesn't change the model's behavior (see Fig. 8.5). This means that realistic phenomenological theory of adaptation/learning process should take into account additional characteristics of a neural cell.

8.2.2 Solutions of fuzzy dynamics model of a neural cell's behavior

The simplified fuzzy model, which has been formulated in Table 8.2, doesn't give a satisfactory description of the adaptation/learning process of a neural cell. So, apparently, the model, which is described in the Table 8.1, is a "minimal" proper phenomenological model of a neuron's behavior. Note, that taking into account *expectation* of environmental reaction on a neuron's response - Θ reveals a difference between the trained and the control (surrounding) neurons, since environmental reactions for their responses are different. Moreover, under instrumental conditioning, environmental reaction for both the trained and control neurons depends on the AP-activity of the trained and surrounding neurons (see Chapter 1) together, so they co-evolute collectively and their dynamics cannot be considered separately. Possibility-measures of the dynamics laws for the trained - $\mathcal{P}(\boldsymbol{V};\boldsymbol{X},\boldsymbol{Y},\boldsymbol{\Theta})$ and for the control - $\mathcal{P}(\boldsymbol{W};\boldsymbol{Y},\boldsymbol{\Theta}',A(X_1))$ neurons should be calculated, now from the Eqs. (8.5)-(8.7) and (8.10)-(8.11). In fact, expression for $\mathcal{P}(\boldsymbol{W};\boldsymbol{Y},\boldsymbol{\Theta}',A(X_1))$ can be obtained from $\mathcal{P}(\boldsymbol{V};\boldsymbol{X},\boldsymbol{Y},\boldsymbol{\Theta})$ by replacement: $\boldsymbol{V} \to \boldsymbol{W}$, $\boldsymbol{X} \to \boldsymbol{Y}$, $\boldsymbol{\Theta} \to \boldsymbol{\Theta}'$, but for the environmental reactions we should use (8.7) instead of (8.6). Note,

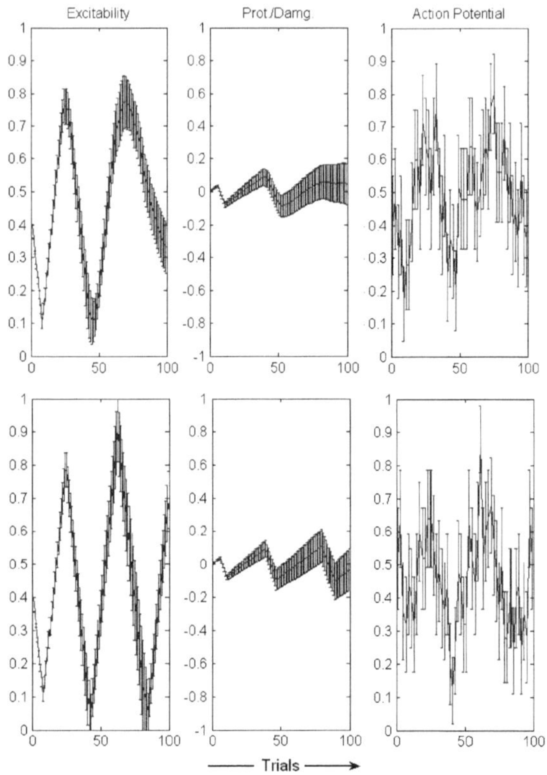

Fig. 8.4. Solution for the simple model. The first row corresponds to the *Most*[,] and the second to the *Max*[,] approximations. Initial values of the variables were: $x_1 = [0.45, 0.5]$; $x_2 = 0$. The bars in the graphs for *excitability* and *damage* show intervals of their values corresponding to the most possible ($\mu = 1$) way of the system's evolution. The last column shows average activity (points) and standard deviation (bars) of the neuron's activity (AP) for 30 repeats of the learning process. AP was calculated as *Most*[,] and *Max*[,] approximations for $p(x_1)$ in (8.2) with $\theta = 0.5$.

that differentiating signal expressions (8.16) instead of (8.6), (8.7) should be used for both neurons.

Since the equalities $\mathcal{P}(\boldsymbol{V}, \boldsymbol{X}, \boldsymbol{Y}, \boldsymbol{\Theta}) = 1$ and $\mathcal{P}(\boldsymbol{W}; \boldsymbol{Y}, \boldsymbol{\Theta}', A(X_1)) = 1$, Eq. (8.11) leads to:

$$\mathcal{T}_{\mathcal{V}}(\tilde{\boldsymbol{V}}, \boldsymbol{X}, \boldsymbol{Y}, \boldsymbol{\Theta}) = \sup_{\boldsymbol{V}} \mathcal{T}_{\mathcal{V}}(\boldsymbol{V}, \boldsymbol{X}, \boldsymbol{\Theta}),$$
$$\mathcal{T}_{\mathcal{V}}(\tilde{\boldsymbol{W}}, \boldsymbol{Y}, \boldsymbol{\Theta}', A(X_1)) = \sup_{\boldsymbol{W}} \mathcal{T}_{\mathcal{V}}(\boldsymbol{W}, \boldsymbol{Y}, \boldsymbol{\Theta}', A(X_1)). \tag{8.20}$$

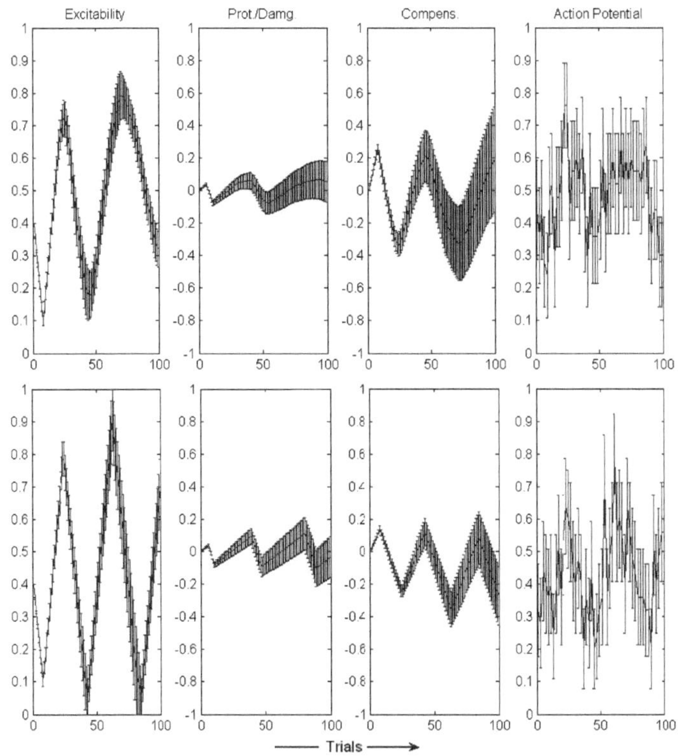

Fig. 8.5. Influence of the compensation mechanism on the instrumental reaction elaboration.

In accordance with our choice of the membership functions for $\boldsymbol{V}$ (see Fig. 8.1), solutions of (8.20) are set-valued, so mutual equations for evolution of the trained and the control neurons will be actually, differential inclusions like (8.17). Eqs. (8.20) corresponds to evolution under conditioning signal. For differentiating signal we will have $\tilde{\boldsymbol{V}}^{dif} = \tilde{\boldsymbol{W}}^{dif}$ and these quantities have to be found from the equation:

$$\mathcal{T}_{\mathcal{V}}(\tilde{\boldsymbol{W}}, \boldsymbol{Y}, \boldsymbol{\Theta}') = \sup_{\boldsymbol{W}} \mathcal{T}_{\mathcal{V}}(\boldsymbol{W}, \boldsymbol{Y}, \boldsymbol{\Theta}'). \tag{8.21}$$

Finally, evolution of the neurons under instrumental conditioning is found from:

$$\begin{gathered} \frac{dX_i}{dt} = \tilde{V}_i^s \; ; \quad \frac{dY_i}{dt} = \tilde{W}_i^s \, , \\ \frac{d\Theta_\alpha^s}{dt} = \tilde{V}_\alpha^s \; ; \quad \frac{d{\Theta'}_\alpha^s}{dt} = \tilde{W}_\alpha^s \, , \\ X_i(0) = x_i, \; \Theta_\alpha^s(0) = \theta_\alpha \; ; \quad Y_i(0) = y_i, \; {\Theta'}_\alpha^s(0) = \theta'_\alpha \, , \end{gathered} \tag{8.22}$$

$$\mu_O(\boldsymbol{x}, \boldsymbol{\theta}_\alpha) = 1\ ; \quad \mu_O(\boldsymbol{y},\ \boldsymbol{\theta}'_\alpha) = 1,$$

where $i = 1, 2, 3$; $\alpha = 1, 2$ and the index $s = \{con,\ dif\}$ corresponds to the conditioning or to the differentiating signals.

The system (8.20)-(8.22) can be solved analytically in an implicit form, but the solution is too immense, in order to be analyzable. So the better way is to solve this system numerically. A corresponding solution for the *Min*-representation of the *AND* connective in the dynamics rules is shown in the Fig. 8.6 and Fig. 8.7 (*Most*[,] and *Max*[,] approximations correspondingly). In order to fix it up to the experimental situation (see Chapter 1) the control neuron activity has been taken some higher (i.e. $\theta_{con} < \theta_{tr}$) than activity of the trained one and conditioning and differentiating signals were alternated randomly.

Comparison of Figs. 8.6 and 8.7 with Fig. 8.4 shows that taking into account of the neuron's *expectation* of an environmental reaction meaningfully change the model's behavior. Instead of joyless oscillations, we see quite a realistic picture of the adaptation process both for the trained and for the control neurons. In spite of uncertainty of description of the neuron's behavior increases with time, the model indicates clear progress of the trained neuron in elaboration of the instrumental reflex, while the control neuron demonstrates habitation at the later stage of the training procedure. The trained neuron quickly recognizes the difference between the environmental reactions on the responses on the conditioning and the differentiating signals. We see also, that the *Max*[,] approximation leads to a sharped picture of the system behavior and reveals more stable neuron response.

It should be emphasized that the model's behavior is quite sensitive to choice of the parameters $a, c, z_{max,min}$ of the membership functions. In particular, as it can be seen in Fig. 8.3 that the membership functions *Positive* - $P(V_2)$ and *Negative* - $N(V_2)$ for velocity of change of the *Damage/Protection* variable x_2 were defined as considerably asymmetric. In Fig. 8.9 one shows what happens with the neuron's behavior if we change their definitions as it is shown on the Fig. 8.8. We see that the ability of the trained neuron to elaborate the instrumental reflex is completely lost, the neuron's damage is overcompensated and the neuron becomes overprotected at the later stage of the adaptation process. This feature of the phenomenological model is quite realistic. Remember that in accordance with definition of the variable x_2 positive value of x_2 corresponds to the neural cell's protection, while the negative one corresponds to the cell's damage. So, positive V_2 describes the velocity of increase of the cell's protection and negative V_2 describes velocity of increase of the cell's damage. Asymmetry in definitions of $P(V_2)$ and $N(V_2)$ means that the neural cell's protection should increase much more slowly than increasing of the cell's damage in order for neural cell to be able to elaborate the instrumental reaction.

In Fig. 8.10 effect of changing of the parameters of the compensation's velocity is shown. In this case the neural cell doesn't lose the ability of the

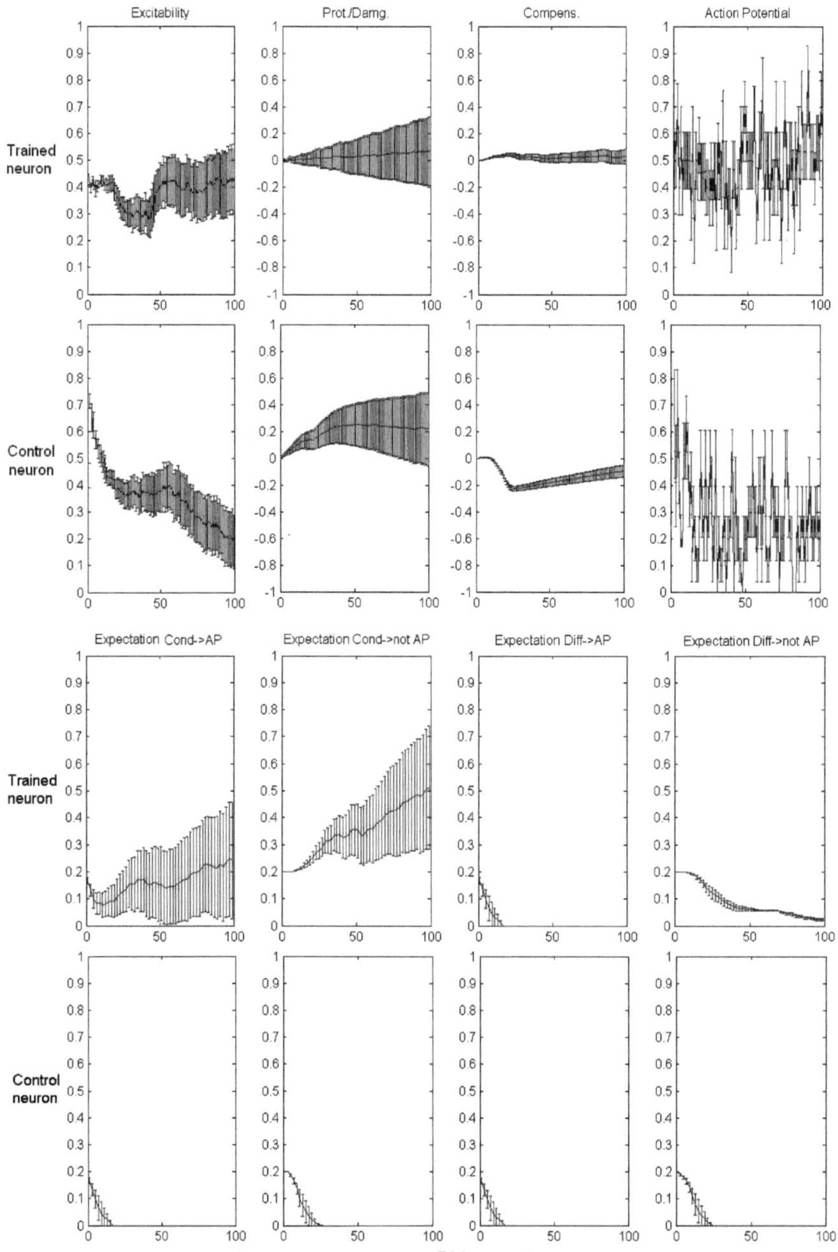

Fig. 8.6. Elaboration of the instrumental conditioning reaction (*Most*[,] approximation). AP's was calculated in accordance with (8.2) with $\theta = 0.5$, $\beta = 2$ for the control and $\theta = 0.7$, $\beta = 2$ for the trained neurons (points - average and bars - standard deviation for 30 repeats is shown).

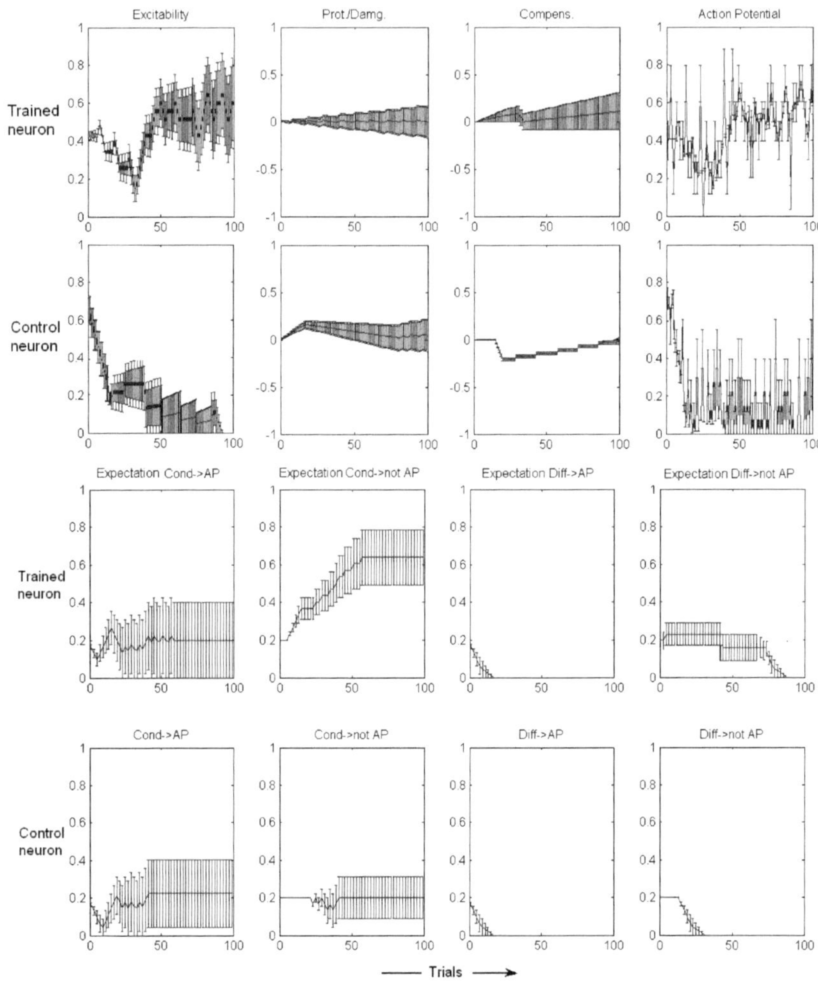

Fig. 8.7. Elaboration of the instrumental conditioning reaction ($Max[,]$ approximation. AP's was calculated in accordance with (8.2) with $\theta = 0.5$, $\beta = 2$ for the control and $\theta = 0.7$, $\beta = 2$ for the trained neurons, average and standard deviation for 30 repeats is shown).

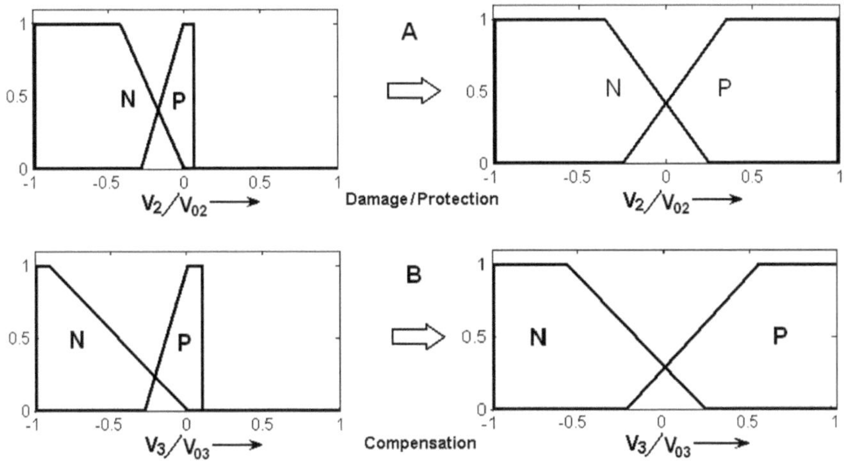

Fig. 8.8. Changing of definition of the membership functions $P(V_2)$, $N(V_2)$ and $P(V_3)$, $N(V_3)$.

instrumental reaction's elaboration, but at the later stage of the training procedure the cell's protection becomes overcompensated and the cell remains damaged despite elaboration of the instrumental reaction. In Fig. 8.11 and Fig. 8.12, elaborations of the classical conditioning and habitation are shown. The dynamics inclusions in these cases are the same as (8.21)-(8.22), but for the environmental reactions the expressions (8.15) and (8.16) should be used (for both neurons). In both cases the more active neuron elaborates then reaction quicker than less active one. Note, that under habitation the neural cell becomes overprotected and compensation mechanism tries to compensate this protection.

How does the representation of *AND* connective influence on the model behavior? In Fig. 8.13 we show the numerical solution of the system (8.20)-(8.22) for *pseudo-arithmetic sum* as the *AND* connective. The function f (see Eq. (6.11) was chosen as $f(w) = \exp(-kw)$, which leads to $T[U, W] = UW$. Qualitatively, the system behavior is still the same as for *Min*-representation of the *AND* connective, but there are some quantitative differences. Apparently, such a situation will take place for another *Q-consistent* representations of $T[U, W]$ as well.

Interestingly, that for this $T[U, W]$ we meet with quite a rare case when bundle of the most preferable trajectories of the evolution is collapsed during the time (this means that sometimes uncertainty of prediction of a system's behavior may decrease during the time).

Presented results allow to conclude that fuzzy-dynamics is a powerful and promising tool that can be successfully used in the phenomenological theory of the neural cell's behavior. As we will argue in the next Chapter, its description

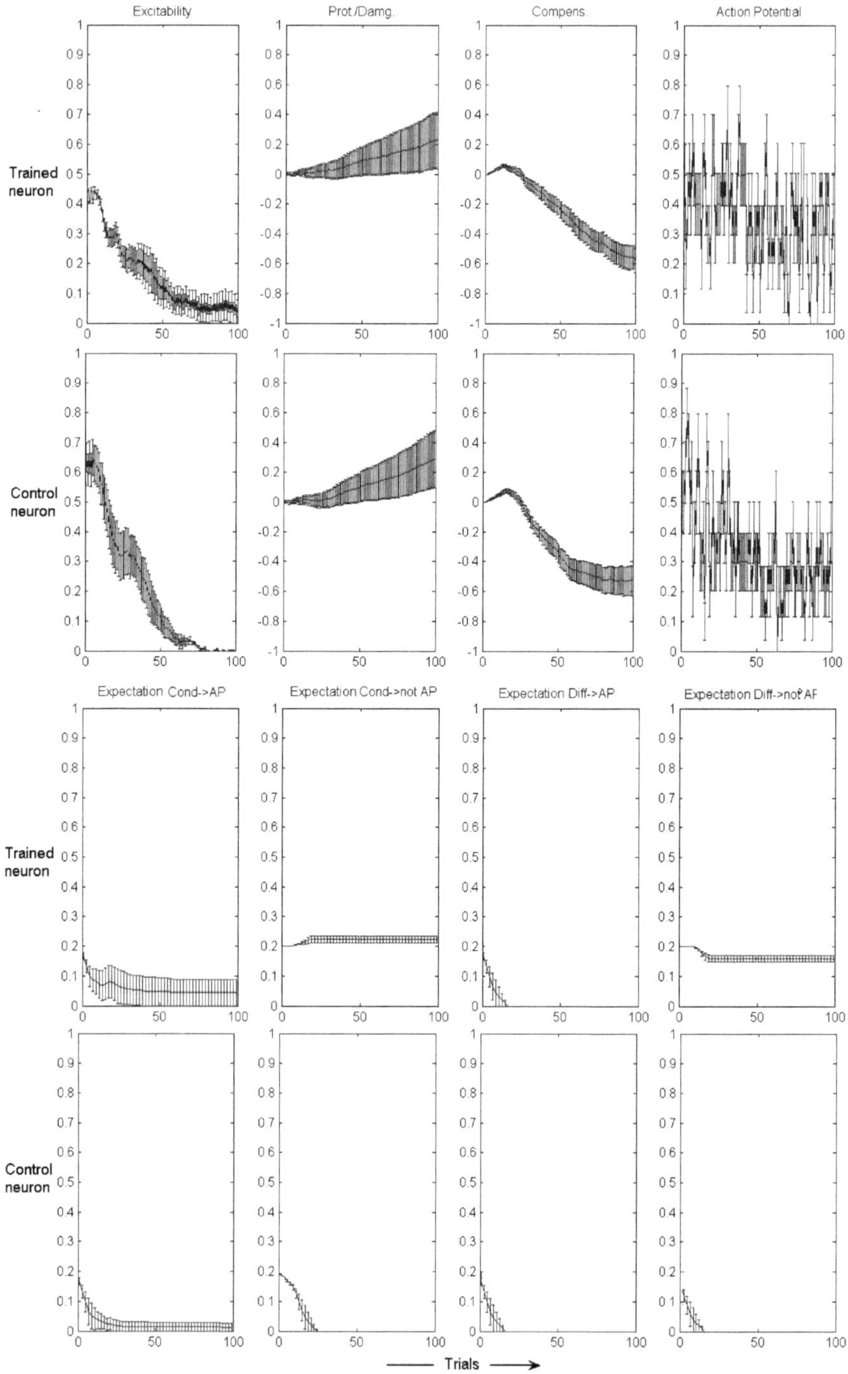

Fig. 8.9. Influence of changing of definition of the membership functions $P(V_2)$ and $N(V_2)$ (Fig. 8.8A) on the system's behavior. Remember that $N(V_2)$ indicates truth value for velocity of increasing of *damage*, while $P(V_2)$ indicates truth value for velocity of increasing of *protection*.

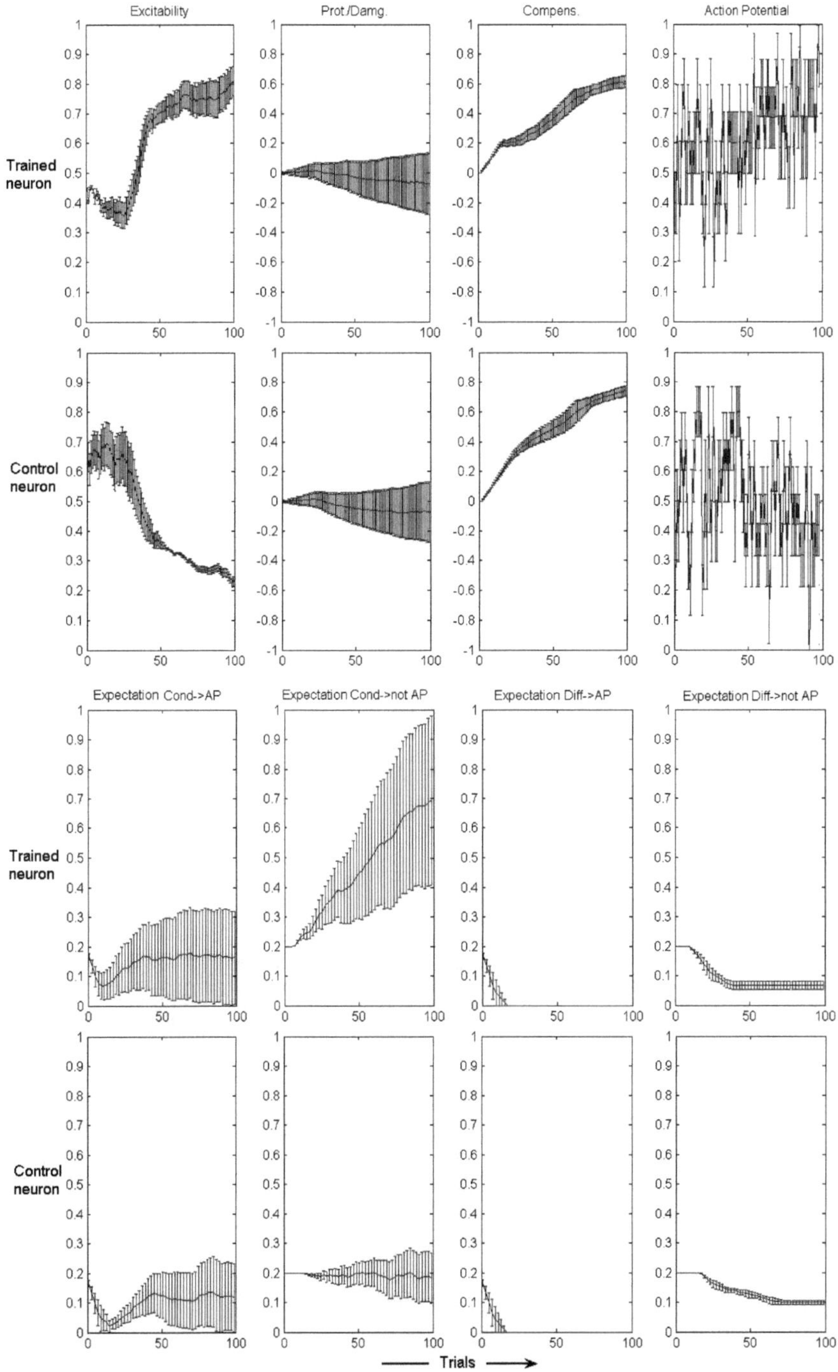

Fig. 8.10. Influence of changing of definition of the membership functions $P(V_3)$ and $N(V_3)$ (Fig. 8.8B) on the system's behavior.Remember that $N(V_3)$ indicates truth value for velocity of compensation of *damage*, while $P(V_3)$ indicates truth value for velocity of compensation of *protection*.

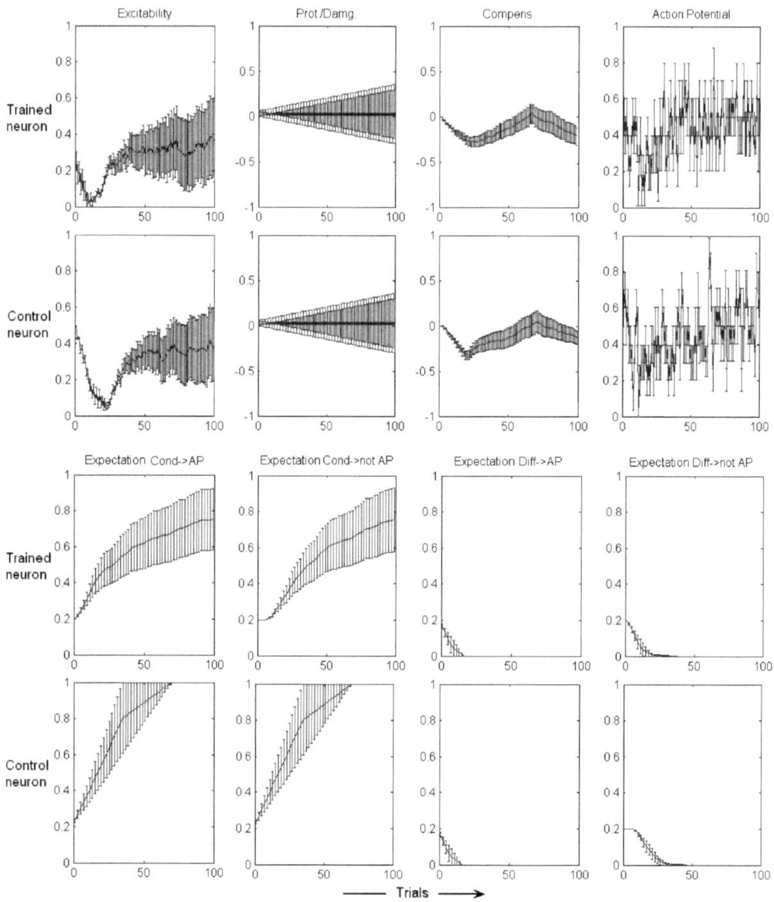

Fig. 8.11. Elaboration of the classical conditioning reaction. (*Most*[,] approximation. AP's was calculated in accordance with (8.2) with $\theta = 0.5$, $\beta = 2$ for the control and $\theta = 0.7$, $\beta = 2$ for the trained neurons, average and standard deviation for 30 repeats is shown).

of the adaptation/learning process is in good agreement with the experimental data at least in first approximation.

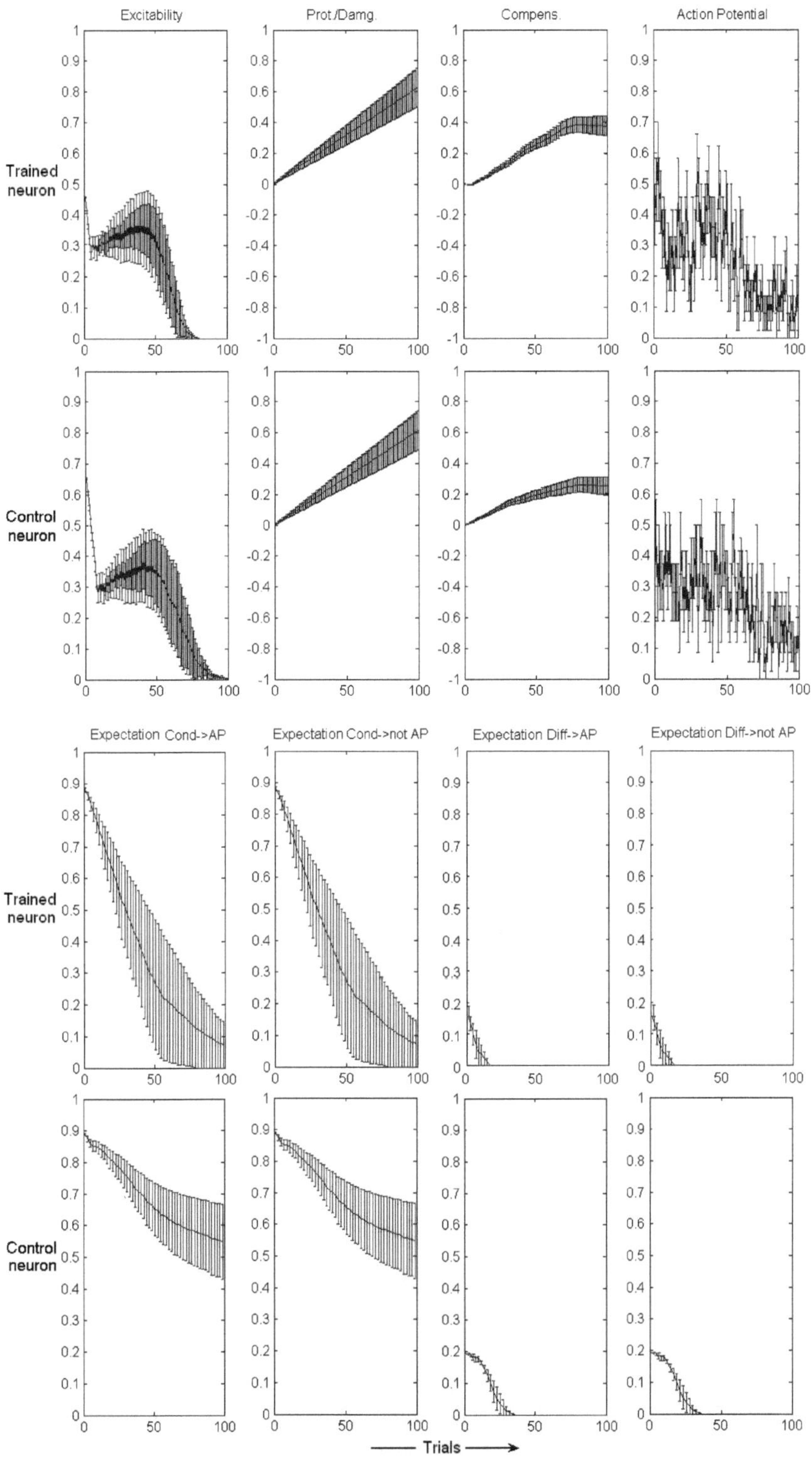

Fig. 8.12. Elaboration of the habitation. (*Most*[,] approximation. AP's was calculated in accordance with (8.2) with $\theta = 0.5$, $\beta = 2$ for the control and $\theta = 0.7$, $\beta = 2$ for the trained neurons, average and standard deviation for 30 repeats is shown).

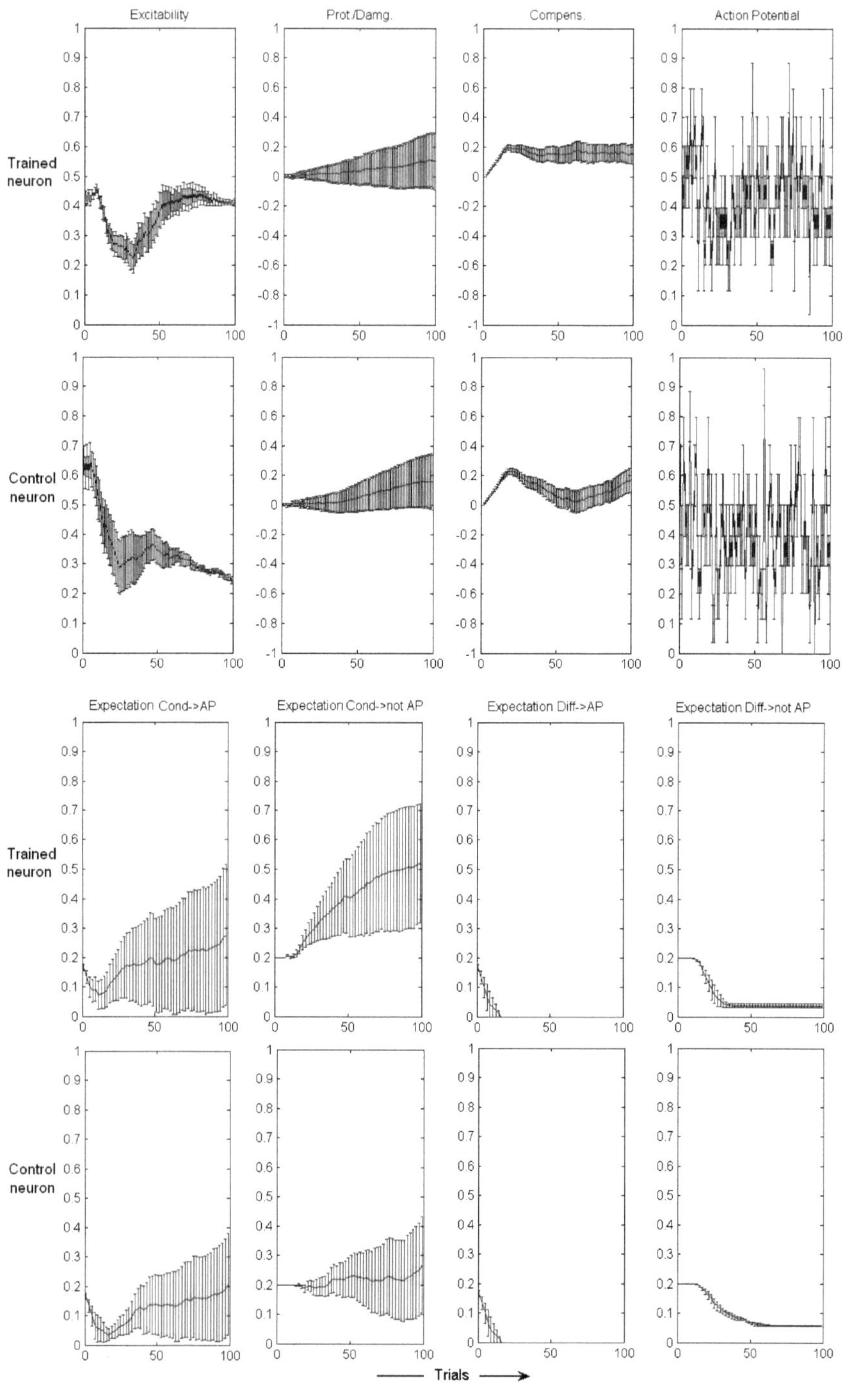

Fig. 8.13. Elaboration of the instrumental conditioning reaction in the model with another representation of the $\bigwedge(AND)$ connective: $T[U, W] = UW$ ($Most[,]$ approximation. AP's was calculated in accordance with (8.2) with $\theta = 0.5$, $\beta = 2$ for the control and $\theta = 0.7$, $\beta = 2$ for the trained neurons; average and standard deviation for 30 repeats is shown).

9

Conclusion: Is real neuron a primary fuzzy unit?

9.1 The operation of memory

The problems of brain and behavior are so difficult, that sometimes even speculative hypotheses are difficult to formulate, even if one imposes as the only constraint: lesser mysticism. The perplexing question is: how are physiological events converted into subjective feeling? As a minimum, these two kinds of events, objective and subjective, are interdependent and therefore a physiological approach is plausible. At the present, there are some disciplines, headed by philosophy, that are dealing with this problem. During the last years, physiologists, leaning on Sir F. Crick's authority, have also begun to delve into this problem.

Anyone who acquaints himself with a neural system is surprised by the complexity of the brain (the same complex organ) and the neuron (the same complex cell). We know too much and comprehend too little. According to prevailing understanding, if we confine ourselves to a textbook level, the brain receives information from receptors, carries out a large volume of calculations and sends results to effectors. A neuron, according to the same textbook, adds up coming excitations and inhibitions and compares the sum with the threshold, though the calculation might be complicated because of neuronal geometry, and sends action potentials in the output. After training, the efficiency of some pathways increases while others decrease, depending on circumstances, and thus brain accumulates memory. If so, why does a brain need thousands of biochemical reactions, why does brain use tens of billions of neurons and glial cells and why do operations of this wonderful construction often endure catastrophic destruction and stay intact, while a short-term drop in metabolism destroys brain function while preserving the construction unbroken?

Two main avenues of neuroscience emphasize a prevailing importance of neural network or neuron. Brain is considered, either as a complex network consisting of simple neurons, or as a very complex network of self-contained neurons with the accent put on a neuron's complexity. Depending on our approach, memory, principally, may be considered as the establishment of new connectivity in the space of the brain or as a code in chemical language.

A relatively simple task, although not a trivial one, is to determine where the particular functions of brain are localized. Some functions seem to be distributed within brain, so that one area may control a great number of functions. Some other functions have concrete address within the brain. The degree of localization seems to be different for ancient and new functions, for fresh and automatic memory, for important and unessential information, for conscious and unconscious memories, etc. Extent of localization grows in high animals, but even local functions have the potential to become delocalized. Likely, delocalization of function is a primary principle in the brain, while advanced brain has specific constructions for recent functions and does not rebuild ancient gadgets.

Therefore, there is a question: inasmuch as the brain perfects its construction in evolution, when it acquires new functions maybe it, too, completes the construction during an individual life, when it acquires new habits? Neurons in various brain regions change their activity correspondingly to a learning procedure and this indicates scattered recording of new information. However, in special experiments, learning may be artificially directed to a restricted population of neurons and even to a single neuron. This means that even if in natural condition many neurons participate in the same function, a single neuron is potentially capable of learning. Moreover, in exceptional cases, a single neuron has power over an entire behavior. The only possible 'connectivity' for a single neuron is connectivity in the world of biochemical reactions.

During learning, some stimuli acquire high biological significance, while other stimuli become insignificant and neurons somehow decide which stimulus is important in the current behavior. Change in synaptic efficacy and excitability are the means for modulation of neuronal activity. If such modifications are persistent and serve as the traces of long-term memory, this means that learning rebuilds the structure of brain. Really, both synaptic efficacy and excitability are changed during learning. Whether these modifications are persistent or transient is now a controversial topic. A persistent mechanism is simple and is therefore attractive, but it may explain only unspecific effects. For instance, if synaptic efficacy (presynaptic supply of neurotransmitter, density of chemoreceptors, etc.) or neuron excitability (membrane potential, input resistance, etc) have changed in a stationary manner, this will exert modification of any signal involving these synapses or neurons, not only the signal that had participated during learning, that is whose biological significance changed as a result of learning. Actually, persistent changes were found only during unspecific forms of learning such as sensitization, long-term potentiation and depression and some forms of learning connected with a stress and therefore also of low specificity.

A specific form of plasticity definitely better corresponds to properties of learning, but unspecific forms also exist. A non-specific form is observed at both levels, behavioral and neuronal. The most primitive, unspecific forms of behavioral plasticity related to fatigue, exhausting of resources and stress. These primitive forms of plasticity might be an evolutional ancestor of learning. Particularly during the beginning of associative learning, a phase of generalization is observed [947] when, together with the change of responses to the CS^+, responses to the stimuli resembling the CS^+ also change. Unspecific alterations take place also during homeostatic reorganizations of neuronal activity.

Behavioral and neuronal responses change selectively during habituation and associative learning. Particularly in some cases, it has been demonstrated that material substrates for synaptic efficacy and excitability are not anchored to specific locations in the brain. Efficacy of the same synapse and excitability of the same neuron flexibly alters, depending on current behavior. Static interconnections and fixed excitability are not relevant for explanation of short-term memory. A functional plasticity is not reduced to structural connectivity. This means that chemical processes, beginning in the activated synapses, interact within a neuron and therefore efficacy of the synapse depends on other simultaneously-activated synapses. Similarly, a neuron displays different excitability in the correspondence to different signals. Neurons evaluate input signals according to their significance, not only according to their strength; afterwards they somehow choose an appropriate excitability and affect their output volley. Thus, neurons are selectively activated or inhibited for one stimulus, but not for another. Such a parsimonious reorganization of neuronal properties extends the possibilities of neurons to participate in various types of activity. Recently it was established that neurons display selective excitability not only as a result of learning. Excitable membrane plasticity is a reason for neuronal reaction selectivity during visual image recognition and during responses to the preferred and non-preferred orientations of the whisker angular deflection.

Many indirect observations illustrate the chemical nature of memory. For instance, stability of memory after severe brain damage with catastrophic impairment of neural network structure demonstrates that a chemical hypothesis may be correct. Information is lost during moderate damage through after a long period of time (weeks or months) is usually recovered. Where else, except in the chemical substratum, may temporarily missed information be maintained? Memory may stay intact while brain morphology is strongly reorganized in the course of individual development, for instance, during insect and amphibian metamorphosis. These facts are difficult to explain without an assumption of the chemical substrate of memory. In addition, experimental evidence exists verifying a close link between learning, memory and intracellular biochemical processes.

Mechanisms of selective forms of plasticity are less transparent compared with unselective plasticity. We developed the detailed scheme of chemical reactions, supposedly occurring in the neural cells and the properties of such a model neuron resemble the properties of a live neuron in respect to elaboration of habituation, classical conditioning and the instrumental reflex (during short-term memory). Neural networks consisting of such model neurons demonstrate an advantage over other models. Although chemical processes in the real neuron may differ from our scheme, our model demonstrates that the system of chemical reactions may explain basic properties of learning without implementation of changes in neuron connectivity.

We and others have demonstrated that chemical alterations do take place during learning. However, we do not know whether the same chemical events encode the same elements of memory in different neurons. It may be that various neurons must unite their scanty resources of memory in order to create a whole image. We cannot suggest a reasonable hypothesis of long-term memory. The idea of a role for the reverse route (nontemplate RNA synthesis and reverse transcription) for recording of the acquired memory into a stable molecule of DNA is preliminary, but it may be tested by the comparison of a diversity of DNA structures in infants and adults.

9.2 The verve of injured neurons

When brain operates, its neurons and glial cells permanently spend and recover their resources and cellular being is in a stream of interaction between damage and repair. Failure of homeostasis immediately leads to damage and death. Neurons and glial cells survive together. They collaborate in maintenance of ion homeostasis, exchange of neurotransmitters and important metabolites, ensure spread of both slow electrical potentials and damage in neural tissue. Damage and recovery of brain tissue are not at all times connected with pathology; they may be associated with normal brain functions. Slow reorganization of behavior accompanied by damage and recovery of neural cells in the high neural centers, connected with behavior, are especially sensitive to damage. Neurons and glia, probably, jointly perform physiological functions connected with long-term alterations of behavior. Neurons execute current control of behavior, while glial cells participate in the time-consuming regulation of cerebral activity. Glial cells participate in specific neurological diseases, in regulation of mood and their activity is related to arousal, levels of motivation and learning, but not to current flow of behavior, as is the activity of neural cells.

Cells avoid damage by means of a homeostatic system, which contains protein sensors, responsible for regulation the particular variables of being. Nevertheless, persistent disturbance of particular characteristics of an injured cell does not lead to death, since homeostasis may tune other variables and thus compensate deleterious effects of the distorted variable. Therefore, homeostasis regulates a certain general quality of a neural cell, for example, the level of damage, but the manner of such regulation is unknown. Such a system of regulation resembles a bit the system of an authority, which strictly brings to light street hooliganism, small stealing, keeps streets clean and punishes infringement of traffic regulation. Besides, the authority makes a great effort against corruption by means of augmentation of taxes.

Our discussion has focused on the subject of an organism and its cells, on the diverse ways that lead from damage to death and on the means by which cells may be defended from death. The total diversity of forms of damage proves to be not too considerable; they roughly correspond to the spatial diversity of brain elements. In Chapter 1 we have shown that the variety of information by which a brain operates exceeds the entire sum of brain elements, neurons and synapses: diversity, through which the network hypothesis operates. The chemical hypothesis corresponds by appropriate variety to a set of memory elements, though the nature of chemical substratum of memory is vague. A chemical nature of memory has a variety of experimental evidences. Memory as a reorganization of a neural network is feebly compatible with experiments. Direct proofs for this hypothesis were revealed only in the example of LTP that was considered as a model of memory. In Chapter 2 we have demonstrated that LTP is too crude a memory paradigm and corresponds more to conditions of cell damage-protection. Thus, the network hypothesis loses its basic proofs. The concept of neurons with static interconnections of fixed or slowly changing efficacy (during learning) is no longer appropriate. However, to reveal that LTP is the result of damage is not the same as to insist on its primitive character. Damage and, especially, homeostatic compensation is anything but simple. Homeostasis is a perfect adjustment for small deviations and perfect management in the case of serious breakages. Indirect hints point us to the apparatus of memory as an amplifier of the power of homeostasis. What is diversity of damage-protection states and does this diversity correspond to a memory set?

The diversity of protective paths may be greater than the diversity of damages, since it corresponds to a variety of metabolic pathways that homeostasis uses for compensation multiplied by the number of cracks. However, the pool of homeostatic pathways may be also not overly huge relative to the pool of memory elements, and then the role of homeostasis in a behavior will be restricted by a correction of the most important decisions. Really, despite a variety of metabolic pathways of homeostasis, only a small part of which we touched, this number does not exceed the entire sum of proteins that a cell can synthesize and only a fraction of the proteins that participate in homeostatic regulation. Therefore, even if we will take into consideration a possible combination of a few proteins for mutual control of complex functions, their overall number cannot reach the overall number of memory elements. The experimental data known at present also do not indicate a large assortment of protective means compared to the assortment of possible behaviors. Certainly, the power of homeostasis would be strongly extended if it could use neuronal memory. All the more, the relationship of homeostasis to the system of neuronal memory is not an entirely speculative hypothesis.

Damage of neural tissue is caused not only by injury of different kinds, but may be evoked by a physiological activity. Particularly, extreme excitation usually causes cell damage, while inhibition protects neurons from death. On the other hand, moderate damage increases excitability of neurons, whereas protection inhibits neurons. Severe damage deregulates ion equilibrium and excitability of a dying neuron bleakly decreases. In these cases, hyperpolarization protects such neurons from injury and paradoxically excites them. This means that homeostatic compensation of damage may lead to generation of output neuronal reactions. Some indirect data indicate a possible correction of homeostasis under the influence of acquired memory; at least, both phenomena are compatible in their sense and time scopes. At the neuronal level, control of ion homeostasis has been shown to undergo behavior-related plastic changes and may be directed toward anticipatory compensation of the factors that lead to a disturbance of the homeostasis. Further, damage triggers general protective mechanisms and this may ameliorate consequences of yet-to-come more severe injury. Yet, protective preconditioning by means of a weak injury actuates unspecific protection. This resembles a simple, non-associative form of learning. Nevertheless, partial selectivity of damage-protection, dependent on experience, is often observed in the experiments. In addition, an acute influence of deleterious and protective factors usually has opposite results in comparison with their chronic action.

Although we may say a little about enrichment of the homeostasis repertoire at the expense of memory, definition of the reverse problem looks more convincing. There are some data concerning the relative participation of homeostatic compensation in the control of behavior. First of all, damage and homeostatic compensation directly affects neuronal activity. On the other hand, the factors affecting cell damage and homeostasis, such as second and retrograde messengers, cytokines, neuropeptides, gap junction regulators, anesthetics, etc. possess a clear psychoactive potential. They relate to anxiety, pain, appetite, epilepsia, insomnia, depression and many other slow changes of awareness, even if their interactions with the brain tissue are subtle and evoke only the threat of damage instead of real injury. Further, learning-related influences of the sensory deprivation to the following behavior during development indicate a connection between damage and behavior. Complex phenomena of memory recovery after brain damage may be also somehow related to a homeostatic compensation.

At the same time, control of even that number of variants of compensational processes, which already have been investigated, is astonishing. Homeostasis maintains optimal values of a large quantity of vital variables. Moreover, if for any of them the deficiency becomes unmanageable, homeostasis alters the set of other variables, serviceability of the brain recovers and afterwards these new optima are maintained. The mechanism of change of priorities for the current set points is unclear, exactly as the mechanism of change of dominated behavior, dependent on context, was unclear in the previous chapter. We did not know, how, after a change of environment, the current quasi steady combination of short-term memory molecules may be modified into a new quasi steady combination of molecules by means of an apparatus of long-term memory. Similarly, we do not know how homeostasis reorganizes its function, depending on forced changes of equilibrium conditions. The formation of a new combination of stable factors by homeostasis one can compare also with the puzzle of coalescence of sprinkled features in a whole image. In order to recognize the combination of features, all traits of the image ought to be integrated. We referred this as a 'binding problem". However, homeostasis solves this problem from another side. In the beginning, the set of new optimal values is not recognized. Recognition is created in the course of compensation and then is supported, if the hazard was removed. In such a fashion, self-assembly of protein units occurs and the same chain of amino acids may be converted into several different steady conformations. May be the image, feeling and recollection are also each time created again from dispersed features? By the way, from the same details is repeatedly assembled something similar, but not always the same. This, on the one hand, may ensure constancy of the sense of Self, but on the other hand, an identity is not always identical. This helps us to decide "binding problem". Nevertheless, this does not clarify "binding problem" for homeostasis, which collects scattered information about an organism's proper functioning and elaborates the strategy for maintenance of being. At the cellular level, it is necessary to comprehend how homeostasis "knows" that the new combination of optimal constants that have been discovered is proper. What does homeostasis really do? A change in excitability is not the sole determinant of neuron damage. There are many other characteristics observed during injury and exposure too many factors hurts cells. Therefore, there is the problem of what parameter, or combination of parameters, we are obliged to accept as a criterion of cell damage. Then there is a much more important problem of what parameter and at which optimal value is an organism (or a brain, or, in a narrow sense, a neuron) regulated in order to avoid death. That is, regulation of what parameter can an organism never substitute for regulation of other parameters, since modification of this parameter is the most important? Death.

Contrarily, an organism modifies other parameters for the sake of the optimal value of just this vital parameter. One may consider that the major trouble of homeostasis is self-preservation. Certainly, this is right, but at the same time, it is necessary to know the criterion of self-preservation. What is the most important flaw in a cell that is the omen of death? One such main entity is the condition of being of homeostasis itself, since after homeostatic failure ruin is immediate. At the beginning of an attack on the given variable, homeostasis tries to return the variable to its optimal magnitude and spends reserves of energy and substances. Let us imagine, for example, that there is too high a Na^+ concentration in the extracellular medium. Na^+ will penetrate into the cell, Na^+, K^+-ATPase begins to pump Na^+ out of the cell and spends intracellular resources. If the injurious factor, in our case the distorted ion environment, was not removed, resources such as energy-rich molecules of ATP become exhausted and other cell variables will suffer. Thus, exhaustion of resources or excessive augmented activity of homeostatic mechanism may serve as the signal of inevitable death. Homeostasis can reveal this signal and tries to reorganize the whole system. Thus, prolonged being in a vicinity of energetic local minimum may serve as a signal of unfavorable circumstances; this is rather an old principle. Now our system may alter the number or sensitivity of Na^+ channels. If the system is simple enough, this is the end of our description.

Let us add a small complication. In the extracellular environment, the cell reveals an excess of glutamate, instead of Na^+. Energetic hunger will force now a decrease in the number or sensitivity of AMPA receptors. If in the basis of homeostasis is put the equilibrium of one general global variable, how may it determine which particular set of variables is in need of reorganization, controlling Na^+ concentration or controlling AMPA receptors? The entire brain, compared with neurons, disposes of additional means for defense. It uses urgent-local (gap junctions) intermittent-distant (synapses) and sluggish-local (retrograde messengers and spread of damage-protection) remedies for numerous threats. Cells defend themselves and other cells (sometimes) from the extracellular environment, while brain organizes behavior for defense from the outer environment. Naturally, it is necessary to take into consideration the properties of the outer environment. Therefore, in spite of complexes of cells, which are capable of individual behaviors, brain operates with goals that are unknown for its cells. We will return to this problem in Chapter 4, but at present will make one important remark. The cell, which in its activity is guided by the state of only one global variable such as shift in the scale of "damage-protection", will behave as the object possessing minimum awareness. It state, is really changed within the scale "better-worse". The approach of death increases cellular efforts to operate within homeostasis and increases energetic exhaustion. If we interpret this as a worsening of the state, we have not any possibility to determine how the cell evaluates its states as "Too little ATP!" or "It is terrible!" In the end, appearance of a global evaluation of the quality of life might be the basis of feeling in multicellular animals.

9.3 Subjective nature of motivation

Motivation is disposed at the thin verge between substance and inner Self. It is an inter-disciplinary notion and it belongs to both physiology and psychology. The problem of physiology is to understand material roots of motivation, while the problem of psychology is to comprehend the consequence of motivational existence, which is the cause of behavior. Going deeper into the physiological problems touches levels of brain, neural centers, neural networks, the cellular level and chemical processes within cells. Expansion of knowledge in psychology relates to sociology, criminology, philosophy and other fields connected with the active attitude to life. At the same time, both material and ideal considerations ought not to become reserved within the limits of only one approach. In particular, the most appealing task of the physiology of motivation is to realize, how material events convert motivation into a subjective sense. Certainly, we may consider a motivation as something like the modulation of neural tissue, which alters an output reaction. However, if one excludes this ultimate goal, one immediately loses the sense of consideration of motivation as a special notion. More simple notions, excitation, inhibition, signal and reflex can approximate a description of behavior. Therefore, we understand a motivation as an organism's subjective attitude to its current or future physiological state, which somehow modulates the generation of actions until an optimal state is attained.

The data of literature and our experiments demonstrate that the capability of a single neural cell within the brain to achieve goal-directed behavior is rather a probable hypothesis. It is not a common rule that each neuron execute a whole cycle of goal-directed behavior: evaluation of an input signal, generation of an output reaction and consumption of the environment reward. On the contrary, this is a rare exception, since the majority of neurons generate goal-directed behavior together, while the closed cycle of a single neuron is usually directed to a subsystem of a whole behavior. However, such exceptional cases have been demonstrated and they are very significant, because they demonstrate the capability of a neuron to generate the whole complex of goal-directed behavior. When cells form a body, they do not become by workpieces of this organism and do not stop to be individuals. Most cell types in the body, at a given time are capable of or perform migration and there is plasticity in cell migration [415].

In particular, a neural cell is not a workpiece of brain, which sums excitation and transmits its predetermined decision to other cells of the brain. It is a primitive organism, living, behaving, suffering and dying within the brain. On the other hand, brain is not a construction designed of neural and glial cells. It is a collective of primitive organisms. Certainly, as a first approximation, a neuron is a workpiece and a brain is the construction. This approximation can explain some features of behavior as a system of signals and reflexes. However, none of the "hard" problems, formulated at the beginning of this book, may benefit from construing of brain as this first approximation. The idea about capability of a neuron to self-contained behavior had been discussed from time to time [1126, 18, 661, 181, 43]. This happened each time that one needed an explanation of some of the 'hard' problems.

Nevertheless, the property of an individual neuron to be as if it were a self-contained system is difficult to detect directly, since neurons are interconnected within the brain and a common decision of brain influences the decision of any neuron. However we may evaluate cell chemotaxis as an example of the simplest goal-directed behavior of the single cell. A motivationally-like activity of single cells can fulfill a complete behavioral cycle with its own goal, action, reward and satiation.

Chemotactic activity has many similar features with motivational behavior. Primary metabolic signals, activating motivational centers, affect chemotaxis as well, and a similar set of intracellular biochemical reactions are involved in motivation and in chemotaxis and both motivation and chemotaxis satisfy by similar means. Motivationally-relevant substances affect chemotaxis in the similar concentrations and with similar dependence on concentration, in accordance with the U-shaped function. Lastly, functions of chemotaxis and motivations coincide. In both cases, the stimulus for goal-directed behavior is a metabolic disturbance, a shortage of nutrients or danger, while removal of this reason causes the behavior to cease. One may suppose that the phenomenon of behavior arose in single-cellular organisms and was perfected in high animals.

Chemotaxis may change after interaction with the environment and these modifications may have an adaptive character. The chemotactic system, even in bacterium, demonstrates desensitization that sometimes possesses features of selectivity. Some authors inform us about the existence of primitive conditioned reflexes in a unicellular animal, although such results need verification. In critical circumstances for survival, single cells may collaborate and unite their efforts in the fruiting body that has a more perfect behavior.

If a neuron possesses the capability to behave as do unicellular animals, neurons within a brain may represent a collective of sentient units. A completely isolated neuron can demonstrate elaboration of habituation and classical conditioning that satisfies to the basic properties of behavioral conditioning. Motivational behavior can be demonstrated in an individual neuron and its homeostasis may serve as a unit of goal-directed behavior.

As a future perspective, it would be important to examine real behavior of living neurons (their partially differentiated predecessors) in tissue culture. Chemotactic movement of such cells will allow simulation of a living neural network by the collective behavior of living neurons. It would also be interesting to compare the chemotaxis of neurons from motivational centers and the movement of other neurons. Furthermore, chemotaxis and chemokinesis of a neuron from primary motivational centers can be examined under the influence of their relative metabolic signals. Instead of the chemotactic substances for given neurons one may use their target neurons. It is no less important to examine the influence of reward-related substances, such as glutamate, GABA, dopamine and opioids on neuronal chemotaxis. Substitution of motivational behavior by chemotaxis of young neurons will allow one to evaluate the capability of neurons to survive and will clarify behavior of neurons within the brain.

There are numerous pieces of evidence that during development of simple motivations, neurons in specific motivational zones undergo injury, while artificially-induced damage augments motivation or causes it to arise it anew. On the other hand, biochemical pathways involved in the rise of motivation are the same alterations of neuronal homeostasis that lead to cell growth and cell damage, and the factors that augment cell damage also provoke motivation. One important reason for both motivations and cell damage is excessive excitation. We may conclude that the origin of elemental motivations is a consequence of transient cell damage.

The brain can identify its current needs since there are specific distinctive features specifying a given motivation. Usually different neurons control different stages of behavior. Nevertheless, each specialized neuron can detect any metabolic flaws, which is necessary to compensate for its survival. Although cell can feel any trouble threatening to its existence, the cell and, especially, the neuron can be tuned for the fine revelation of a specific flaw. To understand the nature of differences between various motivations is much easier than to grasp the enigmatic spirit of motivation as a natural phenomenon. The roles of initial signals for motivational behavior change in relation to pH, temperature, salinity of extracellular environment, osmolality, glucose concentration, reproductive steriods and a shortage of energy. Neurons in different motivational centers are tuned to detect a specific homeostatic discrepancy, so that their sensitivity to damage is attenuated. Sensitive zones for various motivations are localized in different areas of the brain. This sensitivity concerns neurons in the motivational centers, and different neurons reveal different needs and different motivations affect partially dissimilar neuronal populations. One may conclude that apparatuses of specific motivations are to some extent separated in the brain. Different motivations affect partially dissimilar neuronal populations and spatial organization of motivation plays, as we illustrated, a much more essential role than spatial organization of memory. Difference between motivations is easy explained by a spatial separation of their material substrates. Thus, a material difference between motivations does exist and somehow may be evaluated.

With the exception of only a receipt of primary metabolic signals, the mechanisms of different motivations are identical or very similar. Whereas initial signals of different motivations are specific, following metabolic events are the same for various simple motivations. Synaptic pathways connected with activation of motivation are usually excitatory amino acids, which induce excitotoxic damage, whereas satisfaction of motivation is connected with opioids, GABA and dopamine. We may also pick out some motivationally-relevant events that seem to be connected with further development of motivational processes. Evidently, emergence of motivation involves participation of approximately the same metabolic systems, as this is what happens during the crucial reorganization of neuronal states. Developments of damage and motivation pass through the same time scopes. When motivation begins, neurons in motivational centers display pathological excitation, energy stores are exhausted, Na^{+}, K^{+}-ATPase activity reduces, Ca^{2+} channels and other cation channels are activated, intracellular Ca^{2+} is accumulated, temperature in the brain arises, concentrations of cytokines and other inflammatory mediators increases, activities of second messenger cyclic AMP with the participation of stimulatory G_s proteins and retrograde messenger nitric oxide also increase. G proteins also alter their interaction with several types of GABA-, dopamine- and glutamate-receptors and with the calcium-binding proteins. The number of neuronal and glial gap junctions during motivation increases and metabolic coupling between the glia and neurons is also enhanced.

Potential-dependent channels, chemoreceptors, second and retrograde messengers and other means may change excitability without a change in membrane potential or alter conventional dependencies of excitability on membrane potential. This leads to a disturbance of classical properties of excitable membrane in injured neurons and during goal-directed behavior, as we and others have demonstrated in examples of elaboration of the local instrumental reflex, long-term sensitization and during thirst. Homeostasis can affect neuronal activity by various means, not only through membrane potential.

If homeostasis is really capable to so fine a regulation of neuronal responses (transient, forestalled and biologically plausible), this means that it can recognize input signals and have access to memory stores. In order to perform so rapid an on-line control of neuronal reactions, homeostasis must operate in a corresponding tempo. However, we know that homeostatic protection operates slowly. For organization of a new set-point and appreciable reorganization of a function, homeostasis needs days or weeks. And even for recovery of the previous optimal characteristics, without set-point reorganization, it is necessary to spend considerable time. Recovery to baseline of intracellular Na^+ levels, mediated by Na^+, K^+-ATPase, typically requires tens of seconds. Standard glucose-dependent production of energetically rich molecules of ATP also consumes a large amount of time. Access to genetic information is an even more protracted process. Moreover, time for simple molecular diffusion across a cell exceeds latencies of neuronal reactions.

Nevertheless, possibilities for rapid homeostatic modulation of neuronal responses exist. Limitations in the rapidity of chemical reactions have been discussed as applied to memory. We have concluded that dynamic participation of chemical long-term memory in the current behavior is time-consuming and therefore doubtful. Long-term memory should, in accordance with actual circumstances, fertilize short term memory and then a prepared brain may quickly use data that were recently located in the long-term memory. Homeostasis may be somehow participates in a rapid control of behavior using memory mechanism. It is clear, for example that homeostasis cannot use a renovation of energetic stores or genetic information for current control of the cell state. However, a neuron possesses instruments for the rapid modulation of its conditions of being. Changes of functions of potential-dependent channels and other proteins by means of phosphorylation-dephosphorylation may proceed during milliseconds. Ionotropic $GABA_A$ receptors and AMPA glutamate receptors operate also at a fast rate. Diffusion of the small molecules of retrograde messengers at short distances and modulation of gap junctions may be rapid processes, too. Thus, theoretical prohibitions for the existence of on-line homeostatic modulation of neuronal activity are absent, while indirect data support such a possibility. At the present, we cannot contend that a rapid connection of homeostasis and behavior definitely exist, but this problem deserves careful experimental analysis in the future.

Motivation has, evidently, a chemical mechanism. There are a number of motivationally-relevant substances and usually their small concentrations are sufficient in order to affect an initiation of motivations or their satisfaction. Production of motivationally-relevant substances in the motivational center is observed when motivation arises. On the other hand, administration of such substances in specific brain area modulates existing motivations or recharges motivation. Defensive and sexual motivations may be reorganized by substances in extraordinarily low doses; maybe only one molecule per cell is enough. Motivationally-relevant substances act through the same metabolic pathways as a motivational behavior and the same pathways are activated during cell injury. Usually, although not in the every case, a motivationally-relevant substance at physiologically relevant concentrations excites neurons in motivational centers if it increases motivation, whereas it inhibits neurons if the particular substance is rewarding. In addition, motivationally-relevant substances, such as natural motivation and neuronal injury, affect excitability without a direct impact on the membrane potential of neurons. Hence, one may suppose that these substances affect motivation by means of modulation of compensational plasticity and the reorganization of damage-protection of neurons.

Motivation accomplishes an interaction of organism with environment. Necessity in such an interaction arises in two cases. The environment may threaten an organism and it is forced to take measures for escape. Moreover, an organism needs restoration of its inner resources and is forced to apply to the environment to searching for source of its requests.

The direct goal of defensive behavior is support of the physical integrity of the organism. A goal of other motivations is, ultimately, also the maintenance of being. Any elemental motivation, not only a defensive one, is directed to resistance from an unfavorable state of an organism, unfavorable at present or in the predicted future. Therefore, defense is the main goal of behavior. Besides, any motivation executes defensive function in respect to some metabolic disturbance in the organism. It is rather probable that every form of motivational behavior has evolved from defensive motivation. Any reward system of brain is, probably, based on the rewarding mechanism of defensive motivation and this is not strange. Defensive motivation is, evidently, the key system for the maintenance of longevity. However, the primary goals of motivations controlling feeding, drinking, respiration, temperature regulation and the need to sleep are distinct from defense from damage and only unsatisfied simple motivations threaten to damage or lead to death. Initial signals for defensive motivation come from the environment, while other motivations initiate action by means of inner brain signals, which arise after energy exhausting or exhausting of other internal resources and disturbance of homeostasis leading to internal recovery. On the other hand, the urgent source for satisfaction of the defensive motivation is located within the brain: this is the opioid system. At the same time, the urgent sources of satisfaction of other motivations are located outside of brain: food, water, oxygen, heat or rest. Thus, when the reason for behavior is in the environment, the source of temporary satisfaction lies within the organism. Correspondingly, when the cause of behavior lies inside the organism, the source of satisfaction is outside of the organism. However, the inner consequence of these outside rewards is found in the opioid system, since one consequence of initial motivational signals from these motivations also threaten life. In this body of motivations, sexual motivation has a special place. The initial signal for sexual motivation does not connect with a disturbance of homeostasis. It is initiated by an outer signal, as with a defensive motivation does. Furthermore, this outer signal is converted into inner cellular damage by means of the action of sexual hormones, but satisfaction of sexual motivation is completed by the opioid system, too. Mechanisms of sexual and defensive motivation are closed and their motivational centers are located in close proximity in the brain.

Consumption of a reward evokes in the motivational centers processes that are mostly opposite to the event that happened during motivational excitation. A natural reward and rewarding substances usually inhibit neurons in motivational centers, but not neurons outside of these centers. Rewarding substances also protect cells from damage and exert euphoric effects. The final link of any reward, not only avoidance of punishment, is found in opioid system of brain. There is information that the ultimate links of many rewards for other motivations, which were not the object of our consideration, are also connected to the system of endogenous opiates. For example, if one scratches oneself, one receives a dose of endogenous opiates, and when somebody receives winnings, he also feels opiate satisfaction.

It is necessary to emphasize that the correspondence between "excitation-damage-motivation", on the one hand and "inhibition-protection-satisfaction", on the other hand, is not absolute: activation of neurons may lead to inhibition of follower neurons and, consequently, the same impact may induce a complex pattern of excitation and inhibition in a network. However, on the whole, we interpret the body of evidence presented here as showing that motivation and reward are related to the transient damage and recovery of specific brain neurons. We suggest that motivation intervenes in intimate cellular processes resulting in hyperthermia. Note that the same chemical substances, at similar concentrations, affect motivational behavior and cellular damage. Motivational excitations exert the same changes in neuronal states that are observed during necrotic damage, while alterations of neuronal conditions, which inevitably lead to damage, lead to the arousal of motivations. The reversible nature of motivation-related damage suggests that this damage is necrotic, since apoptosis is irreversible.

Elemental biological motivations are directed to recovery of attenuated cellular metabolism or to avoidance from punishment. Therefore, it is natural to suppose that satisfaction of a motivation somewhat protects an animal and, hence, its neurons. Really, rewards of various kinds protect neurons. We have described in Chapter 2 that protection of a cell against damage proceeds by means of homeostatic compensation. Homeostatic compensation may play role in the process of satisfaction of the motivation with a reward. However, between these two forms of protection, homeostatic compensation and reward, there is present a principal difference. Compensation may play a role in the endogenous protection against metabolic disruption, unlike a reward that exerts an exogenous protection. For example, deficiency of energy in a tissue may be compensated for by the homeostasis by minimization of metabolic energy expenditure or by means of a transfer from oxidative metabolism through aerobic respiration to the production of energy by means of anaerobic glycolysis. Protection using a food reward to supply energy can be recovered via consumption of energy rich nutrients, which later are metabolized by the homeostatic machinery. Thus, specific primary signals of many elemental motivations are in either case connected with decrease of energy metabolism. Failure of energy metabolism concerns the feeding motivation, the respiratory motivation, temperature regulation, drug-dependence, the motivation to sleep and, to some extent, drinking. However, energy metabolism failure is not a part of defensive and sexual motivations. Defensive motivation is directly connected with tissue injury, damage because of pain, or with the threat of damage. As well, a starting signal for sexual motivation is an outer gesture of the body or, at least, of the brain.

Currently, a whole complex of data convincingly proves that the cause of motivation is damage to specific neurons in corresponding motivational centers, while satisfaction of motivation protects these neurons. Strong emotional experiences sometimes leave so strong traces in the neural system that they are almost no differ than organic damage.

At the end of 2005, an epidemic of a strange illness occurred in a school in the one of regions of Chechnya. Children and some adults suffered breathing difficulties, convulsions and nausea. Regional physicians supposed group poisoning. The epidemic quickly spread to neighboring schools. A conference of specialist doctors diagnosed this case as mass hysteria. Anxiety-related illness can closely mimic illness caused by poisoning, since anxiety causes the same symptoms as does poisoning. This sad example demonstrates that aversive experience can leave real physical damage in one's body. Unsatisfied motivation is always an aversive experience and it produces transient or even permanent damage in brain. A fatal role of social stresses was demonstrated also in rats. For example, rats subjected to the social-defeat (social conflict) procedure showed an enhanced nociceptive behavior to a subcutaneous administration of formalin 5 d after the last confrontation session [37]. Repeated psychosocial or restraint stress causes atrophy of apical dendrites in pyramidal neurons of the hippocampus, mediated by excitatory amino acids together with adrenal steroids. A single stress session did not produce these changes [777].

We have convincingly argued that transient injury of specific neurons gives rise to the wish to get rid of damage and when one finds the possibility to protect these neurons, one feels pleasure and this becomes the basis of the reward system of brain. Then we feel wishes or satisfactions, being in an aware state. Thus our neurons constantly balance between damage and protection. As a rule, such states of cells are reversible. However, there is the question, "Does this astonishing mechanism only frighten and encourage our neurons?" May be, in reality, nothing threatens our neurons and their drift between being, death and being again is only a happy find of nature, giving us the possibility to discover, in good time that deviation from homeostasis does not aggravate danger or remove a threat. Origin of awareness gives us a general criterion for evaluation of the status of our organism instead of the necessity to regulate each homeostatic parameter separately. However, we know that neuronal damage, being usually reversible, sometime does kill neurons. Although the protective role of reward means that neuronal damage during motivation is transient, neurons are sometimes irreversibly injured, such as during drug-dependence, self-stimulation, stress, and redundant mating. Therefore, our neurons really die from time to time, but here is one consolation. Our neurons die instead of us.

And here a new question arises. May be, our neurons generate motivation and impel us to avoid their damage since there is danger for these particular neurons, but, really, nothing threatens our life: it is necessary only to support our proper functioning. For example, generals sometime send soldiers into battle for the sake of personal career. Do our wishes actually shorten our life, even if we satisfy our motivations soon enough? The response is yes. The lifespan of an organism increases when it reduces its calorie intake [1029, 1379]. The lifespan of the fly [228] and the nematode [564] decreases with an increase in sexual activity. Deletion of half of the cells in a nematode that generate the sperm and eggs increases the lifespan [1029]. A drug used as an antidepressant in humans increases nematode lifespan and its effect on lifespan involves mechanisms associated with lifespan extension by dietary restriction [969]. Anticonvulsants, decreasing superfluous excitations, retard the aging process and extend the life-span of the roundworm [368]. Moreover, people who experience high subjective well-being typically have good health outcomes and longevity [184]. And lastly, duration of life is regulated by the nervous system [157, 1350].

Phenomena motivation and homeostasis are closely connected. The essences of motivations and homeostasis are related to some general and principal property of living tissue. Both motivation and homeostasis recover equilibrium and defend from death. Homeostasis recovers equilibrium at the level of inner resources, whereas motivation interacts with the environment. Although motivation may be an aware feeling, while homeostasis performs its function without participation of consciousness, we cannot consider homeostasis as a more primitive mechanism. When homeostasis is impeded, there is no recovery of tissue from damage by means of endogenous resources and motivational behavior is the only possibility for recovering the proper equilibrium. We have demonstrated that homeostasis may sometimes be connected with completing high neural functions and we may suppose that, when inner resources are exhausted, homeostasis undertakes efforts for organization of goal-directed action, such as the establishment of a new set-point, when maintenance of the previous set-point is unattainable (see Chapter 2). Thus motivation turns out to be the highest form of homeostatic activity and goal directed actions are produced by intensified homeostatic efforts. This is a weighty supposition and it needs careful examination. Our discussion has demonstrated that this idea is in a good accordance with the data known at present. Firstly, both mechanisms perform the same function. Secondly, they initiated by the same set of metabolic disturbances and, thirdly, involve the same biochemical pathways. Further, time courses of development of motivation and damage-protection are close enough. Next, a wide class of chemical substances in approximately the same concentration affects developments of motivations and homeostatic protection. In addition, both mechanisms regulate rather general qualities of cells and their physical integrity, instead of particular variables.

Intimate connection between motivation and being is intuitively plausible. The French writer Honore de Balzac in his novel "The Magic Skin" described a mysterious shagreen skin. Any owner of this skin may fulfill any wish, but at the expense of longevity of his life: "Your wishes will be accurately fulfilled, but at the expense of your life. The compass of your days, visible in that skin, will contract according to the strength and number of your desires." Here is how Honore de Balzac describes the shagreen skin: "A piece of shagreen which was only about the size of a fox's skin, but it seemed to fill the deep shadows of the place with such brilliant rays that it looked like a small comet, an appearance at first sight inexplicable. There is the mark of the seal which they call in the East the Signet of Solomon. The mysterious words were thus arranged, drawing of Sacred characters:

Possessing me thou shalt possess all things.
But thy life is mine, for god has so willed it.
To wish, and thy wishes shall be fulfilled;
But measure thy desires,
According to the life that is in thee.
This is the life, with each wish
I must shrink even as thy own days

9.4 Goal-directed actions

Generation of intentional action is a bright and distinctive feature of living being. A certain mysterious force compels animals suddenly to start to run, swim or fly and this is the main trait demonstrating that the object is alive. Is this mysterious force somehow connected with the homeostatic mechanism, controlling our survival? Recovery of homeostatic equilibrium is an autonomous process, as is intentional action. Error in metabolic equilibrium is normally compensated by homeostasis but if the adaptive response cannot be generated internally by means of compensational homeostatic processes (say, because of a shortage of food, water or oxygen), it must be achieved via interaction with the environment. We have supposed (Chapter 3) that homeostatic compensation not only directly recovered intracellular metabolism, but also indirectly promote recovery of metabolic resources affecting behavior. In this case, an adaptive behavior can assist in avoidance of death. We think that homeostatic compensation advances to generation of behavior, but this still falls short of solving the exceptionally convoluted problem of brain/mind relationships.

Our discussion reveals that the conventional properties of a neuron are disturbed during motivational behavior, whereas the neuron recovers itself (although not always) after homeostatic compensation of damage or after reward obtained from the environment, as a result of goal-directed behavior. The conventional properties of a neuron, such as membrane potential maintenance and spike generation, are disturbed during instrumental learning and this decreases neuronal reaction. A healthy neuron better performs its functions. It therefore cannot be excluded from consideration that homeostatic plasticity controls behavioral plasticity and processes of compensation participate in the organization of behavior. Compensational hyperpolarization protects neurons after damage evoked by excitation and leads to augmentation of the neuronal responses. Development of compensation counteracts cell damage and thus satisfies motivation. Therefore, generation of instrumental reactions may be a consequence of the homeostatic recovery of neuronal properties after metabolic damage or after damage because of a motivational excitation. Disturbance of the injury-protection balance during instrumental learning may turn a neuron into an emissive subject.

Thus, intention to act may be based on homeostatic efforts. When homeostasis cannot manage the inner flaw at the expense of inner resources, it strains efforts to obtain outer resources and this is an incentive to act. Surely, homeostasis knows nothing about supplies in the environment. It simply augments compensation in order to recover equilibrium and this transiently recovers the health of the neuron and promotes action generation. Tension of compensation directs energy to operation. Therefore, brain temperature falls after completing goal-directed actions. The stronger is motivation, the larger is the disturbance in homeostasis, the bigger the compensational efforts and the brawnier the action. These elemental functions may be presented at the level of a single neuron. Further, this driving force for behavior is necessary to direct into the proper channel of action those neurons which will execute the behavior. A single neuron cannot decide what action will be proper and the choice of effective output neurons cannot proceed at the neuronal level.

The broad spectrum of data, obtained by means of various methods is evidence that many properties of brain that promote control of behavior are present already at the level of a neuron. A neuron is capable to distinguish biologically significant and insignificant events, that is, to determine how a given event foreruns the change of the neuron state. In order to accomplish this evaluation in the future, the neuron must memorize direct consequences of actual signals. Moreover, a neuron is capable to evaluate whether, in the past, consequences of a given event depended on its own activity (i.e. after generation or failure of an AP). After such evaluation, the neuron makes a decision if it is beneficial to send its control's command in the given case. These capabilities are founded on a neuron's competence to evaluate the quality of its being and to undertake efforts for normalization of its state. It is important that a neuron displays these unique properties, as a result of own activity and not as a passive reaction to the operational activity of its environment. This neuronal and glial encirclement is not more than the micro-environment within which the neuron lives. Certainly, neuronal behavior depends on its neighbors, but the neuron directs its actions for its own survival. In exceptional cases, the neuron is able to subordinate global brain function under its responsibility. This means that a brain is a well organized society of primitive creatures rather than an ideally adjusted mechanism.

If this scenario is proper, it is necessary to explain how billions of primitive evaluations of the situation are amalgamated into a whole apprehension and how countless elemental actions are united in the entire will and sense of the Self. It is not a trivial task to determine how a distributed system makes a decision. The first important quality that the brain adds up to a facility of a single cell is the ability to choose dominant output action out of some potential possibility.

A neuron has the potential to choose its reaction: to generate or not to generate action potentials in response to a given stimulus within its channel of action. Yet, in order for an organism to choose the direction of its reaction, it has to have the possibility for sending its signal to any of at least two pathways. This task is an arduous one, since a neuron has only output, even if bifurcated. Thus the organism has a choice, while a neuron does not. An association of neurons and glial cells in ensembles may ensure a multiple output.

A brain has a few means for association of cells in a system. These are the means for interaction between neurons: synaptic connection, a collection of ions and some simple substances in the extracellular milieu, retrograde messengers and gap junctions. Synaptic connections attach neurons along the natural spread of signals, retrograde messengers link spatially close cells having corresponding receptors; extracellular milieu (K^+ and Ca^{2+}ions, regenerative secretion of ATP, adenosine and glutamate) affects all closed cells, while gap junctions associate neurons having related functions. Therefore synaptic connections control the temporal sequence of events in the neural system, retrograde messengers keep related metabolic systems under control, ions constituents ensure continuity of the process in neural tissue and gap junctions unite relevant but scattered events in the whole.

Retrograde messengers and gap junctions may unite neurons into an ensemble. These assembling and also second messengers cyclic AMP, Ca^{2+} and inositol trisphosphate in the intracellular milieu synchronize neuron instability. One or a few dominant cells may initiate a reaction, while related cells are involved in this dominant activity. Ensembles may promote the dominance and give the possibility to determine the temporal cooperation of neurons and glial cells, ensuring a unity of action in the given instant. Changeability of neuronal ensembles is determined by dynamic properties of gap junctions. New formation of gap junctions is a rather slow process (hours), but when coupling has already been established, conductivity of gap junctions may change at the millisecond timescale. Considering that gap junctions are involved in motivational behaviors on the one hand and in the processes of damage on the other hand, an arrangement of cellular ensembles by the gap junction could contribute to the development of goal-directed behavior. An ensemble of cells with a united intracellular milieu and electrically coincided compartments can make a general decision, but has multiple outputs.

A cellular ensemble is an example of the collective behavior of individuals in a group. Interaction in a group increases the complexity of the system. Such a system may complete functions that are unknown to cells. For instance, it can make a conclusion from general information provided by several cells and choose the direction of the command. On the other hand, a collective decision is based on individual decisions of cells. Cells within the ensemble are not equal in rights and this may be determined not only by their morphological differences, but also by the degree of complementarities between the individual pieces of information of the cells collected in the past, and also the current informational flow. Unfortunately, collective behavior of neurons is a poorly developed branch of Neuroscience, but we are waiting for rapid progress in this field.

A postulated picture of action generations indicates the way that behavior may be not predetermined in advance. Neuronal damage, which has been induced by motivation, allows cutting the Gordian knot of existence of arbitrary actions. Transient damage of the neurons in the primary motivational centers evidently induces instability of neuronal reactions, which may serve as the basis for a choice. Motivational behavior is directed to overcoming the transient injury of specific cells and it has an erratic component. After consumption of the reward and satisfaction of motivation, non-stability decreases. If the action is proportional to the degree of deterioration and is directed accidentally, this will, on average, lead to a decrease in deterioration. This simple rule ensures simultaneous goal-directed behavior and non-predetermined search.

Causal dependence between stimuli and reactions is not obligatory and existence of non-predetermined choice is in agreement with logic. Brain may emit actions without logical contradictions. We are evaluating this mechanism as more or less plausible, but are not sure that it functions in this simple form, since according to computer modelling, the search is too slow. The search is accelerated for a one-dimensional case, when homeostasis recovers equilibrium for only one general variable or when homeostasis is coupled with memory. In any case, we now may explain the free will paradox. The neural system may produce different goal-directed behaviors in response to the same internal and externals signals and for the same initial conditions. Moreover, brain may generate behavior by its own will, without any visible signal. All artificial objects perform actions in correspondence with a preliminary entered program. Therefore, study of intentional actions during a goal-directed behavior is a tempting task for the designer of artificial life. We hope that this task will be initiated before long.

9.5 Death as an awareness-rising factor

A brain has a number of different functions and it is not completely clear why one function should be completed with consciousness. We have argued that basic grains of behavior, learning and memory, choice of the preferable signal, motivation, reward and intentional behavior are present at the neuronal level. All these items have a subjective coloring and may receive a unified explanation of whether a neuron (and maybe other cells also) feels an elemental sensation. However, what is 'sensation' on the physical background? What does a neuron have to sense? The response is simultaneously simple and complex. A neuron has to reveal mismatch in the variable that characterizes its intactness, or the level of its health.

It would be plausible to consider cellular damage as a primary negative sensation of a cell. This approach reduces the problem of sensation to a cellular level. The very moment of physiological control of damage-protection also may be a moment of psychological sensation. Transient 'damage-suffering' induces compensatory protection and random action generation, which is logically equivalent to a goal-directed search. The will to perform arises as purposeful homeostatic efforts and as long as the value of protection is proportional to the magnitude of the random actions, the sensation of injury is converted into direction to goal. This does not prove free will, but demonstrates consistency of the free will idea. After learning, compensation through modulation of excitability is directed to the causes of damage thus accelerates the search for correct actions. An association of neurons into ensembles may convert primary cellular sensations into feeling and promote to generation of dominant action, which will furnish satisfaction.

Mortality is not an annoying inevitability. Consciousness consists of being between life and death and, in the narrow sense, between rest and depression. Awareness is the consequence of mortality and we feel drives because we are mortal. Approach in homeostatic equilibrium is pleasant. We cannot determine whether or not an injured cell feels a negative sensation. We can only judge cellular behavior such as taxis of a single cell, local instrumental behavior of a neuron and control of entire animal behavior by a single neuron. Nevertheless, cellular sensation is almost as plausible as feeling of a whole organism: in both cases we can only observe behavior.

9.6 Fuzzy dynamics model of neuronal behavior

Behavior of a neuron's model is presented in the Chapter 8 well corresponds to properties of neural cells during learning. This concerns elaboration of habituation, classical and instrumental conditioning. Besides similarities of general dynamics between behavior of neural cell and the model neuron, there are many fascinating details that meet data of biological experiments. It is important that these details were exhibited posterior modelling, but were not laid in the foundation of the model.

In particularly, during learning, parameters of cells are altered in non-monotonous mode, and change in stepwise mode, in agreement with results are shown in Fig. 8.7, which better correspond to the neurobiological data than the results is shown in the Fig. 8.6. Remember, that Fig. 8.7 corresponds to *Max-approximation* for the membership functions of the set-valued arguments (see page 368), while Fig. 8.6 corresponds to the quasi-random approximation of this function. It implies that stochastic elements in the neuron's decision-making process are limited.

Condition $V_{02} \sim 2V_{03}$ on the Fig. 8.3 means that damage should be developed more rapidly than protection[1]. An example of slow development of compensation after damage is the harmful preconditioning, when preliminary deleterious impact exert protective effect to following (through tens of minutes) more strong harmful impact (Chapter 2). This corresponds to results of the computer simulations, since realistic behavior of the model neurons was observed only if protection developed slower than damage and development of compensation of protection was slower than compensation of damage (see Fig. 8.7 and Fig. 8.8B together with the Figs. 8.9, 8.10). Besides, our model has elucidated the reason of appearance of stepwise variability during learning. Distinctive feature of instability is their slow change in time, during sequential trials. The state of neurons usually stays steady during 10-15 minutes, then suddenly changes and again stays stable during several minutes. This is related to change of membrane potential and responses to the signals participating in the current behavior (Chapter 4). Such slow changes may in different neurons proceed not simultaneously, while rapid alterations are synchronous. For instance the neuron may become responsible for output reaction during several trials and then lost this function. Temporary stability of decisions may be connected with formation of neuronal ensembles during aware behavior. An emergence of neuronal ensemble is protracted process. Since quantity of neurons in ensemble is large, an ensemble is stable and its characteristics are steady.

In the real situations protection may be dangerous, because completion of the protection after satisfaction of motivation stops activity, so the necessity in behavior cease to exist. However, augmentation of protection ceases also beneficial physiological activity and evokes narcosis. Further overprotection may lead even to death. Therefore, homeostatic compensation of proposed danger is directed against both the damage and overprotection. It supports precise magnitude of the optimal parameters thus preventing instability in the level of optimal parameters, but damage and protection are non-symmetrical phenomena. Only small damage and protection are symmetrical, since small damage excites cells, while small protection, which is directed against small

[1] Remember that V_{02} designate maximal velocity of the damage development and V_{03} is maximal velocity of development of the compensation.

damage, inhibits them. Nevertheless, even in this simple case increasing of the damage should be more rapid than increasing of its protection. The reason for cell damage is outer factor, while protection is developed at the account of inner resources and needs time for organization of the defense. Therefore, damage should develop relatively rapidly (seconds or minutes: this depends on the severity of injury).

Result $V_{02} > V_{03}$ well correspond to properties of living tissue. We have demonstrated that homeostasis can be, probably, directed against threat of death, not only against damage itself (Chapter 2). Compensational recovery of the normal state of tissue is even more protracted process, than damage (tens of minutes). For instance, during learning animals can differentiate currents signals, participating in the learning procedure, through 7-10 ms (Chapter 1). Necessity to apply to long-term memory takes 0.3-1.0 sec. Artificially (say, chemically) induced motivation injures neural tissue and expands during minutes. Satisfaction of motivation protects neurons and it extends even slower (Chapter 3). During negative learning, a punishment evokes a rapid immediate reaction and evokes deleterious influence in the next learning trial, through 2-3 minutes. This period is close to necrotic reorganization in neuronal electrical activity. Development of compensation after damage take tens of minutes or hours to complete, while development of compensation during negative learning takes 10-15 minutes (Chapters 1,2,4). Such a slow induction of compensation is connected with necessity of chemical alterations in the tissue. Because of compensation is sluggish process and it is developed longer than damage. It is impossible to fulfill precise on-line compensation and therefore protective influences may exceed initial damage and overcompensate neuronal states. In this case endogenous homeostasis-induced damage counteracts overprotection. This property of compensation explains a development of drug addiction: although acute administration of drugs evokes protection and produces euphoria, homeostasis compensates and overcompensates drug-induced protection and rose damage, despair and abstinent syndrome (Chapter 3).

9.7 Fuzzy logic of a neural cell

Multiple biochemical and electrochemical processes, which are initiated by an external signal, determine the way in which the neural cell generates its response. What is the type of logic which is generated by the neural cell's processing of information? Could it be well-approximated by the classical Boolean logic or is a non-classic continuing logic, like fuzzy logic, a much better approximation? We believe that the presented experimental and theoretical results point out the latter possibility. In order to be able to generate fuzzy logic, a physical system should possess several properties:

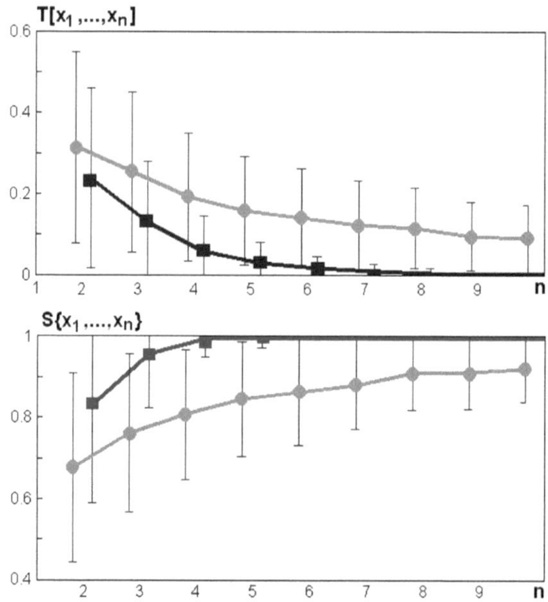

Fig. 9.1. Accumulation of several parameters by the connectives $T[x_1;...;x_n]$ and the connective $S\{x_1,...,x_n\}$. Gray (circle) curves correspond to $S\{x_1,...,x_n\} = \max\{x_1,...,x_n\}$ and $T[x_1;...;x_n] = \min[x_1;...;x_n]$, while black (square) curves correspond to product and restricted summation of the parameters $S\{x_1,...,x_n\} = \min\{1, \sum_{i=1}^{n} x_i\}$. In this example random values of the variables from the interval $[0,1]$ have been used. The curves show average values of $T[x_1;...;x_n]$ and $S\{x_1,...,x_n\}$ for 30 repeats and $n = 2,...,10$, while the vertical bars show the corresponding standard deviation.

:

a. In order to produce a response to an external signal, the system operates with continuous parameters and functions of these parameters[2]. It should be noted, however, that the system's output could be discontinuous.

b. In the process of generating the system's response on an external signal (the process of "decision making"), physical and chemical processes, which are induced by the signal, should enable to aggregate and to accumulate the inner parameters and/or functions of these parameters.

c. In order for information processing to be effective in the case where many parameters (actually, more than 5) are involved in the "decision making" process, the system should be able to compare these parameters or functions of them and to choose the maximal and/or minimal one.

[2] This means that if the parameter x lies in the interval $[x_d, x_u]$, then all values of x in this interval are admissible and that if $|x_1 - x_2|$ is small, then $|f(x_1) - f(x_2)|$ is small as well.

The first feature (a) is correct for a physiological system of the calm brain. Although, the central event in brain activity is spike generation and dependence of AP generation on input signal has a break at the point of threshold, this discontinuity disappears (or at least sharply decreases) during decision-making, as we have demonstrated in Chapter 1. The break in characteristics of excitability also decreases during processes of damage-protection (Chapter 2) and during motivational excitation (Chapter 3). In addition, discrete morphological units of the brain, synapses and neurons, dissolve their individuality in the collective behavior of the related units during decision-making. In these cases, neurons perceive a synaptic pattern, as a whole, whereas neuronal ensembles unite cells into a temporary executive organ. Thus morphological discreteness of the brain is also weakened during decision-making. Therefore the feature of continuity corresponds correctly to the cases where the aware brain makes voluntary actions.

The feature (b) is usually satisfied in the working brain, and this statement is especially correct for the decision-making process. Examples are given in Chapters 1-3. The neuron has a variety of means for interaction with synaptic excitations which arise in intracellular biochemical processes (second and retrograde messengers) within neurons. These biochemical processes inter-relate and their yield exerts direct and indirect influences to potential-dependent channels, chemoreceptors of the cellular membrane, gap junction conductivities and genetic apparatus. More importantly, the routes of these chemical reactions are rearranged during motivational behavior, after learning and after change in the levels of awareness. Neurons, for sure, accomplish decision-making by chemical means. As a result, when a neuron converts the input signal into the output reaction, it enriches the signal qualitatively with the information recorded in memory. Probabilistic logic of molecular stochastic dynamics participates in these processes, but cannot explain the mass of experimental data and vast variety of neuronal behavior.

The last feature (c) is illustrated in Fig. 9.1. Let us denote by $T[x_1;...;x_n]$ accumulation of the parameters $x_1,...,x_n$, which corresponds to the logical connective *AND* that joins the variables, while $S\{x_1,...,x_n\}$ denotes aggregation of the parameters, which corresponds to choice among the alternative variables, i.e. to the logical connective *OR*. Assume, now, that our parameters are some concentrations[3], so $0 \leq x_i \leq 1$. It is obvious that if n is large and $T[x_1;...;x_n]$, $S\{x_1,...,x_n\}$ include any kind of product or summation of the variables[4], for almost all samplings of $x_1,...,x_n$, their product will be close to zero, while the restricted sum will be close to one (see Fig. 9.1). This

[3] or any other normalized, bounded quantities

[4] For example like in the *Probabilistic logic*, where:

$$S\{x_1,...,x_n\} = 1 - \Pi_{i=1}^{n}(1 - x_i), \quad (9.1)$$

$$T[x_1;...;x_n] = \Pi_{i=1}^{n} x_i. \quad (9.2)$$

Here $\Pi_{i=1}^{n} x_i = x_1 \cdot x_2 \cdot ... \cdot x_n$.

means that almost anyone expression, which includes the logical connectives OR and AND, becomes insensitive to the concrete sampling of its variables, if the sampling is long enough[5]. However, only during decision-making the parameters differ from 0 or 1 and the number of effective parameters must be small in order for $T[x_1; ...; x_n]$ and $S\{x_1, ..., x_n\}$ would differ from 0 or 1. So, probabilistic logic is ineffective for complex systems like a neural cell, while fuzzy logic, which is using Maximal/Minimal representation of the logical connectives, demonstrates good robustness even for the long samplings of the variables. An example that shows how Min-Max representation can appear in the system, which is governed by the chemical-like kinetics, is considered in [1074].

The aware brain and its neurons operate in accordance with the feature (c) during decision-making. Behavioral reaction depends on a number of variables. Several simultaneously- activated synaptic influences interact in order that the neuron might evaluate the biological significance of an input signal, and efficacy of particular synapses is not equivalent when it is involved in different patterns (Chapter 1). Besides, many particular parameters of the current image located in different neurons must be amalgamated in the neuronal ensemble, before decision-making can be possible. The image in a neuronal ensemble may emerge as a result of interaction of matching features spreading in a number of neurons, and contribution of neurons in the formation of the ensemble differs (Chapter 4). In addition, activity of specific neurons and even the single neuron may remain dominant and control output activity of the brain. Decision-making based on a large variety of essential variables, when each particular variable may stay decisive, means that a probabilistic logic is, certainly ineffective for decision-making within a neural cell, within an ensemble of neural cells and within the brain.

Dependence of brain behavior on numberless decisive variables is not the single characteristic indicating possible operation of fuzzy logic in the working brain. Goal-directed behavior is an extremely non-stable process and peculiarity of this instability points out to choice of reaction by means of Maximal/Minimal mode. Peculiarities of decision-making clearly indicate fuzzy decisions in the brain. Both animal behaviors and neuronal reactions are extremely unstable. Instability refers to both the type of chosen reactions and the magnitude of the, instability is usually greater. Variability of urgent reactions compared with reactions prepared beforehand, is also higher. Neuronal and behavioral instability is especially irregular during goal-directed behavior, when an animal needs to make decisions during each trial. This instability is not connected with the imprecision of experimental data. The shortage of information, accessible to an animal is not also the reason for instability. Although instability increases during vagueness of a choice, but even when an

[5] "long sampling" means actually the sampling, which is longer than $n \geq 5$. Note, however, that if all x_i are close to 0 or 1, n may be larger.

animal is quite familiar with the right decision, it continues an exploration of the environment and generates trials-and-errors (Chapter 1).

An animal explores the environment and generates probing movements. Training is the series of attempts to achieve a proper result. The animal builds its behavioral strategy on the basis of actual circumstances and past experience. Dependence upon these factors does not explain why behavior is so variable in stationary conditions. There is, probably, some robust factor, which makes animal actions non-predictable. This factor is not an exhibition of the disadvantage of a neural system. Evidently, instability is an essential property of a working brain, since it is larger in higher animals, in high neural centers and during aware reaction (Chapter 4). Does the brain use a natural instability of neuronal activity when it searches a suitable behavior? Perhaps , moreover, special dispersive means increase instability, when the brain makes decision? Particularly, some neurotransmitters, such as dopamine and glutamate can increase dispersion of neuronal reactions. This may be important for organization of probabilistic logic. For instance, it would be possible to suppose that an animal accidentally chooses its action: one of several possible actions.

This is, however, not a correct conclusion. Instability of neuronal and behavioral reactions does not prove that reactions are chosen by chance. Neuronal and macroscopic reactions exhibit instabilities similar in amplitude and synchronous in time. Averaging reactions ought to decrease instability of macro-reactions comparing with neuronal reactions. Nevertheless, non-stable activities of different neurons are coordinated. This coordination cannot be connected with the existence of a common source of variability. Instability is a general property of neural tissue and nobody can isolate instability by means of surgical or pharmacologic manipulations. So, in many cases, the source of brain reaction irregularity can lay in inconsistency of activity at the single neuron level, rather than in the imprecise averaging of activity of the individual neurons.

Therefore it is necessary to explain why alterations of responses are so great, but similar in different, especially, neighboring neurons. If neuronal responses are pre-determined, it would be difficult to explain, how this predetermination may be similar in different neurons. Large instability suggests either a small number of participants in reaction - generation or non-accidental choice. At the same time, poor averaging of neuronal activity creates evidences against probabilistic logic. Averaging of synaptic signals, chemical influences and neuronal activities is not a working tool of the brain during decision-making. Chance processes take place in the brain and they do participate in the organization of behavior, particularly, diffusion of specific substances and choice of complementary molecular interactions. However, random action generation is not a main principle of decision-making.

This conclusion is related to different levels of information processing. We have demonstrated that neurons do not average synaptic influence, when they make a decision to generate spike or not (Chapter 1). Similarly, the brain generates dominant reaction that is not an average reaction of its neurons, but on the contrary corresponds to the reaction of dominant neurons, in a given case - neurons that evaluated their participation in a particular behavior, as the most preferable. At last, in special, favorable, although rare cases, the single neuron is able to subordinate entire behavior under its responsibility (Chapters 1,3). If the averaging is a principle of brain function, this would be impossible, since reaction of a single neuron must vanish at the noise of other neurons. Therefore, averaging cannot be a basic principle of brain or neuron functioning, but absence of averaging is not evidence that actions are chosen accidentally among the preferable actions, since these actions are coordinated in different neurons. This coordination would be easily explained, if neurons choose each time maximal or minimal predictions out of available reactions, that is they use fuzzy logic. Moreover, neurons, probably, generate fuzzy logic and they contain some factor responsible for the origin of instability. Our discussion revealed that this "mysterious" factor is the change of the neuron's state at the point of damage-protection. This is rather probable, since the decision to act is connected with development of critical conditions in the neural system.

During decision-making, neurons solve perilous dilemmas for their survival problems: whether and which of their actions will promote their endurance. Therefore, the most preferable action corresponds to the most secure alternative variant of behavior among minimally acceptable actions, such as, for instance: the reliability of the mechanical chain is determined by the strength of the weak member, and there is the sense to choose that chain, whose poor member is the strongest. Voluntary actions are aware reactions and they emerge in consequence of intensification of homeostatic compensation directed against damage (see Chapter 4). Goal-seeking behavior arises in the second phase of damage-induced decline in neuronal function, when endogenous compensation cannot overcome damage and threats of cell death develop (Chapter 2). Therefore, instability of aware reactions depends on the current activity of the compensational processes that is on figure of merit for the current state of neurons. This evaluation is usually similar for neighboring neurons, especially connected through gap junctions. Thus instability in such adjacent neurons may be close. Instability in the brain is well-controlled and the degree of instability is, perhaps, connected with alteration in the state of damage-protection of brain cells. The cells that are amalgamated in the ensemble may synchronously evaluate their state as injurious or protective. The brain and its neurons make decisions by means of fuzzy logic since they choose the most preferable in a given moment action, but at another time, other decisions may be considered as the most preferable.

9.8 Artificial motivational neurons and feeling robots

The idea of using some of the features of physiological processes in cybernetics is very old [61], but it is still attractive. For example, concepts of self-consciousness and emotion for robotic systems has been discussed recently in [681].

Our approach makes it feasible to design new kinds of artificial neurons: *motivational artificial neurons*, because Fuzzy Rules in Table 8.1 (see Sec. 8.1) are easily computerized.

Artificial motivational neurons (AMNn) seem to be very promising as elements of the "brain of the feeling robot". Like most of the fuzzy systems, AMNn's are easily hybridized with almost any sensors, actions controllers and long-term memory system. Behavior of such robots will be initiated by a general "defensive motivation", expressed as a few artificial drives like energy recuperation, avoidance of injury and aspiration for survival. The robot's main task may be considered as an analog of an animal's sexual motivation, so the robot will enjoy the work. Each of these motivations can be implemented as a block of the AMNn, where the neurons affect their neighbors through gap junctions. Every time the dominant motivation is passed to the robot's knowledge base and inference engine, which performs the current task, and the commands are sent to the Action controllers. So, in a "feeling robot" performance of the main task, aspiration to survive, trial-and-error learning and "instinct of researcher" will be naturally combined.

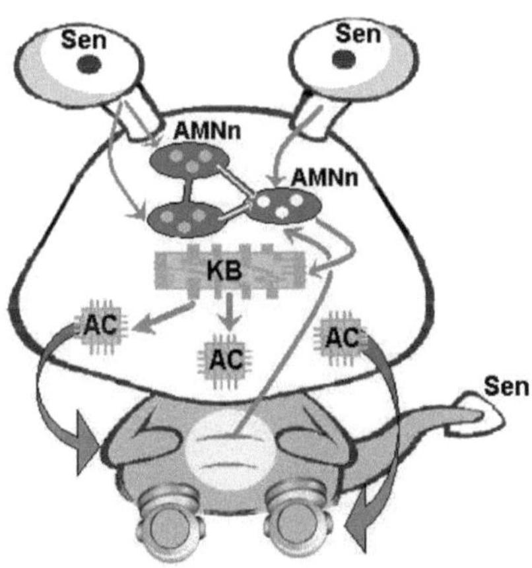

Fig. 9.2. "Feeling robot". AMNn - block of the Artificial Motivational Neurons; subdominated motivational neurons (gray dots) augment the dominant neuron (white dots) (see Chapter 4 for explanation). KB - main Knowledge Base with Inference engine, Sen - Sensors, AC - Action controllers.

If a robot has to act autonomously in poorly predictable and difficult environmental conditions, a motivational paradigm may be significantly more effective than conventional reinforcement learning approaches [1201].

A

Appendix

A.1 *Model of a chemical memory of a neuron

According to the classical notion, a neuron sums up excitation and generates an action potential upon reaching a threshold in conformity with the "all-or-none" principle [815]. The weighted sum is usually further processed by an activation function. This may be a linear or "sigmoid" function. A neural network consisting of such neurons stores information thus changing the efficacy of synaptic connections between neurons. It has many well-known remarkable properties and can solve some problems in line with algorithms of the networks operation. This classical approach is both easier to calculate and easier to analyze than a functioning of complex neuron units.

The simple neural networks have been called "perceptrons". They consist of a single layer or several layers of artificial neurons connected by weights to a set of inputs. Rosenblatt [1041] investigated perceptron learning. He and others explored a network paradigm for pattern recognition after learning (the tuning of the weights). It was necessary to adjust the weights of the network in a way that minimizes error between the network output and the desired output. Such a possibility gives to the method back-propagation of the error. A back-propagation algorithm is a systemic method for training multilayered neural networks [1052]. The input signal is passed through the network and each neuron in the subsequent layer produces a weighted sum and output signal. Back-propagation trains the hidden layers by propagating the output error back through the network layer by layer, adjusting weights at each layer. This algorithm is rather effective, but, evidently, does not correspond to real processes in the brain. Spike back-propagation is observed in dendrites, but this is not the case for axons. Biological neurons cannot interchange as large an amount of information about their states as is required for artificial neurons when implementing the algorithm. Back-propagation is characterized by the recursive way of computing gradients of complex non-linear functions and it is tremendously complicated in large neural networks. Besides, decision-making in real brain is so rapid that the possibility for travelling signals in the tissue is rather restricted. The same objection can be raised also for J.J. Hopfield's idea [559], since relaxation of an entirely interconnected neural network to a local minimum of energy (that is interpreted as an associative recognition) is very slow.

Irrespectively of the physiological relevance of these models, they have some pluses and some disadvantages. The form of algorithms in general is not universal but is determined by the model of the neuron, by the particular network structure and by the salient feature of the practical problem. The tuning is usually too slow and is sometimes accompanied by network paralysis. Memory capacity is low and it is often necessary to create huge networks. Besides, false images may appear and it is difficult to overcome local minima. Deterministic properties of these models impede normal function in an intricate environment. Real brain operates without these disadvantages. Therefore, it is important to bring theory into line with the experimental data.

If a neuron recognizes the pattern of active synapses as a whole, the problem of reaction specificity arises anew. There is no evidence that during normal learning, such as habituation, as well as classical and instrumental conditioning, the efficacy of the synapse changes independently from the activation of other synapses. We have demonstrated the correctness of this notion in the simple experiment [1272]. When repeated presentation of the tactile stimulation of the mollusk body evoked habituation, there was the stimulation of the same part of body but with a reduced intensity: we used this as a rare stimulus. After decrease in the response of the frequent stimulus, a sudden decrease in its strength leaded to a decrease in threshold in the same response (Fig. A.1, left column). At the same time, decrease in the habitual stimulus intensity did not increase the number of spikes in the response, because of failure in spike generation in some responses (Fig. A.1, right). Amplitude of EPSP did not decrease when intensity of the habitual stimuli decreased (not shown in the figure), but dispersion of the responses to a small rare stimulus was augmented: in some responses EPSP increased and in some others the EPSP almost disappeared. Naturally, responses with an AP failure cannot be taken into consideration for evaluation of thresholds, while estimation of EPSP is concerned with responses to AP failure.

In fact, if some of a population of initially activated excitatory synapses evokes the same or larger EPSP than the population as a group, this EPSP cannot be represented as the sum of the potentials generated by each synapse separately, since the whole must be greater than its part. Likewise, large regions of cellular membrane are involved in spike generation, and selective augmentation of excitability in response to reduced input signal also points to recognition of pattern synapses as a whole. Chemical processes initiated in the synapses may perhaps interact in the neuron and thus solve the "binding problem". After this series of experiments, new corroborations were received. When in a mammalian cortex pre-synaptic neurons are activated in synchrony, their effect on a mutual post-synaptic unit may increase by several folds without modification of the individual synapse. Neuronal interactions between neuron

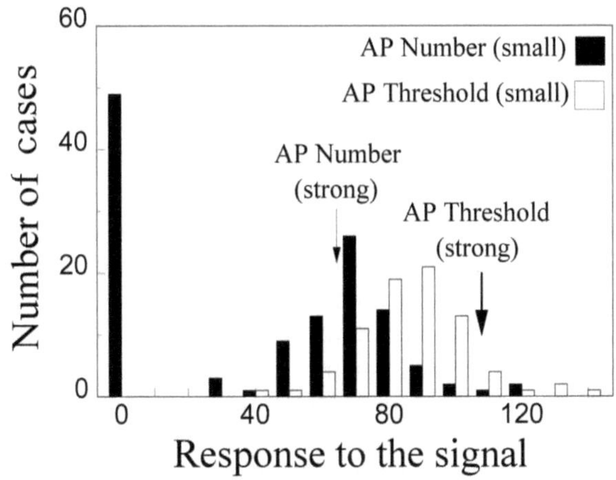

Fig. A.1. Disinhibition in the response to a sudden decrease of value of habitual tactile signal. At the left change in the thresholds during habituation in response to the habitual (light symbols) and rare stimuli (dark symbols). 100% - average value in the response to habitual stimulus. Squares: threshold AP, rhombi: AP number. At the right distributions of AP number and AP threshold for small rare tactile stimulus at the end of habituation. Arrows, corresponding average values for strong habitual stimulus at the end of habituation.

pairs can be modified in relation to behavioral state. The modification may be linked to external events and may occur on a very short time scale, within a fraction of a second. When the same cell is coactivated with another set of neurons, a different part of its connections become effective.

There is no need for each of the synaptic connections to be particularly strong because they become effective through synchronous activity with other neurons [1291]. Synaptic transmission of activity among neurons seems to be very labile and can be changed dynamically. Selective enhancement refers not to a permanent change in synaptic efficacy between neurons but to the transient dynamics of a neuronal interaction which sustains it during activity [1064]. Nevertheless, it was shown [45] that during classical conditioning of the siphon withdrawal reflex in Aplysia the monosynaptic EPSP, which constitutes part of the postsynaptic conditioned response, increases after pairing. This is the first indirect evidence that synaptic efficacy may change independently of the activation of other synapses, and it supports the idea that synaptic plasticity contributes to learning, but does not prove it. After pairing, the sensory neuron evoked enlarged monosynaptic EPSP in the motor neuron only if in the response to the CS^+ this sensory neuron generated an AP [45] and, hence, in the presynaptic neuron the plasticity of excitable membrane may be developed. Besides, the behavior of those synaptic connections that are included in both effective and ineffective signals is unknown. Recently it was convincingly shown that receptorreceptor interactions participate in learning and memory and that a cluster of receptors can work as collective functional units located at the plasma membrane [505]. Functional interactions between synapses complement properties of selective change of excitability during learning and allow the development of a neuronal model that well conforms to experimental data [1044].

Computational models of the information processes occurring at the cellular level have been developed [945]. There are known attempts to describe the mechanisms of intracellular regulation of chemoreceptor sensitivity [919] or the size of a neurotransmitter pool [473] as a basis for associative learning. In the enzymatic model of a neuron [596] information may be stored as a distribution of ferments in the cellular membrane. S. Gerzon and A. Michailov [461] presented the most complete model of non-equilibrium biochemical neuron. However, these models are not capable of distinguishing between different excitabilities relative to different input signals, and recognition of the general common properties of similar synaptic patterns is complicated.

An alternative approach considers a neuron as an object that makes predictions as to the consequences of input signals and generates an output signal in conformity with this prediction [1044]. The final level for decision-making is, apparently, molecular. It is assumed that different levels of threshold are associated with modulation of sodium channels. We now propose a neuronal model that is capable of habituation, classical conditioning and instrumental learning, merges synaptic pattern as a whole and provides a different excitability relative to the different input signals generation after a learning procedure.

Briefly, it would appear reasonable to consider that chemical processes involved in learning begin at chemoreceptors and terminate on excitable membrane channels. The chemical singularity during learning was generated by interactions of second messengers specific for the corresponding excited synapses and for the rewarding input. The kinetics of these supposed chemical reactions can be described by a set of differential equations. The yield of chemical reactions was considered to affect the transition of sodium channels in an open state as described by the Hodgkin-Huxley model. When the thresholds within the response to the CS^+ and CS^- become different, these signals induce different chemical reactions within the neuron.

We have developed a neuronal model that after a learning procedure, exhibits different excitability to stimuli, thus predicting different changes in the environment. Let us consider the neuronal model with N inputs and one output. The signal on each of the N inputs (synapses) is in the initial conditions either 1 or 0, depending on the excitation state of the presynaptic neurons, if an initial memory is absent. Further input signal depends on the memory state of the presynaptic neuron and is continuous.

The properties of excitable membranes in our model were controlled by the supposed chemical reactions occurring in the neural cells. Let us write the equations of the chemical reactions involved in the information recording process and their corresponding differential equations (see explanation in the previous paragraph). Here is the set of the chemical and differential equations corresponding to the memory formation process that is memory recording.

Chemical reactions	Differential equetions
$x_i + x_j \to w_{ij}^+$	$\frac{dx_i}{dt} = -k_1 x_i \sum x_j$
$w_{ij} + Ca^{2+} + R \to W_{ij}^+ + Ca^{2+} + R$	$\frac{dw_{ij}}{dt} = -\Xi w_{ij} + k_1 x_i x_j$
$w_{ij} + Ca^{2+} \to W_{ij}^+ + Ca^{2+}$	$\frac{dW_{ij}^+}{dt} = (k_2 Ca^{2+} R + k_5) w_{ij}$
$w_{ij} + R \to W_{ij}^- + R$	$\frac{dW_{ij}^-}{dt} = (k_3 Ca^{2+} + k_4 R) w_{ij}$
$R \to 0$	$\frac{dR}{dt} = -k_5 R,$

where

$$\Xi = k_2 Ca^{2+} R + k_3 Ca^{2+} + k_4 R + k_5$$

and $k_1, ...k_5$ are the reaction rate constants, while $i, j = 1...N$; N is the number of synapses converging on a neuron. The concentrations of the substances are designated by the same letters as the substances themselves and their values are positive and small. Ca^{2+} ions point to AP generation. The molecules of instantaneous memory w_{ij} interact with the second messenger R and Ca^{2+} ions and form the short-term memory molecules W_{ij}^+ and W_{ij}^-. W_{ij}^+ molecules are synthesized when an AP generation promotes a reward, while W_{ij}^- is connected with punishment.

During the process of reproduction, the W_{ij}^+ molecules support AP generation, while W_{ij}^- molecules reduce it. w_{ij} molecules carry information that synapses i and j (and, correspondingly, pairs of second messengers x_i and x_j) are components of the input pattern, which may correspond to the CS$^+$, CS$^-$ or to be similar to any of them. Similarity between the current signal and the past signal (but we want to remember this is a recent signal concerning current behavior), fixed in short-term memory, is determined by the set of pairs i,j matching two signals, CS$^+$ and CS$^-$. The neuronal response to the input signals depends on the interaction between molecules of instantaneous memory w_{ij} with the short-term memory molecules W_{ij}^+ and W_{ij}^-. The result of these reactions is the synthesis of the substances α^+ and α^- that shift the balance between the substances β^+ and β^-, determining the mean opening time of potential-sensitive channels. Let us write the equations of the chemical reactions involved in the information reproduction process:

The number of substances entering the reactions of α^+ and α^- synthesis is less than the number of the resulting substances. We assume that some agent enters these reactions and that its concentration is large enough and remains practically unchanged in the cell. Therefore, the concentration of this agent can be included in the value of the constant k_7. We also assume that the short-term memory molecules exhibit enzymatic properties during these reactions.

The kinetics of the reactions of information reproduction is described by the following set of chemical and differential equations. Immediately after the arrival of the input signal the neuron assesses the probability of a reward on the basis of accumulated experience, and either generates or does not generate an AP and only then does it receive information on the arrival of a reward. Therefore, the processes of recording and reproduction must be separate in time, and may be realized at the cost of exceeding the rate constants of the reproduction process over the constants of the recording process.

Chemical reactions	Differential equetions
$x_i + x_j + W_{ij}^+ \to x_i + x_j + W_{ij}^+ + \alpha^+$	$\frac{d\alpha^+}{dt} = k_7 \sum W_{ij}^+ x_i x_j - k_8 \alpha^+ \beta^+ \beta^-$
$x_i + x_j + W_{ij}^- \to x_i + x_j + W_{ij}^- + \alpha^-$	$\frac{d\alpha^-}{dt} = k_7 \sum W_{ij}^- x_i x_j - k_8 \alpha^- \beta^+ \beta^-$
$\alpha^+ + \beta^+ + \beta^- \to \beta^+$	$\frac{d\beta^-}{dt} = k_8(\alpha^+ + \alpha^-)\beta^+ \beta^-$
$\alpha^- + \beta^+ + \beta^- \to \beta^-$	$\frac{d\beta^+}{dt} = -\frac{d\beta^-}{dt}$

where k_7, k_8 are the reaction rate constants. The properties of neuronal learning may be changed by varying the model constants. For example, an increase in the k_3 coefficient enhances the ability of the model to develop habituation. A compromise has to be reached between times of the input excitation and the arrival of the reinforcement signal. If the time of arrival of the reinforcement signal t, differs from the optimum, the participation of the reinforcement messengers R in the process of memory consolidation is minimal. A limitation in the amount of reinforcement impedes development of cellular analogs of associative learning. This corresponds to the experimental data [525].

The model describes process habituation, classical conditioning and instrumental reflex. The properties of such a model neuron resemble the properties of a live neuron. Neural networks consisting of such model neurons demonstrate an advantage over other models. The neuron's error depends on the output reaction of the network and does not depend directly on output reactions of a given neuron. It is necessary, however, to tune each neuron during learning. Neurons are tuned by means of an intra-neuronal mechanism. Most of the optimization schemes, such as back-propagation of error, require knowledge about network connections. This means that an artificial neural network, as hardware built from our model neurons, may be easily constructed from standard neuro-chips and it may work without outside control.

For a numerical solution of the set of differential equations, we employed the Euler difference scheme ensuring an accuracy of count by only one step. It is considered to be equal to 0.05 ms. We have described earlier the initial conditions for the set of differential equations and other details [1056, 1055, 1044]. As a result of a one step approximation, we have received parameters α^+, α^- and β^- that control prediction of the reward P and properties of sodium channels.

$$\alpha^- - \alpha^+ = \sum_{ij}^{N}(W_{ij}^- - W_{ij}^+)x_i x_j,$$

$$\beta^- = 1 + \sum_{ij}^{N}(W_{ij}^- - W_{ij}^+)x_i x_j,$$

The values of $W_{ij}^+(0)$ and $W_{ij}^-(0)$ correspond to the state of memory prior to the arrival of the first input signal. We take it as null matrices. The conditions for memory modification (value of memory elements shift) are: $\Delta W_{ij} = \gamma(y - y^*)x_i x_j$, where γ is a positive constant determining the rate of learning, correct output $y^* \in [0, 1]$, y is the output function of the neuron and $\Delta W_{ij} = W_{ij}^- - W_{ij}^+$.

Prediction P is connected with the portion of rewards ($P \leq 0$) or punishments ($P \geq 0$), which followed in the past after AP generation in this neuron, and modulates the threshold level θ, dependently on excess of the positive W_{ij}^+ or negative W_{ij}^- memory, connecting with the given input signal:

$$P = \frac{\sum_{i,j}^{N}(W_{ij}^- - W_{ij}^+)x_i x_j}{\sum_{i,j}^{N}(W_{ij}^- + W_{ij}^+)x_i x_j},$$

which imply that $|P| \leq 1$. Prediction P is related to the statistical significance of difference between the fractions of the rewards, received in the past, after AP generation and AP failure for simultaneous activation of synaptic pairs appearing in the given signal.

Besides that, prediction P regulates an output signal, which is larger when AP generation in the past lead to a reward, so that a neuron transmits prediction P, which is a partially processed signal, into neuronal output. Therefore, heavy neuronal calculations do not disappear for nothing and the neuron sends to target neurons its prediction. We call this process the forward propagation of prediction (see next paragraph).

Active information processing depends on the parameters of Hodgkin-Huxley equations and these parameters are usually stable, but not constant. In many cases, one parameter may be compensated for at the expense of another. An equation describing the correspondence between ion currents and voltage V in the cellular membrane is:

$$C\frac{dV}{dt} + (V - V_0)\rho + I_{Na} + I_K + I_{Ca} = I$$

Here:

$$\begin{aligned} I_{Na} &= G_{Na} m^3 h(V - E_{Na}) \\ I_K &= G_K n^4 (V - E_K) \\ I_{Ca} &= G_{Ca} z \frac{V - E_{Ca}}{1 + \xi[Ca]}, \end{aligned}$$

where C and ρ are the capacitance and resistance of the membrane, G_{Na}, G_K, G_{Ca} and E_{Na}, E_K, E_{Ca} are the maximum conductivities and equilibrium potentials for the sodium, potassium and calcium channels respectively, ξ is the degree of calcium inhibition of the calcium channel and V_0 is the resting potential. The time dynamics of magnitudes h, n and z, figuring in the equations is indicated in [363].

Properties of neural networks depend both on network structure and on a model of neurons, which this network incorporates. The cascade of chemical reactions results in change of the neuronal electrogenesis by means of its action on potential-sensitive channels.

We have considered the effect of regulation of the properties of the voltage-gated Na^+ channels. Regulation allows simulation of the change occurring in the AP parameters associated with neuronal excitability during learning. Na^+ channels are the main target of G protein-coupled receptors, which initiate signaling cascades joined to input-output events in neurons. It has already been shown that properties of sodium channels are modulated [744]. Neuronal activity and membrane depolarization through phosphorylation of Na^+ channels can change their availability. For instance, depolarization rearranges Na^+ channels in a slow inactivated state. In this state Na^+ channels are unavailable and do not open in response to membrane depolarization. Entrance and exit from this state is rather slow and takes seconds, but the process is accelerated after synaptic receptor stimulations and activation proteinkinases A and C [211].

Such processes may change availability of Na^+ channels during a few tens of milliseconds for 5-10% and this amount will explain the magnitude of changes in excitability revealed in the experiments during selective reorganization of thresholds. Rapidity of modulation of Na^+ channels is in agreement with the data concerning a selective decrease in excitability during habituation, when the latency of responses increases. Yet, rapidity of these reorganizations must be a few milliseconds during the selective increase in excitability during classical conditioning, but this may be achieved when Na^+ channels are located near the active synaptic input. In the context of our model, the result of regulation of other channels may also be studied. For example, AP duration may be effectively controlled by the action on the Ca^{2+} channels [1111], but we will consider only participation of Na^+ channels.

Let the substances β^+ and β^- mutually compensate their influence on the channel at the initial state. Then, after the arrival of the input signal the balance between β^+ and β^- is disrupted according to the accumulated memory and, therefore, the probability of AP generation is changed. The dynamics of the Na^+-channel transition into the open and closed states are described by the following equation:

$$\frac{dm}{dt} = k_-(V)m[k_-(V) + k_+(V)\beta^-],$$

where m is the activation gating variable of the Na^+-channel and β^- is determined by the set of differential equations of memory reproduction. k_-, k_+ are the functions of the membrane potential.

The cascade of chemical reactions results in the change of the neuronal electrogenesis by means of its action on the state of potential-sensitive channels. We assume that the substances β^- influence the probability of sodium channel transition into the open state. This effect is realized by the following scheme of Na^+ -channel transition into the open state and inversely:

$$CLOSED \quad \underset{k_+(V)\beta^-}{\overset{k_-(V)}{\rightleftharpoons}} \quad OPENED$$

In order to examine the properties of the model, we performed a computer simulation of the process of development of instrumental learning. Let us have two super-threshold input signals. The neuron receives the reinforcement signal (analog of US) in the case of AP generation after presentation of the given input signal (analog of CS^+). Another input signal is an analog of CS^-. The learning procedure consists of the successive presentations of the CS^+ and US or an isolated (from time to time) presentation of the CS^-. It may be seen that the neuronal response to the presentations of the CS^+ and CS^- is variable. Latency of the responses (estimated as the moment when the membrane potential reaches its maximum value) decreases after presentations of the CS^+, and increases with the CS^- presentations (Fig. 1.33).

During learning, the AP generation threshold in responses to the CS^+ decreased, while in responses to the CS^- it increased until the point of AP blockade. The change in the AP parameters (amplitude, duration and generation threshold) associated with neuronal excitability in the course of the numerical experiment is shown in Fig. A.2. The generation threshold was defined as the difference between the membrane potential value at the point of maximum curvature on the leading edge of AP and the value of the resting potential. The AP amplitude was defined as the maximum deviation of the membrane potential from the resting potential.

The generation threshold decreased in the neuronal responses to the CS^+ during the numerical experiment and the amplitude and duration of the AP increased (Figs. A.2A,A.2B and Fig. A.2C, curve 1). These changes in the AP parameters correspond to the rise in the neuronal excitability. Meanwhile, the amplitude and duration of the AP decreased, and the generation threshold increased in the neuronal response to the CS^- (Figs. A.2A,A.2B and Fig. A.2C, curve 2) that corresponds to the drop in neuronal excitability. Thus, the model exhibits different excitability relative to the different input signals after learning and is in good agreement with the experimental data.

Fig. A.3 depicts the dependencies of the fraction of correctly recognized signals and the necessary number of learning cycles on the quantity of input signals presented. Each point in the graphs is the result of averaging over 100 independent experiments. In each experiment the necessary number of

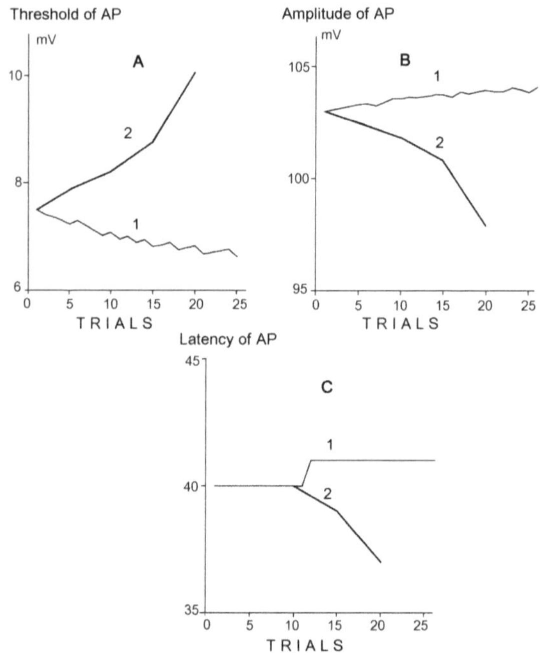

Fig. A.2. The change of the neuronal response during the course of the numerical experiment with 8-synapsed model neuron. The vector 1111000 was used as the CS^+ and the vector 0001111 as the CS^-. The components of vectors correspond to the state of excitation of the neuronal inputs. The CS^- was presented to the model after each fifth presentation of the CS^+. A- Generation threshold, B- Amplitude of AP and C Duration of AP. The AP duration was defined as the interval between its leading and back edges taken at some super-threshold level of the membrane potential. Explanations in the text.

different input signals was randomly generated, each of them also randomly being assigned to one of two classes. On presentation of the signals of only one class, the model neuron received a reward for the generation of an AP.

As Fig. A.3 shows, the curves of the dependence of the percentage of correctly classified signals have a point of inflection. The number of signals at a given point roughly corresponds to the informational capacity of the model. Thus, the model is capable of division into two classes of N^2 images, where N is the number of inputs of the model. The resulting recognition is resistant to damage of 10-20% memory elements.

Taking into account the above considerations, we may formulate the basic principles of neuron function. A neuron evaluates the significance of difference between the number of reinforcements that had been given following the generation and the failure of the spike in the response to the current signal, and excitability of neurons increases if a stimulus with a greater significance

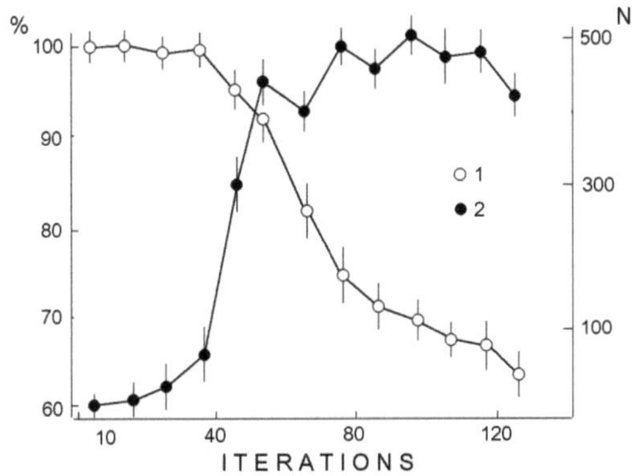

Fig. A.3. Fraction of correctly recognized signals (1) and the necessary number of the trials N (2). Dimensionality of input vector 7-synapses.

acts. Meanwhile, a neuron exhibits the all-or-none principle of spike generation for any given combination of inputs. A neuron calculates the sum of an input signal and compares this sum with the modified threshold; the change in the excitability within the response of a presynaptic neuron is accompanied by changes in the monosynaptic response value in the postsynaptic target neurons.

Chemical reactions happening in real neurons may be distinguished from our description. Nevertheless, our model demonstrates the real possibility of the existence of a chemical scheme for explaining complex behavior of a neuron. Depending on the biological significance of the stimulus, a living neuron can thus regulate its own excitability by changing the electrical properties of its membrane, and can also regulate the efficiency of its synaptic inputs by changing the chemical properties of the membrane. The change in synapse efficiency and neuronal excitability are of short duration and take place immediately after recognition of the stimulus. Thus, perhaps an unstable molecular micro-system controls brain behaviors. Models of chemical processes in a neuron are based on such properties as a perspective means for the creation of self-contained systems. A neuronal net comprising such neurons display excellent tuning, possesses a large volume of memory and continues to function after removal of a part of its elements [1262].

A.2 *An alternative type of fuzzy dynamics equations

In this section we briefly consider the case:

$$\lim_{\Delta_{\boldsymbol{x}} \to \boldsymbol{x}} m(\Delta_{\boldsymbol{x}}, t) = 0. \tag{A.1}$$

In this case a system's state space can be approximated by a Riemann-manifold. Since choice of the coordinates on the manifold is not unique,

$m(\Delta, t)$ should be invariant under appropriate transformations of the coordinates. Therefore, the function $m(\Delta, t)$ could depend only on invariant measure of the domain Δ. Simple such a measure is ρdU, where dU is "volume" of Δ in a given coordinate system, while the multiplier ρ makes the product ρdU invariant under the admissible transformations of the coordinates. Therefore, we can write asymptotically:

$$\lim_{\Delta_{\boldsymbol{x}} \to \boldsymbol{x}} \frac{m(\Delta_{\boldsymbol{x}}, t))}{g(\rho(\boldsymbol{x}, t) dU_{\boldsymbol{x}})} = 1, \tag{A.2}$$

where $g(0) = 0$ and function $g(...)$ is continuous (but not necessary differentiable) near zero. We will assume, also, that ρ is a smooth function of $\boldsymbol{x}$ and t. In physics, the multiplier like ρ is called - density, so we call $\rho(\boldsymbol{x}, t)$: *State Density*. Generally speaking, the function g could depend explicitly on point $\boldsymbol{x}$, but such dependence must be omitted, if we want hold homogeneity of the representations of the logical connectives $\bigvee$ and $\bigwedge$ for the whole state space.

Consider small domains $\Delta_{\boldsymbol{x}} = \Delta_{\boldsymbol{x}+\boldsymbol{\varepsilon}} \bigcup \Delta_{\boldsymbol{x}-\boldsymbol{\varepsilon}}$ with the volumes:

$$dU_{\boldsymbol{x}} = dU_{\boldsymbol{x}-\boldsymbol{\varepsilon}} + dU_{\boldsymbol{x}+\boldsymbol{\varepsilon}}, \tag{A.3}$$

The possibility, that the system is in $\Delta_{\boldsymbol{x}}$, is equal to:

$$g(\rho(\boldsymbol{x}, t) dU_{\boldsymbol{x}}) = S(g(\rho(\boldsymbol{x} - \varepsilon, t) dU_{\boldsymbol{x}-\boldsymbol{\varepsilon}}); g(\rho(\boldsymbol{x} + \varepsilon) dU_{\boldsymbol{x}+\boldsymbol{\varepsilon}})). \tag{A.4}$$

Using (A.3), one has for the smooth $\rho(\boldsymbol{x}, t)$ and small ε:

$$\rho(\boldsymbol{x}) dU_{\boldsymbol{x}} = \rho(\boldsymbol{x} - \varepsilon) dU_{\boldsymbol{x}-\boldsymbol{\varepsilon}} + \rho(\boldsymbol{x} + \varepsilon) dU_{\boldsymbol{x}+\boldsymbol{\varepsilon}} + o(\varepsilon dU).$$

Thus, for the connective $\bigvee$, we have for the small neighboring domains Δ_1, Δ_2:

$$S(g(\rho(\boldsymbol{x}_1) dU_1); g(\rho(\boldsymbol{x}_2) dU_2)) \approx g(\rho(\boldsymbol{x}_1) dU_1 + \rho(\boldsymbol{x}_2) dU_2). \tag{A.5}$$

In order to find out an admissible representation of the possibility of transition, let us assume now that at the time $t - \delta$ the system was in the domain Δ_0 with the possibility 1, and with the possibility 0 in all other places. It is obvious, that the possibility, that at the time t the system will be in a domain $\Delta_{\boldsymbol{x}}$, is equal to the possibility that the system transferred to $\Delta_{\boldsymbol{x}}$ from Δ_0. That is in such a situation the transition possibility - $P_\delta(\Delta_{\boldsymbol{x}}, \Delta_{\boldsymbol{x}_0}, t)$ should be equal to $m(\Delta_{\boldsymbol{x}}, t)$. It is follows (A.2) that asymptotically:

$$m(\Delta_{\boldsymbol{x}}, t) \sim g(\rho(\boldsymbol{x}, t) dU_{\boldsymbol{x}}),$$

so we can write:

$$P_\delta(\Delta_{\boldsymbol{x}}, \Delta_{\boldsymbol{x}0}) \approx g(\Gamma_\delta dU_{\boldsymbol{x}}), \tag{A.6}$$

where Γ_δ depend on $(\boldsymbol{x}_0; \boldsymbol{x}; t)$, but doesn't depend on dU_0.

Now we want to find admissible representation of the connective $\bigwedge$. To do this, consider decomposition of the transition from $\Delta_{\boldsymbol{x}_0}$ to $\Delta_{\boldsymbol{x}}$ during the

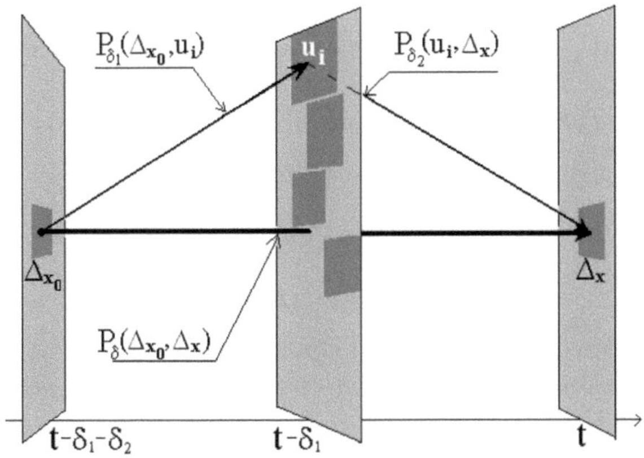

Fig. A.4. Decomposition of the transition possibility in the case $\lim_{\Delta_{\boldsymbol{x}}\to\boldsymbol{x}} m(\Delta_{\boldsymbol{x}}, t) = 0$.

time $\delta = \delta_1 + \delta_2$ on the transitions from $\Delta_{\boldsymbol{x}_0}$ to u_1 during the time $\delta_1 \bigwedge$ from u_1 to $\Delta_{\boldsymbol{x}}$ during δ_2, *or* from $\Delta_{\boldsymbol{x}_0}$ to $u_2 \bigwedge$ from u_2 to $\Delta_{\boldsymbol{x}}$, *or* ... so on, for all possible intermediate domains (see Fig. A.4). In according with (A.6) we have:

$$P_{\delta_1}^i = P_{\delta_1}^i \left(\Delta_{\boldsymbol{x}0}, u_i\right) = g\left(\Gamma_{\delta_1}^i du_i\right),$$
$$P_{\delta_2}^i = P_{\delta_2}^i (u_i, \Delta_{\boldsymbol{x}}) = g\left(\Gamma_{\delta_2}^i dU_{\boldsymbol{x}}\right).$$

Then, for possibility of the "two-fold" transition from $\Delta_{\boldsymbol{x}_0}$ to u_i during the time $\delta_1 \bigwedge$ from u_i to $\Delta_{\boldsymbol{x}}$ during the time δ_2 one obtains:

$$T\left(P_{\delta_1}^i, P_{\delta_2}^i\right) = g\left(F\left(\Gamma_{\delta_1}^i du_i; \Gamma_{\delta_2}^i dU_{\boldsymbol{x}}\right)\right),$$

with some unknown function F. It follows from Fig. A.4 and (A.5) that:

$$P_\delta\left(\Delta_{\boldsymbol{x}0}, \Delta_{\boldsymbol{x}}\right) = S\left(..., T\left(P_{\delta_1}^i, P_{\delta_2}^i\right), ...\right) = g\left(\sum_i F\left(\Gamma_{\delta_1}^i du_i, \Gamma_{\delta_2}^i dU_{\boldsymbol{x}}\right)\right),$$

As $P_\delta = g\left(\Gamma_\delta dU_{\boldsymbol{x}}\right)$ this implies that:

$$\Gamma_\delta dU_{\boldsymbol{x}} = \sum_i F\left(\Gamma_{\delta_1}^i du_i, \Gamma_{\delta_2}^i dU_{\boldsymbol{x}}\right) = \left(\sum_i f\left(\Gamma_{\delta_1}^i du_i\right)\Gamma_{\delta_2}^i\right) dU_{\boldsymbol{x}}.$$

In order that the last sum will converge for any partition $\{du_i\}$, we have:

$$f\left(\Gamma_{\delta_1}^i du_i\right) = const \cdot \Gamma_{\delta_1}^i du_i.$$

Without loss of generality we can put $const = 1$. Therefore:

$$F\left(\Gamma^i_{\delta_1} du_i; \Gamma^i_{\delta_2} dU_{\boldsymbol{x}}\right) = \Gamma^i_{\delta_1} \Gamma^i_{\delta_2} du_i dU_{\boldsymbol{x}}.$$

Thus, the logical connective $\bigwedge$ for the case $\lim_{\Delta_{\boldsymbol{x}} \to \boldsymbol{x}} m(\Delta_{\boldsymbol{x}}, t) = 0$ has to be defined as:

$$T(P_1(\Gamma_1 dU_1); P_2(\Gamma_2 dU_2)) = g\left(\Gamma_1 dU_1 \Gamma_2 dU_2\right). \tag{A.7}$$

If the function g is invertible we can write:

$$S(m_1; m_2) = g(g^{-1}(m_1) + g^{-1}(m_2)), \tag{A.8}$$

$$T(P_1; P_2) = g(g^{-1}(P_1) \cdot g^{-1}(P_2)), \tag{A.9}$$

so in the considered case the logical connectives $\bigvee$ and $\bigwedge$, at least infinitesimally, should be represented as *pseudo-arithmetic sum* and *pseudo-arithmetic product* [829],[937].

Therefore, in the given case, the symbolic expression (7.6) leads to the equation:

$$g\left(\rho\left(\boldsymbol{x}, t\right) d^n \boldsymbol{x}\right) = g\left(\left(\int \Gamma\left(\boldsymbol{x}, \boldsymbol{x}', t; \delta\right) \rho\left(\boldsymbol{x}', t - \delta\right) d^n \boldsymbol{x}'\right) d^n \boldsymbol{x}\right), \tag{A.10}$$

where n is dimension of a system's state space and we put $dU_{\boldsymbol{x}} = d^n \boldsymbol{x}$. Expanding (A.10) with respect to δ leads to the equation:

$$\frac{\partial \rho}{\partial t} + (\boldsymbol{V}\left(\boldsymbol{x}, t\right) \cdot \nabla)\rho = \left\{D\left(\boldsymbol{x}, t\right) \nabla^2 + U\left(\boldsymbol{x}, t\right)\right\} \rho, \tag{A.11}$$

with

$$U\left(\boldsymbol{x}, t\right) = \lim_{\delta \to 0} \frac{1}{\delta} \left(\int \Gamma_\delta\left(\boldsymbol{x} - \boldsymbol{u}, \boldsymbol{x}, t\right) d^n \boldsymbol{u} - 1\right),$$

$$\boldsymbol{V}\left(\boldsymbol{x}, t\right) = \lim_{\delta \to 0} \frac{1}{\delta} \int \Gamma_\delta\left(\boldsymbol{x} - \boldsymbol{u}, \boldsymbol{x}, t\right) \boldsymbol{u} d^n \boldsymbol{u},$$

$$D\left(\boldsymbol{x}, t\right) = \lim_{\delta \to 0} \frac{1}{2\delta} \int \Gamma_\delta\left(\boldsymbol{x} - \boldsymbol{u}, \boldsymbol{x}, t\right) \left|\boldsymbol{u}\right|^2 d^n \boldsymbol{u}.$$

If a system's state space is stratified on M, singly-connected manifolds, this equation is transformed to:

$$\frac{\partial \rho_\sigma}{\partial t} + V^{\kappa,i}_\sigma\left(\boldsymbol{x}, t\right) \nabla_i \rho_\kappa = \left\{D^{\kappa,ij}_\sigma\left(\boldsymbol{x}, t\right) \nabla_i \nabla_j + U^\kappa_\sigma\left(\boldsymbol{x}, t\right)\right\} \rho_\kappa, \tag{A.12}$$

where indexes $\sigma, \kappa = 1, ..., M$ describe the different manifolds.

The coefficients in these equations have the following meaning. The coefficient $D(\boldsymbol{x}, t)$ describes the diffusion, the coefficient $\boldsymbol{V}(\boldsymbol{x}, t)$ describes the flows and the coefficient $U(\boldsymbol{x}, t)$ describes the external fields. For $V_\sigma \equiv 0$, $D = \hbar/2m$ and complex valued $\rho \to \psi = \psi_1 + i\psi_2$, equation (A.12) corresponds to the Schreodinger equation:

$$i\frac{\partial \psi(\boldsymbol{x}, t)}{\partial t} = \left\{-\frac{\hbar}{2m}\nabla^2 + \frac{1}{\hbar}U(\boldsymbol{x}, t)\right\}\psi(\boldsymbol{x}, t),$$

If the system's velocities are strictly restricted: $|\boldsymbol{v}| \leq c < \inf$, it can be shown that $D^{\kappa}_{\sigma} \equiv 0$. Then, for spinor-valued $\rho \to \boldsymbol{\psi} = \psi_{\sigma}$, $\sigma = 0, 1, 2, 3$ Eq. (A.12) corresponds to the Dirac equation:

$$\left\{ i\gamma^0 \frac{\partial}{\partial t} - -i(\boldsymbol{\gamma} \cdot \nabla) \right\} \boldsymbol{\psi}(\boldsymbol{x}, t) = m\boldsymbol{\psi}(\boldsymbol{x}, t).$$

where $\{\gamma^0, \boldsymbol{\gamma}\}$ are Dirac matrixes. For scalar valued ρ and $U = (\nabla \cdot \boldsymbol{V})$ (A.12) leads to the Fokker-Plank equation:

$$\frac{\partial \rho(\boldsymbol{x}, t)}{\partial t} = D\nabla^2 \rho + (\nabla \cdot \boldsymbol{V}(\boldsymbol{x}, t)\rho).$$

All these equations are well known in the Statistical and Quantum Physics, so the case (A.1) is equivalent to the stochastic or quantum approaches. Our discussion so far leads to conclusion that general form of the dynamics equations are determined, in fact, by the causality principal, while the system specificity is included into the state space geometry and into two first moments of the possibility of a local system's movement.

References

1. Abdulla FA, Smith PA (1999) Neuroscience 89:43–60.
2. Abdullaev Y, Posner M (2005) Nat Neurosci 8:1424–1425.
3. Abeles M (2004) Science 304:523–524.
4. Abrams R, Hammel HT (1965) Am J Physiol 208:698–702.
5. Abraham WC, Robins A (2005) TRENDS in Neurosci 28:73–78
6. Adachi N (2006) Masui 55:542–551
7. Adamec RE, Blundell J, Burton P (2003) Progress in Neuro-Psychopharmacology and Biological Psychiatry 27:1243–1267.
8. Afraz SR, Kiani R Esteky H (2006) Nature 442: 692–695
9. Agmo A (1999) Behav Brain Res 105:129–150.
10. Ahern GP, Klyachko VA, Jackson MB (2002) TRENDS in Neurosci 25:510–517
11. Ahlbom E, Prins GS, Ceccatelli S (2001) Brain Res 892:255–262.
12. Aihara H, Li X, Fujiwara S, Matsumura T, Okumura S, Tozaki H, Kanno T, Nishizaki T (2001) Brain Res 922:158–162
13. Aisemberg GO, Gershon TR, Macagno ER (1997) J Neurobiol 33:11–17.
14. Aizenman CD, Akerman CJ, Jensen KR, Cline HT (2003) Neuron 39:831–842.
15. Aizenman CD, Linden DJ (2000) Nat Neurosci 3:109–111.
16. Akirav I, Maroun M (2007) Neural Plast 10:30–73.
17. Alanko L, Porkka–HeiskanenT, Soinila S (2006) Journal of Chemical Neuroanatomy 31:162168.
18. Albe-Fessard D (1951) Scalpel (Brux) 104:799–808.
19. Albe-Fessard D (1971) Zh Vyssh Nerv Deiat Im I P Pavlova 21:509–513. (in Russian)
20. Albright TD, Kandel ER, Posner MI (2000) Curr Opin Neurobiol 10:612–624
21. Aleksandrov YI, Grinchenko YV, Shevchenko DG, Mats VN, Laukka S, Averkin RG (2005) Neurosci Behav Physiol 35:501–509
22. Alex KD, Pehek EA (2007) Pharmacology Therapeutics 113:296–320.
23. Alger BE (2002) Prog Neurobiol 68:247–286
24. Alkon DL (1979) Science 205:810–816.
25. Allen AM, Moeller I, Jenkins TA, Zhuo J, Aldred GP, Chai SY, Mendelsohn FAO (1998) Brain Res Bulletin 47:17–28
26. Allen JW, Mutkus LA, Aschner M (2002) Neurotoxicology 23:693–700.
27. Allen NJ, Barres BA (2005) Curr Opin Neurobiol 15:542–548
28. Alkire MT, Miller J (2005) Prog Brain Res 150:229–244

29. Allen JW, Mutkus LA, Aschner M (2002) Neurotoxicology 23:693–700.
30. Amalric M, Koob GF (1993). Prog Btain Res 99:209–226.
31. Ames A (2000) Brain Res Rev 34:42–68.
32. Amir R, Argoff CE, Bennett GJ, Cummins TR, Durieux ME, Gerner P, Gold MS, Porreca F Strichartz GR (2006) J Pain 7:1–29.
33. Andersen M, Lambert JDC (1999) J Physiol 519:85–100.
34. Anderson JS, Carandini M, Ferster D (2000b) J Neurophysiol 84:909-926
35. Anderson JW, Washburn AV, Ferguson AV (2000a) Neuroscience 100:539–547.
36. Anderson TR, Jarvis CR, Biedermann AJ, Molnar C, Andrew RD (2005) J Neurophysiol 93: 963–979
37. Andre J, Zeau B, Pohl M, Cesselin F, Benoliel JJ, Becker C (2005) Behav Biochem Studies. J Neurosci 25:78967904.
38. Angelino E, Brenner MP (2007) PLoS Comput Biol 3:1751–1759.
39. Angenstein F, Riedel G, Reymann KG, Staak S (1994) NeuroReport 5:381–384.
40. Anis Y (2006) Modulator 21:25–27
41. Ankarcrona M, Dypbukt JM, Bonfoco E, Zhivotovsky B, Orrenius S, Lipton SA, Nicotera P (1995) Neuron 15:961–973
42. Anokhin KV, Mileusnic R, Shamatkina I, Rose SPR (1991) Brain Res 544:101–107.
43. Anochin PK (1974) Uspechi Fisiol Nauk 5:5–76.
44. Anouara Y, Ghzilia H, Grumolotoa L, Montero–Hadjadjea M, Lesageb J, Galasa L, Ait-Alia D, Elkahlounec AG, Vieaub D, Vaudrya H (2006) Frontiers in Neuroendocrinology 27:82–83.
45. Antonov I, Antonova I, Kandel ER, Hawkins RD (2003) Neuron 37:135–147.
46. Antonsen BL, Herberholz J, Edwards DH (2006) J Neurosci 25:3086-3094.
47. Antzoulatos EG, Byrne JH (2004) TRENDS in Neurosci 27:555–560.
48. Aou S, Woody CD, Birt D (1992) J Neurosci 12: 560–569.
49. Arango V, Huang YY, Underwood MD, Mann JJ (2003) J Psychiatr Res 37:375–386.
50. Araujo IA, Gutierrez R, Oliveira-Maia AJ, Pereira P, Nicolelis MAL, Simon SA (2006) Neuron 51:483–494
51. Arban R, Maraia G, Brackenborough K, Winyard L, Wilson A, Gerrard P, Large C (2005) Behav Brain Res 158:123–132.
52. Arhem P, Klement G, Nilsson J (2003) Neuropsychopharmacology 28:40–47.
53. Arkhipov VI (1999) Behav Brain Res 106:39–46
54. Armano S, Rossi P, Taglietti V, Angelo ED (2000) J Neurosci 20:5208–5216.
55. Armstrong CM, Hille B (1998). Neuron 20:371–380.
56. Arnold VI (1978) Mathematical Methods of Classical Mechanics. Graduate Texts in Mathematics, Springer-Verlag, New York.
57. Aron A, Fisher H, Mashek DJ, Strong G, Li H, Brown LL (2005) J Neurophysiol 94:327–337.
58. Arshavsky YI (2001) A new insight. Brain and Cognition 46:414–428.
59. Arthur PG, Lim SCC, Meloni BP, Munns SE, Chan, A, Knuckey NW (2004) Brain Res 1017:146–154.
60. Arumugam TV, Gleichmann M, Tang SC Mattson MP (2006) Ageing Research Reviews 5:165–178.
61. Ashby WR (1956), An Introduction to Cybernetics. Chapmen and Hall, London.
62. Ashmarin IP, Eropkin MJ, Malukova IV (1976) Dokl. AN USSR 229:761–763.

63. Assimakopoulos SF, Vagianos CE, Zervoudakis G, Filose KS, Georgioud C, NikolopoulouV, Scopa CD (2004) Regulatory Peptides 120:185–193.
64. Atkinson AP, Thomas MSC, Cleeremans A (2000) Trends Cogn. Sci. 4 (2000) 372–382.
65. Atrens DM, Menendez JA (1993) Pharmacol Biochem.Behav 46:617–622.
66. Averbeck BB, Lee D (2004) TRENDS in Neurosci 27:225–230.
67. Aubin JP, Cellina A (1984) Differential Inclusions, Springer-Verlag, New York.
68. Auger AP (2003) J Neuroendocrinol. 2003 Jun;15(6):622–627.
69. Azmitia EC (1999) Neuropsychopharmacology 21:33–45.
70. Azouz R (2006) J Neurophysiol 94:2785–2796.
71. Azouz R, Gray CM (1999) J Neurosci 19:2209–2223.
72. Baars BJ, Laureys S (2005) TRENDS Cogn Sci 9:269–276.
73. Badaut J, Hirt L, Price M, Ribeiro MC, Magistretti PJ, Regli L (2005) Brain Res. 1051:117–122.
74. Baev KV (1997) Prog Neurobiol 51:129–166.
75. Bahh BE, Auvergne R, Lere C, Brana C, La Salle GLG, Rougier A (2001) Brain Res 894:209–217.
76. Bai X, Wang H, Li Z, Liu K (2004) J Huazhong Univ Sci Technolog Med Sci 24:68–70.
77. Bailey CH, Kandel ER (1994) Sem Neurosci 6:35–44
78. Baker TL, Fuller DD, Zabka AG, Mitchell GS (2001) Respir Physiol 129:25–35.
79. Balboa MA, Balsinde J (2006) BBA - Molecular and Cell Biology of Lipids 1761:385–391.
80. Ballantyne JC, LaForge KS (2007) Pain 129:235–255.
81. Ballanyi K, Onimaru H, Homma I (1999) Prog Neurobiol 59:583–634.
82. Balthazart J, Ball GF (2006) TRENDS in Neurosci 29:241–249.
83. Banasiak KJ, Xia Y, Haddad GG (2000) Prog Neurobiol 62:215–249.
84. Bandemer H, Nather W (1992) Fuzzy Data Analysis, Kluwer Academic Publisher, London.
85. Bao J-X, Kandel ER, Hawkins RD (1998) J Neurosci 18:458-466.
86. Barco A, Bailey CH, Kandel ER (2006) J Neurochem 97:1520-1533.
87. Barrot M, Wallace DL, Bolaos CA, Graham DL, Perrotti LI, Neve RL, Chambliss H, Yin JC, Nestler EJ (2005) Proc Natl Acad Sci USA 102:8357–8362.
88. Basavappa S, Chartouni V, Kirk K, Prpic V, Ellory JC, Mangel AW (1995) J Neuroscie 15:3662–3666.
89. Bauco P. Rompr PP (2001) Eur J Pharmacol 432:57–61.
90. Baumeister RF, Bratslavsky E, Finkenauer C, Kathleen D (2001) Rev Gen Psychol 5:323–370.
91. Bayley PJ, Frascino JC, Squire LR (2005) Nature 436:550–553.
92. Baylor DA, Lamb TD, Yau KW (1979) J Physiol 288:613–634.
93. Bean BP (2007) Nat Rev Neurosci 8:451–465.
94. Bear MF, Malenka RC (1994) Curr Opin Neurobiol 4:389–399.
95. Bechtholt AJ, Smith R, Raber J, Christophe L (2004) Pharmacol Biochem Behav 77:783–792.
96. Beckman M (2004) Science 304:1888–1889.
97. Beckstead MJ, Grandy DK, Wickman K, Williams JT (2004) Neuron 42:939–946.
98. Begley S (2007) How the brain rewires itself. Time Friday, Jan. 19
99. Behar TN, Smith SV, Kennedy RT, McKenzie JM, Maric I, Barker JL (2001) Cerebral Cortex 11:744–753.

100. Behnisch T, Reymann KG (1994) Neurosci Lett 165:23–26.
101. Beinfeld MC (2001) Peptides 22:1197–1200.
102. Belelli D, Lambert JJ (2005) Nat Rev Neurosci 6:565–575.
103. Belenkov NJ (1965) Conditional reflex and vegetative centers of brain. Medicine, Moscov (in Russian)
104. Belmeguenai A, Vaudry H, Leprince J, Vivet B, Cavelier F, Martinez J, Louiset E (2000) Neuroendocrinology 72:379–391.
105. Beltowski J, Marciniak A, Wjcicka G, Grny D (2003) Acta Biochim Pol 50:103–14.
106. Benedetti F, Amanzio M (1997) Prog Neurobiol 52:109–125.
107. Benington JH, Frank MG (2003) Prog Neurobiol 69:71–101.
108. Bennett MVL, Zukin RS (2002) Neuron 41:495–511.
109. Bennett MVL, Zukin RS (2004) Neuron 43:495–511.
110. Beraudi A, Bruno V, Battaglia G, Biagioni F, Rampello L, Nicoletti F, Poli A (2007) Exp Brain Res 84 544–522.
111. Bernabeu R, Princ F, de Stein ML, Fin C, Juknat AA, Batile A, Izquierdo I, Medina JH (1995) NeuroReport 6:516–518.
112. Berrendero F, Mendizabal V, Robledo P, Galeote L, Bilkei-Gorzo A, Zimmer A, Maldonado R (2005) J Neurosci 25:1103–1112.
113. Berridge KC (2003) Brain Cogn 52:106–128.
114. Berridge KC (2004) Physiol Behav 81:179–209.
115. Berridge KC, Robinson TE (1998) Brain Res Rev 28:309–369.
116. Beshers SN, Fewell JH (2001) Annu Rev Entomol 46:413–440.
117. Besson JM (1999) Lancet 353:1610–1615.
118. Beyenburg S, Stoffel-Wagner B, Bauer J, Watzka M, Blumcke I, Bidlingmaier F, Elger CE (2001) Epilepsy Res 44:141–153.
119. Bezzi P, Volterra A (2001) Curr Opin Neurobiol 11:387–394.
120. Bhandari V, Elias JA (2006) Free Radical Biology and Medicine 41:4-18.
121. Bhargava HN (1994) Pharmacol Rev 46:293–324
122. Bhargava HN, Yousif DJ, Matwyashyn GA (1983) Gen Pharmacol 14:565–570.
123. Bhattacharjee A, Kaczmarek LK (2005) TRENDS in Neurosci 28:422–428.
124. Bi G-Q, Rubin J (2005) TRENDS in Neurosci 28:222–228.
125. Bianchi AL, Denavit-Saubie M, Champagnat J (1995) Physiol Rev 75:1–45.
126. Bickler PE, Fahlman CS (2004) Neuroscience 127:673–683.
127. Biegon A, Fry PA, Paden CM, Alexandrovich A, Tsenter J, Shohami E (2004) Proc Natl Acad Sci USA 101:5117–5122.
128. Bielajew C, Trzcinska M (1994) Behav Brain Res 61:29–35.
129. Bielau H, Mawrin C, Krell D, Agelink MW, Trbner K, Davis R, Gos T, Bogerts B, Bernstein H-G, Baumann B (2005) Brain Res 1039:43–52.
130. Bienengraeber MW, Weihrauch D, Kersten JR, Pagel PS, Warltier DC (2005) Vascular Pharmacology 42:243–252.
131. Billington CJ, Levine AS (1992) Curr Opin Neurobiol 2:847–851.
132. Bird DM (2004) Curr Opin in Plant Biol 7:372–376.
133. Blanke O, Mohr C (2006) Brain Res Rev 50:184–199.
134. Blech-Hermoni Y, Kiyatkin EA (2004) Pharmacol Biochem Behav 77:823–837.
135. Bliss TVP, Lomo T (1973) J Physiol 232:331–356.
136. Bloch GJ, Eckersell C, Mills R (1993) Brain Res 620: 259–268.
137. Blondeau N, Widmann C, Lazdunski M, Heurteaux C (2002) Neuroscience 109:231–241.

138. Bloodgood BL, Sabatini BL (2005) Science 310:866–899.
139. Blumenfeld H (2005) Prog Brain Res 150:271–286.
140. Blumenfeld R (2006) Biophys J 91:1970-1983.
141. Bodnar RJ, Klein GE (2006) Peptides 27:3391-3478.
142. Boehm S, Harvey RJ, Holst van A, Rohrer H, Betz H (1997) J Physiology 504:683–694.
143. Bogdanov AV, Galashina AG (1997) Behav Physiol 27:97–104.
144. Boghossian S, Jourdan D, Dacher M, Alliot J (2001) Mech Ageing Devel 122:1825–1839.
145. Boix F, Sandor P, Nogueira PJC, Huston JP, Schwarting RKW (1995) Neuroscience 64:1045–1055.
146. Boldyrev AA, Johnson P, Wei Y, Tan Y, Carpenter DO (1999) Neurosci Lett 263:169–172.
147. Bolyard LA, Van Looy JV, Vasko MR (2000) Pain 88:277–285.
148. Bongard MM, Losev IS, Smirnov MS (2006) Model of behavior :ANIMAL: In: Redko VG (ed) From model of behavior to artificial intellect. ComKniga, Moscow, pp 61–81.
149. Bono de M, Maricq AV (2005) Annu Rev Neurosci 28:451–501.
150. Borlongan CV, Stahl CE, Redei E, Wang Y (1999) Neuroreport 10:3501–3505.
151. Borlongan CV, Su TP, Wang Y (2001) Cell Transpl. 10 (2001) 53–58.
152. Borroni B, Montagna M, Cerri G, Baldissera F (2005) Brain Res 1065:115-124.
153. Borst JG, Lodder JC, Kits KS (1994) J Neurophysiol 71:639-655.
154. Bou-Flores C, Berger AJ (2001) J Neurophysiol 85:1543–1551.
155. Boules M, Fredrickson P, Richelson E (2006) Peptides 27:2523–2533.
156. Bouchon-Meunier B (1997) Aggregation and Fusion of Imperfect Information, Springer-Verlag, NY.
157. Boulianne GL (2001) Mech Ageing Devel 122:883–894.
158. Boutilier RG, St-Pierre J (2000) Comparative Biochemistry and Physiology - Part A 126:481–490.
159. Bozzi Y, Borrelli E (2006) TRENDS in Neurosci 29:167–173.
160. Braet K, Cabooter L, Paemeleire K, Leybaert L (2004) Biology of the Cell 96:79–91.
161. Branchereau P, Champagnat J, Denavit-Saubie M (1994) J Neurophysiol 70:25842595.
162. Bray D (1995) Nature 376:307–312.
163. Brecht M, Schneider M, Sakmann B, Margrie TW (2004) Nature 427:704–710.
164. Bredeloux P, Costentin J, Dubuc I (2006) Behav Brain Res 175:399407.
165. Brickley SG, Revilla V, Cull-Candy SG, Wisden W, Farrant M (2001) Nature 409:88–92.
166. Briggman KL, Abarbanel HDI, Kristan WBJr (2005) Science 307:896–901.
167. Brivanlou IH, Warland DK, Meister M (1998) Neuron: 20:527–539.
168. Broberger C, McCormick DA (2005) J Neurosci 25:1664–1673.
169. Brooks PJ (2000) Neurochem Int 37:403–412.
170. Brown PL, Wise RA, Kiyatkin EA (2003) J Neurosci 23:3924-3929.
171. Bruel-Jungerman E, Davis S, Rampon C, Laroche S (2006) J Neurosci 31:5888-5893.
172. Brun P, Mastrotto C, Beggiao E, Stefani A, Barzon L, Sturniolo GC, Palu G, Castagliuolo I (2005) Am J Physiol Gastrointest Liver Physiol 288:621–629.
173. Brundin L, Petersn A, Bjrkqvist M, Trskman-Bendz L (2007) J Affective Disorders 100:259–263.

174. Brunetti L, Di Nisio C, Orlando G, Ferrante C, Vacca M (2005) Int J Immunopathol Pharmacol 18:201–212.
175. Brzozowski T, Konturek PC, Konturek SJ, Pajdo R, Duda A, Pierzchalski P, Bielanski W, Hahn EC (1999) Eur J Pharmacol 374:263–276.
176. Brzozowski T, Konturek PC, Konturek SJ, Pierzchalski P, Bielanski W, Pajdo R, Drozdowicz D, Kwiecien S, Hahn EG (2000) Digestion 62:126–142.
177. Buchman TG (2002) Nature 420:246–251.
178. Buckner RL, Petersen SE (1996) Sem Neurosci 8:47–55.
179. Buhler AV, Choi J, Proudfit HK, Gebhart GF (2005) Pain 114:285–294.
180. Bukauskas FF, Verselis VK (2004) Biochimica et Biophysica Acta (BBA) - Biomembranes 1662:42–60.
181. Bullock TH (1958) Parameters of integrative action of the nervous system at the neuronal level. Exp Cell Res 14:323–337.
182. Bucolo C, Drago F, Lin LR, Reddy VN (2006) Neuroreport 17:287–291.
183. Buresova O (1964) Psychopharmacologia 5, 255–263.
184. Burgdorf J, Panksepp J (2006) Neurosci Biobehav Rev 30:173–187.
185. Burrone J, O'Byrne M, Murthy VN (2002) Nature 420:414–418.
186. Burroughs S, French D (2007) Current Anaesthesia Critical Care 18:34-41.
187. Burton MD, Kazemi H (2000) Respir Physiol 122:111–121.
188. Busel BI, Kniga AP (1991) Neurofiziologia 23:564–574.
189. Busselberg D, Bischoff AM, Paton JF, Richter DW (2001) Pflugers Arch 441:444–449.
190. Butler TL, Kassed CA, Pennypacker KR (2003) Brain Res Bull 59:339–351.
191. Butnariu D (1994) On Triangular Norm-Based Propositional Fuzzy Logic, preprint Haifa University, Haifa, Israel.
192. Butovas S, Hormuzdi SG, Monyer H, Schwarz C (2006) J Neurophysiol 96:1227–1236.
193. Buzsaki G, Kaila K, Raichle M (2007) Neuron 56:771–783.
194. Byrne WL and 22 others (1966) Science 153:658–659.
195. Cabanac M (1971) Science 173:103–110.
196. Cabanac M (1979) Quarterly Rev Biol 54:1–29.
197. Cabanac M (1992) J Theor Biol 155:173–200.
198. Cabanac M (2002) Behavioural Processes 60:69–83.
199. Caillard O, Moreno H, Schwaller B, Llano I, Celio MR, Marty A (2000) Proc Natl Acad Sci USA 97:13372-13377.
200. Caldwell JD (2002) Neurosci BioBehav Rev 26:13–30.
201. Campfield LA, Smith FJ (1990) Systemic factors in the control of food intake: evidence for patterns as signals: In: Stricker EM (ed) Handbook of behavioral neurobiology. Vol.10, Neurobiology of food and fluid intake, Plenum press, N-Y, (1990) pp 183–206.
202. Canas AFM (2000) An R Acad Nac Med (Madr) 117:855–869.
203. Cannon WB (1928) Physiol Rev 9:399–431.
204. Cannon SC (2002) Neuron 34:853–854.
205. Cao R, Hasuo H, Ooba S, Akasu T, Zhang X (2006) Neurosc Lett 401:136–141.
206. Cardinal RN, Everitt BJ (2004) Curr Opin Neurobiol 14:156–162.
207. Cardinal RN, Parkinson JA, Hall J, Everitt BJ (2002) Neurosci BioBehav Rev 26:321–352.
208. Carmena JM, Lebedev MA, Henriquez CS, Nicolelis MAL (2005) J Neurosci 25:10712-10716.

209. Carmignoto C (2000) Prog Neurobiol 62:561–581.
210. Carpentier P, Foquin A, Rondouin G, Lerner-Natoli M. de Groot DM. Lallement G (2000) Neurotoxicology 21:521–540.
211. Carr DB, Day M, Cantrell AR, Held J, Scheuer T, Catterall WA, Surmeier DJ (2003). Neuron 39:793–806.
212. Carrascal L, Nieto-Gonzalez JL, Cameron WE, Torres B, Nunez- Abades PA (2005) Brain Res Rev 49:377–387.
213. Carrasco GA, Van de Kar LD (2003) Eur J Pharmacol 463:235–272.
214. Carraway RE, Plona AM (2006) Peptides 27:2445–2460.
215. Castagliuolo I, Wang CC, Valenick L, Pasha A, Nikulasson S, Carraway RE, Pothoulakis C (1999) J Clin Invest 103:843-849.
216. Castelfranchi C (2001) Cognitive System Research 2:5–38.
217. Castellucci V, Pinsker H, Kupferman J, Kandel ER (1970) Science 167:1745–1748.
218. Castrn E (2005) Nat Rev Neurosci 6:241–245.
219. Catterall WA (2000) Neuron 26:13–25.
220. Caudle RM (2006) Theoretical Biology and Medical Modelling 3:1742–1782.
221. Centonze D, Marfia GA, Pisani A, Giacomini P, Bernardi G, Calabresi P (2001) Prog Neurobiol 63:687–696.
222. Cepinskas G, Rui T, Kvietys PR (2002) Free Radical Biology and Medicine 33:433–440.
223. Chalmers DJ (2002) What is the neural correlate of consciousness? In: Metzinger T (ed) Neural correlates of consciousness. MIT, Cambridge, MA, pp 121–137.
224. Chance WT, Sheriff S, Peng F, Balasubramaniam A (2000) Neuropeptides 34:167–172.
225. Chang J-Y, Janak PH, Woodward DJ (1998) J Neurosci 18:3098–3115.
226. Chang J-Y, Sawyer SF, Lee R-S, Woodward DJ (1994) J Neurosci 14:1224–1244.
227. Chang LK, Putcha GV, Deshmukh M, Johnson EM (2002) Biochimie 84:223-231.
228. Chapman T, Liddle LE, Kalb JM, Wolfner MF, Partridge L (1995) Nature 373:241–245.
229. Chaudhuri A, Kolodecik TR, Gorelick FS (2005) Am J Physiol Gastrointest Liver Physiol 288:235–243.
230. Chen BM, Grinnell AD (1995) Science 269:1578-1580.
231. Chen L, Yung KK, Yung WH (2004) Neuroscience 125:853–859.
232. Chen S-H, Cheung RTF (2002) Brain Res 927:138–143.
233. Chen Z-F, Schottler F, Arlinghaus L, Kassell NF, Lee KS (1996) Brain Res 708:82–92.
234. Cheng G, Gaob B-X, Verbny Y, Ziskind-Conhaim L (1999) Brain Res 845:224–231.
235. Cheron G, Servais L, Dan B, Gall D, Roussel C, Schiffmann SN (2005) Prog Brain Res 148:165–180.
236. Chester JA, Cunningham CL (2002) Alcohol 26:131–143.
237. Choe ES, Wang JQ (2002) Int J Mol Med 9:3–10.
238. Choi Y-H, Furuse M, Okimura J-I, Denbow DM (1994) Brain Res 654:163–166.
239. Choi Y-H, Hartzell D, Azain MJ, Baile CA (2004) Physiol Behav 77:1–4.
240. Christie JM, Bark C, Hormuzdi SG, Helbig J, Monyer H, Westbrook GL (2005) Neuron 46:761–772.

241. Chung S, Civelli O (2006) Neuropeptides 40:233–243.
242. Ciccarelli R, Ballerini P, Sabatino G, Rathbone MP, D'Onofrio M, Caciagli F, Di Iorio P (2001) Int J Devel Neurosci 19:395–414.
243. Clancy AN, Zumpe D, Michael RP (2000) Horm Behav 38:86–93.
244. Clatworthy AL, Walters ET (1993) J Neurophysiol 70:1187–1194.
245. Cleary LJ, Hammer M, Byrne JH, (1989) Insight into the cellular mechanisms of short-term sensitization in Aplysia. In: Alan R (ed) Perspectives in neural systems and behavior. Liss, Inc pp 1015–1019.
246. Cleary LJ, Lee WL, Byrne JH (1998) J Neurosci 18:5988–5998.
247. Clement Y, Chapouthier Y (1998) Neurosci BioBehav Rev 22:623–633.
248. Codispoti M, Ferrari V, Bradley MM (2006) Brain Res 1068:213–220.
249. Cohen DA, Pascual-Leone A, Press DZ, Robertson EM (2005) Proc Natl Acad Sci USA 102:18237–18241.
250. Cohen H, Kaplan Z, Matar MA, Loewenthal U, Kozlovsky N, Zohar J (2006) Biol Psychiatry 60:767-776.
251. Cohen JD, Blum KI (2002) Reward and Decision. Neuron 36:193–198.
252. Cohen MR, Newsome WT (2004) Curr Opin Neurobiol 14:169–177.
253. Cohen MV, Yang X-M, Downey JM (2006) Cardiovascular Research 70:223–239.
254. Colmers WE, Bleakman D (1994) Trends Neurosci 17:373–379.
255. Companac e, Debanne D (2007) Arch Ital Biol 145:277–287.
256. Condes-Lara M, Veinante P, Rabai M, Freund-Mercier MJ (1994) Brain Res 637:277–286.
257. Connors BW, Long MA (2004) Annu Rev Neurosci 27:393–418.
258. Conradt L, Roper TJ (2005) TRENDS in Ecol Evol 20:449–456.
259. Consoulas C, Duch C, Bayline RJ, Levine RB (2000) Brain Res Bull 53:571–583.
260. Contreras JE, Snchez HA, Vliz LP, Bukauskas FF, Bennett MVL, Sez JC (2004) Brain Res Rev 47:290–303.
261. Cook DG, Carew TJ (1989) J Neurosci 9:3115–3122.
262. Corbett D, Thornhill J (2000) Brain Pathol. 10:145–152.
263. Corbetta M, Kincade MJ, Lewis C, Sapir A (2005) Nat Neurosci 8:1603–1610.
264. Corry J (1986) TV: a Soviet-Donahue summit. The New York Times January 4.
265. Costa FF (2007) Gene 386:1–10.
266. Costa FG, Filhob RF, Felicio LF (2001) Eur J Pharmacol 428:97–103.
267. Cotman CW, Nieto-Sampedro M (1984) Science 225:1287–1292.
268. Cottrell GT, Ferguson AV (2004) Regul Pept 117:11–23.
269. Cowburn RF, Marcusson JO, Eriksson A, Wiehager B, O'Neill C (1994) Brain Res 633:297–304.
270. Cowen MS, Lawrence AJ (1999) Prog Neuro-Psychopharmacol Biol Psychiatr 23:1171–1212.
271. Coyle JT, Duman RS (2003) Neuron 38:157–160.
272. Cox CL, Metherate R, Ashe JH (1994) Synapse 16:123–136.
273. Crair MC, Malenka RC (1995) Nature 375:325-328.
274. Craner MJ, Damarjian TG, Liu S, Hains BC Lo AC, Black JA, Newcombe J, Cuzner ML, Waxman SG (2004) Glia 49:220-229.
275. Crawford AC, Fettiplace R (1985) J Physiol 364:359–379.
276. Crick FH (1996) Nature 379:485–486.

277. Crick F, Koch C (1998) Cerebral Cortex 8:97–107.
278. Croll RG (2003) Brain Res Rev 43:266–274.
279. Cromwell HC, Anstrom K, Azarov A, Woodward DJ (2005) Brain Res 1043:12–23.
280. Crown ED, Ye Z, Johnson KM, Xu ZY, McAdoo DJ, Westlund KN, Hulsebosch CE (2005) Neurosci Lett 384:139–144.
281. Cruikshank SJ, Weinberger NM (1996) Brain Res Rev 22:191–228.
282. Cunningham ST, Kelley AE (1992) Brain Res Bull 29:675–680.
283. Cunnington R, Windischberger C, Moser E (2005) Human Movement Science 24:644-656.
284. Cunnington R, Windischberger C, Robinson C, Moser E (2006) NeuroImage 29:1294–1302.
285. Currie PL, Coscina DV (1996) Brain Res 737:238–242.
286. Cutolo M, Silli A, Barone A, Seriolo BS (1993) Clin Exp Rheumatol 11:331–339.
287. Daglish MRC, Nutt DI (2003) Eur Neuropsychopharmacol 13:453–458.
288. Damalama M, Swann J (1993) Neuropeptides 24:327–334.
289. Damasio AR, Grabowski TJ, Bechara A, Damasio H, Ponto LLB, Parvizi J, Hichwa RD (2000) Nat Neurosci 3:1049-1056.
290. Dan Y, Poo M-M (2006) Physiol Rev 86:1033–1049.
291. Daoudal G, Debanne D (2003) Learn Mem 10:456–465.
292. Daprati E, Nico D, Franck N, Sirigu A (2003) Consciousness and Cognition 12:670–683.
293. Dauge V, Lena I (1998) Neurosci BioBehav Rev 22:815–825.
294. Davidowa H, Albrecht D, Gabriel HJ, Zippel U (1995) Brain Res Bull 36:533–537.
295. Davies P, Anderton B, Kirsch J, Konnerth A, Nitsch R, Sheetz J (1998) Prog Neurobiol 55:651–658.
296. Davis GW (2006) Annu Rev Neurosci 29:307–323.
297. Davis M, Rainnie D, Cassel M (1994) Trends in Neurosci 17:208–214.
298. Davis RL (2005) Annu Rev Neurosci 28:275–302.
299. Deadwyler SA, Hayashizaki S, Cheer J, Hampson RE (2004) Neurosci Biobehav Rev 27:703–711.
300. Dehaene S, Naccache L (2001) Cognition 79:1-37.
301. Dehaene S, Changeux JP (2005) PLoS Biol 3:1–41.
302. De Hert SG (2004) Curr Opin Anaesthesiol 17:57–62.
303. Delacour J, Houcine O, Costa JC (1990) Neuroscience 34:1-8.
304. De La Fuente M, Carrasco M, Del Rio M, Hernanz A (1998) Neuropeptides 32:225–233.
305. De Leoa JA, Tawfika VL, LaCroix-Fralisha ML (2006) Pain 122:17–21.
306. Del Rio M, De La Fuente M (1994) Regul Pept 49:185–193.
307. Demb JB (2007) Neuron 55:179–186.
308. Demir E, Dickson BJ (2005) Cell 121:785–794.
309. Deneubourg J-L, Lioni A, Detrain C (2002) Biol Bull 202:262–266.
310. Deng PY, Porter JE, Shin HS, Lei S (2006) J Physiol 577:497–511.
311. Dermietzel R (1998) Brain Res Rev 26:176–183.
312. Desai NS, Rutherford LC, Turrigiano GG (1999) Nat Neurosci 2:515-520.
313. Dessi F, Charriaut-Marlanguc C, Ben-Ari Y (1994) BrainRes 650:49–55.
314. De Wied D (1979) Acta Endocrinol 91:416–418.

315. De Wied D (1999) Eur J Pharmacol 375:1–11.
316. DeZazzo, J Tully T (1995) Trends Neurosci 18:212–218.
317. Dhein S (2002) Peptides 23:1701–1709.
318. Dhein S (2004) Cardiovasc Res 62:287–298.
319. Dhingra NK, Freed MA, Smith RG (2005) J Neurosci 25:8097–8103.
320. Di Marzo V, Vial D, Sokoloff P, Schwartz JC, Piomelli D (1993). J Neurosci 13:4846–4853.
321. Di Paolo E (2003). Adaptive Behavior 10:243–263.
322. Diamond P (1999) IEEE Trans Fuzzy Systems 7:734–740.
323. Diamond P (2000) IEEE Trans Fuzzy Systems 8:583–590.
324. Diamond P, Watson P (2000) Appliied Mathematics Letters 13:31–35.
325. Dietz D, Wang H, Kabbaj M (2007) Behav Brain Res 181:287–291.
326. Dilon PF, Root-Bernstein RS (1997) J Theor Biol 188:481–493.
327. Dib-Hajj SD, Fjell J, Cummins TR, Zheng Z, Fried K, LaMotte R, Black JA, Waxman SG (1999) Pain 83:591–600.
328. Dirnagl U, Simon RP, Hallenbeck JM (2003) Trends in Neurosci 26:248–254.
329. Dobrenis K, Chang HY, Pina-Benabou MH, Woodroffe A, Lee SC, Rozental R, Spray DC, Scemes E (2005) J Neurosci Res 82:306–315.
330. Docherty RJ, Farrag KJ (2006) Neuropharmacology 51:1047–1057.
331. Dodd R, Beckmann AM, Davidson MS, Wilce PA (2000) Neurochem Int 37:509–533.
332. Doetsch F, Hen R (2005) Curr Opin Neurobiol 15:121–128
333. Dominguez JM, Gil M, Hull EM (2006) J Neurosci 26:1699-1703.
334. Dopico AM, Chu B, Lemos JR, Treistman SN (1999) Neurochem Int 35:103–106.
335. Drago F, Busa L (1997) Brain Res 779:135–139.
336. Dragunow M, de Castro D, Faull RLM (1990) Brain Res 527:41–54.
337. Droge W (2002) Physiol Rev 82:47–95.
338. Du J, Creson TK, Wu L-J, Ren M, Gray NA, Falke C, Wei Y, Wang Y, Blumenthal R, Machado-Vieira M, Yuan P, Chen G, Zhuo M, Manji HK (2008) J Neurosci 28:68–79.
339. Dubinsky JM, Oxford GS (1985) Proc Natl Acad Sci USA 82:4282–4286.
340. Dublineau I, Dudoignon N, Monti P, Combes O, Wysocki J, Grison S, Baudelin C, Griffiths NM, Scanff P (2004) Can J Physiol Pharmacol 82:103–113.
341. Dudai Y, Eisenberg M (2002) Neuron 44:93–100.
342. Duman RS (2005) Neurobiology of Aging 26:88–93.
343. Dunn AJ (1993) Ciba Found Symp 172:226–239.
344. Dunzendorfer S, Wiedermann CJ (2000) Ann Med 32:429–439.
345. Durkovic RG (1975) Physiol Behav 14:297–311.
346. Dubois D, Pride H (1991) Fuzzy Sets and Systems 40:143–202.
347. Dwivedi Y, Mondal AC, Rizavi HS, Conley RR (2005) Biol Psychiatry 58:315–324.
348. Dyakonova TL, Michaltsev EI (1983). J Physiol (USSR) 69:795–802. (in Russian)
349. Eagleman DM (2004) Science 303:1144–1146.
350. Ebenezer IS (2002) Eur J Pharmacol 441:79–82.
351. Eberhardt O, Schulz JB (2003) Toxicol Lett 139:135–151.
352. Edeline J-M (1998) Prog Neurobiol 57:165–224.
353. Edwards JCW (2005) J Consciousness studies 12:60–76.

354. Eger II MD EI (2005) Seminars in Anesthesia, Perioperative Medicine and Pain 24:89–90.
355. Egorov AV, Muller W (1999) Neurosci Lett 261:77–80.
356. Ehrlich GJ, Rabjohna RR, Routtenberger A (1977) Pharmacol Biochem Behav 6:169–177.
357. Elliott CR, Marsh B, Bacon A, Pope R, Vanderplank P, Wynick D (2004) Proc Natl Acad Sci USA 101:5105–5110.
358. Emamghoreishi M, Li PP, Schlichter L, Parikh SV, Cooke R, Warsh JJ (2000) Biol Psychiatry 48:665-673
359. Emery DL, Royo NC, Fischer I, Saatman KE, McIntosh TK (2003) J Neurotrauma 20:1271–1292.
360. Endres M, Laufs U, Liao JK, Moskowitz MA (2004) Trends in Neurosci 27:283–289.
361. Engel AK, Singer W (2001) Trends Cogn Sci 5:16–25.
362. Engel D, Jonas P (2005) Neuron 48:405–417.
363. Epstein IR, Marder E (1990) Biol Cybern 63:23–32.
364. Ercan F, Akici A, Ersoy Y, Hrdag C, Erin N (2006) Reg Peptides 133:82–89.
365. Ermak G, Kelvin JA Davie KJA (2002) Mol Immunol 38:713–721.
366. Estap N, Steckler T (2001) Behav Brain Res 121:39–55.
367. Etienne AS, Maurer R, Berlie J, Reverdin B, Rowe T, Georgakopoulos J, Sguinot V (1998) Nature 396:161–164.
368. Evanson K, Huang C, Yamben I, Covey DF, Kornfeld K (2005) Science 307:258–262.
369. Everitt BJ (1990) Neurosci BioBehav Rev 14:217–232.
370. Everitt BJ, Dickinson A, Robbins TW (2001) Brain Res Rev 36:129–138.
371. Eytan D, Brenner N, Marom S (2003) J Neurosci 23:9349–9356.
372. Eytan D, Minerbi A, Ziv N, Marom S (2004) J Neurophysiol 92:1817–1824.
373. Fadda F, Rossetti ZL (1998) Prog Neurobiol 56:385–431.
374. Fan Y, Fricker D, Brager DH, Chen X, Lu H-C, Chitwood RA, Johnston D (2005) Nat Neurosci 8:1542–1551.
375. Farber I (2005) J Consci Stud 12:77–95.
376. Farel PB, Glanzman DL, Thompson RF (1973) J Neurophysiol 36:1117–1130.
377. Feldman DE, Brecht M (2005) Science 310:810-815.
378. Feldman JL, Mitchell GS, Nattie EE (2003) Annu Rev Neurosci 26:239–266.
379. Felicio LF, Mazzini BK, Cacheiro RJR, Cruz TN, Florio JC, Nasello AG (2001) Peptides 22:1299–1304.
380. Fellin T, Pascual O, Gobbo S, Pozzan T, Haydon PG, Carmignoto G (2004) Neuron 43:729–743.
381. Felzen B, Berke G,Gardner P, Binah O (1997) Pflgers Arch Eur J Physiol 433:721-726.
382. Ferguson AV, Washburn DLS (1998) Prog Neurobiol 54:169–192.
383. Ferretti A, Caulo M, Gratta CD, Matteo RD, Merla A, Montorsi F, Pizzella V, Pompa P, Rigatti P, Rossini PM, Salonia A, Tartaro A, Romani GL (2005) NeuroImage 26:1086–1096.
384. Fetler L, Amigorena S (2005) The Microglia Patrol Science 309:392-393.
385. Fetz E, Barker MA (1973) J Neurophysiol 36:179–204.
386. Figenschou A, Hu GU, Storm JF (1996) Eur J Neurosci 8:211–219.
387. Figiel M, Allritz C, Lehmann C, Engele J (2007) Mol Cell Neurosci 35:130–137.
388. Figueroa SE, Lpez E, Arce C, Oset-Gasque MJ, Gonzlez MP (2005) Brain Res 1034:25–33.

389. Fischer M, Kaech S, Knutti D, Matus A (1998) Neuron 20:847-854.
390. Fiscus RR (2002) Neurosignals 11:175–190.
391. Fitch CL, de Sousa SM, O'Day PM, Neuber TA, Plantilla CM, Spencer M, Yarfitz S, Apte D, Hurley JB (1993) Cellr Signal 5:187–201.
392. Flint DJ, Knight CH (1997) J Mammary Gland Biol Neoplasia 2:41–48.
393. Flohr H (1995) Behav Brain Res 71:157–161.
394. Flohr H (1995a) Neuropsychologia 33:1169–1180.
395. Flood DG (1993) Neurobiology of Aging 14:649–654.
396. Flood DG, Coleman PD (1993) Neurobiology of Aging 14:575–587.
397. Flood JF, Landry DW (1981) Pharmacol Biochem Behavior 15:289–296.
398. Fonseca R, Nagerl UV, Morris RGM, Bonhoeffer T (2004) Neuron 44:1011–1020.
399. Forrester JW (1961) Industrial Dynamics. MIT Press, Cambridge, Mass.
400. Fontella FU, Cimarosti H, Crema LM, Thomazi AP, Leite MC, Salbego C, Gonalves CAS, Wofchuk S, Dalmaz C, Netto CA (2005) Brain Res Bull 66:443–450.
401. Foo H, Helmstetter FJ (1999) Pain 83:427–431.
402. Fowler CJ (2003) Brain Res Rev 41:26–43.
403. Frahm C, Haupt C, Witte OW (2004) Neuroscience 127:341–346.
404. Franzen R, Bouhy D, Schoenen J (2004) Neurosci Lett 361:76-78.
405. Franklin JL, Rodriguez-Sanz C, Juhasz A, Deckwerth TL, Johnson EM (1995) J Neurosci 15:643-664.
406. Franco R, Pacheco R, Lluis C, Ahern GP, O'Connell PJ (2007) Trends in Immunol 28:400–407.
407. Franks NP, Honore E (2004) Trends in Pharmacol Sci 26:601–608.
408. Franks NP, Lieb WR (1998) Toxicol Lett 100:1-8.
409. Frassoni C, Radici C, Spreafico R, de Curtis M (1998) Neuroscience 83:229–237.
410. Freed WJ Dillon-Carter O, Kleinman JE (1993) Exp Neurol 121:48–56.
411. Freeman JH, Nicholson DA (1999) Brain Res 833:225–223.
412. Freeman WJ (2006) Int J Psychophysiol 60:149–161.
413. Fricker D, Varheugen JAH, Miles R (1999) J Physiol 517:791–804.
414. Frick A, Magee J, Johanston D (2004) Nat Neurosci 7:126–135.
415. Friedl P (2004). Curr Opin Neurobiol 16:14–23
416. Friedman JE, Haddad GG (1994) Brain Res 663:329–334.
417. Friedman Y, Sandler U (1994) A New Approach to Fuzzy Logic. In: Proc 12th IAPR Inter Conf on Pattern Recognition, Jerusalem,Israel.
418. Friedman Y, Sandler U (1996) Fuzzy Set and Systems 84:61–74.
419. Friedman Y, Sandler U (1997) Inter J of Chaos Theory and Application 2(3–4):5–21.
420. Friedman Y, Sandler U (1997) Fuzzy Dynamics as Alternative to Statistical Mechanics. In: Proc Intern Conf on Fuzzy Logic and Application - FUZZY'97, Zichron Yaakov, Israel.
421. Friedman Y, Sandler U, (1977) Fuzzy Dynamics : A new way description of uncertainty in complex systems. In: Proc 7th World Congress on Fuzzy Set and Systems - IFSA'97, Prague, CR.
422. Friedman Y, Sandler U (1997) Fuzzy Dynamics of Immune Systems: T-cell maturation. In: Proc 7th World Congress on Fuzzy Set and Systems - IFSA'97, Prague, CR.

423. Friedman Y, Sandler U (1998) Fuzzy Logic as a Natural Basis for Dynamics Theory. In: Proc 7th Inter Conference on Information Processing and Management of Uncertainty - IPMU'98, Paris.
424. Friedman Y, Sandler U (1998) Prediction based on Fuzzy state Dynamics. In: Proc 7th Inter Conference on Information Processing and Management of Uncertainty - IPMU'98, Paris.
425. Friedman Y, Sandler U (1998) Fuzzy Logic and Dynamics Theory of Complex Systems. In: Proc 6th Europian Congress on Intelligent Techniques and Soft Computings - EUFIT'98, Aachen.
426. Friedman Y, Sandler U (1999) Fuzzy Set and Systems 106:61–74.
427. Friedman Y, Sandler U, (1999) Aggregation Operator and Fuzzy Dynamic Equations: Different Dynamics for Different Operators. What is an optimal choice? In: Proc 7th Europian Congress on Intelligent Techniques and Soft Computings - EUFIT'99, Aachen.
428. Fries 2005) TRENDS in Cognitive Sci 9:474–480.
429. Frost WN, Kandel ER (1995) J Neurophysiol 73:2413–2427.
430. Fry M, Ferguson AV (2007) Physiol Behav 91:413–423.
431. Frye CA (2001) Brain Res Rev 37:201–222.
432. Fryer RM, Auchampach JA, Gross GJ (2002) Cardiovascular Research 55:520–525.
433. Fu XW, Nurse CA, Wang YT, Cutz E (1999) J Physiol (Lond) 514:139–150.
434. Fuentealba P, Steriade M (2005) Prog Neurobiol 75:125-141.
435. Fujiki M, Kobayashi H, Inoue R, Goda M (2004) Journal of Neurotrauma 21:459–470.
436. Fujimoto JM, Arts KS, Rady JJ, Tseng LF (1990) Neuropharmacology 29:609–617.
437. Fung ML, Croning MDR, Haddad GG (1999) Neurosci Lett 275:41–44.
438. Funk GD, Smith JC, Feldman JL (1995) J Neurophysiol 70:1497–1515.
439. Fyhn M, Hafting T, Treves A, Moser MB, Moser EI (2007) Nature 446:190–194.
440. Gabrilovac J (1993) Int J Immunopharmacol 15:557–565.
441. Gahr A, Garcia-Segura LM (1996) Brain Res 712:69–73.
442. Gainutdinov KL, Chekmarev LJ and Gainutdinova TH. (1998) Neuroreport, 1998, 9:517–520.
443. Gainutdinov R, Premont RT, Bohn LM, Lefkowitz RJ, Caron MG (2004) Annu Rev Neurosci 27:107–144.
444. Gainutdinova TK, Andrianov VV, Gainutdinov KL (2004) Neurosci Behav Physiol 34:731–733.
445. Galvez R, Weiss C, Weible AP, Disterhoft JF (2006) J Neurosci 31:6062-6068.
446. Gandhi CC, Matzel LD (2000) J Neurosci 20:2022-2035.
447. Garcia SI, Pirola CJ (2005) Regulatory Peptides 128:239-246.
448. Garcia-Dorado D, Inserte J, Ruiz-Meana M, Gonzalez MA, Solares J, Julia M, Barrabes JA, Soler-Soler J (1997) Circulation 96:3579–3586.
449. Garcia-Dorado D, Ruiz-Meana M, Padilla F, Rodriguez- Sinovas A, Mirabet M (2002) Cardiovasc Res 55:456–465.
450. Garcia-Minoz M, Segal DS, Patino SJ, Young SJ, Kuczenski R, Groves PM (1996) Brain Res 720:131–138.
451. Garrido JJ, Arahuetes RM, Hernanz A, Dela Fuente M (1992) Regulatory Peptides 41:27–37.
452. Gasanov UG (1992) Zh Vyssh Nerv Deiat 42:854–863.

453. Gasull X, Liao X, Dulin MF, Phelps C, Walters ET (2005) J Neurophysiol 94:2218–2230.
454. Gaumond I, Arsenault P, Marchanda S (2005) Brain Res 1052:105–111.
455. Gebber GL, Das M, Barman SM (2005) Brain Res 1046:216–223.
456. Gebke E, Muller AR, Jurzak M, Gerstberger R (1998) Neuroscience 85:509–520.
457. Geldwert D, Norris JM, Feldman IG, Schulman JJ, Rayport S (2006) BMC Neurosci 7:53–69.
458. Genedani S, Bernardi M, Ottaviani E, Franceschi C, Leung MK, Stefano GB (1994) Peptides 15:203–206.
459. Geocadin RG, Malthora AD, Tong MS, Seth A, Moriwaki G, Hanley DF, Thakor NV (2005) Brain Res 1064:146–154.
460. Gerard C, Gerard NP (1994) Curr Opin Immunol 6:140–145.
461. Gerzon SA, Michailov AS (1986) Dokl Acad Nauk USSR 291:160–163.
462. Gether U (2000) Endocrinol Rev 21:90–113.
463. Ghafour-Rashidi Z, Dermenaki-Farahani E, Aliahmadi A, Esmaily H, Mohammadirad A, Ostad SN, Abdollahi M (2007) Pesticide Biochem Physiol 87:261-270.
464. Ghersi G, Chen W, Lee EW, Zukowska Z (2001) Peptides 22:453–458.
465. Giachino C, Canalian, Capone F, Fasolo A, Alleva E, Rivama A, Cirulli F, Peretto P (2007) Neuroscience 145:568-578.
466. Giacino JT (2005) Prog Brain Res 150:381–395.
467. Gibson RM, Rothwell NJ, Le Feuvre RA (2004) The Veterinary Journal 168:230–237.
468. Giese KP, Peters M, Vernon J (2001) Physiol Behav 73:803–810.
469. Gilby KL, Kelly ME, McIntyre DC, Robertson HA (2005) Neuroscience 131:785–792.
470. Gilby WL, Sydserff SG, Robertson HA (2005a) Brain Res 1035:196–205.
471. Gillette R, Huang R-C, Hatcher N, Moroz LL (2000) Proc Natl Acad Sci USA 97:3585–3590.
472. Gingrich B Liu Y, Cascio C, Wang Z, Insel TR (2000) Behav Neurosci 114:173-183.
473. Gingrich KJ, Byrne JH (1987) J Neurophysiol 57:1705–1715.
474. Gisabella B, Rowan MJ, Anwyl R (2003) Neuroscience 121:297–305.
475. Gladkova AI (1999) Uspechi Fiziologocheskih Nauk 30:97–105. (in Russian)
476. Glanzman DL (1995) Trends in Neurosci 18:30–35.
477. Glassman R (1969) Annu Rev Biochem 38:387–400.
478. Glassman RB, Malamut BL (1977) Biosystems 9:257–268).
479. Glezer II, Hof PR, Morgane PJ (1998) Journal of Chemical Neuroanatomy 15:203–237.
480. Gohar O (2006) Modulator 21:2–8.
481. Gold PE (2003) Neurobiol Learn Mem 80:194–210.
482. Goldberg AL, Amaral LAN, Hausdorff JM, Ivanov PC, Peng C-K, Stanley HE (2002) Proc Natl Acad Sci USA 99:2466–2472.
483. Goldberg GS, Valiunas V, Brink PR (2004) Biochimica et Biophysica Acta (BBA) - Biomembranes 1662:Pages 96–101.
484. Goldberg I, Harel M, Malach R (2006) Neuron 50:329–339.
485. Goldin PR, Hutcherson CAC, Ochsner KN, Glover GH, Gabrieli JDE, Gross JJ (2005) NeuroImage 27:26–36.

486. Goldman SA, Luskin MB (1998) Trends in Neurosci 21:107–114.
487. Goldstone RL, Janssen MA (2005) TRENDS in Cogn Sci 9:424–430.
488. Larisa B. Goncharova, Alexander O. Tarakanov (2007) Brain Res Rev 55:155–166.
489. Gondolfo F, Li C-SR, Benda BJ, Schioppa CP, Bizzi E (2000) Proc Natl Acad Sci USA 96:191–196.
490. Gotikh BP, Kraevskii AA, Kukhanova MK (1974) Usp Biol Khimii 15:24–39.
491. Gottesmann C (2004) Brain Res Rev 45:230–249.
492. Gould E, Vail N, Wagers M, Gross CG (2001) Proc Natl Acad Sci USA 98:10910–10917.
493. Gould TD, Quiroz JA, Singh J, Zarate CA, Manji HK (2004) Mol Psychiatry (2004) 9, 734-755.
494. Grabisch M, Nguen HT, Walker EA (1995) Fundamentals of Uncertainty Calculi with Applications to Fuzzy Inference, Kluwer Academic Publisher, London.
495. Grammatopoulos TN, Ahmadi F, Jones SM, Fariss MW, Weyhenmeyer JA, Zawada WM (2005) Brain Res 1045:64–71.
496. Greco B, Edwards DA, Michael RP, Clancy AN (1998) Neuroendocrinology 67:18–28.
497. Greenspan RJ, van Swinderen B (2004) TRENDS in Neurosci 27:707–711.
498. Griffin JD (2004) J Therm Biol 29:327–331.
499. Griggio MA, Luz J, Carvalho SMT (1992) Brazilian J Med Biol Res 25:205–208.
500. Gross CG (2002) The Neuroscientist 8:512–518
501. Gross-Isseroff R, Biegon A, Voet H, Weizman A (1998) Neurosci BioBehav Rev 22:653–661.
502. Grosskreutz J, Lin C, Mogyoros I, Burke D (1999) J Physiol 518:301–314.
503. Groves PM, Thompson RF (1970) Psychol Rev 1970:77:419–450.
504. Gudermann T, Kalkbrenner F, Schultz G (1996) Annu Rev Pharmacol Toxicol 36:429–459.
505. Guidolin D, Fuxe K, Neri G, Nussdorfer GG, Agnati LF (2007) Brain Res Rev 55:119-133.
506. Guo XW, Wang XL, Gao ZL (2005) Zhonghua Xin Xue Guan Bing Za Zhi 33:1006–1009.
507. Gwag BJ, Canzoniero LMT, Sensi SL, Demaro JA, Koh JY, Goldberg MP, Jacquin M, Choi DW (1999) Neuroscience 90:1339–1348.
508. Hackley SA, Boelhouwer AJW (1997) The more or less startling effects of weak prestimulation revisited: prepulse modulation of multicomponent blink reflexes. In: Lang PJ, Simons RF, Balaban P (eds.) Attention and orienting: sensory and motivational processes. Mahwah NJ: Lawrence Erlbaum pp 205–227.
509. Haefely WE, Martin JR, Richards JG, Schoch P (1993) Can J Psychiatry 38:102–108.
510. Hai T, Hartman MG (2001) Gene 273:1–11.
511. Haist F, Gore JB, Mao H (2001) Nat Neurosci 4:1139–1145.
512. Halassa MM, Fellin M, Haydon PG (2007) Trends in Molecular Medicine 13:54–63.
513. Halperta AG, Olmsteada MC, Beninge RJ (2003) Pharmacol Biochem Behav 75:173–179.
514. Halter JA, Carp JS, Wolpaw JR (1995) J Neurophysiol 73:867–871.

515. Hammer M (1997) Trends in Neurosci 20:245–252.
516. Hao J, de Vaca SC, Pan Y, Carr KD (2006) Brain Res 1087:123-133.
517. Hao Y, Yang JY, Guo M, Wu CF, Wu MF (2005) Brain Res 1040:191–196.
518. Harper C, Matsumoto I (2005) Curr Opin Pharmacol 5:73–78.
519. Harris DA, True HL (2006) Neuron 50:353–357.
520. Harrison LM, Kastin AJ, Zadina JE (1998) Peptides 19:1503–1630.
521. Hasenstaub A, Shu Y, Haider B, Kraushaar U, Duque A, McCormick DA (2005) Neuron 47:423-435.
522. Hask G, Pacher P, Vizi ES, Illes P (2005) Trends in Pharmacological Sciences 26:511–516.
523. Hatanaka K, Kawata H, Toyofuku T, Yoshida K (2004) Jpn Heart J 45:1007–1019.
524. Hawkins RD (1981) J Neurophysiol 45:327–336.
525. Hawkins RD, Carew TG, Kandel ER, (1986) J Neurosci 6:1695–1701.
526. Haydon PG, Carmignoto G (2006) Physiol Rev 86:1009–1032.
527. Hayes RL, Dubner R, Hoffman DS (1981) J Neurophysiol 46:428–443.
528. Hazelwood RL (1993) Proc Soc Exp Biol Med 202:44–63.
529. Headrick JP, Peart J (2005) Vascular Pharmacology 42:271-279.
530. Hebb DO (1949) The organization of behavior. Wiley
531. Heidel E, Davidowa H (1998) Neuropeptides 32: 423–429.
532. Heiss WD, Graf R (1994) Curr Opin Neurol 7:11–19.
533. Henderson LP, Penatti CAA, Jones BL, Yang P, Clark AS (2006) Neuroscience 138:793–799.
534. Hendricks JC, Williams JA, Panckeri K, Kirk D, Tello M, Yin JC, Sehgal A (2001) Nat Neurosci 4:1108–1115.
535. Herb (2005). The missing link in cognition: origins of self-reflective consciousness. Oxford University Press, London
536. Hernandez PJ, Sadeghian K, Kelley AE (2002) Nat Neurosci 5:1327-1331.
537. Hernandez PJ, Kelley AE (2004) Learn Mem 11:748–754.
538. Hertz L, Zieke HR (2004) TRENDS in Neurosci 27:735–743.
539. Herrero JF, Headley PM (1995) Anesthesiology 82:267–275.
540. Herring TL, Slotin IM, Baltz JM, Morris CE (1998) Am J Physiol 274:272–281.
541. Hestrin S, Galarreta M (2005) TRENDS in Neurosci 28:304–309.
542. Hirano T, Simbulan D, Kumazawa T (1994) Brain Res 665:54–62.
543. Hirota K (1981) Fuzzy Sets and Systems 5:31–462.
544. Holsmann S, Oku Y, Zhang W, Richter DW (2000) Eur J Neurosci 12:856–862.
545. Hnasko TS, Sotak BN, Palmiter RD (2005) Nature 438:854–857.
546. Hochman S, McCrea DA (1994) J Neurophysiol 71:1468–1479.
547. Hochner B, Klein M, Schacher S, Kandel ER (1986) Proc Natl Acad Sci USA 83:8794–9798.
548. Hodgkin AL, Huxley AF (1952) J Physiol 117:500–544.
549. Hodgson TM, Crow T (1992) Brain Res 570:267–271.
550. Hofer SB, Mrsic-Flogel TD, Bonhoeffer B, Hbener M (2006) Nat Neurosci 9:127132
551. Hoheisel U, Sander B, Mense S (1995) Neurosci Lett 188:143–146.
552. Holland PC, Gallagher M (2004) Curr Opin Neurobiol 14:148–155.
553. Hollerman JF, Schultz W (1998) Nat Neurosci 1:304–309.
554. Holmes GM, Bresnahan JC, Beattie MC (2001) Physiol Behav 74:57–64.
555. Hlscher C, Gigg J, O'Mara SM (1999) Neurosci BioBehav Rev 23:399–410.

556. Holschneider DP, Yang J, Sadler TR, Nguyen PT, Givrad TK, Maarek J-MI (2006) NeuroImage 29:1344–1358.
557. Holsmann S, Oku Y, Zhang W, Richter DW (2000) Eur J Neurosci 12:856–862.
558. Hoop B, Beagle JL, Maher TJ, Kazemi H (1999) Respiration Physiology 118:117–129.
559. Hopfield JJ (1982) Proc Nat Acad Sci USA 79:2554–2558.
560. Horiguchi T, Snipes JA, Kis B, Shimizu K, Busija DW (2005) Brain Res 1039:84–89.
561. Horn G (2004) Nat Rev Neurosci 5:108–121.
562. Houten JV (1994) Trends in Neurosci 17:62–75.
563. Houweling AR, Brecht M (2008) Nature 452:65–69.
564. Hsin H, Kenyon C (1999) Nature 399:362–366.
565. Hsu M, Bhatt M, Adolphs R, Tranel D, Colin F (2005) Science 310:1680–1683.
566. Hu S, Chao CC, Hegg CC, Thayer S, Petterson PK (2000) J Psychopharmacol 14:238–243.
567. Huang R, Bossut DF, Somjen GG (1997) J Neurophysiol 77:2349–2359.
568. Hung T, Nguyen, Walker E (2006) A First Course in Fuzzy Logic CRC Press, London, NY.
569. Huff RM (1996) Cell Signal 8:453–459.
570. Hullermeir E (1997) Int J Uncertainty Fuzziness and Knowledge-Based Systems 5:117–137.
571. Hunt GE, Atrens DM (1992) Pharmacol Biochem Behav 42:563–577.
572. Hunter MD, Eickhoff SB, Miller TW R, Farrow TFD, Wilkinson ID, Woodruff PWR (2006) Proc Natl Acad Sci USA 103:189–194.
573. Hussy N, Deleuze C, Desarmenien MG, Moos F (2000) Prog Neurobiol 62:113–134.
574. Hyden H (1967) Behavior, neural function, and RNA. Progress in nucleic acid research and molecular biology 6:187–218.
575. Ikeda M, Ikeda-Sagara M, Okada T, Clement P, Urade Y, Nagai T, Sugiyama T, YoshiokaT, Honda K, Inou (2005) Neuroscience 130:1029–1040.
576. Immler S, Moore HD, Breed WG, Birkhead TR (2007) PLoS ONE 2:e170
577. Ingram SL, Williams JT (1994) Neuron 13:179–186.
578. Inoue S, Matsuzawa T (2007) Current Biology 17:1004–1005.
579. Inserte J, Garcia-Dorado D, Hernando V, Barba I, Soler-Soler J (2006) Cardiovasc Res 70:364–373.
580. Ishiwata T, Saito T, Hasegawa H, Yazawa T, Kotani Y, Otokawa M, Aohara Y (2005) Brain Res 1048:32–40
581. Izquierdo I, Medina JH (1995) Neurobiology of learning and memory 63:19–32.
582. James W (1890) The principles of psychology, http:// psychclassics.yorku.ca /James /Principles /index.htm
583. Janmey PA (1998) Physiol Rev 78:763–781.
584. Jasper HH (1995) Adv Neurol 66:97–114.
585. Jayakar SS, Dikshit M (2004) Int J Nurosci 114:695–734.
586. Jester JM, Campbell LW, Sejnowski T.J. (1995) J Physiol 484:689–705.
587. Jiang RG, Eyzaguirre C (2006) Brain Res 1076:198–208.
588. Johansen FF (1993) Acta Neurol Scand Suppl 150:1–32.
589. Johansson S, Arhem P (1994) Proc Natl Acad Sci USA 91:1761–1765.
590. Johnston GAR (1991) Sem Neurosci 3 (1991) 205–210.
591. Jonas P, Bischofberger J, Fricker D, Miles R (2004) Trends in Neurosci 27:30–40.

592. Julius D, Basbaum AI (2001) Nature 413:203–210.
593. Kaila K (1994) Prog Neurobiol 42:489–537.
594. Kaleva O (1990) Fuzzy Sets and Systems 35:389–396.
595. von Kamke E (1959) Differentialgleichungen. 6.Verbesserte auflage. Leipzig.
596. Kampfner RR, Conrad M (1983) Bull Math Biol 45:931–968.
597. Kandel ER (1998) Am J Psychiatry 155:457–470.
598. Kandler K, Katz LC (1995) Curr Opin Neurobiol 5:98–105.
599. Kandler K, Katz LC (1998) J Neurosci 18:1419-1427.
600. Kaneko H, Tache Y, Kusugami K (2002) J Gastroenterol 37:128–132.
601. Kanter ED, Haberly LB (1993) J Neurosci 13:2477-2482.
602. Kapuscinski J, Ardelt B, Piosik J, Zdunek M, Darzynkiewicz Z (2002) Biochem Pharmacol 63:625–634.
603. Karpiak SE, Rapport MM (1979) Behav Neural Biol 27:146–157.
604. Kask A, Harro J, Horsten S, Redrobe JP, Dumont Y, Quirion R (2002) Neurosci BioBehav Rev 26:259–283.
605. Kato A, Sakuma Y (2000) Brain Res 862:90–102.
606. Kato K, Clifford DB, Zorumski CF (1993) Neuroscience 53:39–47.
607. Katsaras AK and Liu D B (1977) J Math Anal Appl.,58:135–146.
608. Katsura N, Torigoe F, Hayashida S, Honda T, Tsujimura A, Ohkuma S (2005) Brain Res 1039:211–215.
609. Katsuura G, Asakawa A, Inui A (2002) Peptides 23:323–329.
610. Katz LM, Young AS, Frank JE, Wang Y, Park K (2004) Brain Res 1017:85–91.
611. Katz PS, Clemens S (2001) TRENDS in Neurosci 24:18–24.
612. Kausik S, Lindquist S, Kandel ER (2003) Cell 115:879–891.
613. Kawai Y, Gemba M (1996) Jpn J Pharmacol 72:67–70.
614. Kelley AE (2004) Neurosci BioBehavior Rev 27:765–776.
615. Kelley AE, Smith-Roe SL, Holahan MR (1997) Proc Natl Acad Sci USA 94:12174–12179.
616. Kemenes G, Staras K, Benjamin PR (1997) J. Neurophysiol 78:2351–2362.
617. Kennedy AL, Harakall SA, Lynch SW, Braas KM, Hardwick JC, Mawe GM, Parsons RL (1998) J Autonom Nerv Syst 71:190–195.
618. Khan S, Spudich JL, McCray JA, Trentham DR (1995) Proc Natl Acad Sci USA 92:9757–9761.
619. Khisti RT, Boyd KN, Kumar S, Morrow AL (2005) Brain Res 1049:104–111.
620. Khodorov BI (1972) Problem of Excitability. Plenum Press, N-Y
621. Kiehn O, Tresch MC (2002) Trends in Neurosci 25:108–115.
622. Kien J, Altman JS (1995) Prog Neurobiol 45:361–372.
623. Kikuchi S, Kim SU (1993) J Neurosci Res 36:558–569.
624. Killcross S, Robbins TW, Everitt BJ (1997) Nature 388:377–380.
625. Kilner JM, Vargas C, Duval S, Blakemore S-J, Sirigu A (2004) Nat Neurosci 7:1299-301
626. Kim CH, Oh Y, Chung JM, Chung K (2001) Mol Brain Res 95:153–161.
627. Kim HI, Kim TH, Song J-H (2005) Brain Res 1045:134–141.
628. Kim JJ, Jung MV (2006) Neurosci Biobehav Rev 30:188–202.
629. Kim ML, Lee JH, Park SO, Hong KW, Kim CD, Kim KY, Lee WS (2006) Pharmacol Res 54:261-267.
630. Kim M-S, Cheong Y-P, So H-S, Lee K-M, Kim T-Y, Oh J, Chung Y-T, Son Y, Kim B-R, Park R (2001a) Biochem Pharmacol 61:779–786.
631. Kinkeada B, Bindera EB, Nemeroff CB (1999) Biological Psychiatry 46:340–351.

632. Kirkwood A, Lee H-K, Bear MF (1995) Nature 375:328-331.
633. Kirsch J (1999) Curr Opin Neurobiol 9:329–335.
634. Kircher T, David A (2003) The self in neuroscience and psychiatry. Cambridge Univ Press, Cambridge.
635. Kiyatkin EA (2005) Brain Res Rev 50:27–56.
636. Kiyatkin EA, Brown PL (2004) Brain Res 1005:101–116.
637. Kiyatkin EA, Brown PL (2005) Eur J Neurosci 22:930–938.
638. Kiyatkin EA, Rebec GV (2001) Neuroscience 102:565–580.
639. Kiyatkin EA, Wise RA (2002) J Neurosci 22:1072–1080.
640. Kleim JA, Freeman JH, Bruneau R, Nolan BC, Cooper NR, Zook A, Walters D (2002) Proc Natl Acad Sci USA 99:13228–13232.
641. Klein M, Kandel ER (1979) Proc Natl Acad Sci USA 27:3512–3515.
642. Klepeis VE, Cornell-Bell A, Trinkaus-Randall V (2001) J Cell Sci 114:4185–4195.
643. Kloeden PE (1982) Fuzzy Sets and Systems 7:275–296.
644. Klotz KN, Jesaitis AJ (1994) Bioassays 16:193–198.
645. Kodama T, Kimura M (2002) Peptides 23:1673–1681.
646. Koch C, Bernander O, Douglas RJ (1995) J Comp Neurosci 2:63–82.
647. Koch C, Segev I (2000) Nat Neurosci 3:1171–1177.
648. Koch C, Tsuchiya N (2007) Attention and consciousness: two distinct brain processes. TRENDS in Cogn Sci 11:16–22.
649. Koenigs M, Huey ED, Raymont V, Cheon B, Solomon J, Wassermann EM, Grafman J (2007) Nat Neurosci 11:232–237.
650. Kofke WA, Garman RH, Stiller RL, Rose ME, Garman R (1996) Anesth Analg 83:1298–1306.
651. Kofman O, Einat H, Cohen H, Tenne H, Shoshana C (2000) J Neural Transm 107:241–253.
652. Koh G, Suh KS, Chon S, Oh S, Woo JT, Kim SW, Kim JW, Kim YS (2005) Arch Biochem Biophys 438:70–79.
653. Kohidai L (1999) Acta Biol Hung 50:375–394.
654. Kohsaka A, Bass J (2007) TRENDS in Endocrinology and Metabolism 18:4–11.
655. Kolenda H, Steffens H, Hagenah J, Schomburg ED (2003) Neurosci Res 47: 357–366.
656. Koob GF, Moal ML (2001) Neuropsychopharmacology 24:97–129.
657. Kopelman MKD (2002) Brain 125:2152-2190.
658. Korobkova E, Emonet T, Vilar JMG, Shimizu TS, Cluzel P (2004) Nature 428:574–578.
659. Korte M, Carroll P, Wolf E, Brem G, Thoenen H (1995) Proc Natl Acad Sci USA 92:8856–8860.
660. Korz V, Frey JU (2005) J Neurosci 25:7393-7400.
661. Kostjuk PG, Semenjutin IP (1961) J Physiol USSR 47:678–686. (in Russian)
662. Kotlyar BI (1977) Plasticity of neural cells. Pb Moskow State Univeesity, Moscow (in Russian)
663. Kotlyar BI, Pivovarov AS (1989) J Higher Nerv Act 39:195–214. (in Russian).
664. Kotlyar BI, Yeroshenko TM (1971) Physiol Behav 4:609–615.
665. Kovac MP, Davis WJ (1980) J Neurophysiol 43:469–487.
666. Kow L-M, Commons KG, Ogawa S, Pfaff DW (2002) Brain Res 935:87–102.
667. Kow L-M, Easton A, Pfaff DW (2005) Brain Res 1043:124–131.
668. Kowall NW, Beal MF, Busciglio J, Duffy LK, Yankner BA (1991) Proc Natl Acad Sci USA 88:7247-7251

669. Koyama C, Kamimura S (2003) Physiol Behav 80:267–272.
670. Kozhedub RG (1993) J Higher Nervous Activity 43:509–513. (in Russian)
671. Krampfl K (2007) Clinical Neurophysiology 118:62–63.
672. Krawitz S, Fedirchuk B, Dai Y, Jordan LM, McCrea DA (2001) J Physiol 532.1:271–281.
673. Kreibich AS, Blendy JA (2005) Int Rev Neurobiol 65:147–178.
674. Krezel W, Dupont S, Krust A, Chambon P, Chapman PF (2001) Proc Natl Acad Sci USA 98:12278–12282.
675. Krishna MT, Jacob MH (2001) Proc Natl Acad Sci USA 98:3631–3632.
676. Krnjevic K, Reiffenstein RJ, Topert N (1981) Neuroscience 6:2465–2484.
677. Krncke K-D, Fehsel K, Kolb-Bachofen V (1997) Nitric Oxide 4:107–120.
678. Kronitiris-Litowitz JK, Walters ET, McAdoo DJ (1994) J Comp Physiol 163:671–675.
679. Kruger THC, Haake P, Hartmann U, Schedlowski M, Exton MS (2002) Neurosci BioBehav Rev 26:31–44.
680. Kruglikov RI (1981) Neurochemical mechanisms of learning and memory. Nauka, Moscow
681. Kubota N at al (2001) Self-Conscioumess and Emotion for a Pet Robot with Structured Intelligence. In Proc of the Joint 9th IFSA World Congress and 20th NAFIPS Intrnational Conference, 2786–2791.
682. Kujawa SG, Liberman MC (1999) J Neurophysiol 82:863–873.
683. Kukreja RC, Salloum F, Das A, Ockaili R, Yin C, Bremer YA, Fisher PW, Wittkamp M, Hawkins J, Chou E, Kukreja AK, Wang X, Marwaha VR, Xi L (2005) Vascular Pharmacology 42:219–232.
684. Kulik A, Brockhaus J, Pedarzani P, Ballanyi K (2002) Neuroscience 110: 541–554.
685. Kulkosky PJ, Allison CT, Mattson BJ (2000) Alcohol 20:87–91.
686. Kullmann DM (1994) Neuron 12:1111–1120.
687. Kunes S (2000) Curr Opin in Neurobiol 10:58–62.
688. Kunimatsu T, Asai S, Kanematsu K, Zhao H, Kohno T, Misaki T, Ishikawa K (1999) Brain Res 831:273–282.
689. Kupfermann I (1994) Curr Opin Neurobiol 6:869–876.
690. Kuptsov PA, Pleskacheva MG, Voronkov DN, Lipp Kh-P, Anokhin KV (2006) Neurosci Behav Physiol 36:341–350.
691. Kurihara K, Nakanishi N, Ueha T (2000) Am J Physiol Cell Physiol 279:1516–1527.
692. Kuriyama H. Hotta M. Wakabayashi I, Shibasaki T (2000) Neurosci Lett 282:109–112.
693. Kvolik S, Glavas-Obrovac L, Bares V, Karner I (2005) Life Sci 77:2369–2383.
694. Kwak S, Weiss JH (2006) Curr Opin Neurobiol 16:281–287.
695. Ladebya R, Wirenfeldta M, Garcia-Ovejerob D, Fengera C, Dissing- Olesena L, Dalmaua I, Bente Finsen B (2005) Brain Res Rev 48:196–206.
696. Lahor-Beltra R, Hameroff SR, Dayhoff JE (1993) BioSystems 29:1–16.
697. Laming PR, Kimelberg H, Robinson S, Salm A, Hawrylak N, Mller C, Roots B, Ng K (2000) Neurosci Biobehav Rev 24:295–340.
698. Lampe PD, Lau AF (2004) Int J Biochem Cell Biol 26:1171–1186.
699. Lampl I, Reichova I, Ferster D (1999) Neuron 22:361–374.
700. Landau L (1937) Phys Zs Sowjet 11:26–44.
701. Landisman CE, Connors BW (2005) Science 310:1809–1813.

702. Lang F, Busch GL, Ritter M, Volkl H, Waldegger S, Gulbins E, Haussinger D (1998) Physiol Rev 78:247–285.
703. Lang R, Gundlachc AL, Koflera B (2007) Pharmacology Therapeutics 115:177–207.
704. Las L, Stern AA, Nelken I (2005) J Neurosci 25:1503–1513.
705. Lashley KS (1929) Brain mechanisms and intelligence: a quantitative study of injuries to the brain. University of Chicago Press, Chicago.
706. Lassen MB, Brown JE, Stobbs SH, Gunderson SH, Maes L, Valenzuela CF, Ray AP, Henriksen SJ, Steffensen SC (2007) Brain Res 1156:46-58.
707. Lau HC, Rogers RD, Ramnani N, Passingham RE (2004) NeuroImage 21:1407-1415.
708. Laureys S (2005) TRENDS in Cogn Sci 9:556–559.
709. Laurini RN, Arbeille B, Gemberg C, Akoka S, Locatelli A, Lansac J, Arbeille P (1999) Eur J Obstetrics Gynecol Reprod Biol 86:15–22.
710. Leal-Campanario R, Delgado-Garcia JM, Gruart A (2006) Proc Natl Acad Sci USA 103:10052–10057.
711. LeBeau FEN, Traub RD, Monyer H, Whittington MA, Buhl EH (2003) Brain Res Bull 62:3–13.
712. Lechner HA, Baxter DA and Byrne JH. (2000) J Neurosci 20:3377–3386.
713. Lechner HA, Byrne JH (1998) Neuron 20:355–358.
714. Ledoux VA, Wooley CSJ (2005) J Neurosci 25:971–976.
715. Lee R-S, Koob GF, Henriksen SJ (1998) Brain Res 799:317–322.
716. Lee SM, Clemens MG (2004) Shock 22:82–87.
717. Lee SW, Stanley BG (2005) Brain Res 1023:1–8.
718. Leedham JA, Kong J-Q, Taylor DA, Johnson SM, Fleming WW (1992) J Pharmacol Exp Ther 263:15–19.
719. Leick V, Bog-Hansen TC, Juhl HA (2001) J Membr Biol 181:47–53.
720. Leon LR (2005) J Therm Biol 30:273–288.
721. Leranth C, Nitsch R, Deller T, Frotscher M (1992) Epilepsy Res 7:49–64.
722. Leresche N, Asprodini E, Emri Z, Cope DW, Crunelli V (2000) Neuroscience 98:513–522.
723. Leszczuk M, Gandhi C, Muzzio JA, Matzel LD (1997) Modification of action potential kinetics following associative conditioning of Hermissenda contributes to synaptic facilitation of B photoreceptors. Abstr.Neurosci. 526.5:1339
724. Levary R (1990) International Journal of System Sciences 21:1701–1707.
725. Levi AJ, Issberner J (1996) J Physiol 493:19–37.
726. Levin BE (2006) Physiol Behav 89:486-489.
727. Lewis DI, Coote JH (1993) Brain Res 614:1–9.
728. Lewis KE, O'Day DH (1996) J Euk Microbiol 43:65–69.
729. Lewis M, Todd D (2005) J Consci Stud 12:210–235.
730. Leznik E, Llins R (2005) J Neurophysiol 94:2447-2456.
731. Li Q, Roberts AC, Glanzman DL (2005) J Neurosci 25:5623–5637.
732. Li Z, Ferguson AV (1994) Neuroscience 55:197–207.
733. Lian X-Y, Stringer JL (2004) Brain Res 1012:177–184.
734. Liao D, Hessler NA, Malinow R (1995) Nature 375:400–404.
735. Liberman EA, Minina SV (1996) BioSystems 38:173–177.
736. Liberzon I, Zubieta JK, Fig LM, Phan KL, Koeppe RA, Taylor SF (2002) Proc Natl Acad Sci USA 99:7084–7089.
737. Libet B (2006) Prog in Neurobiol 78:322-326.

738. Lichtman JW, Colman H (2000) Neuron 25:269–278.
739. Lien R, Mishra OP, Graham E, Papadopoulos MD, Anday EK (1994) Brain Res 637:249–254.
740. Lin XY, Glanzman DL (1997) J Neurophysiol 77:667-674.
741. Lin YJ, Greif GJ, Freedman JE (1993). Molecular Pharmacology 44:907–910.
742. Linda R, Mills LR, Velumian AA, Agrawal SK, Theriault E, Fehlings MG (2004) NeuroImage 21:1069–1082.
743. Linden AM, Schoepp DD (2006) Drug Discovery Today: Therapeutric Strategies 3:507–517.
744. Linden D (1994) Learn Mem 1:121–128.
745. Linden DJ, Smeyne M, Connor JA (1993) Neuron 11:1093–1100.
746. Linden DJ (1999) Neuron 22:661–666.
747. Linden R, Martins RAP, Silveira MS (2005) Progress in Retinal and Eye Research 24:457–491.
748. Liou AKF, Clark RS, Henshall DC, Yin X-M, Chen J (2003) Prog Neurobiol 69:103-142.
749. Lipton SA (1999) Cell death and differentiation 6:943–951.
750. Lisman J, Lichtman JW, Sanes JR (2003) Nat Neurosci Rew 4:926–929.
751. Lisman J, Spruston N (2005) Nat Neurosci 8:839-841.
752. Litvinov EG, Logunav DG (1979) J Vish Nevn Deyat 29:284–294. (in Russian)
753. Liu B, Du L, Kong L-Y, Hudson M, Wilson BS, Chang RC, Abel HH, Hong JS (2000) Neuroscience 97:749–756.
754. Liu B, Qin L, Yang SN, Wilson BC, Liu Y, Hong JS (2001a). J Pharmacol Exp Ther 298:1133–1141.
755. Liu PK, Grossman RG, Hsu CY, Robertson CS (2001) TRENDS in Neurosci 24:581–588.
756. Liu Q-s, Pu1 L, Poo M-m (2005) Nature 437:1027–1031.
757. Liu Y, Wang ZX (2003) Neuroscience 121:537-544.
758. Liu YC, Salamone JD, Sachs BD (1997) J Neurosci 17:5245-5253.
759. Lledo P-M, Saghatelyan A (2005) TRENDS in Neurosci 28:248–254.
760. Logunov DB, Tsitolovsky LE (1984) J High Nerv Activity 34:575–577. (in Russian).
761. Lopes J, Huisamen B, Lochner A (2006) J Mol Cell Cardiol 40:998–999.
762. Lorenzetti FD, Mozzachiodi R, Baxter DA, Byrne JH (2006) Nat Neurosci 9:17-19.
763. Louiset E, Belmeguenai A, Desrues L, Leprince J, Tononi MC, Vaudry H (2005) Annals of the New York Academy of Sciences 1040:131–136.
764. Lowe G, Gold GH (1995) Proc Natl Acad Sci USA 92:7864–7868.
765. Lu R, Wang W, Uzzau S, Vigorito R, Zielke HR, Fasano A (2000) J Neurochem 74:320–326.
766. Lu YM, Mansuy IM, Kandel ER, Roder J (2000a) Neuron 26:197–205.
767. Lubar I, Middaugh L (1971) J Compar Physiol Psychol 76:26–30.
768. Luft AR, Buitrago MM, Ringer T, Dichgans J, Schulz JB (2004) J Neurosci 24:6515-6520.
769. Lukowiak K, Cotter R, Westly J, Ringseis E, Spenser G and Syed N. (1998) J Exp Biol 201: 877–882.
770. Lynch WJ, Taylor JR (2005) Eur J Neurosci 22:1214–1220.
771. Lyubkin M, Durand DM, Haxhiu MA (1997) J Neurophysiol 78:2475–2482.
772. Ma X, Tong Y-G, Schmidt R, Brown W, Payza K, Hodzic L, Pou C, Godbout C, Hokfelt T, Xu Z-QD (2001) Brain Res 919:169–174.

773. Machado-Vieira R, Lara DR, Souza DO, Kapczinski F (2002) Medical Hypotheses 58:297–304.
774. Mackowiak M, Markowicz-Kula K, Fijal K, Wedzony K (2005) Brain Res 1055:149–155.
775. MacLean DB, Luo L-G (2004) Brain Res 1020:1–11.
776. Maeda N, Shimomura L, Kishida K, Nishizawa H, Matsuda M (2002) Nat Medicine 8:731–737.
777. Magarios AM, Verdugo JMG, McEwen BS (1997) Proc Natl Acad Sci USA 94:14002–14008.
778. Maghazachi AA (2000) Int J Biochem Cell Biol 32:931–943.
779. Magoski NS, Bulloch AGM (2000) J Neurobiol 42:410-423.
780. Magoul R, Dubourg P, Benjelloun W, Tramu G (1993) J Chem Neuroanatomy 6:419–429.
781. Maia TV, Cleeremans A (2005) Trends in Cognitive Sciences 9:397–404.
782. Maier I (1993) Plant Cell Env 16:891–907.
783. Makarewicz D, Duszczyk M, Gadamski R, Danysz W, Lazarewicz JW (2006) Neurochem Int 48 485–490.
784. Malatynska E, Knapp RJ (2005) Neurosci BioBehav Rev 29:715–737.
785. Manahan-Vaughan D, Behnish T, Reymann KG (1999) Neuropharmacology 38:487–494.
786. Mani SK, Allen JM, Rettori V, McCann SM, O'Malley BW, Clark JH (1994) Proc Natl Acad Sci USA 91:6468–6472.
787. Maniotis AJ, Chen CS, Ingber DE (1997) Proc Natl Acad Sci USA 94:849–854.
788. Mann JJ, Oquendo M, Underwood MD, Arongo V (1999) J Clin Psych 60:113–116.
789. Mantle D, Eddeb F, Areni K, Snowden C, Mendelow AD (2000) Clin Chimica Acta 301:41–53.
790. Marban E, Yamagishi T, Tomaselli GF (1998) J Physiol 508:647–657.
791. Marder E, Goaillard J-M (2006) Nat Rev Neurosci 7:563–575.
792. Marder E, Prinz AA (2002) BioEssays 24:1145-1154.
793. Maren S, Baudry M (1995) Neurobiology of learning and memory 63:1–18.
794. Marois R, Ivanoff J (2005) TRENDS in Cogn Sci 9:296–305.
795. Marois R, Yi D-J, Chun MM (2004) Neuron 41:465–472.
796. Martin I, Grotewiel MS (2006) Mechanisms of Ageing and Development 127:411–423.
797. Martin RL, Lloyd HGE, Cowan AI (1994) Trends in Neurosci 17:251–257.
798. Martin S, Vincent JP, Mazella J (2003) J Neurosci 23:1198–1205.
799. Martin-Soelch C, Leenders KL, Chevalley AF, Missimer J, Kunig G, Magyar S, Mino A, Schultz W (2001) Brain Res Rev 36:139–149.
800. Martnez AD, Eugenn EA, Branes MC, Bennett MVL, Saez JC (2002) Brain Res 943:191–201.
801. Marks JD, Bindokas VP, Zhang X-M (2000) Developmental Brain Res 124:101–116.
802. Mas M (1995) Neurosci BioBehav Rev 19:261–277.
803. Masuko S, Nakajima S, Nakajima Y (1992) Neuroscience 49:347–364.
804. Mattson MP (1998) Trends Neurosci 20:53–57.
805. Matsumoto Y, Watanabe T (2002) Neurosci Lett 322:67–69.
806. Maubach KA, Martin K, Smith DW, Hewson L, Frankshun RA, Harrison T, Seabrook GR (2001) Neuropharmacology 40:806–817.

807. Maurice T, Casalino M, Lacroix M, Romieu P (2003) Pharmacol Biochem Behav 74:869–876.
808. Maurice T, Gregoire C, Espallergues J (2006) Pharmacol Biochem Behav 84:581–597.
809. Maus MV, Reilly SC, Clevenger CV (1999) Endocrinology 140:5447–5450.
810. Maximova OA, Balaban PB (1983) The neuronal mechanisms of the plasticity of behavior. Nauka, Moscow (in Russian)
811. Mayford M, Kandel ER (1999) TIG 15:463–470.
812. Mayr B, Montminy M (2001) Nat Rev Mol Cell Biol 2:599–609.
813. McBride WJ (2002) Pharmacol Biochem Behav 71:509–515.
814. McCarthy L, Szabo I, Nitsche JF, Pintar JI, Rogers TJ (2001) J Neuroimmunol 114:173–180.
815. McCulloch WS, Pitts W (1943) Bull Math Biophys 5:115–133.
816. McConnell JV (1962) J Neuropsychiatry 3:42–48.
817. McEachern JC, Shaw CA (1996) Brain Res Rev 22:51–92.
818. McEchron MD, Green EJ, Winters RW, Nolen TG, Schneiderman NM, Cale PM (1996) J Neurosci 16:1273–1283.
819. McEven BS (1994) Sem Neurosci 6:239–246.
820. McEven BS (2001) Clin Neurosci Res 1:19–34.
821. McGaugh JL (2004) Annu Rev Neurosci 27:1–28
822. McKenna KE (2000) Neurosci BioBehav Rev 24:535–540.
823. McLay RN, Pan W, Kastin AJ (2001) J Peptides Peptides 22:2181–2255.
824. McMahon SB, Cafferty WBJ, Marchand F (2005) Exp Neurol 192:444–462.
825. Meckler LB (1967) Nature 215:481–484.
826. Meech RW, Mackie GO (1993) J Neurophysiol 69:884–893.
827. Meerson FZ (1975) Physiol Rev 55:79–123.
828. Klement EP, Mesiar R, Pap E (2000) Triangular Norms, Trends in Logic 8, Kluwer Academic Publisher, Dordrecht.
829. Meisar R, Rybarik J (1995) Fuzzy Set and Systems 74:365–369.
830. Klement EP, Mesiar R (2005) Logical, Algebraic, Analytic and Probabilistic Aspects of Triangular Norms, Elsevier.
831. Meldolesi J (2001) Prog Neurobiol 65:309–338.
832. Meldrum BS (2000) J Nutrition 130:1007–1015.
833. Meliza CD, Dan Y (2006) Neuron 49:183–189.
834. Meme W, Calvo CF, Froger N, Ezan P, Amigou E, Koulakoff A, Giaume C (2006) FASEB J 20:494–496.
835. Menini A, Picco C, Firestein S (1995) Nature 373:435–437.
836. Mercuri NB, Bonci A, Johnson SW, Stratta F, Calabresi P, Bernardi G (1994) J Neurophysiol 71:1165–1173.
837. Mermelstein PG, Becker JB, Surmeier DJ (1996) J Neurosci 16:595–604.
838. Miguel TT, de-Souza RLN (2006) Brain Res 1076:42–48.
839. Mikhailova MV, Mayeux PR, Jurkevich A, Kuenzel WJ, Madison F, Periasamy A, Chen Y, Cornett LE (2007) Mol Endocrinol 21:2178–2188.
840. Millan MJ (1999) Prog Neurobiol 57:1–164.
841. Miller NE (1969) Science, 1969, 163:434–445.
842. Miller RR, Berk AM (1977) J Exp Psychol 3:343–356.
843. Mills JC, Lee VM-Y, Pittman RN (1998) J Cell Sc 111:625–636.
844. Mills LR, Velumian AA, Agrawal SK, Theriault E, Fehlings MG (2004) NeuroImage 21:1069–1082.

845. Milner B, Squire LR, Kandel ER (1998) Neuron 20:445–468.
846. Minerbo G, Albeck D, Goldberg E, Lindberg T, Nakari M, Garritano J, Smock T (1994) Exp Brain Res 97:444–450.
847. Ming G-L, Wong ST, Henley J, Yuang X-B, Song NC, Spitzer M-M (2002) Nature 417:411–418.
848. Minghetti L, Levi G (1998) Prog Neurobiol 54:99–125.
849. Mironov SL, Richter W (2000) Brain Res. 869 (2000) 166–177.
850. Miserendino MJD, Nestler EJ (1995) Brain Res 674:299–306.
851. Misra K, Pandey SC (2006) Neuropsychopharmacology 31:1406–1419.
852. Mitchum RD, Kiyatkin EA (2004) Brain Res 1000:110–122.
853. Miura T, Ohnuma Y, Kuno A, Tanno M, Ichikawa Y, Nakamura Y, Yano T, Miki T, Sakamoto J, Shimamoto K (2004) Am J Physiol: Heart Circ Physiol 286:214-221.
854. Mizumoto M, Tanaka K, (1976) Information and Control 30:312–340.
855. Mizumoto M (1989) Fuzzy Sets and Systems 31:217–242.
856. Mizutani A, Murakami K, Okajima K, Kira S, Mizutani S, Kudo K, Takatani J, Goto K, Hattori S, Noguchi T (2005) Shock 24:281–287.
857. Mochizuki T, Klerman EB, Sakurai T, Scammell TE (2006) Am J Physiol Regul Integr Comp Physiol 291:533–540.
858. Moffett SB (1995) Prog Neurobiol 46:289–330.
859. Monje ML, Phillips R, Sapolsky R (2001) Brain Res 911:37–42.
860. Montgomery JM, Pavlidis P, Madison DV (2001) Neuron 29:691–701.
861. Moran TH (2000) Nutrition 16:858–865.
862. Morimoto Y, Bart RD, Pearlstein RD, Dexter F, Warner DS (1997) J Neurosurg Anesthesiol 9:316–323.
863. Morley JE, Flood HF (1991) Life Sci 49:707–711.
864. Motosaka K, Koganezawa M, Narikawa S, Furuyama A, Shinozaki K, Isono K, Shimada I (2007) J Comp Physiol A Neuroethol Sens Neural Behav Physiol 193:279–283.
865. Moyer JR, Thompson LT, Disterhoft JF. (1996) J Neurosci 16:5536–5546.
866. Mzsik G, Peidl Z, Szolcsnyi J, Dmtr A, Hideg K, Szekeres G, Kardi O, Hunyady B. (2005) Inflammopharmacology 13:139–149.
867. Mulkey DK, Stornetta RL, Weston MC, Simmons JR, Parker A, Bayliss DA, Guyenet PG (2004) Nat Neurosci 7:1360–1369.
868. Mungarndee SS, Baldwin BA, Chindadoungratana C, Kotchabhakdi N (2002) Brain Res 925:204–212.
869. Murai Y, Nakashima T, Miyara S, Kiyohara T (1998) Neurosci Res 30:35–41.
870. Murata T, Maruoka N, Omata N, Takashima Y, Igarashi K, Kasuya F, Fujibayashi Y, Wada Y (2007) Prog Neuropsychopharmacol Biol Psychiatry 31:848–857.
871. Murav'eva EV, Anokhin KV (2006) Zh Vyssh Nerv Deiat 56:274–281. (in Russian)
872. Musallam S, Corneil BD, Greger B, Scherberger H, Andersen RA (2004) Science 305:258–262.
873. Muzur A (2005) J Theor Biol 233:103–118.
874. Nagy JI, Dudek FE, Rash JE (2004) Brain Res Rev 47:191–215.
875. Nahmias E, Morris ST, Nadelhoffer T (2004) J Consci Stud 11:162–179.
876. Nakagaki T, Yamada H, Gota T (2000) Nature 407:470–475.
877. Nakagawa Y, Iwasaki T (1995) Brain Res 686:70–76.

878. Nakase T, Naus CCG (2004) Biochimica et Biophysica Acta 1662:149-158.
879. Narita M, Mizoguchi H, Nagase H, Suzuki T, Tseng LF (2001) Brain Res 913:170–173.
880. Nader K, Schafe GE, Le Doux JE (2000) Nature 406:722–726.
881. Nargeot R, Baxter DA, Byrne JH (1997) J Neurosci 17:8093–8105.
882. Nargeot R, Baxter DA, Byrne JH (1999a) J Neurosci 19:2247–2260.
883. Nargeot R, Baxter DA, Byrne JH (1999b) J Neurosci 19:2261–2272.
884. Nazaroff GJ (1973) J Math Anal Appl 41:478–485.
885. Nedergaard M, Ransom B, and Goldman SA (2003) Trends in Neurosci 26:523–5302.
886. Nelson TE, Ur CL, Gruol DL (2005) Brain Res 1048:69–79.
887. Nestler EJ (2005) Nat Neurosci 8:1445–1449.
888. Nestler EJ, Carlezon VA (2006) Biological Psychiatry 59:1151–1159.
889. Nevo N, Reiss TA, Levy R, Bayewitch M, Vogel Z (2000) Neuropharmacology 39:364–371.
890. Nicol RA (2003) Philos Trans R Soc Lond B Biol Sci 358:721–726.
891. Nicolaidis S (2006) Metabolism 55:24–29.
892. Nicolaidis S, Even P (1992) Int J of Obesity 16:31–41.
893. Nicolelis MA, Baccala LA, Lin RC, Chapin JK (1995) Science 268:1353–1358.
894. Nicoll RA, Malenka RC (1995) Nature 373:115–118.
895. Nie Z, Schweitzer P, Roberts AJ, Madamba SG, Moore SD, Siggins GR (2004) Science 303:1512–1514.
896. Nie Z, Yuan X, Madamba SG, Siggins GR (1993) J Pharmacol Exp Ther 266:1705–1712.
897. Nishino S, Fujiki N (2007) Expert Opin Ther Targets 11:37–59.
898. Niswender ED, Morton GJ, Walter HS Rhodes CJ Myers AG, Scwartz MW (2001) Nature 413:794–795.
899. Nodgren RA, Woodruff DS, Bick MD (1970) Physiol Behav 5:1169–1174.
900. Novk V, Perfilieva I, Mockor J (1999) Mathematical Principles of Fuzzy Logic, Springer.
901. Nunez JL, McCarthy MM (2002) Endocrinology 144:2350–2359.
902. Oberheim NA, Wang X, Goldman S, Nedergaard M (2006) Trends in Neurosci 29:547–553.
903. Obuchowicz E, Krysiak R, Herman ZS (2004) Neurosci BioBehav Rev 28:595–610.
904. O'Doherty JP, Deichmann R, Critchley HD, Dolan RJ (2002) Neuron 33:815–826.
905. Ogawa WN, Baptista V, Aguiar JF, Varanda WA (2005) Neuroscience 130:309–315.
906. Okano K, Kuraishi Y, Satoh M (1993) Biol Pharmaceutic Bull 16:861–865.
907. Okazaki M, Takeda R, Haji A, Yamazaki H (2001) Brain Res 914:34–47.
908. O'Keefe J, Nadel L (1978) The hippocampus as a cognitive map. Oxford University Press, London
909. O'Leary FA, Byrne JH, Cleary LJ (1995) J Neurosci 15:3519-3525.
910. Olson GA, Olson RD, Kastin AJ (1993) Peptides 14:339–378.
911. Olson VG, Zabetian CP, Bolanos CA, Edwards S, Barrot M, Eisch AJ, Hughes T, Self DW, Neve RL, Nestler EJ (2005) J Neurosci 25:5553–5562.
912. Omholt SW, Kefang X, Andersen O, Plahte E (1998) J Theor Biol 195:339–350.
913. Onn S-P, Grace AA (2000) J Neurosci 20 2332–2345.

914. Onodera H, Kogure K, Ono Y, Igarashi K, Kiyota Y, Nagaoka A (1989) Neurosci Lett 98:101–104.
915. Oomen SP, Hofland LJ, Lamberts SW, Lowenberg B, Touw IP (2001) FEBS Lett 503:163–167.
916. Oquendo MA, Placidi GPA, Malone KM, Campbell C, Keilp J, Brodsky B, Kegeles LS, Cooper TB, Parsey RV, Van Heertum RL, Mann JJ (2003) Arch Gen Psychiatry 60:14-22.
917. Ordieres MGL, Arnaiz GRLA (2000) Peptides 21:571–576.
918. Ordieres MGL, Arnaiz GRLA (2005) Regulatory Peptides 129:177–182.
919. Orpwood RD (1990) J Theor Biol 143:145–162.
920. Osborne LC, Lisberger SG, Bialek W (2005) Nature 437:412–416.
921. Osborne NN, Casson RJ, Wood JPM, Chidlow G, Graham M, Melena J (2004) Progress in Retinal and Eye Research 23:91–147.
922. O'Steen WK, Bare DJ, Tytell M, Morris M, Gower DJ (1990) Brain Res 534:99–105.
923. Otani S, Connor JA (1995) J Neurophysiol 73:2596-2601.
924. Otori T, Friedland JC, Sinson G, McIntosh TK, Raghupathi R, Welsh FA (2004) J Neurotrauma 21:707–718.
925. Ouardous M, Lacaille JC (1995) J Neurophysiol 73:199–205.
926. Oyamada M, Oyamada Y, Takamatsu T (2005) Biochimica et Biophysica Acta (BBA) Biomembranes 1719:6–23.
927. Ozawa H, Kushiya E, Takahashi Y (1980) Neurosci Lett 18:191–194.
928. Ozawa S, Kimura N (1979) Proc Natl Acad Sci USA 76:6017–6020.
929. Paban V, Jaffard M, Chambon C, Malafosse M, Alescio-Lautier B (2005) Neuroscience 132:13–32.
930. Pace RW, Mackay DD, Feldman JL, Del Negro CA (2007) J Physiol 582:113–125.
931. Pacheco MA, Stockmeier C, Meltzer HY, Overholser JC, Dilley JE, Jope RS (1996) Brain Res 723:37–45.
932. Padovan CM, Guimaraes FS (2004) Pharmacol Biochem Behav 77:15–19.
933. Palva S, Linkenkaer-Hansen K, Ntnen R, Palva JM (2005) J Neurosci 25:5248-5258.
934. Pan Y, Zhang H, Acharya AB, Patrick PH, Oliver D, Morley JE (2005) Brain Res 1043:195–204.
935. Pandey SC (2004) Pharmacology Therapeutics 104:47–58.
936. Pandey SC, Roy A, Zhang H, Xu T (2004) J Neurosci 24:5022–5030.
937. Pap E (1995) Null-Additive Set Function. Kluwer Ac Publ, Dordrecht-Boston-London.
938. Papp R, Gnczi M, Kovcs M, Seprnyi G, Vgh A (2007) Cardiovasc Res 74:396–405.
939. Paredes RG, Karam P, Highland L, Agmoi A (1997) Pharmacol Biochem Behav 58:291–298.
940. Parent CA (2004) Curr Opin Cell Biol 16:4–13.
941. Parenti R, Campisi A, Vanella A, Cicirata F (2002) Arch Ital Biol 140:101–108.
942. Park E, Velumian AA, Fehling MG (2004) J Neurotrauma 21:754–774.
943. Paton JA, Nottebohm FN (1984) Science 225:1046–1048.
944. Paton JJ, Belova MA, Morrison SE, Salzman CD (2006) Nature 439:865–870.
945. Paton RC (1993) BioSystems 29:63–75.
946. Paulsen O, Sejnowski TJ (2006) J Neurosci 26:1661–1662.

947. Pavlov IP (1927) Conditioned reflexes: an investigation of the physiological activity of the cerebral cortex. Oxford University Press; London
948. Pawlak R, Magarinos AM, Melchor J, McEwen B, Strickland S (2003) Nat Neurosci 6:168-174.
949. Payne RS, Goldbart A, Gozal D, Schurr A (2004) Brain Res 1029:195–199.
950. Paz R, Natan C, Boraud T, Bergman H, Vaadia E (2005) J Neurosci 25:10941-10951.
951. Paz R, Vaadia E (2004) J Physiol (Paris) 98a:331–348.
952. Paz R, Wise SP, Vaadia E (2004) TRENDS in Neurosciences 27:496–503.
953. Pearse DD, Pereira FC, Marcillo AE, Bates ML, Berrocal YA, Filbin MT, Bunge MB (2004) Nat Med 10:610–616.
954. Pecia S, Berridge KC (2005) J Neurosci 25:11777-11786.
955. Pekary AE, Stevens SA, Sattin A (2006) Neurochem Int 48:208–217.
956. Pena F, Tapia R (2000) Neuroscience 101:547–561.
957. Pena F, Ramirez J-M (2004) J Neurosci 34:7549–7556.
958. Penfield W (1958) Proc Natl Acad Sci USA 44:51–66.
959. Penland S, Hoplight B, Obernier J, Crews FT (2001) Alcohol 24:45–54.
960. Peoples LL, Uzwiak AJ, Gee F, West MO (1999) Brain Res 822:231–236.
961. Peracchia C (2004) Roles of calcium, pH and calmodulin. BBA - Biomembranes 1662:61–80.
962. Pereda AE, Faber DS (1996) J Neurosci 16:983-992.
963. Pereda AE, Rash JE, Nagy JI, Bennett MVL (2004) Brain Res Rev 47:227–244.
964. Perea G, Araque A (2002) J Physiol (Paris) 96:199–207.
965. Perentonis SJ, Lisboa JG (1992) IEEE Tr on NN3:241–251.
966. Prez-Otaoa I, Ehlers MD (2005) Trends in Neurosci 28:229–238.
967. Perez-Velazquez JL, Valiante TA, Carlen PL (1994) Brain Res 649:305–309.
968. Pessoa L (2008) Nat Rev Neurosci 9:148–158.
969. Petrascheck M, Ye X, Buck LB (2007) Nature 450:553–557.
970. Pesonen U, Huupponen R, Rouru J, Koulu M (1992) Brain Res Molecular Brain Res 16:255–260.
971. Petrnyi GG (2002) Transplant Immunology 10:91–100.
972. Petrie RXA, Reid IC, Stewart CA (2000) Pharmacol Ther 87:11–25.
973. Petrovic B, Cupic D (1979) Neurosci Lett 13:186–189.
974. Phan KL, Wager T, Taylor SF, Liberzon I (2002) NeuroImage 16:331–348.
975. Phillips MI, Sumners C (1998) Regul Peptides 78:1–11.
976. Piazza PV, Moal ML (1998) Trends in Pharmacol Sci 19:67–74.
977. Pierce RC, Kumaresan V (2006) Neurosci BioBehav Rev 30:215–238.
978. Pike CJ (2001) Brain Res 919:160–165.
979. Ping A, Chun ZX, Xue XY (2005) Brain Res 1059:105–112.
980. Pinker S (2007) The mystery of consciousness. Time, Friday, Jan. 19.
981. Pistis M, Muntoni ML, Pillolla G, Perra S, Cignarella G, Melis M, Gessa GL (2005) Neuroscience 131:465–474.
982. Pivovarov AS, Valoushkova VI (1980) Nauch Dokl Vyssh Shkoly, Biol Nauki 7:55–63.(in Russian)
983. Plotnikoff NP (1975) Method of enhancing memory and learning. Pat USA, C1 424–177 (A61 c 27/00) N 3865934 Publ 11.02.
984. Pomfrett CJD (2005) Anaesthesia and intensive care medicine 6:221–216.
985. Pompeia C, Lima T, Curi R (2003) Cell Biochem Funct 21:97–104.
986. Polakiewicz RD, Schieferl SM, Gingras AC, Sonenberg N, Comb MJ (1998) J Biol Chem 273:23534–23541.

987. Popescu IR, Frost WN (2002) J Neurosci 22:1985–1993.
988. Power AE (2004) Proc Natl Acad Sci USA 101:1795–1796.
989. Prange AJ (1992) Annals of N-Y Acad Sci 668:298–306.
990. Prangishvili DA, Bibilashvili RS (1979) Usp Biol Khimii 20:5–14.
991. Prast H, Philippu A (2001) Prog Neurobiol 64:51–68.
992. Pratt KG, Aizenman CD (2007) J Neurosci 27:8268–8277.
993. Prehn JH, Lippert K, Kreiglstein J (1995) Eur J Pharmacol 292:179–189.
994. Prendergast MA, Harris BR, Mayer S, Holley RC, Hauser KF, Littleton JM (2001) Neuroscience 102:75–85.
995. Prendergast MA, Harris BR, Mayer S, Littleton JM (2000) Alcohol Clin Exp Res 24:1583–1592.
996. Pribram K (2003) Int J Psychophysiol 48:115–131.
997. Priebe NJ, Ferster D (2005) Neuron 45:133–145.
998. Prigogine I, Stengers I (1984) Order out of chaos. Bantam Books, N-Y
999. Ptak K, Pasquale ED, Monteau R (1999) Development Brain Res 114:217–227.
1000. Przyklenk K, Maynard M, Whittaker P (2006) Am J Physiol Heart Circ Physiol Am J Physiol Heart Circ Physiol 291:2008–2012.
1001. Ptak K, Pasquale ED, Monteau R (1999) Development Brain Res 114:217–227.
1002. Pulvirenti L (1993) Functional Neurol 8:433–440.
1003. Quirk GJ, Repa JC, LeDoux JE (1996) Neuron 15:1029–1039.
1004. Quirk MC, Blum KI, Wilson MA (2001) J Neurosci 21:240–248.
1005. Quiroga RQ,. Reddy L, Kreiman G, Koch C, Fried I (2005) Nature 435:1102–1107.
1006. Rabinovich MJ (1975) Completed function of the brain. Medicine, Moscow
1007. Racin RJ, Wilson D, Teskey GC, Milgram NW (1994) Brain Res 637:73–82.
1008. Raggenbass M (2001) Prog Neurobiol 64:307–326.
1009. Rajkowska G (2000) Biol Psychiatr 48:766–777.
1010. Rakic P, Cameron RS, Komuro H (1994) Curr Opin Neurobiol 4:63–69.
1011. Rami A, Volkmann T, Winckler J (2001) Exp Neurol 170:297–304.
1012. Ransom B, Behar T, Nedergaard M (2003) Trends in Neurosci 26:520–522.
1013. Rao AP, Sikdar SK (2004) Brain Res 1022:126–136. bibitemRao et al. 1993Rao BS, Desiraju T, Raju TR (1993) Brain Res 627:216–224.
1014. Rauca C, Schafer K, Hollt V (1999) Naunyn Schmiedebergs Arch Pharmacol 360:633–638.
1015. Raupach T, Ballanyi K (2004) Brain Res 1017:137–145.
1016. Raveh L, Weissman BA, Cohen G, Alkalay D, Rabinovitz I, Sonego H, Brandeis R (2002) Neurotoxicology 23:7–17.
1017. Ray DL, Cattaert D (1999) J Neurosci 19:1473-1483.
1018. Reaux A, Gallatz K, Palkovits M, Llorens-Cortes C (2002) Neuroscience 113:653–662.
1019. Rector DM, Topchiy IA, Carter KM, Rojas MJ (2005) Brain Res 1047:45–55.
1020. Redko VG (2005) Evolution, neural networks, intellect. Comkniga, Moscow. (in Russian)
1021. Reeve AJ, Walker K, Urban L, Fox A (2000) Neurosci Lett 295:25–28.
1022. Reich CG, Mason SE, Alger BE (2004) J Neurophysiol 91:239–247.
1023. Reichenberg A, Yirmiya R, Schuld A, Kraus T, Haack M, Morag A, Pollmcher T (2001) Arch Gen Psychiatry 58:445-452.
1024. Rekling JC, Funk GD, Bayliss DA, Dong X-W, Feldman JL (2000). Physiol Rev 80:767–852.

1025. Remaury A, Vitaa N, Gendreau S, Jung M, Arnone M, Poncelet M, Culouscou JM, Le Fur G, Soubri P, Caput D, Shire D, Kopf M, Ferrara P (2002) Brain Res 953:63–72.
1026. Rhodes JS, Crabbe JC (2005) Curr Opin in Pharmacol 5:26–33.
1027. Rick M, Dijkhuizen RM, Beekwilder JP, van der Worp HB, Brain Res 840:194–205.
1028. Riedel G (1996) TRENDS in Neurosci 19:219–224.
1029. Riddle DL (1999) Nature 399:8–9.
1030. Rijen van HVM, Veen van TAB, Hermans MMP, Jongsma HJ (2000) Cardiovascular Research 45:941-951.
1031. Ring BD, Scully S, Davis CR, Barker MB, Cullen MJ, Pelleymounter MA, Danilenko DM (2000) Endocrinology 141:446–449.
1032. Rizk NN, Rafols J, Dunbar JC (2005) Brain Res 1053:1–9.
1033. Roberts AC, Glanzman DL (2003) Trends in Neurosci 26:662–670.
1034. Robinson TE, Kolb B (2004) Neuropharmacology 47:33-46.
1035. Rogan MT, Staubli UV, LeDoux JE (1997) Nature 390:604-607.
1036. Roldan G, Cobos-Zapian G, Prado-Alcala Q (2001). Behav Brain Res 121:173–179.
1037. Rolls ET (2000) Neuron 27:205–218.
1038. Romo R, Salinas E (2001) Annu Rev Neurosci 24:107–137.
1039. Rose SP (2000) Learn Mem 7:1–17.
1040. Rose S (1995) Molecules and mind. Open University Press.
1041. Rosenblatt F (1961) Principles of neuro-dynamics. Spartan D.C. (ed) Washington
1042. Rosenbrock H, Koros E, Bloching A, Podhorna J, Borsini F (2005) Brain Res 1040:55-63.
1043. Rossokhin AV, Saakyan J (1992) Biophysics 37:969–973.
1044. Rossokhin AV, Tsitolovsky LE (1997) Biosystems 40:159–167.
1045. Rothman SM (1994) Sem Neurosci 6:315–322.
1046. Rothwell NJ, Relton JK (1993) Cerebrovascular and Brain Metabolism Reviews 5:178–198.
1047. Rouach N, Koulakoff A, Giaume C (2004) Neurochemistry International 45:265–272.
1048. Rouach N, Tenc M, Glowinski J, Giaume C (2002) Proc Natl Acad Sci USA 99:1023–1028.
1049. Routtenberg A, Rekart JL (2005) Trends in Neurosciences 28:12–19.
1050. Rudolph U, Antkowiak B (2004) Nat Rev Neurosci 5:709–720.
1051. Ruiz-Ortega M, Lorenzo O, Suzuki Y, Ruperez M, Egido J (2001) Curr Opin Nephrol Hypertens 10:321–329.
1052. Rumelhat DE, McClelland JL (1986) Parallel distributed processing: explorations in the microstructure of cognition. MIT Press 1,2, Cambridge
1053. Rupprecht R, Michele F, Hermann B, Striohle A, Lancel M, Romeo E, Holsboer F(2001) Brain Res Rev 37:59–67.
1054. Rybnikova EA, Khozhai LI, Tyulrsquokova EI, Glushchenko TS, Sitnik NA, Pelto-Huikko M, Otellin VA, Samoilov MO (2005) Neurosci Behav Physiol 35:383–387.
1055. Saakyan YZ, Rossokhin AV, Tsitolovsky LE (1993) Biofiziks 38:471–476.
1056. Saakyan YZ, Tsitolovsky LE (1990) Biofizika 35:518–519, (in Russian).
1057. Sacerdote P (2006) Palliat Med 20:9–15.

1058. Sacerdote P, Ruff MR, Pert CB (1988) Peptides 9:29–34.
1059. Saez JC, Spray DC, Nairn AC, Hertzberg E, Greengard P, Bennet MV (1986) Proc Natl Acad Sci USA 83:2473–2477.
1060. Sahu A, Carraway RE, Wang YP (2001) Brain Res 888:343–347.
1061. Sahu A, Metlakunta AS (2005) J Neuroendocrinol 17:720–726.
1062. Sakaguchi T, Kuno M, Kawasaki K (1999) Neurosci Lett 274:135–138.
1063. Sakurai T (2007) Nat Rev Neurosci 8:171–81.
1064. Sakurai Y (1999) Neurosci BioBehav Rev 23:785–796.
1065. Sakata Y, Fujioka T, Chowdhury GMI, Nakamura S (2000) Brain Res 871:271–280.
1066. Salameh A, Dhein S (2005) Biochimica et Biophysica Acta 1719:36-58.
1067. Salamone JD, Cousins MS, Bucher S (1994) Behav Brain Res 65:221–229.
1068. Salamone JD, Correa M, Mingote SM, Weber SM (2005) Curr Opin Pharmacol 5:34–41.
1069. Saleh TM, Kombian SB, Zidichouski JA, Pittman QJ (1997) Neuroscience 77:23–35.
1070. Salganik RI, Shumskaya IA, Tomsons VP (1981) Dokl Akad Nauk USSR 256:1269–1273.
1071. Salinska EJ, Chaudhury D, Bourne RC, Rose SPR (1999) Neuroscience 93:1507–1514.
1072. Snchez-Alvarez R, Tabernero A, Medina JM (2005) Brain Res 1039:189–198.
1073. Sanderson NSR, Weissler E, Crews D (2005) Brain Res 1052:236–239.
1074. Sandler U (1997) Neurocomputing 14:41–62.
1075. Sandler U (2002) Evolution of Fuzzy Systems and Dynamics Theory. In: Proc of the World Congress on Computational Intelligence, FUZZ-IEEE2002, Honolulu, Hawaii, US.
1076. Sandler U (2002) Fuzzy theory of evolutionary processes: Two types of dynamics. In: Proc 9th Inter Conference on Information Processing and Management of Uncertainty IPMU2002, Annasy, France.
1077. Sandler U, Tsitolovsky L (1997) Fuzzy Dynamics of Brain: the Model of Instrumental Reflex. In: Proc 7th World Congress on Fuzzy Set and Systems - IFSA'97, Prague, CR.
1078. Sandler U, Tsitolovsky L (1998) Fuzzy Dynamics of a Motivated Behavior. In: Proc 7th Inter Conference on Information Processing and Management of Uncertainty - IPMU'98, Paris.
1079. Sandler U, Tsitolovsky L (2001) Fuzzy Sets and Systems 121/2:237–245.
1080. Sandler U, Tsitolovsky L (2004) Motivational Processes and Fuzzy Logic of the Brain: Can a robot have drives and feel love for a task? In: Proc Inter Conference on Fuzzy Systems FUZZ-IEEE 2004, Budapest, Hungary.
1081. Sandler U, Tsitolovsky L (2005) Inter J of Comp Cogn (IJCC) 3:91–108.
1082. Sandler U, Tsitolovsky L (2006) WSEAS Trans on Systems 5:2025–2032.
1083. Saper CB, Chou TC, Elmquist JK (2002) Neuron 36:199–211.
1084. Saraga F, Ng L, Skinner FK (2006) J Neurophysiol 95:1669–1682.
1085. Saransaari P, Oja SS (2000) Amino Acids, 19 (2000) 509–526.
1086. Sarbadhikari SN, Chakrabarty K (2001) Med Eng Physics 23:445–455.
1087. Sawamura H, Orban, GA, Rufin V (2006) Neuron 49:307–318.
1088. Saxe R, Carey S (2006) Acta Psychol (Amst) 123:144–165.
1089. Scammell TE, Saper C (2005) Nat Neurosci 8:1286–1288.
1090. Schaffer S, Takahashi K, Azuma J (2000) Amino Acids 19:527–546.

1091. Schallert T, Jones TA (1993) J Neural Transpl Plast 4:193–198.
1092. Schinelli S, Paolillo M, Corona GL (1994). J Neurochem 62:944–949.
1093. Schioth HB, Watanobe H (2002) Brain Res Rev 38:340–350.
1094. Schoenbaum G, Chiba AA, Gallagher M (1999) J Neurosci 19:1876–1884.
1095. Scholz AT, Cooper JC, Madison DM (1973) Olfactory impriting in coho salmon, behavioral and electrophysiological evidence. In: Proc 16th Conf Great Lakes Researches Hiron. Ohio p.73
1096. Schubert P, Ogata T, Marchini C, Ferroni S (2001) Mechanisms of Ageing and Development 123:47–57.
1097. Schultz JEJ, Gross GJ (2001) Pharmacol Ther 89:123–137.
1098. Schultz W (2007) Trends in Neurosci 30:203–210.
1099. Schwanke U, Konietzka I, Duschin A, Li X, Schulz R, Heusch G (2002) Am J Physiol Heart Circ Physiol 283:1740–1742.
1100. Screurs BG, Gusev PA, Tomsic D, Alkon DL, Dhi T (1998) J Neurosci 18:5498–5507.
1101. Schmahl C, Bohmer G (1997) Brain Res 746:151–163.
1102. Schmidt AP, Lara DR, Souza DO (2007) Pharmacology Therapeutics 116:401–416.
1103. Schroeter M, Jander S (2005) Neuromolecular medicine 7:183–195.
1104. Schrter A, Andrabi SA, Wolf G, Horn TFW (2005) Brain Res 1060:1–15.
1105. Schulingkamp RJ, Pagano TC, Hung D, Raffa RB (2000) Neurosci BioBehav Rev 24:855–872.
1106. Schulteis G, Markou A, Cole M, Koob GF (1995) Proc Natl Acad Sci USA 92: 5880–5884.
1107. Schultz JEJ, Gross GJ (2001) Pharmacol Ther 89:123–137.
1108. Schultz W (2002) Neuron 36:241–263.
1109. Schwartz M, Butovsky O, Brck W, Hanisch U-K (2006) TREND in Neurosci 29:68–74.
1110. Schwartz AB (1994) Curr Opin Neurobiol 4:840–846.
1111. Scroggs RS, Fox AP (1992) J Neurosci 12:1789–1801.
1112. Seeburg P (2000) Genetically altered synaptic function and plasticity. Society for Neuroscience, Washington, DC.
1113. Seeley TD, Visscher PK, Passino KM (2006) American Scientist 94:220–225.
1114. Seidemann E, Meilijson I, Abeles M, Bergman H, Vaadia E (1996) J Neurosci 16:752-768.
1115. Seiger A (2007) Journal of Psychiatric Research 41:537–552.
1116. Sergent C, Baillet S, Dehaene S (2005) Nat Neurosci 8:1391-1400.
1117. Sergent C, Dehaene S (2004) Psychol Sci 15:720-728.
1118. Servais L, Bearzatto B, Delvaux V, Nol E, Leach R, Brasseur M, Schiffmann SN, Guy C (2005) Brain res 1055:171–179.
1119. Shaikh AG, Finlayson PG (2005) Hearing Research 201:70-80.
1120. Sharma P (1997) Neurochem Int 30:337–344.
1121. Sharaf G, Marom S (2001) J Neurosci 15:8782–8788.
1122. Sharf R, Lee DE, Ranaldi R (2005) Brain Res 1033:179–185.
1123. Shen K-F, Crain SM (1994) Brain Res 636:286–297.
1124. Sheriff S, Chance WT, Fischer JE, Balasubramaniam A. (1997) Mol Pharmacol 51:597–604.
1125. Sheriff S, Chance WT, Iqbal S, Rizvi TA, Xiao C, Kasckow JW, Balasubramaniam A (2003) Peptides 24:245–254.

1126. Sherrington C (1937) Man on his nature. MacMillan, Edinburgh, New York
1127. Shewmon DA, Holmes GL, Byrne PA (1999) Developmental Medicine Child Neurology 41:364-374.
1128. Shi T–JS, Holmberg K, Xu, Z-QD, Steinbusch H, de Vente J, Hkfelt T (1998) Pain 78:171–180.
1129. Shi W-X, Rayport S (1994) J Neurosci 14:4548–4560.
1130. Shi W, Khler T, Zusman DR (1994) J Bacteriol 176:696–701.
1131. Shi W, Zusman DR (1994) J Bacteriol 176:1517–1520.
1132. Shigenaga M, Gimeno C, Ames B (1989). Proc Natl Acad Sci U S A 86 (24):697-701.
1133. Shima K, Hoshi E, Tanji K (1996) J Neurophysiol 76:2115–2119.
1134. Shimura T, Yamamoto T, Shimokochi M (1994) Brain Res 640:215–222.
1135. Shin J, Koch C (1999) Neural Computation 11:1893–1913.
1136. Shishido Y, Furushiro M, TanabeS, Shibata S, Hashimoto S, Yokokura T (1999) Eur J Pharmacol 372:135–142.
1137. Shizgal P (1997) Curr Opin Neurobiol 7:198–208.
1138. Shouman B, Fontaine RH, Baud O, Schwendimann L, Keller M, Spedding M, Lelievre V, Gressens P (2006) Br J Pharmacol 148:442–451.
1139. Shulman RG, Rothman DL, Behar KL, Hyder F (2004) TRENDS in Neurosci 27:489–495.
1140. Shultz DE, Sosnik R, Ego V, Haidarllu S, Ahissar e (2000) Nature 403:549–553.
1141. Shvirkov VB (1995) Introduction in the objective psychology. Neuronal basis of psychic. Institute of Psychology, RAN, Moscow
1142. Silberstein SD (2006) TRENDS in Pharmacological Sciences 27:410–415.
1143. Silva IV, Caruso-Neves C, Azeredo IM, Carvalho TL, Lara LS, de Mello MC, Lopes AG (2002) Arch Biochem Biophys 406:183–189.
1144. Simic G, Lucassen PJ, Krsnik Z, Kruslin B, Kostovic I, Winblad B, Bogdanovi N (2000) Exp Neurol 165:12–26.
1145. Sinz EH, Kofke WA, Garman RH (2000) Anesth Analg 91:1443–1449.
1146. Sirotkin AV, Nitray J (1993) The Journal of steroid biochemistry and molecular biology 46:573–577.
1147. Sivaci R, Kahraman A, Serteser M, Sahin DA, Dilek ON (2006) Clinical Biochemistry 39:293–298.
1148. Sjstrm PJ, Turrigiano GG, Nelson SB (2001) Neuron 32:1149–1164.
1149. Sklair-Tavron L, Shi W-X, Lane SB, Harris HW, Bunney BS, Nestler EJ (1996) Proc Natl Acad Sci USA 93:11202–11207.
1150. Skyba DA, Radhakrishnan R, Bement BKH, Sluka KA (2004) Drug Discovery Today: Disease Models 1115–119.
1151. Sliwa L (2001) Arch. Androl. 46 (2001) 135–140.
1152. Smith CUM (2006) Brain and Cognition 61:181–188.
1153. Smith DA, Browning M, Dunwiddie TV (1993) Brain Res 608:259–265.
1154. Soest PF, Kits KS (1998) J Neurophysiol 79:1619–1632.
1155. Sohl G, Maxeiner S, Willecke K (2005) Nat Rev Neurosci 6:191–200.
1156. Soffin EM, Evans ML, Gill CH, Harries MH, Benham CD, Davies SH (2002) Neuropharmacology 42:127–133.
1157. Sokolov EN, Grechenko TN (1974) Neirofiziologiya 8:192–204.
1158. Sokolov EN, Nezlina NI (2004) Neurosci Behav Physiol 34:847–857.
1159. Sokolova I, Lester HA, Davidson N (2006) J Neurophysiol 95:2570–2579.

1160. Somekava S, Fukuhara S, Nakaoka Y, Fujita H, Saito Y, Mochizuki N (2005) Circ Res 97:655–662.
1161. Song XS, Cao JL, Xu YB, He JH, Zhang LC, Zeng YM (2005) Acta Pharmacol Sin 26:789–798.
1162. Soreni N, Apter A, Weizman A, Tufeled OD, Leschiner S, Karp L, Gavish M (1999) Biol Psychiatry 46:484–488.
1163. Sourjik V, Berg HC (2004) Nature 428:437–441.
1164. Spanswick S, Smith MA, Mirshamsi S, Routh VH, Ashford MLJ (2000) Nat Neurosci 3:757-758.
1165. Sperelakis N. (1998) Cell Physiology. Academic Press, London.
1166. Spitzer NC (1999) Nat Neurosci 2:489–491.
1167. Stalnaker TA, Franz TM, Singh T, Schoenbaum G (2007). Neuron 54:51-58.
1168. Stanley BG, Willett III VL, Donias HW, Ha LH, Spears LC (1993) Brain Res 630:41–49.
1169. Stanley S, Wynne K, McGowan B, Bloom S (2005) Physiol Rev 85:1131–1158.
1170. Staras K, Kemenes G, Benjamin PR (1998) J Neurophysiol 79:3030–3040.
1171. Stefano GB, Fricchione GL, Slingsby BT, Benson H (2001) Brain Res Rev 35:1–19.
1172. Steiger A. (2007) Journal of Psychiatric Research 41:537–552.
1173. Stein L, Xue BG, Belluzzi JD. (1994) Journal of Experimental Analysis of Behavior 61:155–168.
1174. Stein DG (2001) Trends in Neurosci 24;386–391.
1175. Steinmetz JE (2000) Behav Brain Res 110:13–24.
1176. Steller H (1995) Science 267:1445–1449.
1177. Stelzer A, Simon G, Kovacs G, Rai R (1994) Proc Natl Acad Sci USA 91:3058–3062.
1178. Stemmler M, Koch C (1999) Nat Neurosci 2:521–527.
1179. Steriade M (2005) Trends in Neurosci 28:317–324.
1180. Stern EA, Jaeger D, Wilson CJ (1998) Nature 394:475-478.
1181. Stevens CF, Wang Y (1993) Nature 364:147-149.
1182. Stobbs SH, Ohran AJ, Lassen MB, Allison DW, Brown JE, Steffensen SC (2004) J Pharmacol Exp Ther 311:282–289.
1183. Stone EA, Zhang Y, John S, Filer D, Bing G (1993) Brain Res 603:181–185.
1184. Storozhuk VM (2005) Neurosci Behav Physiol 35:655–665.
1185. Stuart G, Sakman B (1995) Neuron 15:1065–1076.
1186. Stuart EB, Thompson JM, Rhees RW, Lephart ED (2001) Brain Res Developmental Brain Res 129:125–133.
1187. Stubbe J, Tian J, He A, Sinskey AJ, Lawrence AG, Liu P (2005) Annu Rev Biochem 74:433–480.
1188. Stys PK (2005) J Neurol Sci 233:3–13.
1189. Su TP (2000) J Biomed Sci 7:195–199.
1190. Sublette ME, Russ MJ, Smith GS (2004) Bipolar Disord 6:95–105.
1191. Sudakov KV (1993) Integr Physiol Behav Sci 28:396–407.
1192. Suhayl D-J (2007) Neurology 68:13–21.
1193. Sun K, Ferguson AV (1997) Am J Physiol 272:1625–1630.
1194. Sun R (2001) Cognitive Systems Research 2:1–3.
1195. Supinder S, Bedi AS, Salim A, Chen S, Glanzman DL (1998) potential role of cAMP. J Neurophysiol 79:1371–1383.
1196. Sur M, Rubenstein JLR (2005) Science 310:805–810.

1197. Surmeier DJ, Day JDM, Wang Z, Shen W (2007) Trends in Neurosci 30:228–235.
1198. Susanna C, Benn SC, Woolf CJ (2004) Nat Rev Neurosci 5:686–700.
1199. Sutcliffe G, Lecea L (2002) Nat Rev Neurosci 3:339–349.
1200. Sutor B, Hablitz JJ (1989) Brain Res Devel Brain Res 46:155–160.
1201. Sutton RS, Barto AG (1998), Reinforcement Learninig. MIT Press, Cambridge, MA.
1202. Suzuki M, Beuckmann CT, Shikata K, Ogura H, Sawai T (2005) Brain Res 1044:116–121.
1203. Svarnic OE, Alexandrov YI, Gavrilov VV, Grinchenko YV, Anokhin KV (2005) Neuroscience 136:33–42.
1204. Svensson L, Bergquist J, Wenneras C (2004) Microbes Infect 6:256–264.
1205. Svoboda KR, Fetcho JR (1996) J Neurosci 16:843–52.
1206. Szechtman H, Woody E (2004) Psychol Rev 111:111-127.
1207. Szelenyi Z (2001) Peptides 22:1245–1250.
1208. Szlufcik K, Bultynck G, Callewaert G, Missiaen L, Parys JB, De Smedt H (2006) Cell Calcium 39:325-336.
1209. Taheri S, Zeitzer JM, Mignot E (2002) Annu Rev Neurosci 25:283–313.
1210. Takano K, Asano S, Yamashita N (1994). Am J Physiol 266:318–325.
1211. Tambiev AE, Medvedev SD (2005) Neurosci Behav Physiol 35:643–647.
1212. Tan PSP, Potas JR, Killinger S, Horiuchi J, Goodchild AK, Pilowsky PM, Dampney RAL (2005) Brain Res 1036:70–76.
1213. Tanaka E, Yamamoto S, Kudo Y, Mihara S, Higashi H (1997) J Neurophysiol 78:891–902.
1214. Tanaka K, Nogawa S, Nagata E, Ito D, Suzuki S, Dembo T, Kosakai A, Fukuuchi Y (2000) Exp Neurol 161:462–471
1215. Tanaka M, Cummins TR, Ishikawa K, Black JA, Ibata Y, Waxman SG (1999) Proc Natl Acad Sci USA 96:1088–1093.
1216. Tanapat P, Hastings NB, Rydel TA, Galea LA, Gould E (2001) J Compar Neurol 437:496–504.
1217. Tang P, Xu Y (2002) Proc Natl Acad Sci USA 99:16035–16040.
1218. Tang YP, Shimizu E, Dube GR, Rampon C, Kerchner GA, Zhuo M, Liu G, Tsien JZ (1999) Nature 401:63-69.
1219. Teodorescu HNL, Brezuliani A (1997) Intern Journal of Chaotic Theory and Applications 2(1):17–44.
1220. Tarasiuk A, Grossman Y, (1991) Respir Physiol 86:369–379.
1221. Tarek M, Saleh TM, Connell BJ, Legge C, Cribb AE (2004) Brain Res 1018:119–129.
1222. Taube JS, Schwartzkroin PA (1988) J Neurosci 8:1632–l644.
1223. TenBroek EM, Lampe PD, Solan JL, Reynhout JK, Johnson RG (2001) J Cell Biol 155:1307–1318.
1224. Thoby-Brisson M, Ramirez J-M (2000). J Neurosci 20:5858–5866.
1225. Thompson AM (2000) TRENDS in Neurosci 23:305–312.
1226. Thompson AM, Swant J, Gosnell BA, Wagner JJ (2004) Neuroscience 127:177–185.
1227. Thompson RJ, Zhou N, MacVicar BA (2006) Science 312:924–927.
1228. Tilders FJ, DeRijk RH, Van Dam AM, Vincent VA, Schotanus K (1994) Psychoneuroendocrinology 19:209–232.
1229. Tirassa P, Aloe L, Stenfors C, Turrini P, Lundeberg T (1999) Proc Natl Acad Sci USA 96:6473–6477.

1230. Tirelli E, Laviola G, Adriani W (2003) Neurosci Biobehav Rev 27:163–178.
1231. Tisa LS, Adler J (1995) Proc Natl Acad Sci USA 92:10777–10781.
1232. Toda N, Ayajiki K, Okamura T (2005) Pharmacology Therapeutics 106:233–266.
1233. Toga AW, Thompson PM (2005) Annu Rev Neurosci 28:1–23
1234. Tolkunov BF, Orlov AA, Afanas'ev SV (1997) Neurosci Behav Physiol 27:297–302.
1235. Tomasulo RA, Ramirez JJ, Steward O (1993) Proc Natl Acad of Sci USA 90:11578–11582.
1236. Tomatis C, Pla AF, Munaron L (2007) Cell Calcium 41:261–269.
1237. Tononi G, Edelman GM (2000) Brain Res Rev 31:391–400.
1238. Toppila J, Niittymki P, Porkka-Heiskanen T, Stenberg D (2000) Pharmacol Biochem Behav 66:721–727.
1239. Torup L, Borsdal J, Sager T (2003) Neurosci Lett 351:173–176.
1240. Tracey DJ, Romm MA, Yao NNL (1995) Brain Res 669:245–254.
1241. Trapp S, Luckermann M, Kaila K, Ballanyi K (1996) NeuroReport 7:2000–2004.
1242. Traub R (1997) Brain Res 778:34–42.
1243. Travis KA, Jonson AK (1993) Am J Physiol 264:1200–1205.
1244. Tremblay L, Schultz W (1999) Nature 398:704–708.
1245. Tremblay L, Schultz W (2000) J Neurophysiol 83:1877–1885.
1246. Triller A, Choquet D (2005) Trends in Neurosci 28:133–139.
1247. Trotsky L (1991) Portraits of revolutionaries. Felshtinsky JG (ed), Moscow worker, Moscow.
1248. Truong BG, Magrum LG, Gietzen DW (2002) Brain Res 924:1–9.
1249. Tsetlin ML (1969) Theory of automation and modeling of biological systems. Nauka. Moscow (in Russian)
1250. Tsitolovskaya LA (1974) J Higher Neural Act 24:536–543. (in Russian)
1251. Tsitolovskaya LA (1981) J Higher Neural Act 31:276–283. (in Russian)
1252. Tsitolovskaya LA, Ryabchikova NA (1988) Procceedings of High School 33:72–82. (in Russian)
1253. Tsitolovsky LE (1973) Proc Akad Sci USSR 209:1247–1248. English translation: 1973 Consultant Bureau, a division of Plenum Publishing. Doklady Biol Sci Section 182–183.
1254. Tsitolovsky LE (1979) Chem Life 18:20–25. (in Russian)
1255. Tsitolovsky LE (1986) Usp Physiol Sci 17:83–103. (in Russian)
1256. Tsitolovsky LE (1997) J Biol Syst 197:301–323.
1257. Tsitolovsky LE (2005) Brain Res Rev 49:566–594.
1258. Tsitolovsky LE, Babkina NV (1992) Brain Res 595:67-73.
1259. Tsitolovsky LE, Babkina NV (2002) Brain Res 946:104-118.
1260. Tsitolovsky LE, Babkina NV, Shvedov A (2004) Neurobiol Learn Mem 81:82–95.
1261. Tsitolovsky LE, Guselnikov VI (1974) Biol Nauki 10:36–47. (in Russian)
1262. Tsitolovsky LE, Kovalenko M (1993) Structure independent learning in feed-forward networks. In: The 10th Israeli sumposium on artificial intelligence. IAAI. Proceedings. pp 273–282.
1263. Tsitolovsky LE, Kraevskii AA (1984) Neurosci Behav Physiol 14:121–127. (in Russian)
1264. Tsitolovsky LE, Orlov IV (1979) Dokl Akad Nauk USSR 248:766–769. (in Russian)

1265. Tsitolovsky LE, Pivovarov AS (1973) Physiol J USSR 59 236–259.(in Russian)
1266. Tsitolovsky LE, Pivovarov AS (1975) Neurophysiologia 7:324–326. (in Russian)
1267. Tsitolovsky LE, Pivovarov AS (1976) Neurophysiologia 8:22–28. (in Russian)
1268. Tsitolovsky LE, Sandler U (2006) WSEAS Trans on Systems and Control 1:192–199.
1269. Tsitolovsky LE, Sandler U (2006) WSEAS Transactions on Systems and Control 19:192–199.
1270. Tsitolovsky LE, Shvedov A (1997) Brain Res 745:271–282.
1271. Tsitolovsky LE, Shvedov A (1998) NeuroReport 9:1–7.
1272. Tsitolovsky LE, Tsaturyan OI (1980) Neurosci Behav Physiol 10:333–339.
1273. Tsoy EL (2007) (Personal communication).
1274. Tsuda I (1991) World Futures 32:167–184.
1275. Tsuda I (1992) Neural Networks 5:313–326.
1276. Tully T, Cambiazo V, Kruse L (1994) J Neurosci 14:68–74.
1277. Turrigiano GG, Nelson SB (2000) Curr Opin Neurobiol 10:358–364.
1278. Turrigiano GG, Nelson SB (2004) Nat Rev Neurosci 5:97–1007
1279. Tzounopoulos T, Kim Y, Oertel D, Trussell L (2004) Nat Neurosci 7:719–725.
1280. Tzschentke TM (1998) Progress in Neurobiology 56:613–672.
1281. Uhlhaas PJ, Singer W (2006) Neuron 52:155-168.
1282. Ulanovsky N, Las L, Farkas D, Nelken I (2004) J Neurosci 24:10440–10453.
1283. Ukai M, Maeda H, Nanya Y, Kameyama T, Matsuno K (1998) Pharmacol Biochem Behav 61: 247-252.
1284. Ukhtomsky AA (1923) J Physiol (Russia) 6:31–45.
1285. Ullrich O, Merker K, Timm J, Tauber S (2007) J Neuroimmunol 184:127–135.
1286. Ungar G, Desiderio DM, Parr W (1972) Nature 238:198–202.
1287. Ungerleider LG (1995) Science 270:769–775.
1288. Urani A, Romieu P, Roman FJ, Yamada K, Noda Y, Kamei H, ManhTH, Nagai T, Nabeshima T, Maurice T (2004) Eur J Pharmacol 20:151–161.
1289. Urban NN, Barrionueve G (1998) Proc Natl Acad Sci USA 95:11450–11455.
1290. Ure J, Baudry M, Perassolo M (2006) J Neurol Sci 247:1-9.
1291. Vaadia E, Ahissar E, Bergman H, Lavner Y (1991) Correlated activity of neurons: a neural code for higher brain function, In: Kruger J (ed) Neuronal cooperativity, Springer- Verlag, Berlin, 249-279.
1292. Vaadia E, Haalman E, Abeles M, Bergman H, Prut Y, Slovin H, Aersten A (1995) Nature 373:515–518.
1293. Vaccarino AL, Kastin AJ (2001) Peptides 22:2257–2328.
1294. Valenstein ES (1988) Surg Neurol 30:75–76.
1295. Valtcheva R, Stephanova E, Jordanova A, Pankov R, Altankov G, Lalchev Z (2003) Chemico-Biological Interactions 25:191–200.
1296. Valverde O, Mantamadiotis T, Torrecilla M, Ugedo L, Pineda J, Bleckmann S, Gass P, Kretz O, Mitchell JM, Schutz G, Maldonado R (2004) Neuropsychopharmacology 29:1122–1133.
1297. Vanderschuren LJMJ (1997) Neurosci BioBehav Rev 21:309–326.
1298. Vastola BJ, Douglas LA, Varlinskaya EI, Spear LP (2000) Physiol Behav 77: 107–114.
1299. Veening JG, Coolen LM (1998) Behav Brain Res 92:181–193.
1300. Velasco RT (2002) J Neurosci Res 67:406–410.
1301. Verbnyi YI (1990) Neurosci Behav Physiol 18:440–442.

1302. Veronesid MC, Yarda M, Jacksonb J, Lahiria DK, Kubeka MJ (2007) Brain Res 1128:79–85.
1303. Vetulani J, Nalepa I (2000) Eur J Pharmacol 405:351–363.
1304. Vibert N, Bantikyan A, Babalian A, Serafin M, Muhlethaler M, Vidal P-P (1999) Neuroscience 94:1–5
1305. Vinod KY, Hungund BL (2006) Trends in Pharmacol Sci 27:539–545.
1306. Visscher PK, Camazine S (1999) Nature 397:400–401.
1307. Voisin DL, Bouque CW (2002) Trends in Neurosci 25:199–205.
1308. Volgushev M, Pernberg J, Eysel UT (2002) J Physiol 540:307–320.
1309. Voronin LL, Kozhedub RE (1971) J Vysh Nervn Deyat 21:997–1005.
1310. Vrontou E, Nilsen SP, Demir E, Kravitz EA, Dickson BJ (2006) Nat Neurosci 9:1469–1471.
1311. Vvedensky NE (1953) Excitation, inhibition and narcosis (1901), In: Complete collection of works, USSR, L. Vol. 4 .
1312. Wafford KA (2005) Curr Opin in Pharmacol 5:47–52.
1313. Wager TD, Phan KL, Liberzon I, Taylor SF (2003) NeuroImage 19:513–531.
1314. Waikar SS, Thalhammer IG, Raymond SA, Huang JH, Chang DS, Strichartz GR (1996) Brain Res 721:91–100.
1315. Walton M, Sirimanne E, Williams C, Gluckman B, Dragunow M (1996) Mol Brain Res. 43:21–29.
1316. Wang H, Hu Y, Tsien JZ (2006) Prog Neurobiol 79:123–135.
1317. Wang H-G, Lu F-M, Jin I, Udo H, Kandel ER, de Vente J, Walter U, Lohmann SM, Hawkins RD, Antonova I (2005) Neuron 45:389–403.
1318. Wang G-X, Li G-R, Wang Y-D, Yang T-S, Quyang Y-B (2001) Life Sci 69:2801–2810.
1319. Wang JW, Richardson SR, Thayer SA (1995) Neurosci Lett 186:139–144.
1320. Waraczynski MA (2006) Neurosci BioBehav Rev 30:472–496.
1321. Warner A, Schuhler A, Bennett GW, Ebling FJP, Brameld JM (2006) Frontiers in Neuroendocrinology 27:155–156.
1322. Warshavsky VI, Pospelov DA (1984) Orchestra plays without a conductor. Physical-Mathematical Ed. Nauka, Moscow (in Russian)
1323. Washburn DLS, Beedle AM, Ferguson AV (1999) Neuroscience 93:349–359.
1324. Watanabe T, Nakagawa T, Yamamoto R, Maeda A, Minami M, Satoh M (2002) Jpn J Pharmacol 88(4):399–406.
1325. Watkins LR, Cobelli DA, Mayer DJ (1982) Brain Res 243:119–132.
1326. Watkins LR, Milligan ED, Maier SF (2001) Trends in Neurosci 24:450–455.
1327. Weeks JC (2003) Prog Neurobiol 70:421–442.
1328. Wegner DM (2002) The illusion of conscious will. Bradford Books/MIT Press, Cambridge
1329. Weinbroum AA, Flaishon R, Sorkine P, Szold O, Rudick V (1997) Drug Saf 17:181–196.
1330. Weiner N (1958) Nonlinear problems in random theory. John Wiley, NY.
1331. Welie I, Hooft JA, Wadman WJ (2004) Proc Natl Acad Sci USA 101:5123–5128
1332. Wentlandt K, Samoilova M, Carlen PL, El Beheiry H. (2006) Anesth Analg. 102:1692–1698.
1333. Wettschureck N, Offermanns S (2005) Physiol Rev 85:1159–1204.
1334. Whim MD, Moss GWJ (2001) Neuron 30:37–50.
1335. Whitlock JR, Heynen AJ, Shuler MG, Bear MF (2006) Science 313:1093–1097.

1336. Whittington MA, Traub RD, Jefferys JGR (1995) Nature 373:612-615.
1337. Wierenga CJ, Ibata K, Turrigiano GG (2005) J Neurosci 25:2895–2905.
1338. Wiesenfeld-Hallin Z, Xu X-J (2001) Eur J Pharmacol 429:49–59.
1339. Wilent WB, Conteras DC (2005) J Neurosci 25:2983–2991.
1340. Wilkie DM, Willson RJ, Carr JAR (1999) Neurosci BioBehav Rev 23:451–455.
1341. Williams G, Bing C, Cai XJ, Harrold JA, King PJ, Liu XH (2001) Physiol Behav 74:683–701.
1342. Willner P (1978) Anim Learn Behav 6:249–257.
1343. Windels F, Kiyatkin EA (2006) Brain Res 1086:104-116.
1344. Wise RA (1987) Psychobiology 15:7–20.
1345. Wise RA (2002) Neuron 36:229–240.
1346. Wise PM (2003) Prog Neurobiol 69:181–191.
1347. Wise PM, Dubal DB, Wilson ME, Rau SW, Bottner M, Rosewell KL (2001) Brain Res Rev 37:313–319.
1348. Wise SP, Murray EA (2000) Trends in Neurosci 23:271–276.
1349. Witkowski G, Szulczyk P (2006) Brain Res 1094:92-106.
1350. Wolkow SA (2002) Trends in Neurosci 25:212–216.
1351. Wolpaw JR (1997) Trends in Neurosci 20:588–594.
1352. Won JS, Im YB, Singh AK, Singh I (2004) Free Radic Biol Med. 37:1834–1844.
1353. Wood SJ, Toth M (2001) Mol Neurobiol 23:101–119.
1354. Woods SC, Ramsay DS (2000) Behav Brain Res 110:175–182.
1355. Woody CD, Black-Cleworth P (1973) J Neurophysiol 36:1104–1116.
1356. Woody CD, Zotova E, Gruen E (2000) Brain Res 868:56–65
1357. Woolcock K, Specht SC (2006) Life Sci 78:1653–1661.
1358. Wright WG, Carew TJ (1995) J Neurosci 15:790–797.
1359. Wright WG, Glanzman DL (2004) Synaptic plasticity depends upon synaptic milieu: observations from co-cultures of Aplysia neurons. Abstr Neurosci 568.16.
1360. Wu G, Hallin RG, Ekedahl R (1996). Brain Res 742:225–238.
1361. Wu J-Y, Tsau Y, Hans-Peter H, Cohen LB, Tang AC (1994) J Neurosci 14:1366–1384.
1362. Wu J, Holstein JD, Upanhyay G, Lin DT, Conway S, Muller E, Lechleiter JD (2007) J Neurosci 27:6510–6520.
1363. Wu Q, Zhao Z, Shen P (2005) Nat Neurosci 8:1350–1355.
1364. Wu Z-Q, Chen J, Chi Z-Q, Liu J-G (2007a) Mol Pharmacol 71:519–530.
1365. Wyse ATS, Bavaresco CS Reis EA, Zugno AI, Tagliari B, Calcagnotto T, Netto CA (2004) Physiol Behav 80:475–479.
1366. Xiang ZM, Bergold PJ (2000) Brain Res 881:77–87.
1367. Xu J, Kang N, Jiang L, Nedergaard M, Kang J (2005) J Neurosci 25:1750-1760.
1368. Xu M, Kim CJ, Neubert MJ, Heinricher MM (2007) Pain 127:253-262.
1369. Xu ZC, Gao T-M, Ren Y (1999) Biol Sign Resept 8:294–308.
1370. Yadid G, Nakash R, Deri I, Tamar G, Kinor N, Gispan I, Zangen A (2000) Prog Neurobiol 62:353–378.
1371. Yager JY (2004) Seminars in Pediatric Neurology 11:31–46.
1372. Yager RR, Zadeh LA (1992) An Introduction to Fuzzy Logic Applications in Intelligent Systems, Kluwer Academic Publisher, London.
1373. Yamada H, Shimoyama N, Sora I, Uhl GR, Fukuda Y, Moriya H, Shimoyama M (2006) Brain Res 1083:61–69.
1374. Yamamoto T, Saitoa O, Shono K, Aoe T, Chiba T (2003) Neurosci Lett 347:183–186.

1375. Yamanaka A, Tsujino N, Hara J, Tominaga M, Goto K, Yanagisawa M, Sakurai T (2002) Regulation of orexin neurons by peripheral nutritional signals: roles of leptin, ghrelin and glucose. Abst Soc Neurosci Washingron DS No 776.8.
1376. Yanamoto H, Xue T-H, Miyamoto S, Nagata I, Nakano Y, Murao K, Kikuchi H (2004) Brain Res 1019:178-188.
1377. Yang LM, Rinke R, Korbmacher C (2006) J Biol Chem 281:9859–9868.
1378. Yang P, Iglesias PI (2007) Systems and Control Letters 55:329- 337.
1379. Yeoman MS, Faragher RGA (2001) Biogerontology 2:85–97.
1380. Yokoyama C, Okamura H, Ibata I (1993) Brain Res Bull 30:551–559.
1381. Yonehara N, Kudo C, Kamisaki Y (2003) Brain Res 963:232–243.
1382. Yoshimura N, de Groat WC (1997) J Physiol 503:269–276.
1383. Yoshioka N, Sakurai M (1995) NeuroReport 6:333–336.
1384. Youssef FE, Addae JI, Stone TW (2006) Brain Res 1073–1074:183–189.
1385. Yu C, Derdikman D, Haidarliu S, Ahissar E (2006) PLoS Biology 4:819–825.
1386. Yu SP (2003) Prog Neurobiol 70:363–386.
1387. Yu S-S, Wang M, Li X-M, Chen W-H, Chen J-T, Wang H-L, Ruan D-Y (2007) BMC Dev Biol 7:51.
1388. Yu WH (1990) Neuroendocrinology 52:595–599.
1389. Yuan J, Lipinsky M, Degterev A (2003) Neuron 40:401–413.
1390. Yule KA, White SR (1999) Exp Lung Res 25:267–273.
1391. Yuste R, Nelson DA, Rubin WW, Katz LC (1995) Neuron 14:7–17.
1392. Zadeh LA (1965) Inform and Control 8:338–353.
1393. Zadeh LA (1968) J Math Anal Appl 23:421–427.
1394. Zadeh LA (1971) Inform Sci 3:177–200.
1395. Zadeh LA (1973) IEEE Trans Systems Man Cybernet SMC 3:28–44.
1396. Zadeh LA (1974) On the analysis of large scale systems. In: Gottinger H (edt), Systems Approaches and Environment Problems. Vandenhoeck and Ruprecht, Gottingen.
1397. Zadeh LA (1975) Synthese 30:407–428.
1398. Zadeh LA (1975) Inf Sci 8:199–249; *ibid* 8:301–357; *ibid* 9:43–80.
1399. Zadeh LA (1976) Inter J Man-Machine Studies 8:249–291.
1400. Zadeh LA (1978) Fuzzy Sets and Systems 1:3–28.
1401. Zadeh LA (1979) Fuzzy sets and information granularity. In: Gupta M, Ragade R, Yager R (eds.), Advances in Fuzzy Set Theory and Applications. North-Holland Publishing Co, Amsterdam.
1402. Zadeh LA (1979) A theory of approximate reasoning. In: Hayes J, Michie D, Mikulich LI (eds.), Machine Intelligence 9. Halstead Press, New York.
1403. Zadeh LA (1981) Possibility theory and soft data analysis. In: Cobb L, Thrall RM (eds.) Mathematical Frontiers of the Social and Policy Sciences. Westview Press, Boulder, CO.
1404. Zadeh LA (1982) Test-score semantics for natural languages and meaning representation via PRUF. In: Rieger B (edt) Empirical Semantics, Brockmeyer, Bochum Germany.
1405. Zadeh LA (1983) Comput Math 9:149–184.
1406. Zadeh LA (1983) J Semantics 3:253–272.
1407. Zadeh LA (1984) Precisiation of meaning via translation into PRUF. In: Vaina L, Hintikka J (eds) Cognitive Constraints on Communication. Reidel, Dordrecht.

1408. Zadeh LA (1986) Outline of a computational approach to meaning and knowledge representation based on the concept of a generalized assignment statement. In: Thoma M, Wyner A (eds) Proceedings of the International Seminar on Artificial Intelligence and Man-Machine Systems. Springer, Heidelberg.
1409. Zadeh LA (1996) Multiple-Valued Logic 1:1–38.
1410. Zadeh LA (1997) Fuzzy Sets and Systems 90:111–127.
1411. Zadeh LA (1998) Soft Comput 2:23–25.
1412. Zadeh LA (1999) IEEE Trans Circuits Systems 45:105–119.
1413. Zadeh LA (2002) J Statist Plann Inference 105:233–264.
1414. Zadeh LA (2004) Data Knowledge Eng 50:291–304.
1415. Zadeh LA (2004) AI Mag 25(3):74–91.
1416. Zadeh LA (2005) Inf Sciences 172:1–40.
1417. Zadeh LA (2006) Computational Statistics and Data Analysis 51:15–46.
1418. Zaslavsky GM, Saagdeev RZ (1988) Introduction to Non-linear Physiscs. Nauka, Moscow.
1419. Zapara TA, Ratushniak AS, Stark MV (1988) Zh Vysh Nervn Dejat 33:140–146. (in Russian)
1420. Zarrindasta MR, Fattahib Z, Rostamib P, Rezayof A (2005) Pharmacol Biochem Behav 82:1–10.
1421. Zemke D, Smith JL, Reeves MJ, Majid A (2004) Neurotoxicology 25:895–904.
1422. Zastrow M (2001) Parkins Rel Disorder 7:265–271.
1423. Zhan RZ, Fujiwara N, Taanaka E, Shimoji K (1998) Brain Res 780:86–94.
1424. Zhang G, Wang L, Liu H, Zhang J (2004) Brain Res 1028:225–232.
1425. Zhang J, Gibney GT, Zhao P, Xia Y (2002) Am J Physiol Cell Physiol 282:1225–1234.
1426. Zhang S, Boyd J, Delaney K, Murphy TH (2005) J Neurosci 25:5333–5338.
1427. Zhang X-F, Zhu CZ, Thimmapaya R, Choi WS, Honore P, Scott VE, Kroeger PE, Sullivan JP, Faltynek CR, Gopalakrishnan M, Shieh C-C (2004a) Brain Res 1009:147–158.
1428. Zhang XF, Hu XT, White FJ (1998) J Neurosci 18:488–498.
1429. Zhang Y, Deng P, Li Y, Xu ZC (2006) J Neurophysiol 95:1537–1544.
1430. Zhao AZ, Huan JN, Gupta S, Pal R, Sahu A (2002) Nat Neurosci 5:727–728.
1431. Zheng S, Zuo Z (2005) Brain Res 1054:143–151.
1432. Zhou Q, Poo M (2004) Trends in Neurosci 27:378–383.
1433. Zhou Y-D, Theoden I, Netoff TI, Sen K, White JA (2005) Proc Natl Acad Sci USA 102:19121–19125.
1434. Zhu B, Herbert J (1997) Brain Res 778:206–214.
1435. Zhu L, Zhao T, Li H, Zhao H, Wu L, Ding A Fan W, Fan M (2005) Brain Res 1055:1–6.
1436. Zhu PJ, Lovinger DM (2006) J Neurophysiol 96:433–441.
1437. Zigmond RE (2001) Trends in Neurosci 24:494–496.
1438. Zilber-Gachelin NF, Chartier MP (1973) J Exp Biol 59:359–381.
1439. Zimmermann M (2001) Eur J Pharmacol 429:23–37.
1440. Ziv I, Lustig C, Markovich S, Susswein AJ (1991) Behav Neural Biol 56:148–169.
1441. Zotova E, Woody CD, Gruen E (2000) Brain Res 868:66-78.
1442. Zucker RS (1972) J Neurophysiol 35:621–637.
1443. Zufall F (2005) Neuron 46:693–694.

Index

A

List of symbols

$AMPA, NMDA$	agonists to glutamate receptors, fast and slow synaptic transmission
AP	action potential
ATP	adenosine triphosphate (energy rich molecule)
CS^+	conditioned stimulus
CS^-	discriminated stimulus
US	unconditioned stimulus
$cAMP, cGMP$	cyclic nucleotides (second messengers)
$EPSP$	excitatory postsynaptic potential
$GABA$	γ-aminobutyric acid (inhibitory neurotransmitter)
$Glutamate$	excitatory neurotransmitter
K^+, Na^+, Ca^{2+}	potassium, sodium and calcium ions
LTD	long-term depression
LTP	long-term potentiation
$Na^+, K^+ - ATPase$	the sodium pump
NO	nitric oxide (retrograde messenger)
DNA	deoxyribonucleic acid
RNA	ribonucleic acid

$\bigcap$	intersection of the sets or domains
$\bigcup$	union of the sets or domains
$\bigwedge$	logical connective *AND*
$\bigvee$	logical connective *OR*
$T[\mu_A;\mu_B]$	mathematical representation of the logical connective *AND*, for example $T[\mu_A;\mu_B] = \min(\mu_A;\mu_B)$
$S\{\mu_A,\mu_B\}$	mathematical representation of the logical connective *OR*, for example $S\{\mu_A,\mu_B\} = \max(\mu_A;\mu_B)$
$\in$	belongs to, for example, $V \in \Omega$
$\partial/\partial t$	partial derivative with respect to t
d/dt	total derivative with respect to t
∇	gradient
$\sup_x$	supremum (maximal value) with respect to x
$\int_a^b$	definite integral
$(\boldsymbol{x}\cdot\boldsymbol{y})$	scalar product of the vectors $\boldsymbol{x}$ and $\boldsymbol{x}$
$\boldsymbol{x}$	(bold $\boldsymbol{x}$) vector $\boldsymbol{x} = \{x_1, ..., x_n\}$
λ	Lagrange-multiplier
$\hat{\Gamma}$	matrix Γ
$\hat{\Gamma}^\top$	Hermitian-transposied matrix
$P([a,b])$	function on interval-valued argument

B

List of definitions

Action potential (AP) is a short electrical wave that travels along the cellular membrane of a neuron.
Apoptosis is programmed cell death or cellular suicide.
Axon is an output branch of a neuron, connected pre and postsynaptic cells by means of synaptic connections.
Chemical synapse is specialized junctions through which the neurons signal to other neurons, muscle, glial cells and to glands.
Chemoreceptors are protein channels in cellular membrane, which become permeable for ions after interaction with the neurotransmitter.
Chemotaxis is directional migration of cells in response to a gradient of a chemical stimulus.
Conditioned stimulus (CS^+) is signal using as a predictive of future important events (US).
Dendrites are branches of neuron accepting input signals from synapses.
Deoxyribonucleic acid (DNA) is a nucleic acid polymer consisting of nucleotide that contains the genetic instructions used protein producing.
Discriminated stimulus (CS^-) is never associated with the US and is used as a control of specificity of learning.
Excitatory postsynaptic potential (EPSP) is a depolarization of postsynaptic membrane caused by the flow of positive ions through synapse. EPSP promotes AP generation.
γ-aminobutyric acid (GABA) is inhibitory neurotransmitter.
Gap junctions connect cytoplasm of adjacent cells. Gap junctions allow small ions, molecules and currents to pass between cells.
Glutamate is excitatory neurotransmitter.
G proteins are GTP-binding proteins that sense substances outside the cell and activate metabolic pathways inside the cell. GTP - one of the nucleotides.
Homeostasis is the property of living beings, which regulates its internal state in order to maintain a stability of life.
Inhibitory postsynaptic potentials result from the flow of negative

ions into the cell or positive ions out of cells. It reduces AP generation.
Long-term potentiation (LTP) is enhancement of neuronal responses after frequent or intensive stimulation.
Membrane potential is the difference of potentials between intracellular and extracellular environments.
Set is a well-defined collection of objects, which is considered as a whole.
Domain is the distinguished part of an abstract space. In mathematic analysis, a domain is an open connected subset of a vector space.
Fuzzy set is a set, whose elements have degrees of membership.
Membership function represents fuzzy set of a universe. The membership function $\mu_A(x)$ quantifies the grade of membership of the element x to the fuzzy set A.
Gradient of a scalar function is a vector field which points in the direction of the greatest rate of increase of the scalar function, and whose magnitude is the greatest rate of change. The gradient is denoted by ∇. By definition, the gradient is a vector field whose components are the partial derivatives of $f(\boldsymbol{x})$. That is:

$$\nabla f = \left\{ \frac{\partial f}{\partial x_1}, \ldots, \frac{\partial f}{\partial x_n} \right\}.$$

Differential inclusion is a generalization of the concept of ordinary differential equation:

$$\frac{d\boldsymbol{X}}{dt} \in \boldsymbol{F}(t, \boldsymbol{X}(t)),$$

where $\boldsymbol{F}(t, \boldsymbol{X})$ is a domain rather than a single point.

MIX
Papier aus verantwortungsvollen Quellen
Paper from responsible sources
FSC® C105338

If you have any concerns about our products,
you can contact us on
ProductSafety@springernature.com

In case Publisher is established outside the EU,
the EU authorized representative is:
Springer Nature Customer Service Center GmbH
Europaplatz 3, 69115 Heidelberg, Germany

Printed by Libri Plureos GmbH
in Hamburg, Germany